Broadcasting in America

A Survey of Electronic Media

NINTH EDITION

Broadcasting in America

A Survey of Electronic Media

NINTH EDITION

Sydney W. Head
Late of the University of Miami

Thomas Spann
University of Nebraska

Michael A. McGregor
Indiana University

HOUGHTON MIFFLIN COMPANY Boston New York

Sponsoring Editor: Adam P. Forrand

Editorial Assistant: Brigitte Maser

Associate Project Editor: Patricia English

Editorial Assistant: Cecilia Molinari

Senior Production/Design Coordinator: Sarah L. Ambrose

Manufacturing Manager: Florence Cadran

Marketing Manager: Stephanie Jones

Cover design: Harold Burch, Harold Burch Designs, NYC
Cover illustration: *ABS 29+ KRE013*, Hugh Kretschmer

Chapter Opening Photo Credits
xxii MSNBC; **18** AP/Wide World Photos; **54** AP/Wide World
Photos; **86** © 1999 Mark Brown/Phototech Imaging; **124**
Photri; **164** © Jay Strauss Photography/Courtesy Priceline.com;
206 Richard Howard for WGBH Boston; **226** Greg Gorman,
Randy Tepper, © 1998 Fox Broadcasting Company/Photofest;
248 Photofest; **290** © Stewart Cohen/Index Stock Imagery; **316**
© James Carroll/Stock Boston; **342** Federal Communications
Commission; **372** Susan Walsh/AP–Wide World Photos;
406 © Robert Knowles/Photo Researchers.

Printed in the U.S.A.

Library of Congress Catalog Card Number: 00-133821

ISBN: 0-618-054197

123456789-DOW-04 03 02 01 00

BRIEF CONTENTS

········
LIST OF EXHIBITS

PREFACE

The first edition of *Broadcasting in America* appeared in 1956. Founding author Sydney Head (1913–1991) launched a text that has helped introduce thousands to the study of broadcasting by viewing electronic media in a broad academic perspective and treating media in context, both as products of contemporary social forces and as social forces in their own right. Later co-authors Christopher Sterling and Lemuel Schofield carried on this vision, and we hope that vision continues in this Ninth Edition. Although their names do not appear on this edition, we gratefully acknowledge that much of what is good about this text is directly attributable to Chris and Lem, and we thank them.

The Ninth Edition of *Broadcasting in America* follows the same basic format as the Eighth Edition. Like the Eighth Edition, this book is more streamlined than the Seventh Edition but contains more information than the two "brief" editions. Although no format is perfect, this approach appears to satisfy most of the needs of the book's loyal users.

massive media convergence (cable television and telephony, for example), conglomeration (the merger of CBS and Viacom, for example), and consolidation (one company owning more than 800 radio stations, for example)
- the movement to digital delivery systems, including implementation of the changeover from analog to digital transmission of television broadcasts
- changes in FCC licensing procedures, notably the abandonment of comparative hearings in favor of spectrum auctions and the court's rejection of FCC equal employment opportunity policies
- the coming of age of direct broadcast satellite (DBS) as a major competitor to cable television systems
- the breathtaking growth of the Internet as an important new player that promises to influence broadcasting and other established media dramatically in the new century

COVERAGE OF RECENT DEVELOPMENTS

Monumental changes in the electronic media have occurred in the two and one-half years since publication of the Eighth Edition. As the media landscape evolves, both industries and consumers—not to mention instructors and students—struggle to keep abreast of this ever-changing environment. And the pace of change increases every year. To keep this text current, we have covered the most significant developments occurring since the publication of the Eighth Edition, including the following:

- the continuing "fallout" from the passage of the Telecommunications Act of 1996, including

CHANGES IN THE NINTH EDITION

This edition of *Broadcasting in America* continues to cover all electronic media, with expanded consideration of emerging mass media such as the Internet and telephone systems. We have also updated and expanded our discussion of newer delivery systems, including local multipoint distribution service (LMDS), digital audio radio service (DARS), and home consumer electronics. The Ninth Edition also provides expanded treatment of digital television (DTV) broadcasting and the evolution of digital cable services.

- Chapter 1 introduces students to the exciting changes taking place in electronic media as we enter a new century. It examines the emerging information infrastructure and the accelerating convergence of various electronic media forms, while also introducing new concepts and vocabulary explored in more detail in later chapters.

- Chapter 2 reviews the history of television and radio, including new exhibits featuring the contributions of Vladimir Zworykin, Philo T. Farnsworth, and Edwin Armstrong.
- Chapter 3 explores the ongoing conversion of cable from analog to digital, the renewed interest in wireless cable technologies, and new media that give consumers increasing control over programming and other electronic communication services.
- Chapters 4 and 5 retain the text's traditional overview of conventional electronic media technology but also provide an expanded discussion of how digital media work, with a special emphasis on DTV transmission and reception. Chapter 5 concludes with a revised and expanded introduction to the Internet. These chapters explore the promise, but also the problems, associated with these new technologies.
- Chapter 6 highlights recent changes in the organization and ownership of the electronic media and recounts the financial successes (and failures) of the media business.
- Chapter 7 updates the programming changes and financial situation of noncommercial media and provides a retrospective look at one of public television's most important figures.
- Chapters 8 and 9 introduce traditional programming basics and vocabulary but also deal with new challenges created by an expanding number of new delivery technologies and the increasing vertical integration of production and distribution controlled by huge media conglomerates. The increase in the number of broadcast networks even as their audiences decline, the growing dominance of cable and DBS programming, and the rapidly emerging influence of the Internet on programming are given greater attention in this edition.
- Chapter 10 presents information on a new data collection system being tested by Arbitron. The latest media use figures also are listed.
- Chapter 11 includes a discussion of the media coverage of President Bill Clinton's relationship with White House intern Monica Lewinsky and his resulting impeachment and trial, as well as public reaction to the coverage.
- Chapter 12 covers new licensing procedures and the demise and rebirth of the FCC's equal employment opportunity guidelines.
- Chapter 13 updates the government's continuing efforts to regulate content in the media—whether it be the Internet, radio, or cable television. The chapter also expands coverage of copyright issues in a digital age.
- Finally, Chapter 14, "Global View," includes various comparative data showing how media resources are unevenly distributed around the world.

ANCILLARY SUPPORT

As with the Eighth Edition, the Ninth will be supported by an *Instructor's Resource Manual*, a computerized test bank, and a web site.

ACKNOWLEDGMENTS

Many people deserve our thanks and appreciation. Walter Gantz, Susan Eastman, Herb Terry, Ron Osgood, and Don Agostino, all colleagues of Mike McGregor at Indiana University, provided information, encouragement, and support for this project. Department of Telecommunications graduate students Rita Zajacz and Smita Saraswat contributed research help, especially for Chapters 6 and 11. Special thanks go to Professor McGregor's students in Introduction to Telecommunications Industry and Management for pointing out spots in the text that were unclear, imprecise, or in need of updating. The University of Nebraska–Lincoln Department of Broadcasting faculty (Richard Alloway, Laurie Thomas Lee, Peter Mayeux, Jerry Renaud, and Larry Walklin) deserve medals for their forbearance while Tom Spann ducked assignments and hid in the shadows working on this edition. Peter Mayeux merits special thanks for going through the Eighth Edition (and many chapters of the Ninth) with the critical eye that only those who know him can fully appreciate. Chapter 5 benefited from the review and comments provided by Laurie

Thomas Lee. Thanks, too, to those University of Nebraska students who were involuntarily subjected to the Eighth Edition. Their moans and groans, though painful at times, helped identify areas that required our attention.

Several people in the industry and government deserve our thanks as well: Suzanne Owen, (WTIU, Bloomington, Indiana) and Christina Kuzmych (WFIU, Bloomington, Indiana) provided insights for Chapter 7. A flock of engineers at Nebraska Educational Telecommunications, most notably Herman Siegl, Paul Sautter, and George Frederick, helped with the technical chapters addressing broadcast technology and DTV. Chief engineers Rex Greenwell (WDSE-TV, Duluth, Minnesota) and Brent Haun (KOLN-KGIN, Lincoln–Grand Island, Nebraska) also contributed to the technical discussions. Richard Bates (Time Warner Cable) provided insights regarding network, local, and syndicated television programming. Gary Fries (Radio Advertising Bureau) and Marcia L. De Sonne (National Association of Broadcasters) offered helpful insights regarding interactivity and the effects of the Internet on traditional electronic media. Lori Chang at the National Cable Television Association handled repeated calls requesting statistical information with efficiency and courtesy. Jim Roberts at ABC was especially helpful in clarifying conflicting information about network radio. Finally, Roger Holberg, Jim Shook, Rebecca Willman, and Gary Kalagian, all of the Federal Communications Commission, provided timely information concerning legal and regulatory matters.

On a more personal note, Mike McGregor would like to thank Debby McGregor for encouraging him to get into the world of academics in the first place. He would like to thank his sons, Geoff and Dan, for their encouragement and humor, and his wife, Celeste, for her understanding, patience, and support.

Finally, we would like to thank the following reviewers for their comments and suggestions:

Dr. Barbara Moore
University of Tennessee

Professor George Bagley
University of Central Florida

Dr. David L. Barner
Westminster College

Professor Robert Hartwig
Cuesta College

Tom Spann
Mike McGregor

CHAPTER

1

Introducing Electronic Media

CHAPTER CONTENTS

You already know something about American electronic mass media even before you pick up this book. Like most of your friends, you probably use radio, television, cable, video cassette recorders, CDs, and perhaps networked computers daily—to the point where they take up a substantial amount of your time. So why, then, read a textbook on something readily available in your life? What can you possibly learn that you don't already know? Well, ask yourself these questions:

- Where did today's media giants come from? How, for example, did CBS, MTV, and Ted Turner get their start?
- How are programs made, how are they paid for, who chooses them, and why?
- Do we watch the same kinds of programs our parents did? Is anything really "new"?
- Just how accurate are those ratings we hear about? ("Nobody ever asks me what I watch!")
- How much does government control what we see and hear? Should it control more? Or less?
- Why do we have a system of electronic media so different from those in most other countries?
- And what is this thing called digital television? How does it differ from regular television? Does it have anything to do with the Internet?

Studying electronic media in the new century is an increasingly exciting thing to do. Rarely has such dramatic and rapid change occurred in the ways people communicate. Reports of new developments appear almost daily. Yet many of us take all this as matter-of-fact evolution. After all, most young people in America have grown up with computers, video games, and multichannel television, and they take these artifacts quite for granted. But staying on top of what some regard as a media revolution, with all of its twists and turns, its subtleties and implications, is a challenge for any serious student of communication.

1.1   THE INFORMATION SUPERHIGHWAY

Some who write about the burgeoning world of electronic media have found it useful to refer to futuristic systems of high-capacity, high-speed paths across the country and around the globe as the *information superhighway* (see Exhibit 1.a).

What It Means

One organization that explores such matters defines the information superhighway as

> a system of pathways that will provide unlimited information—news, entertainment, data, personal communications—to the people who want it. That information will be available widely, easily and in mixtures of text, audio and video. Consumers will be able to send and receive information and choose when they do it. . . . [W]hether by computer, TV, phone or a hybrid of them, people will be able to "travel" the superhighway to get news, shop, socialize, work, attend school, and have fun without having to leave home. (*The Freedom Forum*, May 1994: 4)

Creative adopters of this metaphor have both expanded and subdivided it, suggesting that the superhighway has on-ramps and off-ramps, potholes and traffic jams, tolls and bridges and tunnels, speed limits or *no* speed limits, supersidewalks (for those who wish to move at a slower pace), even roadkill (to describe the unsuccessful would-be traveler).

Like any wonderfully descriptive metaphor, this one, too, has become a cliché through overuse. To avoid it, writers have fashioned substitutes and variations: *infobahn, I-way, info-highway, electronic highway, data highway, communicopia*—the list is endless. Detractors refer to the *info-hypeway*; visionaries speak of a *field of dreams* ("If you build it, they will come").

Although the first use of the term remains obscure, Al Gore claims that he coined "information superhighway" in 1978. While seeking the Democratic Party's presidential nomination early in 1999, Gore told a reporter that he had taken the political initiative in creating the Internet, first as a senator and later as vice president. Former Republican vice president Dan Quayle quickly countered by saying that if Gore invented the Internet, he invented the spell checker. Senate Majority Leader Trent Lott joined the partisan mirth by revealing himself as the father of the paper clip.

Political bickering aside, scholars generally credit Gore with providing strong support for Internet development and helping to popularize the term "information superhighway," a symbolic extension of the national interstate highway system his father, also a U.S. senator, helped build a generation earlier. It was during the Clinton-Gore administration that the federal government laid out its expectations for this superhighway as a key to general economic improvement and individual well-being. These expectations appeared in a 1993 report that referred to the *National Information Infrastructure* (NII), an expression that sounds somehow more official.

The government's interest did not go unnoticed by the nation's news media. According to a Freedom Forum study, selected newspapers, magazines, and broadcasters employed the phrase, or one of its variations, nearly 3,000 times between January 1992 and February 1994. Still, the *Miami Herald* reported three months later (while further refining the metaphor) that

> [T]he information superhighway may be one of the buzzwords of the year, but most people are still in the computerized ditch. A Louis Harris poll found that two out of three Americans had not read or heard anything about [it]. . . . Even among the 34 percent who knew the term, fewer than half said they understood it well. All the same, 27 percent thought the electronic network an excellent idea. (*Miami Herald*, 9 May 1994: 7A)

Today it is difficult to believe that there is anyone in America who remains completely unaware of the concept. Indeed, it is increasingly difficult to find any corner of the world where some vestige of the information superhighway does not extend, if only in the form of an unpaved trail. By the end of 1999, e-mail, that most simple and useful form of commerce on the information superhighway, had reached the world's most remote inhabited island, Tristan da Cunha, about 2,000 miles off the coast of South Africa.

Use of computers and the Internet by Americans increased dramatically during the late 1990s. In January 2000, the Nielsen//NetRatings service estimated that almost 123 million Americans had access to the Internet in their homes. Nielsen estimated that the average home user spent more than nine hours a month on-line. Today more than 98 percent of American households have television sets, and many homes have three or more sets. Early in the year 2000, about 68 percent of those homes were connected to cable, providing dozens of channels. Another approximately 12 percent received TV programs and more by digital *direct-to-home* (DTH) satellite transmission. Cell phones are so common that many teachers place warnings in course syllabi forbidding their in-class use, and some municipalities concerned about public safety have ordinances restricting cell phone use by drivers. (Nielsen//NetRatings, 2000)

In spite of these impressive developments, at the dawn of the new century, relatively few Americans appreciate fully the range of capabilities inherent in the new communications environment made possible by technology. Development of the information superhighway depends not only on the scope and pace of technological advancement, but also on consumer demand for what technology can provide. Add to this mix the influence of corporate economics and government regulation, and the picture becomes fuzzier still. The range of possibilities and the dizzying rate of technological advancement leave both media professionals and scholars uncertain about how electronic media may evolve.

Convergence
• • • • • • • •

In spite of the media's mind-numbing rate of change, perhaps the clearest part of the picture concerns technology. Looking at the information superhighway strictly from a technological perspective, the obvious trend is toward *convergence*—the coming together of, and blurring of lines between, what previously had been essentially discrete communication forms (broadcast, cable, telephony, computers, mail). But this convergence is not only of technologies, of "hardware," of means of transmission and delivery, but also of content. When complete, it will be *seamless* (without interruption) and *transparent* (with separate elements undetectable by users).

Convergence results in what most people call *multimedia* (although some, in what seems a bit of a reach, refer to *telemedia* or even *telepower*)—the combination of "more than one medium, where the media can include speech, music, text, data, graphics, fax, image, video, and animation. A key point is that different media are integrated—that is, they are linked, synchronized, and commonly controlled. Connections are set up in an integrated way. The media can also be interactive or non-interactive" (Mayo, 1992). Multimedia, in turn, serves as a foundation for the superhighway (there's that metaphor again). Exhibit 1.b, on p. 6, offers one concept of what a future home multimedia setup may look like.

EXHIBIT 1.a

The Information Superhighway

Navigators on the information superhighway may need a roadmap to find their way, as content providers—such as the examples represented on the left side of this exhibit—deliver their services (many of them interactive) via cable systems, telephone companies, and satellites; over wires and fiber optics; and through the air to individuals, businesses, and governmental and educational institutions. Interactivity provided by the World Wide Web and other components of the information superhighway blur the traditional distinction between content providers and content receivers. Individuals and institutions depicted on the right side of this exhibit increasingly create, as well as receive, content.

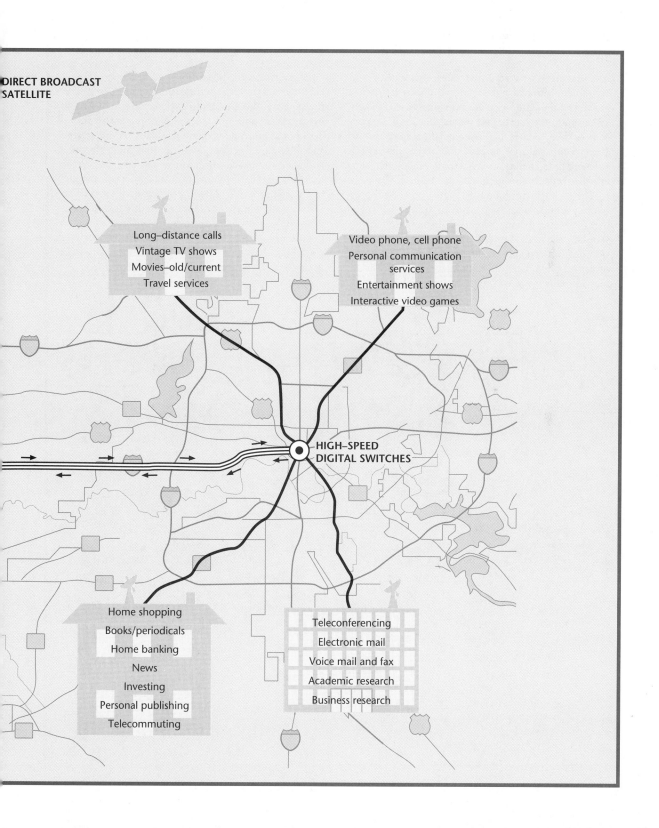

DIRECT BROADCAST
SATELLITE

Long–distance calls
Vintage TV shows
Movies–old/current
Travel services

Video phone, cell phone
Personal communication
services
Entertainment shows
Interactive video games

HIGH–SPEED
DIGITAL SWITCHES

Home shopping
Books/periodicals
Home banking
News
Investing
Personal publishing
Telecommuting

Teleconferencing
Electronic mail
Voice mail and fax
Academic research
Business research

EXHIBIT 1.b

• • • • • • • • • • •

Tomorrow's Multimedia Home

Manufacturers reported that sales of home theater equipment reached almost $8.6 million in 1998, a 5 percent increase over the previous year (CEMA, 1999). Industry sources estimated that about 20 percent of U.S. homes had a home theater system in 1999, compared to 4 percent only five years earlier. Sales are predicted to reach $12 million in 2000, a figure that reflects considerable optimism. The typical cost of such systems (about $3,000) is a major barrier to adoption. In addition, inclusion of a DTV set as part of a home theater system would increase the cost significantly. In the future the number of communication devices available will increase dramatically, and prices may fall. The well-equipped multimedia home may contain many of the items pictured here. The potential size of the projection, wall-mounted HDTV video screen (five feet across or even more), as well as the capacity of digital multitrack sound systems could force multimedia use to take over an entire room. The home multimedia center features an integrated and fully digital set of devices to offer multiple video and audio inputs through integrated decoders at the flick of a universal remote button. An important part of any such room will be interactive audio and video services allowing consumers to "talk back" to program producers. Home units for interactive video telephone and personal communication services (PCS) may be a part of such rooms. Separate devices in this exhibit may be integrated into a few boxes.

Divergence
• • • • • • • •

Some futurists tend to assume a technological determinist's perspective: "If you build it, they will come." But other observers suggest that the superhighway is not inevitable, or at least that it will not achieve wide implementation for many years. By the end of the 1990s, even those most enthusiastic about the promise of the information superhighway were conceding that progress might not be as rapid in all areas as some expected, in spite of impressive advances in the second half of the 1990s.

Detractors point to *technophobia*—the fear of or lack of interest in technology on the part of many Americans. Even in 1999, almost a quarter of all U.S. households did not subscribe to what most young people regard as a necessity—the variety of TV programming provided by cable or by direct-to-home satellites. Although on-screen programming makes operating a VCR very simple, anecdotes still abound describing the millions of units that mindlessly blink "12:00," awaiting their owner's electronic education.

The proliferation of home computers and the use of the Internet at home are criteria sometimes used to measure the pace of progress on the information superhighway. Clearly progress is being made, and rather rapidly. Data from Harris Interactive surveys conducted late in 1999 suggested that more than 50 percent of U.S. adults were using computers at home, and almost as many were using computers at work. The Harris Interactive data indicated that of the approximately 56 percent of all adults who used the Internet, about 46 percent accessed it from their homes. Statistics vary a bit from study to study, but there is little doubt that the number of homes with computers connected to the Internet increased dramatically during the late 1990s. (Harris Interactive Surveys)

Unfortunately, progress along the information highway as measured by computer and Internet use at home is not uniform across all demographic groups. In 1999, the National Telecommunications and Information Administration (NTIA) reported that America has a significant *digital divide*, separating information "haves" and "have nots." The NTIA found that households enjoying incomes above $75,000 were more than 9 times as likely to have a home computer and more than 20 times as likely to have Internet access as those in the lowest income bracket. Black and Hispanic households were only one-third as likely to have home Internet access as

those of Asian/Pacific Islander descent, and about two-fifths as likely as white households. The NTIA found that the home-Internet access gap between white and Hispanic houshesholds and white and black households actually *increased* between 1997 and 1998. The gap between those with the highest and lowest education levels had increased 25 percent; the divide between those with the lowest and the highest incomes had increased almost 30 percent. (National Telecommunications and Information Administration, 1999)

The transition from analog to digital television (DTV) is also seen as a criterion for measuring the pace of progress. The Federal Communications Commission (FCC) in 1996 mandated that the nation's more than 1,500 full-power TV stations convert entirely to digital transmission within a decade, by the end of 2006. Congress later modified the FCC's demand, but let stand the requirement that TV stations convert to digital. In early 2000, the National Association of Broadcasters reported that there were fewer than 120 digital stations on the air (National Association of Broadcasters, 1999). In April 1999, the Consumer Electronics Manufacturers Association (CEMA) reported that only about 26,000 DTV sets had been shipped to dealers. CEMA did not report how many of those sets remained unsold on dealers' shelves. (Consumer Electronics Manufacturers Association, 1999) Although many believe that the pace of the transition will increase dramatically, few Americans were watching DTV as the new century began, and most industry observers believe that analog TV will be around well beyond the year 2006.

Amid such bleakness, however, there are glimmers of hope. One of the happier divergences related to the adoption of new technology centers on age. Young people are far less likely to suffer from technophobia than their elders. The U.S. Education Department's National Center for Educational Statistics (NCES) estimated that 95 percent of the nation's public schools had Internet access in 1999, but there are indications that only a fraction of the nation's teachers are integrating the technology into their instruction. Jokes about students teaching their teachers (and parents) how to use computers and other new technologies have become a cliché, but one supported by research. But the strong interest among young people in traveling the information superhighway bodes well for the future, provided economic and other barriers do not pose serious roadblocks. (U.S. Education Department's National Center for Education Statistics, 1998)

In the 1990s the government sometimes sent conflicting messages. On the one hand, the Clinton administration in 1993 pledged $5 billion to help offset the costs of developing software and equipment necessary to build the superhighway. On the other hand, many blamed government for the 1994 collapse of the historic combination of Bell Atlantic Corporation (a Regional Bell [telephone] Operating Company, or RBOC) and Tele-Communications, Inc. (TCI, at that time the nation's largest cable operator)—a union that offered high hopes of jumpstarting the ride toward the superhighway. Valued at some $33 billion, the merger fell apart soon after the FCC ordered cutbacks in fees cable companies could charge their subscribers.

But four years later, the U.S. government quickly endorsed the merger of TCI and AT&T, the nation's largest provider of long-distance telephone service. In 1999, Viacom and CBS announced plans to merge in a deal some estimated to be worth about $36 billion. Early in the new century, "new media" giant America Online (AOL) shocked the industry by announcing it would acquire Time Warner, an "old media" giant, in what some estimated was a deal worth more than $180 billion. These mergers and acquisitions, and others like them, became attractive following the passage of the Telecommunications Act of 1996 and associated FCC actions that dramatically reduced media ownership restrictions in an effort to encourage competition and the development of new technologies.

By the late 1990s, however, critics were claiming that the government's efforts had produced little competition and, in some ways, had discouraged the development of the information superhighway. These critics said that government policy too often encouraged large media companies to grow even larger and more profitable by expanding their traditional businesses rather than moving into risky new areas. Critics were also concerned increasingly about all media channels and control over content falling into the hands of a few corporate giants. At the end of the decade, some people were calling for the government to take a more active role in regulating the development of the information superhighway; others were calling for even less government involvement.

1.2   THE PLAYERS

In spite of concerns about lack of consumer demand, the danger of economic risks, and the stability of government policies, industry leaders cannot ignore the pressures to move forward toward implementation of the information superhighway in some form. Each industry continues to explore the best route it might take. Although, as noted earlier, distinctions among media grow cloudier, a better understanding of the convergence phenomenon requires a closer look at the major players.

Broadcast

Traditional radio and television run the greatest risk of being stalled on the on-ramp to the information superhighway. With their single, one-way-only channels, they do not fit as easily into a multichannel, interactive future. But traditional radio and television are not dinosaurs yet. Nor need they become extinct if, as now seems the case, they adopt a twenty-first-century outlook and adjust their operations accordingly.

The national TV broadcast networks continue to be the most-watched sources of information and entertainment. Although competition from other sources has eroded their audience, broadcast network affiliates still accounted for about half of all TV viewing in the late 1990s. The broadcast networks remain the principal providers of national news, major sporting events, and broad-based entertainment, as well as the best way for national advertisers to reach a critical mass of consumers. But the advent of interactivity may dramatically change networks from single-channel program *distributors* to multichannel *programmers*.

In fact, the concept of "channels" may disappear altogether as viewers select from menus of programs sorted by content (sitcom, drama, nature) rather than source (ABC, CBS, NBC). So may program "schedules" become obsolete when subscribers can, whenever they want, order a replay of last night's *CBS Evening News* or yesterday's episode of *General Hospital*.

The networks have already moved into cable and direct broadcast satellite (DBS). Examples include NBC's CNBC and MSNBC; ABC's interest in

ESPN, ESPN2, ESPNews, and Lifetime; CBS's TNN and CMT; and Fox's FX, fXM, and Fox News Channel. In 1995, CBS became the first major broadcast television network to set up its own *home page* (a dedicated site for information) on the Internet (see Exhibit 1.c).

The broadcast networks have considerable momentum going into the new millennium as the brand-name nationwide distributors of America's most popular TV programming. The outlook for individual broadcasting stations is less clear. The traditional broadcast model features stations operating as outlets for a single channel of programming flowing in one direction only. In a universe of videotape recorders, CDs, and both cable systems and direct-to-home (DTH) satellites offering many dozens of program channels, broadcast stations approaching the information superhighway using this traditional model do so at a distinct technological disadvantage.

TV stations will have new tools with which to enter the fray once the transition to higher-resolution pictures and digital video compression (allowing multiple channels to operate in the space formerly occupied by only one) is complete. Radio stations also will enjoy a more competitive position once they convert to digital signal transmission technology.

One key to broadcasting's survival is consolidation—owning several stations in a market, each targeting a different audience, or owning many stations spread across many markets. The number of stations under consolidated ownership increased dramatically during the 1990s, with radio and TV station deals topping $25 billion in 1996, three times as much as the previous year. In 1999, radio and TV station sales amounted to more than $33 billion (*Broadcasting & Cable*, 14 February, 2000:32). One 1999 sale involved the purchase of AMFM, Inc., by Clear Channel Communications, a deal that brought more than 800 radio stations under the control of a single company.

A second key to survival is diversification into other program delivery technologies. Television broadcasters look forward to offering additional program and data services, opening up entirely new revenue streams, as more efficient use of their channels becomes possible with the implementation of digital technologies.

Renewed commitment to two founding concepts may also help station owners cope with the new mass communication environment:

- *Localism.* Lifestyles today are not what they were in earlier times, and people tend not to "put down roots" as deeply. But stations can serve the unique needs of their communities far better than a national program delivery system can. The popularity of local TV newscasts and the growth of regional cable networks indicate there is a demand for community-centered programming.
- *Universal access.* Americans are accustomed to getting entertainment and information over the air at no additional direct expense once they purchase a receiver. Broadcasting's appeal as a "free" service is strong, and, for many Americans, any delivery system charging a subscription fee is comparatively less attractive. Broadcasters have used the mantra of "free service" effectively in gaining political support, although only about a fifth of U.S. households now receive TV signals directly from broadcast stations.

Cable

Today more people receive their TV programs by cable than by over-the-air broadcast. But cable also offers much more than television.

Many cable operators are experimenting with, or are actually offering, interactive systems that permit subscribers to order movies or TV program reruns whenever they want (video-on-demand) or almost when they want (near-video-on-demand), to play video games, and to order merchandise without calling a toll-free number. As technologies advance, as systems replace coaxial cable with more efficient optical fiber, applications will expand even further.

Using video compression, some cable program services (HBO, for example) are already *multiplexing* their signals, sending out multiple channels. Nearly every day someone announces plans for a new service, specializing in everything from animals and antiques, to cowboys and golf, to parents, and to women's sports. Some systems have added new digital channels, telephony, and high-speed Internet access. But during much of the 1990s, most cable companies focused heavily on defending themselves

EXHIBIT 1.c

CBS On-line

All major networks have entered the superhighway, as this sample screen from CBS illustrates.

Source: Photo from CBS Worldwide, Inc.

against the new threat of direct-to-home satellite competition and preparing for the transition to digital delivery of popular video programming rather than expanding aggressively into new services. As Microsoft computer billionaire turned cable mogul Paul Allen said in mid-1999, "In terms of deploying new technology, things in the cable industry don't happen overnight." (*Broadcasting & Cable*, 21 June 1999: 33)

So, again, basic questions of regulation and economics—not technology—will determine which services will survive. How many subscribers will pay to get them? How many advertisers will pay to have commercials carried by them? Because cable already has a direct, physical connection to the homes it serves, however, it is a front-runner on the information highway.

Although still only a speck in cable's rear-view mirror, direct-to-home satellite service is an increasingly important player in the delivery of popular television programming. Although lacking cable's established market position and a direct, physical connection to the home that encourages interactivity, satellite delivery offers exceptional picture quality and many more program options than most cable systems. Satellite delivery is, of course, most attractive in areas not served by cable, but many users prefer the quality of service offered by satellite even when cable is available. Satellite delivery promises to become an even larger player as local broadcast signals become an integral part of the service.

Telephone

For years telephone companies (*telcos*) satisfied themselves by delivering *plain old telephone service* (POTS), with occasional bursts of modern technology such as call waiting. All that has changed. Now telephony is moving ahead—as either competitor to or partner with cable and others—to build the superhighway.

In 1992 the FCC began relaxing restrictions on telcos, allowing them to provide video services over their networks in competition with cable companies. This led to some interesting ventures. Deals came from everywhere, strange alliances were forged, and bizarre competitions began: Nynex (another RBOC) backed Viacom (a major cable operator and program supplier) in the latter's acquisition of entertainment giant Paramount; Time

Warner (operator of the local cable system) and Rochester (New York) Telephone (an independent telephone company) entered into an agreement whereby each would compete head-to-head with the other, in the same city, in both cable and telephone services; and Southwestern Bell (yet another RBOC) bought two Washington, D.C., cable companies and later sought to compete in the telephone business in that area, which was already served by Bell Atlantic. The head begins to swim.

In passing the Telecommunications Act of 1996, Congress eliminated all restrictions on telcos providing cable services in their local service areas. But telcos, legally freed to compete with cable and other multichannel video providers, began having second thoughts, at least in the short term. The economic risks of entering a new market area, unexpected technical difficulties with transitional video delivery systems, and renewed interest in expanding telephony services combined to slow expansion into video delivery. In 1995 three RBOCs (Bell Atlantic, Pacific Telesis, and Nynex) formed Tele-TV to spearhead their entry into broadband wireline networks and wireless cable. The next year the telcos dramatically scaled back support of Tele-TV and canceled plans to launch new wireless cable systems. The announcement was symptomatic of the telephone companies' on-and-off approach to the video and interactive media market. Although renewing their commitment to offer a full range of video and interactive services, telcos acknowledged that these services will evolve more slowly as they upgrade their telephone networks.

The 1998 acquisition of TCI by AT&T to form AT&T Broadband & Internet Services (AT&T BIS) renewed speculation that this new cable–telephone corporate entity would provide the long-awaited catalyst needed for dramatic developments. The name of the new company strongly suggests development across a wide range of information services, and in announcing the deal, AT&T officials made it clear that the principal motivation behind the acquisition was to gain access to the local telephone and data delivery market via TCI's existing cable systems.

But the telcos have many advantages in the battle for dominance on the information superhighway even without a cable partner. Like cable, the telephone industry already has a physical connection to the home. Indeed, telcos serve far more homes than do cable operators. Telephone companies have an advantage in that their services are traditionally interactive, while cable's service is based, like broadcasting's, on a one-way model. Phone companies will benefit from advancements in technologies, such as digital subscriber line (DSL), that promise to deliver digital-based video and data services using existing twisted-pair phone lines. This will allow phone companies to compete with cable in providing new services without incurring the prohibitive costs of a systemwide rebuild.

In addition to their huge financial base, what gives telephone companies—at least for a time—an advantage over cable in building the superhighway is their experience with sophisticated *switches* that direct signals to and from designated locations and that are essential to any full-scale interactive service (see Exhibit 5t, p. 158). Offsetting this advantage to some degree is cable's experience in programming.

Computers
• • • • • • • •

Regardless of whether program delivery is by cable systems or telephone companies or both, computers will play a central role in building and operating the superhighway. Indeed, electronic media already employ them in a variety of ways. For example, the familiar set-top *converter box* that facilitates multichannel cable is a computer. Television stations and cable systems use computers to insert commercials and program material at prescribed times. High-volume *video servers* that store and release programs for video-on-demand are computers. Personal computers have handled audio, video, and data since the early 1990s. Now the PC is merging with the TV (and the radio receiver as well) to combine broadcast, cable, and DBS reception with computing functions in a single appliance (see Exhibit 1.d).

The computer industry, however, sees itself as a far more important player in the new information age. Software companies such as Microsoft want to build the operating systems that serve as the data highway's traffic cops, controlling the flow of information to and from each viewer's screen. *TV Guide* and others are perfecting electronic navigators that tell viewers what's on television and where to find it and that provide either instant access for direct viewing or allow easy recording of selected programs for later use.

On-screen electronic program guides (EPGs) will include software agents that can search out programming according to type (sports, news, drama, music) following the viewer's preferences and record it automatically. The EPG also may become, in effect, a portal to the television universe and include computer-like interactive features such as e-mail and e-commerce. But in the case of interactive EPGs, as is often true for technological advances generally, technical capability may be racing ahead of consumer education. One survey found that fewer than a third of the people questioned knew what an on-screen interactive program guide was. (CTAM, 1999)

But there's more. PC owners now receive audio and video programming directly on-line through the World Wide Web. The broadcast networks and many individual stations have web sites providing interactive "streaming" (real-time delivery) of their on-air signals and/or prerecorded supplementary multimedia materials. The technical quality, especially the video portion, does not yet equal that of broadcast, DBS, or cable, but it is improving rapidly.

The form that future home communication equipment may take remains unclear. Some say all communication and information functions—telephone, fax, radio, television, multimedia disc-based players/recorders, even home security and utility management—will merge into a single system. The "loaded" PC/TV is a step in that direction, and Exhibit 1.d suggests how such a highly integrated stand-alone device might be configured. It is relatively easy and inexpensive to add digital TV (DTV) broadcast reception capability to such PCs, and some observers think that more people will initially watch DTV broadcasts on their computers rather than purchase the much more expensive DTV sets currently available.

Other observers, however, believe that people will prefer having a greater number of less expensive, more specialized, and easier-to-use communication appliances rather than a "loaded" PC. For these consumers the PC industry promises a basic PC/TV appliance that looks like a conventional TV and functions with a simple remote control. As the decade began, manufacturers were responding to consumer interest in a variety of computer-based communication appliances that provide specialized services in a smaller, more portable, less expensive, and easier-to-use package than the conventional multifunction PC.

It is possible, of course, that the consumer will want it all: a big-screen, high-definition DTV set in the living room as part of a home multimedia center (see Exhibit 1.b), a basic PC/TV in the bedroom,

EXHIBIT 1.d
· · · · · · · · · · ·
Multimedia PC

Today personal computers are at the heart of media convergence. The modern PC/TV combines the features of a conventional TV or radio receiver with the interactive power of a multimedia computer. Additional features not shown in this illustration include, among others, a video camera and microphone for teleconferencing, a video capture card (which allows frame-grabbing from a TV or VCR for use on the In-

ternet or in video editing), a color printer (ink-jet or laser) for providing a hard copy of on-screen materials, a removable high-capacity storage disk drive, and a scanner for capturing images off a piece of paper or from a photographic negative or slide.

MONITOR
The monitor displays the image created by the video card. As prices fall, future monitors will be flat-screen displays rather than the tube-based device illustrated here.

VIDEO CARD
The video card translates the instructions given by your software into images on your monitor. The video card, not the monitor, determines graphics speed and the number of colors the monitor can display.

DVD-ROM DRIVE
The DVD-ROM drive is capable of holding 4.7 to 17 gigabytes. It can also play CD-ROMs and audio CDs.

FLOPPY DISK DRIVE
The floppy disk drive receives removable disks of various sizes (most of which are no longer "floppy") on which data can be stored.

FAX/MODEM CARD
The fax/modem card for Internet access and for sending and receiving faxes.

PHONE CARD
The phone card for the answering machine.

RADIO/TV CARD
The radio/TV receiver card.

SPEAKERS
The latest sound cards produce CD-quality sound, so it's worth having a good set of speakers on which to hear it.

HARD DRIVE
The hard drive is the computer's main storage device, holding the operating system software, applications, and any files you create.

RAM
RAM (random access memory) is the main type of memory in a computer and provides the microprocessor with a workspace.

MICROPROCESSOR
The microprocessor or central processing unit (CPU) chip is the brain of the system. Basically, the better the microprocessor, the more powerful the machine. The power of microprocessors increases rapidly even as prices remain constant or decrease.

SOUND CARD
The sound card converts the digital information generated by your software into stereo sound. It also lets you record sound onto your computer from a microphone or other sound source.

a high-powered PC/TV with a full array of computer peripherals in the home office, and many less sophisticated communications appliances scattered throughout the house.

Newspapers

Some people say that no electronic device can ever duplicate the convenience and utility of the newspaper. Still, many in that industry have wisely elected not to take a chance.

Newspaper giant Knight-Ridder first tried electronic publishing in 1978. But its Viewtron system proved too costly and slow, and folded in 1986. Undaunted by this experience, Knight-Ridder has joined with other major publishers including Gannett, Times Mirror, and the Tribune Company in exploring ways to use computers and telecommunications for news delivery.

The *Chicago Tribune* in 1992 became the first to offer an electronic newspaper—*Chicago Online*—as a national service. Within two years at least 60 dailies, including most of the nation's biggest newspapers, were offering or working on electronic services of some sort. Some newspapers provide the content of their print editions on-line. Others find it more advantageous to use their on-line presence to complement rather than duplicate their hard-copy publications. In addition, many on-line newspaper sites offer audio and video clips of newsmakers, just as the broadcast and cable news networks do. This is not surprising, because print and broadcast organizations, increasingly under consolidated ownership, are sharing their resources. Newspapers typically allow users to search various on-line databases and often provide helpful information about community events, the weather, sports, entertainment, traffic conditions, financial markets, and other items of interest, in addition to links to the company's other media properties (see Exhibit 1.e).

Newspapers also may provide *audiotext* services, whereby the public calls a published number and selects from a menu of items that include news and sports headlines, weather, even samples from recently released musical recordings.

Some newspaper companies have long since expanded into other media areas. The Tribune Company, for example, owns and operates 22 major-market television stations, including national superstation WGN-TV. The company also owns radio stations in Chicago and Denver, and oper-

EXHIBIT 1.e

Media Convergence On-line

This web page provides direct links to three Atlanta mass media properties owned by Cox. The page features links to information appealing to a wide range of interests and provides opportunities to shop on-line, helping to ensure frequent return visits.

Source: © 2000 Cox Interactive Media, Inc.

ates two 24-hour cable news channels: CLTV (ChicagoLand Television) News and Central Florida News 13. Tribune Entertainment, a subsidiary of Tribune Broadcasting, develops and distributes first-run syndicated TV programming, including *Earth: Final Conflict; Malibu, CA;* and

EXHIBIT 1.f
. . . .

Riding in Style

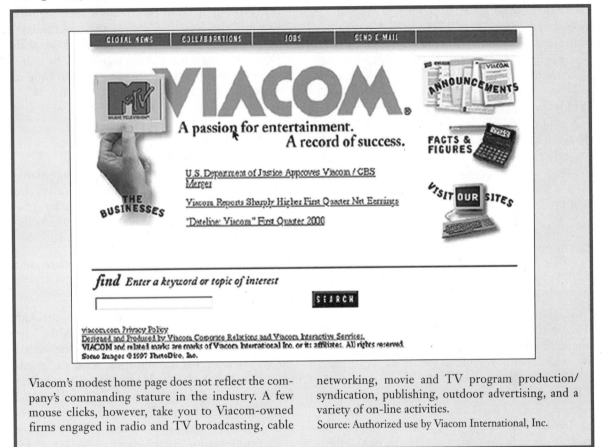

Viacom's modest home page does not reflect the company's commanding stature in the industry. A few mouse clicks, however, take you to Viacom-owned firms engaged in radio and TV broadcasting, cable networking, movie and TV program production/syndication, publishing, outdoor advertising, and a variety of on-line activities.

Source: Authorized use by Viacom International, Inc.

NightMan. Tribune holds an equity position in America Online and other leading Internet companies, and invests as well in The WB Television Network (The WB) and the TV Food Network. In March 2000, Tribune Company announced plans to merge with the Times Mirror Company.

Interestingly, many Americans consider Tribune a "newspaper company" in spite of its broadcasting properties, and think of Rupert Murdoch's News Corporation as a "broadcasting company" because it owns the Fox Network. However, Murdoch's company encompasses not only broadcast stations, broadcast networks, and newspapers but also direct broadcast satellites, cable networks, book publishing, and on-line news, information, and gaming services. Viacom is another example of a colossal multimedia company simultaneously traveling many lanes of the information superhighway (see Exhibit 1.f).

It is important to remember, however, that not every vehicle pulling onto the information superhighway reaches its destination. The roadside is littered with breakdowns. In 1995, for example, 9 major publishers owning about 225 newspapers formed the well-publicized New Century Network (NCN), linking the home pages of affiliated publications at a central site. By early 1997, NCN had more than 80 participating newspapers, and nearly all the nation's major newspapers were expected to be accessible through NCN by 1998. But in the spring of 1998, NCN sputtered to a halt. Reports blamed the venture's demise on disagreements among the controlling publishers regarding business objectives and organizational structure. NCN is only one example of how economics and issues relating to regulation and control may play a more critical role in shaping the future of electronic media than will technology.

1.3 SOME ESSENTIAL TERMS

Given all this, one might well argue that this book's title, *Broadcasting in America*, is misleading. Indeed, its content is not limited to America but rather examines other national systems as well. Nor is it limited, in precise terms, to broadcasting. The trouble is, electronic sources of information and entertainment have grown so varied that, quite frankly, no single word describes them all. The FCC and others use the phrase "broadcasting and electronic media" to distinguish radio, television, and related services such as cable television from common carriers. These terms need exploration here to show how classifications result in legal and financial consequences.

Common Carriers vs. Electronic Mass Media

One way to determine whether a communication service is an electronic mass medium is to ask who has responsibility for its content. Telephone companies are the best example of common carriers. A telephone company may neither edit conversations nor restrict who will make use of its facilities. A broadcast station does both.

Electronic mass media select programs, performers, and speakers and edit what they say and do. In fact, many have a legal *responsibility* to do so. In exchange for assuming this responsibility, electronic mass media come under the protection of the First Amendment to the Constitution, which guarantees freedom from government interference with "speech" and "the press." Operators of common carriers, however, supply communication facilities such as telephones and satellites without assuming any responsibility for what is normally transmitted on those facilities. Having no "speaking role," they have no need to invoke constitutionally protected freedom to speak at will by telephone or satellite.

These fundamental differences between electronic mass media and common carriers have several other practical results, usually involving tradeoffs—each advantage tends to be offset by a disadvantage. The public can demand, as a right, equal access on equal terms to a common carrier, such as a telephone system. No one can assert such a right to use a broadcasting station or cable channel (except for political candidates, who are entitled to *equal opportunities* to use broadcast and cable television facilities). It is easier to obtain a common carrier license than one for a broadcasting station or cable system. The government, however, usually controls common carrier profit levels and the fees they charge their customers. Electronic mass media (except cable, as we shall see) may charge whatever the traffic will bear and may make as much profit as they can, so long as competition prevails.

Broadcasting Defined

Many assume that *any* program or message transmitted over the air, or even over cable, constitutes "broadcasting." But to qualify as broadcasting, programs or messages must pass specific tests. As defined legally, to *broadcast* means

> to send out sound and pictures by means of radio waves through space for reception by the general public.

The phrases "by means of radio waves through space" and "reception by the general public" are crucial: Cable television systems do not broadcast through space, but rather they send programs through cable. And scrambled over-the-air services delivered on a subscription basis are not intended for the "general public," but are meant to be received only by paying subscribers.

Communication "by means of radio waves" includes many nonbroadcast services, such as CB radio, police calls, taxi dispatching, beepers, and mobile telephones. However, these services operate *point-to-point* (from a source to one intended recipient) or *point-to-multipoint* (from a source to more than one specific recipient). Anyone with a suitable receiver can intercept such transmissions, but they are not intended "for reception by the general public" and therefore do not count as broadcasting.

The verb to *broadcast* was adopted to distinguish the new radio communication method from the previous point-to-point orientation of radiotelegraphy and radiotelephony. *Broadcasting* expresses the idea of scattered dissemination to anonymous, undefined destinations (listeners or viewers). The term comes from the farmer's act of hand-sowing grain by *casting it broadly*. The sower lets seeds fall where they may. (Students should be careful to remember, by the way, that the past tense of the verb *broadcast* is also *broadcast*—not *broadcasted*.)

Hybrid Services
• • • • • • • • • • •

Such broad distinctions between common carriers and electronic mass media become less clear with new services that have characteristics of both (referred to here as *hybrids*). Cable television, for example, at first occupied a particularly ambiguous position. Most cable systems operate much like common carriers. They enjoy local monopolies and exercise no editorial control over most of their programs, which come from broadcast stations and cable program networks. Should these factors put cable operators in the same legal basket as common carriers?

Or, since cable systems carry broadcast programs and sometimes originate their own programs in the manner of broadcasters, should they be subject to the same laws as broadcast stations? The question was debated for three decades until Congress finally resolved it in 1984 by recognizing cable as a service that is neither a common carrier nor a broadcaster but a mixture of the two.

Cable operators continue to claim the same First Amendment rights as broadcasters. For example, they maintain that for a city to impose content limitations on a cable system it franchises would violate the cable owner's right of free speech. This claim's legality has been challenged, but so far the courts have upheld it. At the same time, however, cable is subject to limited governmental rate regulation (see Section 12.8, p. 359).

Direct broadcast satellite (DBS) services offer another hybrid example. Satellites usually function as common carriers, but direct-to-home satellite transmission seems equivalent to regular broadcasting, with the station located in space instead of on the ground. However, DBS program providers scramble their signals to prevent anyone other than paying subscribers from receiving them. This barrier to reception by the public puts DBS services in the point-to-multipoint rather than broadcast category. The FCC settled the question in 1987 when it defined DBS transmissions intended only for paying subscribers as *nonbroadcast* services.

Even the telephone, the classic example of a common carrier, has become increasingly hybridized. It first took on a mass entertainment information role when it offered "dial-it" services, sometimes called *mass announcement services*. For a price, listeners can hear, and in some cases participate in, audio "programs" by dialing numbers with special prefixes (usually 1-800 or 1-900).

Public concern over pornographic dial-it services brings into question the common carrier's traditional hands-off policy with regard to message content. It is also possible for one company to provide both common carrier and mass media services, as the entry of AT&T into the cable business (see p. 8) illustrates.

As telephone companies move more and more into multichannel audio and video delivery of entertainment programs and information services—both their own and those of others—they become less and less distinguishable from cable systems. Indeed, telcos have the characteristics of both common carriers (when delivering switched voice/data) and electronic mass media (when delivering cable-like video services).

1.4   BACK TO BASICS
• •

The information superhighway metaphor proves most useful in describing those converging technologies that serve as conduits or delivery systems. It is less helpful in discussions of content providers: program producers and information suppliers.

Cable operators, telephone companies, even computer hardware and software manufacturers want to build the superhighway and control what travels over it. But, in the final analysis, corporate successes in the information age will be determined not by *how* but rather by *what* they deliver. Most people simply have no interest whatsoever in whether they receive information and entertainment through a wire (or possibly *two* wires—one for a cable system and one for the telephone company) or over the air (by traditional broadcast or by direct broadcast satellite), or whether the picture appears on a traditional television receiver or on a computer screen. What they *do* have an interest in is

- the variety and quality of the programs and services they receive
- how easy it is for them to receive them
- how much they have to pay for the package

The student of electronic communication, however, must understand how it has developed, how it works, how it is controlled, how it affects and is affected by its several audiences, and—as we will see in the next chapter—how it all began.

CHAPTER 2

From Radio to Television

CHAPTER CONTENTS

As guideposts in broadcasting's development, six dates stand out:

- 1896, original Marconi wireless patent
- 1920, inception of radio broadcasting
- 1927, passage of Radio Act (most of which carried over to 1934 Communications Act)
- 1948, television network broadcasting begins
- 1975, first home VCR is sold and HBO announces plans to use a satellite for program distribution, setting the stage for cable's expansion
- 1996, FCC adopts technical standards making the transition from analog to digital television broadcasting possible; Congress passes Telecommunications Act

In addition, both world wars helped define important technical phases in radio's development:

- From 1914 to 1918, World War I stimulated rapid evolution of wireless technology and set the stage for radio's debut.
- From 1939 to 1945, World War II temporarily delayed television's emergence, though wartime research developed ultrahigh frequencies that TV's postwar expansion would require. Today's electronic media, however, are based on even earlier cultural and technological precedents.

2.1   CULTURAL PRECEDENTS

Public appetite for mass entertainment developed decades before radio. Popular newspapers, home phonographs, and motion-picture theaters all encouraged a mass media habit, making it easier for broadcasting to achieve success in a very short time.

Urbanization

These older media grew out of fundamental social changes encouraged by the Industrial Revolution (roughly 1740–1850). For centuries most people worked on farms or in related agricultural roles. But industry, based initially on steam power, increasingly drew people away from the land to live and work in cities.

These urban populations became the target of what we now call *mass media*—technology-based means of communication that reach large numbers of people, delivering news and entertainment that most people find interesting and at a price they can afford.

Penny Press

Urban concentration and increasing literacy and leisure time all helped to transform newspapers. Originally aimed primarily at a wealthy and educated elite, they became a product for the masses. The "penny press" signaled this transformation when, in 1833, the *New York Sun* began a trend toward mass-interest, mass-produced papers. Copies sold for a penny, first in the thousands and eventually in the hundreds of thousands. The penny press exploited news of everyday events, sensational crimes, gossip, human-interest stories, sports, and entertaining features—all presented in a breezy style that contrasted with the flowery essay approach of the past. Popular newspapers appealed across lines of class, gender, age, political party, and cherished beliefs. By the 1890s some mass-oriented newspapers had circulations of more than a million.

Vaudeville

Vaudeville was a popular form of entertainment featuring "variety" acts in which performers sang, danced, played musical instruments, and did comedy routines. The first U.S. vaudeville house opened in Boston in 1883, and vaudeville remained immensely popular well into the 1920s, at its peak selling more tickets than all other forms of entertainment combined. New York City alone once had 37 vaudeville houses. The development of motion pictures, especially the "talkies" in the late 1920s, spelled the end of American vaudeville.

But vaudeville developed an audience for forms of entertainment that reappeared later in network radio programming. Vaudeville also served as a training ground for early radio performers such as Jack Benny, Gracie Allen, George Burns, and Milton Berle, all of whom also migrated successfully from radio to TV. Berle's *Texaco Star Theater* on NBC TV was phenomenally popular during the late 1940s and early 1950s, earning Berle the nickname "Mr. Television." The TV audience was small in 1948, but more than 90 percent of the viewers watched Berle's show.

The Phonograph
• • • • • • • • • •

Around the turn of the century, owning a phonograph (often housed in a handsome wooden cabinet) accustomed people to investing in a piece of furniture that brought entertainment into the home. By the end of World War I, on the eve of radio's introduction, some 200 phonograph manufacturers turned out more than 2 million players each year.

But records used acoustic recording methods and were little different from those that Thomas Edison invented in 1878. Radio, in contrast, continued improving its audio quality. By 1930, radio provided better sound than phonographs, as well as "live" music, and it was free, an important consideration during the Great Depression. Broadcasting companies later bought failing phonograph firms, shared radio's advanced audio technology, and brought them back to life. When TV challenged radio in the 1950s, radio broadcasting and the recording industry formed a close interdependent relationship.

Like vaudeville, the phonograph helped create a popular demand for music that radio then supplied, contributing to radio's rapid growth after 1920.

Motion Pictures
• • • • • • • • • •

Like the phonograph industry, movies were well established by the time broadcasting began in the 1920s. Moviegoers returned weekly for serial dramas, which eventually had their counterparts in broadcasting. Movies also benefited from radio's electronic technology.

The lack of synchronized sound—the precise matching of sound and action—long stymied progress toward acceptable "talkies." They finally began in earnest in 1928, with several rival sound systems competing for acceptance. One of these sound-on-film systems had been developed by RCA, the owner of the first national radio network—an example of the many links between broadcasting and motion pictures in the 1920s, long before television brought the two visual media into an even closer relationship.

In addition to technological contributions, motion pictures also helped establish the "visual grammar" later adopted by television. Shooting and editing techniques developed by motion-picture pioneers such as Edwin S. Porter, David Wark Griffith, Sergei Eisenstein, and many others not only helped shape popular tastes regarding visual presentations but also provided a ready guide for early TV producers. Many of the cinematic conventions in today's television programs date back to the birth of the motion-picture industry.

2.2 TECHNOLOGICAL PRECEDENTS

The "penny press," the phonograph, and the motion-picture industry evolved from technological advances in mass communication, but broadcasting's technological precedents focused on point-to-point communication, which may have delayed broadcasting's development. It is also worth noting that broadcasting's technological precedents were electronic, whereas earlier mass media—newspapers, phonographs, and motion pictures—employed mostly mechanical technologies.

The Telegraph
• • • • • • • • • •

The British first developed electrical telegraphy in the 1820s as an aid to early railroad operations. An American artist/inventor, Samuel F. B. Morse, conducted extensive telegraph experiments in the 1830s—at a time when westward expansion put a high premium on rapid communications over long distances. Morse made significant telegraph improvements, including a receiver that recorded messages on strips of paper. He and a partner devised the American Morse code to translate numbers and letters into the dots and dashes of telegraphic language. A slightly modified version called the Continental or International Code served for early prebroadcast wireless communications.

In partnership with the federal government, Morse installed the first operational U.S. telegraph line from Washington to Baltimore in 1844. Three years later the government sold its interest to private investors, retaining only the right to regulate telegraphic services. By the end of the Civil War, Western Union had emerged as the dominant telegraph company.

The decision to sell, or privatize, the telegraph started the United States on a telecommunications path different from that of most other countries.

Typically, governments own and operate national telecommunications systems under Post and Telegraph (PTT) administrations, which often also had some control over broadcasting. However, in recent years PTT monopolies have come under increasing attack as countries adopt the more efficient American model of privately owned, competitively operated telecommunication services.

Communications over transatlantic telegraph cables began in 1866 (after a short-lived attempt in 1858), enabling the exchange of information between Europe and North America in minutes instead of weeks. Newspapers immediately seized on both submarine and land telegraph as a means of sharing information in ways not previously possible. The telegraph made news agencies such as Havas (France, 1835), Reuters (Great Britain, 1851), and the Associated Press (United States, 1856) more important players in the newspaper and magazine business.

The telegraph had a revolutionary influence because it allowed communication with distant locations at a pace literally millions of times faster than anything before it. The shrinking of the Earth into an electronically interconnected "global village" began with the telegraph.

The Telephone

Inventors next turned their attention to wire transmission of speech itself, seeking a way to eliminate tedious encoding and decoding of telegraph messages. Investigators worked simultaneously in many nations on developing what became the telephone. Several appeared on the verge of a solution when Alexander Graham Bell delivered his designs to the U.S. Patent Office in 1876, beating fellow American Elisha Gray by only a few hours. Gray contested Bell's patent, but Bell won in the courts. Bell organized the original Bell Telephone Company a year later, when he secured a second essential patent. Though control over Bell's patents soon passed to others who went on to develop the company known today as AT&T (until 1994, American Telephone & Telegraph Company), Bell's name still means "telephone" to many people.

AT&T developed a "long lines" network connecting local exchanges with one another, soon owning the more important local companies. In 1913, AT&T ensured its supremacy in long distance by acquiring a license to a crucial electronic invention, the Audion (see Section 2.3), which made coast-to-coast telephone service possible. AT&T's control of long-distance voice communication by wire had an important bearing on broadcasting, which after 1927 depended on such wire links for network operations.

Three giant manufacturing firms would also play key roles in radio broadcasting: Western Electric (AT&T's manufacturing arm), General Electric (GE), and Westinghouse. Together they dominated the electrical industry and helped shape its participation in early wireless communication.

2.3 WIRELESS COMMUNICATION

Simultaneous with the development of the telephone, other inventors were working on wireless systems. American dentist Mahlon Loomis transmitted electrical signals almost twenty miles in 1866, and six years later received a patent for his wireless telegraph system. Other Americans tinkering with wireless transmission included William Henry Ward, Nathan B. Stubblefield, and even the "Wizard of Menlo Park," Thomas Edison. France, Germany, Great Britain, and other nations have similar lists of people experimenting with wireless communication in the late 1800s.

Hertzian Waves

Many early experimenters succeeded by "accident" but lacked the understanding of scientific principles required for systematic development of their discoveries. Others, such as Edison, became distracted by projects they considered more rewarding. Their failure to develop a viable commercial wireless communication system was not due to a lack of published information about the subject.

In 1873, James Clerk Maxwell published a paper in which he theorized that an invisible form of radiant energy—electromagnetic energy—must exist. He described it mathematically, foreseeing that it resembled light radiating out from a source in the form of waves.

In 1888, German physicist Heinrich Hertz reported his brilliant laboratory experiments that proved conclusively the validity of Maxwell's theory. Hertz generated electromagnetic waves, transmitted them down the length of his laboratory, detected them, measured their length, and showed that they had other wavelike properties similar to those of light.

In effect, Hertz demonstrated radio. But he sought only to verify a scientific theory, not to develop a new way of communicating. Indeed, when asked whether his discovery could be used for communication, Hertz provided theoretical explanations of why it could not. It remained for other experimenters to communicate with what they called Hertzian waves.

"The Right Releasing Touch"

A young Italian experimenter, Guglielmo Marconi, is credited with developing radio as a viable means of communication. He supplied "the right releasing touch," as a Supreme Court justice put it in upholding Marconi's primacy in a later patent suit (US, 1942: 65). Stimulated by Hertz's paper, Marconi experimented with Hertzian wave signals in the early 1890s, first sending them across the space of an attic, then for greater distances on the grounds of his father's estate.

It was characteristic of Marconi's aggressive personality that he decided his work was "ready for prime time" when he succeeded in transmitting Hertzian waves a mere two miles. He patriotically offered it to the Italian government, which couldn't see the potential. Still only 22, Marconi next went to England, where in 1896 he registered his patent. He soon launched his own company to manufacture wireless-telegraphy equipment and to offer wireless services to ships and shore stations.

To a remarkable degree Marconi combined an inventor's genius with that of a business innovator. As inventor he persisted tirelessly, never discouraged even by hundreds of failed attempts at solving a problem. In 1909 Marconi shared the Nobel Prize in physics (with Germany's Ferdinand Braun) for achievements in wireless telegraphy. As a business manager, Marconi demonstrated a rare entrepreneurial talent plus a flair for effective

 EXHIBIT 2.a
• • • • • • • • • •

Guglielmo Marconi (1874–1937)

Marconi is shown here with some of the early wireless apparatus he used to demonstrate his "miraculous" invention. He is holding a paper tape in his right hand on which a radiotelegraphic message is inscribed in Morse code, a recording technique borrowed from wire telegraphy. Marconi was quick to borrow ideas from other inventors if doing so improved his devices. He was equally quick to patent whatever he could and then to agressively defend his patents.

Source: AP/Wide World Photos.

public relations. In the early 1900s, to prove his system's value, he repeatedly staged dramatic demonstrations of wireless to skeptical officials, scientists, investors, and equipment buyers. Exhibit 2.a shows Marconi with some of the equipment used in these demonstrations. Realizing the valuable publicity that bridging the Atlantic with radio signals would create for his company, Marconi established a receiving station on Signal Hill, 600 feet above the harbor of St. John's, Newfoundland. On December 12, 1901, he heard three short dots—Morse code for the letter *S*— transmitted from England.

Among his business ventures, Marconi's American branch had a decisive influence on the development of broadcasting in America. Founded in 1899, American Marconi developed a virtual monopoly on U.S. wireless communication, owning 17 land stations and 400 shipboard stations by 1914. All these facilities used a wireless extension of the telegraph principle—point-to-point communication between ships and shore stations, between ships at sea, and to a lesser extent between countries.

This limited application of radiotelegraphy dominated the first two decades of radio service because the relatively crude Marconi equipment could transmit only Morse code dots and dashes. Telephones could have been used as microphones to transmit wireless speech, but methods had not yet been found to impress such complex information onto radio waves.

The Vacuum Tube

The solution to this and related radio problems came with the development of an improved *vacuum tube*, capable of electronically amplifying the changing volume and pitch of speech. American inventor Lee de Forest followed up on research leads suggested by Edison's 1883 electric lamp and the original two-element vacuum tube (*diode*) that Ambrose Fleming patented in 1904.

De Forest's crucial improvement consisted of adding a third element to a vacuum tube, turning it into a *triode*. He positioned his new element, a tiny grid or screen, between the diode's existing filament and plate. A small voltage applied to the grid could control with great precision the flow of electrons from filament to plate. Thus a weak signal could be enormously amplified yet precisely modulated. De Forest first experimented with his triode, or *Audion*, in 1906.

It took more than a dozen years to develop the Audion and the new circuits to go with it—and much of the necessary work was done by others. De Forest spent much of his time defending his Audion patents in court. Ambrose Fleming claimed de Forest had infringed his 1904 diode patent. Other British inventors, including H. J. Round and C. S. Franklin, claimed credit for developing electronic amplification devices. In America, General Electric's Irving Langmuir and AT&T's Harold D. Arnold also made claims against de Forest.

The most debated issue involved the *regenerative* or *feedback circuit*, a revolutionary design allowing the Audion to generate radio signals in transmitters and greatly improve receiver sensitivity and selectivity. Both de Forest and Edwin Armstrong filed patents for this critically important application of the amplifying vacuum tube in 1914. The engineering community thought Armstrong had a better understanding of the principles involved, but the Supreme Court ruled in favor of de Forest.

In spite of these legal challenges, de Forest sold seven of his Audion patents to AT&T in 1913. AT&T used these patents to develop vacuum-tube repeaters (amplifiers in telephone lines), achieving the first coast-to-coast telephone service two years later.

The Audion marked a great leap forward into the modern age of electronics. Electrons (particles of energy smaller than atoms) could be controlled. This transforming ability to manipulate electrons would revolutionize not only communications but also virtually all science and industry.

Commercial Uses of Wireless

Into the 1920s, wireless firms made money by supplying telegraphic communication among ships, between ships and shore stations, and between continents. Overland wireless-telegraphy services had less appeal because telephone and telegraph lines satisfied existing needs.

Wireless offered unique advantages to maritime commerce. In fact, Marconi chose international yacht races for some early public wireless demonstrations. Wireless also gained invaluable publicity from its lifesaving role in maritime disasters, one as early as 1898. Each year the number of rescues increased. Exhibit 2.b describes the most famous of them all.

Radio had commercial possibilities as a long-distance alternative to submarine telegraph cables, which were fragile and enormously costly to install. On the eve of the U.S. entry into World War I, General Electric developed its *alternator*, a new type of radio-frequency generator, to send transatlantic messages reliably—a major improvement in long-distance radio communication. But the alternator was a huge and inefficient device compared to tubes. During the 1920s, vacuum-tube transmitters displaced alternators, and radio became a strong competitor of submarine telegraph cables. Tubes enabled use of the shortwave (high-frequency) portion of the spectrum—more efficient than low frequencies previously used for long-distance communication. A sharp rise in transatlantic radio traffic followed.

EXHIBIT 2.b
. . . .

The *Titanic* Disaster, April 1912

The photo above left shows the *Titanic* as the ship looked embarking on its maiden voyage. To the right is a photo showing the *Titanic* wireless room.

A luxury liner advertised as unsinkable, the *Titanic* struck an iceberg and sank in the Atlantic on her maiden voyage from Britain to the United States in April 1912. One heroic Marconi radio operator stayed at his post and went down with the ship; the second operator survived. Some 1,500 people died—among them some of the most famous names in the worlds of art, science, finance, and diplomacy—in part because each nearby vessel, unlike the *Titanic*, had but one radio operator (all that was then required), who had already turned in for the night. Only by chance did the operator on a ship some 50 miles distant hear distress calls from the *Titanic*. It steamed full speed to the disaster site, rescuing about 700 survivors.

Radiotelegraphy's role as the world's only thread of contact with the survivors aboard the rescue liner *Carpathia* as it steamed toward New York brought the new medium of wireless to public attention as nothing else had done. Subsequent British and American inquiries revealed that a more sensible use of wireless

(such as a 24-hour radio watch) could have decreased the loss of life. Because of such findings, the *Titanic* disaster influenced the worldwide adoption of stringent laws governing shipboard wireless stations. The *Titanic* tragedy also set a precedent for regarding the radio business as having a special public responsibility. This concept carried over into broadcasting legislation a quarter of a century later.

The *Titanic* continues to exert an almost eerie influence on American culture today. Discovery of the wreck on the ocean floor in 1985 by researchers from the Woods Hole Oceanographic Institution led by Dr. Robert D. Ballard renewed interest in the tragedy. Dozens of books, TV documentaries, and a blockbuster award-winning motion picture in 1997 followed the discovery.

Sources: *Titanic* photo, Corbis-Bettmann; *Titanic* wireless room, FPG International.

Military Wireless
. . . .

Naturally, navies took an interest in military applications of wireless from the outset. The British Admiralty used Marconi equipment as early as 1899, during the Boer War, but with little success. Technology advanced rapidly during the next decade,

however, and wireless telegraphy became an important military weapon.

In April 1917, as America entered World War I, the U.S. Navy saw wireless as a threat to national security. Enemy spy agents could (some did) use radio to send information about ship movements, for example. Therefore, the navy took over all wireless stations in the country, commercial and amateur alike. It dismantled most, operating a few for its own training and operational needs.

The Army Signal Corps also used radio, as did the fledgling Air Service. Some 10,000 soldiers and sailors received training in wireless. After the war they formed a cadre of amateur enthusiasts, laboratory technicians, and electronics manufacturing employees. They helped popularize radio, creating a ready-made audience for the first regular broadcasting services that began two years after the war ended.

To mobilize America's wireless resources for war more rapidly, the navy decreed a moratorium on patent lawsuits over radio inventions. This allowed manufacturers to pool their patents and use any patent needed. Patent sharing and the pressure of war encouraged companies to move forward with innovations they might not have risked during peacetime.

By the end of the war, big business had developed a stake in wireless and was ready to branch out into new applications. AT&T had added wireless rights to its original purchase of telephone rights to de Forest's Audion. General Electric owned patents for its powerful alternator and had the ability to mass-produce vacuum tubes. Westinghouse, also a vacuum tube producer, joined in seeking new ways to capitalize on wireless. But before any of these firms could move forward, they had to come to terms with the navy, which also had claims on wireless.

2.4 EMERGENCE OF BROADCASTING

Throughout the early 1900s, eager experimenters constantly sought an effective means of *radiotelephony*, the essential precursor of broadcasting.

Early Radiotelephone Experiments

In 1906, Reginald Fessenden made the first known radiotelephone transmission resembling what we would now call a broadcast. Using an ordinary telephone as a microphone and an alternator to generate radio energy, Fessenden made his historic

transmission on Christmas Eve from Brant Rock, on the Massachusetts coast south of Boston. He played a violin, sang, read from the Bible, and played a phonograph recording. Ship operators who picked up the transmission far out at sea could hardly believe they were hearing human voices and musical sounds from earphones that previously brought them only the monotonous drone of Morse code.

In 1907, hard on the heels of Fessenden, de Forest made experimental radiotelephone transmissions from downtown New York City. Some of his equipment appears in Exhibit 2.c.

De Forest, of course, used continuous waves produced by his Audion. A year later, in Great Britain, H. J. Round used yet another device, the *arc transmitter*, to produce the continuous waves needed for voice transmission. Round sent voice messages a distance of 50 miles, but clearly de Forest's tube technology was superior to either Fessenden's alternator or Round's arc transmitter for radiotelephony. In 1915, AT&T engineers connected 500 tubes together and transmitted voice successfully, even if only experimentally, from Arlington, Virginia, to the Eiffel Tower in Paris.

By 1916 de Forest had set up in his Bronx home an experimental transmitter over which he played phonograph recordings—and even aired election returns in November of that year, anticipating by four years the opening broadcast of KDKA. With some justification, de Forest called himself *Father of Radio*, the title of his 1950 autobiography.

Government Monopoly: The Road Not Taken

Though World War I ended in November 1918, the navy did not relinquish control of radio facilities for another 18 months. The critical decisions made at this time profoundly affected the future of broadcasting in America.

Was radio too vital to entrust to private hands? The navy thought so, and it supported a bill in Congress late in 1918 proposing, in effect, to make radio a permanent government monopoly. Despite strong arguments from navy brass at the hearings, the bill failed. Thus radio took the road of private enterprise in the United States, though in many other countries governments remained in charge as

EXHIBIT 2.c
. . . .
Lee de Forest (1873–1961)

It was, by sort of this equipment and through these two headphones that the voice of Madame Farrar was heard at Brooklyn Navy Yard in October 1907 — She being the first woman to sing over The Wireless Telephone

Frank E. Butler

De Forest's invention of the amplifying vacuum tube, the Audion, was critical to the development of modern electronic communication, including broadcasting. The shipboard receiver also pictured is typical of the type used to receive de Forest's experimental radiotelephone transmissions in 1907.

Source: De Forest photo from Corbis-Bettmann; transmitter photo from Smithsonian Institution, negative #79-11558.

radio expanded in the postwar years. Only recently has the global trend been toward privatization of at least some aspects of broadcasting.

Origin of RCA
. . . .

The restoration of private station ownership in 1920 would have meant returning most commercial wireless-communication facilities in the United States to foreign control. Disturbed at the prospect of British Marconi (parent of American Marconi) consolidating its U.S. monopoly by purchasing exclusive rights to GE's alternator, the navy strongly opposed the deal.

GE's alternator was the most advanced technology for intercontinental wireless, and the U.S. government had learned during the recent war how ridiculously easy it was for enemy nations to slash undersea cables. The government feared losing a vital national defense technology to a foreign-controlled corporation. American wireless manufacturers simply feared Marconi would renew aggressive legal actions to protect patents shared during the war, limiting their ability to maximize profits in a booming postwar economy. Caught in a squeeze play between the U.S. government and the giants of American industry, the British-based company could neither expand nor effectively operate. With tacit government approval, Owen Young, head of GE's legal department, quietly negotiated the purchase of all American Marconi stock in a "semi-friendly" takeover.

In October 1919, GE created a subsidiary to take over and operate American Marconi—the Radio Corporation of America (RCA), shortly to become the premier force in American broadcasting. Under RCA's charter, all of its officers had to be Americans, and 80 percent of its stock had to be in American hands.

Westinghouse and AT&T joined GE as investors in the new corporation. AT&T sold its interest in 1923, but RCA remained under GE and

EXHIBIT 2.d

Sarnoff and Paley: Giants of Early Broadcasting

Both network broadcasting pioneers came from immigrant Russian families, but there the similarity ceases. Sarnoff (left) rose from direst poverty, a self-educated and self-made man. In contrast, Paley (right) had every advantage of money and social position. After earning a degree from the Wharton School of Business at the University of Pennsylvania in 1922, he joined his father's prosperous cigar company.

The differences between Sarnoff and Paley extended to their personalities and special skills. Sarnoff was a "technician turned businessman, ill at ease with the hucksterism that he had wrought, and he did not condescend to sell, but Bill Paley loved to sell. CBS was Paley and he sold it as he sold himself" (Halberstam, 1979: 27).

Sarnoff had been introduced to radio by way of hard work at the telegraph key for American Marconi, Paley by way of leisurely DX (long-distance) listen-

ing: "As a radio fan in Philadelphia, I often sat up all night, glued to my set, listening and marveling at the voices and music which came into my ears from distant places," he recalled (Paley, 1979: 32).

Paley's introduction to the business of radio came through sponsoring of programs. After becoming advertising manager of his father's cigar company in 1925, Paley experimented with advertisements on WCAU in Philadelphia. Impressed with the results, he explored getting into the radio business, and late in 1928 took over the struggling CBS network.

Both men were highly competitive and pitted their companies against each other for 40 years before Sarnoff's retirement in 1969. Sarnoff died in 1971; Paley lived until 1990.

Source: Photos from Corbis-Bettmann.

Westinghouse control until 1932, when an antitrust suit forced them to make RCA an independent corporation. Thus it remained for some seven decades until, in an historical irony, GE bought RCA back again (see Section 3.6, p. 77).

If RCA became broadcasting's leading corporation, RCA executive David Sarnoff became

broadcasting's leading corporate figure. From 1930, when he became president, to 1969, when he retired as board chair, he played a major role in the creation of network radio, the evolution of television, and the conversion of television to color. Today a topflight research laboratory bears his name. Exhibit 2.d details his career and that of his chief rival.

EXHIBIT 2.e
• • • • • • • • • •

Conrad's 8XK and Its Successor, KDKA

Dr. Frank Conrad (left) and his amateur radio setup typified the improvised wireless stations operated by inventors and experimenters. It contrasts with the first KDKA transmitter facilities (right), with which the Harding-Cox election returns were broadcast on November 2, 1920.

Source: Conrad photo from UPI/Corbis-Bettmann; KDKA photo CBS Photo Archive.

Cross-Licensing Agreements
• • • • • • • • • • • • • • • • • • •

RCA and its parent companies each held important radio patents, yet each found itself blocked by patents the others held. From 1919 to 1923, AT&T, GE, Westinghouse, RCA, and other minor players worked out a series of cross-licensing agreements, modeled after the navy-run wartime patent pool. Under these agreements, each company had its own slice of the electronics manufacturing and services pie. Each parent company and RCA could build equipment for its own use, but AT&T got exclusive rights to build transmitters for sale to "outsiders." Both GE and Westinghouse held exclusive rights to build receivers sold through RCA. These arrangements were mutually advantageous given the state of wireless communication at the time.

Within a few years, however, these carefully laid plans fell into disarray because of the astonishingly rapid growth of a new use for radiotelephony—broadcasting.

The First Broadcasting Station
• • • • • • • • • • • • • • • • • • •

Stations in many nations claim to be the world's oldest broadcasting station. Certainly, there were stations transmitting forms of information and entertainment we now consider broadcasting following World War I. But these stations typically operated sporadically and often the "program" was just something to make a noise, the real purpose of the transmission being to test the equipment.

Many believe the first broadcasting station began as 8XK, an amateur radiotelephone station operated in a Pittsburgh garage by Westinghouse engineer Frank Conrad. Dr. Conrad's equipment (Exhibit 2.e) was crude, with various components connected by a jumble of exposed wires.

In the fall of 1919 Conrad fell into the habit of transmitting recorded music, sports results, and bits of talk in response to requests from other amateurs. These informal transmissions (hardly "programs" in the formal sense) built up so much interest that newspapers began to comment on them.

Managers at the Joseph Horne Company, a Pittsburgh department store, noticed a growing public interest in wireless and decided people might buy receiving sets to pick up Conrad's "broadcasts." They installed a demonstration receiver and ran a newspaper display ad in September 1920: "Air Concert 'Picked Up' by Radio Here . . . Amateur Wireless Sets made by the maker of the Set which is in operation in our store, are on sale here $10.00 up."

Horne's ad caught the eye of Westinghouse vice president H. P. Davis, who immediately saw a novel merchandising tie-in: The company could increase demand for its receivers and promote Westinghouse's name by regularly transmitting programs. Westinghouse adapted a transmitter, installed it in a makeshift shack on the company's tallest building in East Pittsburgh, and at 6:00 P.M. on November 2, 1920, began broadcasting as KDKA.

By voluntary international agreement, all U.S. broadcasting stations have an initial *K* or *W* in their call signs. Although there are exceptions, such as KDKA, most stations east of the Mississippi River have *W* calls, and those to the west have calls starting with *K*. Some pioneer stations, such as WWJ (Detroit) and WHA (Madison, Wisconsin), received three-letter calls and retained them over the years. Most stations have four-letter calls. Because KDKA's opening coincided with Election Day, 1920, the maiden broadcast reflected public interest in presidential vote results. KDKA offered brief news reports of election returns, fed to the station by telephone from a newspaper office, alternating with phonograph and live banjo music. The announcer, Leo Rosenberg from Westinghouse's public relations department, repeatedly asked anyone listening to contact the company. After the election, KDKA began a regular daily hour-long schedule of music and talk.

KDKA met five criteria that qualify it as the oldest U.S. station still in operation, despite many other claims based on earlier experiments. KDKA (1) used radio waves (2) to send out uncoded signals (3) in a continuous, scheduled program service (4) intended for the general public and (5) was licensed by the government to provide such a service (Baudino & Kittross, 1977). However, no broadcasting licenses as such existed at the time; KDKA had the same kind of license as ship-to-shore radiotelegraphic stations.

Unhampered by competing signals, KDKA's nighttime signal could be picked up at great distances. Newspapers in the United States and Canada printed the station's program schedule. As other stations came on the air, some observed a "silent night" once a week to enable listeners to receive weak "DX" (long-distance) signals from far-off stations whose transmissions would be otherwise drowned out by local stations.

Westinghouse created KDKA to spur demand for Westinghouse receivers, but many early listeners tuned in on homemade sets. As early as 1906, experimenters discovered that certain minerals (crystals) had an amazing ability to detect Hertzian waves when touched in "hot spots" with a thin metallic wire, popularly called a "cat whisker." Almost anyone could assemble a complete "crystal set" for about $10. Best of all, crystal sets required no electrical power for their operation.

In its first year of operation, KDKA pioneered many types of programs that later became standard radio fare: orchestra music, church services, public-service announcements, political addresses, sports events, dramas, and market reports. But KDKA lacked one now-familiar type of broadcast material—commercials. Westinghouse paid KDKA's expenses to promote sales of its own products. The company assumed that each company wanting to promote its wares over the air would operate its own station.

2.5   BROADCASTING BECOMES AN INDUSTRY

Westinghouse did not have the field to itself for long. Department stores, newspapers, educational institutions, churches, and electrical equipment dealers all soon operated their own stations. In early 1922 the new industry gathered momentum; by May more than 200 stations had been licensed, and the number passed 500 early in 1923.

But the number of broadcasting stations did not necessarily increase from year to year during the 1920s. Exact numbers are hard to determine, but some sources show fewer stations on the air at the start of 1924 than had been operating at the start of 1923, for example. The declining numbers in some years probably reflect the fact that some pioneer station owners discovered building and operating a broadcasting station was more expensive than they had anticipated. Receiver ownership, however, increased dramatically and consistently. In 1922 about 60,000 households had a radio; only two years later more than a million had radios.

Conflicting Approaches

RCA entered the field in 1921, purchasing WJZ, a second station that Westinghouse built to reach the New York market. WJZ assumed responsibility for producing its own programs, as had KDKA. But when AT&T put WEAF on the air in New York City in 1922, it took a different approach. The telephone company explained that it would furnish no programs whatsoever. AT&T perceived broadcasting as another *common carrier*—merely a variation of telephony. In a 1922 press release about WEAF, AT&T explained its approach:

> Just as the [telephone] company leases its long distance wire facilities for the use of newspapers, banks, and other concerns, so it will lease its radio telephone facilities and will not provide the matter which is sent out from this station. (Quoted in Banning, 1946: 68)

Strange as it seems today, AT&T could not see the difference between a telephone booth and a broadcasting station. It soon became clear, however, that filling radio schedules entirely with leased time would not work. As with print media, advertisers would have to foot the bill. Advertisers initially, however, had no idea how to prepare programs capable of attracting listeners. To fill its schedule, the telephone company found itself getting into show business after all—a decidedly uncomfortable role for a regulated monopoly bent on maintaining a serious and dignified public image.

Soon the industry broke into two conflicting groups with different ideas about the way broadcasting would work. The "Radio Group" consisted of Westinghouse, GE, and RCA; the "Telephone Group" was made up of AT&T and its Western

Electric subsidiary. In the end, it turned out that each group was partly right.

The Telephone Group correctly foresaw that economic and technical limitations made it impractical for every company wanting to communicate with the public to operate its own radio station. Instead, each station would need to make its services available to many different advertisers. AT&T miscalculated, however, in emphasizing primarily those interested in sending messages instead of those receiving them. Public goodwill had to be earned with listenable programs to pave the way for acceptance of advertising. Here the Radio Group's strategy of providing a program service prevailed. It took about four years for these conflicting approaches to sort themselves out.

Radio Advertising

AT&T's WEAF called the sale of commercial time *toll broadcasting*, an analogy to telephone long-distance toll calls. It first leased facilities for a toll broadcast in August 1922 when a Long Island real estate firm paid for 10 minutes of time to extol the advantages of living in an apartment complex in New York. AT&T would not allow mentioning anything so crass as price.

Advertising on something as "personal" as radio was considered a bit scandalous. Many radio pioneers, including Lee de Forest and David Sarnoff, initially abhorred the practice. Potential advertisers must have had similar thoughts, because net income from advertising after four months was only about $5,000. Broadcasting was not making a large contribution to AT&T's profits.

In 1923 the first weekly advertiser appeared on WEAF, featuring a musical group the client called "The Browning King Orchestra"—which ensured frequent mention of the client's name, although the script carefully avoided mentioning that Browning King sold clothing. By 1928, under pressure of rising station operating costs and advertiser demand, more blatant advertising had become acceptable.

Although AT&T employees had to help advertisers create their programming during the early days of WEAF, relatively few broadcasters developed programming and production skills, leaving a vacuum that advertising agencies filled. They took over commercial program production, introducing the concept of sponsorship. Unlike most modern advertisers, early sponsors did more than simply advertise—they also owned the programs that served as vehicles for their advertising messages. Later on, during the height of network radio's popularity, advertisers and their agencies controlled most major entertainment shows, a surrender of control that broadcasters lived to regret.

Networks

AT&T found a long-lasting broadcast role in its interpretation of the RCA cross-licensing agreements, claiming exclusive rights to interconnect radio stations with its telephone lines. Thus at first only AT&T-owned WEAF could set up a network. It began in 1923 with the first permanent station interconnection—a telephone line between WEAF in New York City and WMAF in South Dartmouth, Massachusetts. By October 1924 AT&T had a regular network of six stations and assembled a temporary coast-to-coast hookup of 22 outlets to carry a speech by President Calvin Coolidge. In broadcasting's early years, it was common practice to call interconnected stations carrying the same program a chain. The terms *chain* and *network* mean the same thing.

Denied the use of AT&T telephone interconnection for its rival network, the Radio Group's stations turned to Western Union telegraph lines. Intended only for the simple on/off pulses of Morse code, the Western Union wires could not carry music and voice with the fidelity of telephone lines, which were specifically designed for audio. Nevertheless, in 1923 WJZ formed a wire link to a station in Washington, D.C., and by 1925 had succeeded in organizing a network of 14 stations.

Cross-Licensing Revisited

The growing demand for broadcasting equipment, especially receivers, upset the careful balance that the cross-licensing agreements among Westinghouse, GE, AT&T, and RCA had created. The agreements had not included broadcasting, which was seen to be of only minor importance. Further, a federal suit alleged that the cross-licensing agreements violated antitrust laws by aiming to control the manufacture and sale of all radio equipment. The suit added urgency to the need for change. At the same time, AT&T concluded that its original concept of broadcasting as a branch of telephony was mistaken.

Accordingly, in 1926 the partners in the cross-licensing agreements redefined the parties' rights to use their commonly owned patents and to engage in the various aspects of the radio business. Briefly, AT&T refocused on telephony, selling WEAF and its other broadcasting assets to the Radio Group for $1 million, agreeing not to manufacture radio receivers. RCA won the right to manufacture receivers while agreeing to lease all network relays from AT&T. It may appear that this concession was small compensation for AT&T, but such was not the case. Radio, and later television, networking became a highly profitable business, a business monopolized by AT&T for many decades.

This 1926 agreement had a defining influence on the future of broadcasting in America. As long as two powerful groups of communications companies fought over basic concepts, broadcasting's economic future remained uncertain. The 1926 agreements removed that uncertainty.

Origin of NBC

A few months after the 1926 settlement, the Radio Group created a new RCA subsidiary, the National Broadcasting Company (NBC)—the first American company organized specifically to operate a broadcasting network. NBC's 4½-hour coast-to-coast inaugural broadcast in late 1926 reached an estimated 5 million listeners and was reported to have cost $50,000 to produce. It took another two years, however, for coast-to-coast network operations to begin on a regular basis.

In 1927 RCA expanded NBC into two semiautonomous networks, the Blue and the Red. WJZ (later to become WABC) and the old Radio Group network formed the nucleus of the Blue; WEAF (later to become WNBC) and the old Telephone Group network formed the nucleus of the Red.

History suggests NBC's networks took their names from the red and blue lines drawn on a map showing interconnecting links among affiliates. Of the two networks, NBC-Red was predominant, with more powerful stations, more popular programming, and a larger income from advertising. NBC-Blue affiliates were weaker stations, or those in less populous cities, and the Blue network offered more public-service programming. Until the 1940s NBC had both a Red and a Blue network affiliate in most major cities, allowing NBC to program two stations in the largest markets during network radio's golden age. NBC's simple Red and Blue designations provide an example of product "branding" long before the concept became a staple of modern broadcast marketing.

Origin of CBS

In 1927, soon after NBC began, an independent talent-booking agency, seeking an alternative to NBC as an outlet for its performers, started a rival network. It went through rapid changes in ownership, picking up the name Columbia Phonograph Broadcasting System as a result of a record company's investment. In September 1928, William S. Paley purchased the "patchwork, money-losing little company," as he later described it. When he took over, CBS had only 22 affiliates. Paley quickly turned the failing network around with a new affiliation contract. In his autobiography a half-century later he recalled:

> I proposed the concept of free sustaining service. . . . I would guarantee not ten but twenty hours of programming per week, pay the stations $50 an hour for the commercial hours used, but with a new proviso. The network would not pay the stations for the first five hours of commercial programming time. . . . [T]o allow for the possibility of more business to come, the network was to receive an option on additional time. And for the first time, we were to have exclusive rights for network broadcasting through the affiliate. That meant the local station could not use its facilities for any other broadcasting network. I added one more innovation which helped our cause: local stations would have to identify our programs with the CBS name. (Paley, 1979: 42)

Paley's canny innovations became standard practice in network contracts. From that point on, CBS never faltered, and Paley eventually rivaled Sarnoff as the nation's leading broadcasting executive (see Exhibit 2.d).

2.6 GOVERNMENT REGULATION

One final piece remained to complete the structure of broadcasting in America—national legislation to impose order on the new medium. With the inception of telegraph systems in the 1840s, most governments recognized the need for both national and international regulation to ensure their fair and efficient operation.

The first international conference devoted to wireless communication issues took place in Berlin in 1903, only six years after Marconi's first patent. It dealt mainly with the Marconi company's refusal to exchange messages with rival maritime wireless systems. The conference emphasized the importance of subordinating commercial interests during emergencies. Marconi's aloof style of operation, aggressive commercial practices, and obvious desire to achieve a wireless monopoly had not made his company popular in international circles. Three years later another Berlin Convention reaffirmed the 1903 provisions, encouraged maritime nations to require wireless equipment on more ships, and established *SOS* as the international distress call.

1912 Radio Act

Congress confirmed American adherence to the Berlin rules with a 1910 wireless act requiring radio apparatus and operators on most ships at sea. In the wake of the April 1912 *Titanic* disaster, the wireless act was modified to require at least two operators on most ships. A few weeks later, Congress passed the Radio Act of 1912, the first comprehensive American legislation to govern land-based stations, requiring federal licensing of all radio transmitters. Specifically, the act required the secretary of commerce to grant licenses to U.S. citizens on request, leaving no basis to reject applications. Congress had no reason to anticipate rejections; it presumed that all who needed to operate radio stations could do so.

The 1912 act remained in force until 1927—throughout broadcasting's formative years. The law worked well enough for the point-to-point services it was designed to regulate. Point-to-multipoint broadcasting, however, introduced demands for spectrum space never imagined in 1912. Unregulated growth of broadcast stations in the mid-1920s soon created intolerable interference. Secretary of Commerce Herbert Hoover called on licensees to restrict themselves to certain frequencies, limit their transmitter power, and operate only at certain times.

An ardent believer in free enterprise, Hoover hoped that the new broadcasting business would discipline itself without the need for government regulation. To that end, he called a series of four annual national radio conferences in Washington, DC. At the first in 1922, about 20 broadcast engineers attended; by the fourth in 1925, the number of participants had increased to 400 and included station managers and attorneys.

Each year interference grew worse, and, to Hoover's chagrin, broadcasters increasingly demanded he impose regulations the industry was unable to impose on itself. Recognizing that broadcasting could not survive without greater discipline imposed by someone, Hoover agreed, although he said he found it remarkable that any industry would request greater government regulation.

Finally, in 1926 things came to a head when a federal court ruled that Hoover had no right (under the 1912 act) to establish regulations. He had tried to sue a Chicago station for operating on unauthorized frequencies at unauthorized times. The court found in favor of the station, remarking that, lacking specific congressional authorization in the law, Hoover could not "make conduct criminal which such laws leave untouched" (F, 1926: 618). In less than a year, 200 new stations took advantage of the government's inability to enforce licensing rules. President Coolidge urged Congress to pass a new law to regulate broadcasting. As he put it, "the whole service of this most important public function has drifted into such chaos as seems likely, if not remedied, to destroy its great value" (Coolidge, 1926). Early in 1927 Congress finally complied.

1927 Radio Act
• • • • • • • • •

The Radio Act of 1927 adopted most of the recommendations of Hoover's Fourth Radio Conference. Not a government move to limit free enterprise,

the 1927 act provided what most broadcasters wanted. It established a temporary Federal Radio Commission (FRC) to bring order into broadcasting. The technical and licensing problems continued, however, and Congress eventually made the Commission permanent.

The FRC's first duty was to eliminate chaotic interference among broadcasting stations. The FRC quickly closed stations operating on frequencies reserved for other nations under international agreements and eliminated portable broadcasting transmitters. It specified transmitter power, frequency, and operating hours for the remaining stations, and increased the number of frequencies available for AM broadcasting. The FRC also established "clear channel" frequency assignments, greatly reducing interference while ensuring that people without a local station could enjoy radio programs at night, when signals on these frequencies travel long distances.

These actions would have been ineffective without the muscle to back them up. The FRC stepped up its inspections of stations and began checking broadcast signals closely for technical violations. An elaborate "constant frequency monitoring station" built during 1929–1930 near Grand Island, Nebraska, became the focal point of the FRC's monitoring activities.

The courts helped by interpreting the Radio Act language requiring stations to operate in the "public interest, convenience, or necessity" in ways that allowed the fledgling FRC to act decisively. These factors combined to reassure investors and advertisers that broadcasting would develop in an orderly fashion.

 • • • • • • • • • • • • • • • • • •

2.7 DEPRESSION YEARS, 1929–1937

Two years after passage of the 1927 act, the Depression settled on the nation. A third of all American workers lost their jobs. National productivity fell by half. None of today's welfare programs existed then to cushion the intense suffering that unemployment, poverty, and hunger caused. Given the depressed state of the economy and the FRC's housecleaning, the number of ra-

EXHIBIT 2.f
• • • • • • • • • •

Growth in Number of Radio Stations, 1920–2000

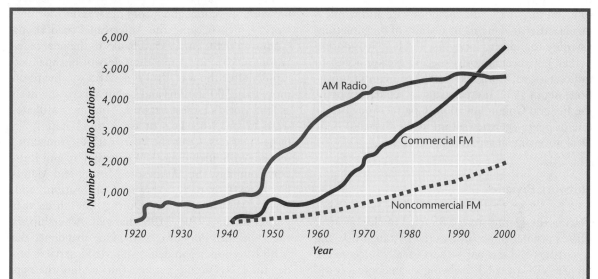

Note that significant downtrends in the growth curves occurred in 1930s AM, when the FRC imposed order on the pre–Radio Act chaos, and in 1950s FM, when its initial promise seemed not to be paying off. The sharp upward trend in the AM growth curve in the late 1940s occurred after the removal of World War II's restraints on consumer goods. In the mid-1990s, the number of commercial FM stations first matched, then surpassed, the number of AM stations.

Source: Adapted from *Stay Tuned: A Concise History of American Broadcasting* by Christopher H. Sterling and John M. Kittross. ©1990 by Wadsworth Publishing Company, Inc. Reprinted by permission of Wadsworth, Inc. 2000 data from FCC.

dio stations on the air actually decreased (see Exhibit 2.f).

Role of Broadcasting
• • • • • • • • • • • • •

In these difficult years, radio entertainment came into its own as the only free (after one owned a receiver), widely available distraction from the grim daily struggle to survive. Listener loyalty became almost irrational, according to broadcast historian Erik Barnouw:

> Destitute families, forced to give up an icebox or furniture or bedding, clung to the radio as to a last link of humanity. In consequence, radio, though briefly jolted by the Depression, was soon prospering from it. Motion picture business was suffering, the theater was collapsing, vaudeville was dying, but many of their major talents flocked to radio—along with audiences and sponsors. (Barnouw, 1978: 27)

Roosevelt and Radio
• • • • • • • • • • • • •

President Franklin D. Roosevelt, coming into office in 1933, proved to be a master broadcaster, the first national politician to exploit the medium to its full potential in presidential politics. His distinctive delivery soon became familiar to every listener who tuned in to his "fireside chats," the term used to suggest the informality, warmth, and directness of these presidential radio reports to the people.

Roosevelt's political opponents also used radio effectively. Louisiana Senator Huey Long and Catholic priest Charles E. Coughlin roundly criticized FDR's policies and drew large radio audiences during the early 1930s. Roosevelt, however, emerged as undisputed master of the new medium. By the 1940s many Americans said they considered radio a more important source of political information than newspapers.

Roosevelt made a significant impact on communication regulation as well. In 1934 the president asked Congress to establish a single communication authority to pull together separate federal responsibilities for regulating both wire and wireless media. After making its own study of the situation, listening to industry lobbyists, and largely ignoring a small but vocal minority seeking greater access to radio stations, Congress passed the Communications Act of 1934. It created a new regulatory body, the Federal Communications Commission (FCC), one of many agencies set up in the flurry of New Deal government activity.

Network Rivalries

During the Depression, CBS tried to chip away at NBC's privileged position as the first network. Big advertisers and star performers usually selected the bigger NBC over CBS when given a choice. It did little good for CBS to build up successful programs: "We were at the mercy of the sponsors and the ad agencies," wrote Paley. "They could always take a successful show away from us and put it on NBC" (Paley, 1979: 174).

The Mutual Broadcasting System (MBS) began in 1934 with a premise different from that of CBS or NBC. In the early 1930s only two large radio stations lacked network affiliation: they were WGN-Chicago and WOR-New York. In 1934 they formed a cooperative network organization with WXYZ-Detroit and WLW-Cincinnati. The four stations pooled some of their own programs, most notably *The Lone Ranger*, to form the nucleus of a network schedule. But MBS attempts to expand into a national network were frustrated because most stations had already committed themselves to NBC or CBS. MBS complained to the FCC about this virtual monopoly of the older networks.

The FCC began an in-depth investigation of network practices. After three years of study, the Commission issued its *Chain Broadcasting Regulations* in May 1941, so sweeping in scope that NBC and CBS argued the new rules would devastate the industry. The FCC sought to give network affiliates more control over their own schedules and the networks less clout. Though the networks fought the rules all the way to the Supreme Court, in 1943 the Court upheld the FCC—a watershed ruling affirming the constitutionality of the commission's rule-making powers (US, 1943).

The decision forced NBC to wrap up its dual network operation, and it sold the weaker Blue network, which became the American Broadcasting Company (ABC) in 1945. Thanks to the new regulations, MBS and the other networks expanded rapidly after the war. By 1948, MBS was heard on more than 500 stations and began advertising itself as "the world's largest network," but most affiliates were in small markets or were small stations in larger markets. Over the years various companies owned MBS, including the General Tire and Rubber Company, the Minnesota Mining and Manufacturing Company, Amway Corporation, and, finally, Westwood One. During one three-year period in the 1950s, MBS changed ownership six times. In April 1999, Westwood One abandoned the MBS label and began delivering MBS's services under the CNN Radio banner, thus ending the network's 65-year struggle for survival.

2.8 EARLY RADIO PROGRAMS

As detailed in Section 2.1, early radio drew on familiar sources for program material, especially vaudeville acts, some of which proved readily transferable to the new medium.

Comedy

The first network radio entertainment program to achieve widespread popularity was a primetime, five-days-a-week situation comedy, *Amos 'n' Andy*. Charles Correll ("Andy") and Freeman Gosden ("Amos") came to radio from vaudeville as a typical song-and-patter team. At one point they were working simply for food, and reports suggest that the station's management thought they were overpaid. Then a Chicago station manager suggested that they try their luck at a radio-comedy series. The two white performers struck a responsive chord in prevailing American culture when they developed a black-dialect show in ghetto English, featuring the ups and downs of the "Fresh Air Taxicab Company of America, Incorporated."

Amos 'n' Andy became radio's first big network hit in the early 1930s. Traffic stopped across the country and movies halted in midreel at 7:00 P.M. for the nightly 15 minutes of chuckles over the antics of Amos, Andy, the Kingfish, Lightnin', Madam Queen, and a host of minor characters, most of whom the versatile Correll and Gosden played themselves.

Such an impersonation of African Americans by white actors using stereotyped dialect and situations based on ghetto poverty would not even be considered today. As early as 1931 a Pittsburgh newspaper called on the FCC to ban the series, alleging racism, but its defenders argued that many blacks seemed to enjoy the program just as much as whites did. CBS dropped a later television version (with black actors) in 1953 because of NAACP (National Association for the Advancement of Color People) opposition, but the series continued in syndication until 1966.

Comedy-variety programs and situation comedies reached their greatest popularity near the end of the Great Depression, just before the beginning of World War II. Comedy became a kind of general anesthetic to ease the pain caused by years of economic hardship and the growing specter of global conflict. Many popular programs featured real-life husband-and-wife combinations: Jack Benny and Mary Livingstone (*The Jack Benny Program*), George Burns and Gracie Allen (*Burns and Allen*), Jim and Marian Jordan (*Fibber McGee and Molly*), and Fred Allen and Portland Hoffa (*The Fred Allen Show*). Ventriloquism and radio may seem incompatible, but NBC had a successful comedy-variety program featuring popular ventriloquist Edgar Bergen and his dummies. (Modern audiences may be more familiar with Bergen's daughter Candice, star of TV's *Murphy Brown*.)

Music
• • • •

Radio relied heavily on music from the beginning. In the mid-1930s more than half of all radio programming consisted of music, three-quarters of it carried on a sustaining (nonsponsored) basis.

Large stations often hired their own musical groups, while networks supported symphony orchestras. Besides having their own large studios in which groups could perform, networks and some stations aired performances originating live outside the studios. Often these "big band remotes" originated in the ballrooms of famous metropolitan hotels. In its early years, CBS devoted a quarter of its entire schedule to music. NBC began regular broadcasts of the Metropolitan Opera in 1931 and hired Arturo Toscanini out of retirement to head the NBC Symphony Orchestra in the late 1930s.

Musicians and composers welcomed such live-performance opportunities, but many stations could afford only recordings. In a single playing, recordings might reach more listeners than would hear live performances in a year. This rapid consumption via radio alarmed musical creators and performers. The musicians' union tried to keep recordings of all kinds off the air, and copyright holders demanded heavy payment for performance rights.

Under copyright law, playing a record in public for profit (as in a broadcast) constitutes a "performance." As such, it obligates the user (in this case a radio station) to pay copyright holders (who may include composers of the music, lyricists, and music publishers) for performing rights. These music copyright holders rely on music-licensing organizations to act on their behalf in monitoring performances and collecting copyright fees for the use of both live and recorded music. The first such U.S. organization, the American Society of Composers, Authors, and Publishers (ASCAP), dates back to 1914.

Almost simultaneously with broadcasting's birth, ASCAP began making demands for substantial payments for using musical works in its catalog, whether broadcast live or from records. These demands threatened stations with an unexpectedly heavy financial burden. In 1923 they formed the National Association of Broadcasters (NAB) to negotiate ASCAP's demands on an industry-wide basis. Over the years the NAB has successfully championed the broadcasters' position, especially on the regulatory front in Washington, D.C. The NAB continues this role in the new century.

In 1937, when ASCAP announced yet another fee increase, broadcasters decided to boycott

ASCAP music and later created their own co-operative music-licensing organization, Broadcast Music, Inc. (BMI). Eventually, BMI built up a comprehensive library. Competition from BMI moderated ASCAP demands, but licensing conflicts have been a cause of lawsuits ever since.

Recordings

Musicians were not the only ones to oppose airing recordings. In 1922 the U.S. secretary of commerce prohibited large stations from using recordings, arguing that records only duplicated programming listeners could receive without radio. Five years later the FRC issued orders that all stations identify as "mechanical reproductions" any programming on phonograph records to avoid having the public think the program was live. Networks banned recordings because they considered their ability to distribute live broadcasts a unique asset but made exceptions for especially newsworthy events, such as Herb Morrison's emotional description of the fiery *Hindenburg* crash in 1937. Concerns about audio quality also fueled the networks' opposition to recordings, especially for entertainment programming.

Early discs ran at 78 revolutions per minute (rpm), allowing time for only three or four minutes on a side. In 1929 broadcasters began using better-quality 16-inch ETs (electrical transcriptions). ETs revolved at 33⅓ rpm, permitting 15 minutes to a side. They were used in radio program syndication and for specially recorded subscription music libraries on which many stations relied.

Network opposition to recorded news programming was swept away by World War II. This opened the door for recorded programming generally. ABC dropped its recording ban in 1946 to lure Bing Crosby away from powerful NBC. The laid-back crooner hated the tension and risks of real-time broadcasting, which were compounded by the need to repeat each live program in New York a second time for the West Coast to compensate for time zone differences. Crosby invested in a little-known company, Ampex, which developed magnetic-tape recording based on German wartime technology. As soon as broadcast-quality audiotape recorders became available, ABC agreed to let Crosby break the network ban by recording his weekly prime-time program. The other networks

soon followed ABC's lead, appreciating the flexibility and control such technology offered.

Daytime Drama

Drama became a staple of daytime network programming in the 1930s. Many of these dramas were called "soap operas" for their typical sponsors. Soaps targeted female listeners, and scripts relentlessly reflected social values and gender roles considered appropriate for women during the period. "Adventurous" women working outside the home were often depicted as glamorous but sadly unfulfilled. According to the scripts, true happiness came only to dutiful wives, mothers, and homemakers. Program titles such as *When a Girl Marries* and *Young Widder Brown* suggest the general themes. The archetypical heroine, and the name of a popular soap, was *Ma Perkins*. Racial and ethnic stereotyping occurred mercilessly, even down to the naming of characters. Heroes usually had sturdy Anglo-Saxon names, but those given to villains had a "foreign" ring.

News

News depends both on outside sources of supply and on technology. Most news came from press associations or *wire services* (so called because they distributed news by telegraph for years). Fearing the immediacy offered by radio, newspapers calculated that they could limit radio's competition by denying broadcasters access to major news agencies. NBC had inaugurated regular 15-minute nightly newscasts by Lowell Thomas in 1930 on its Blue network—a sign that radio might soon seriously compete with newspapers.

In late 1933, newspaper publishers, hoping to limit radio news, forced the networks (which were ill-prepared to gather their own news) to accept use of a special Press-Radio Bureau designed to funnel just enough news to radio to tempt listeners to purchase newspapers for more details.

The Press-Radio Bureau never worked effectively. For one thing, the press-radio agreement did not cover news commentaries, so many newscasters became instant commentators. In 1935, United Press broke the embargo on unrestricted release of news to radio, joined soon by International News

Service (these two merged in 1958 to form United Press International [UPI]). The Press-Radio Bureau finally expired, unmourned, in 1940 when the Associated Press began to accept radio stations as members. The press associations later acquired more broadcasters than publishers as customers and began offering services especially tailored for broadcast stations, including audio feeds ready to go directly on the air.

Excesses
• • • • • •

Reducing interference, a technical issue, initially spurred government's entry into broadcast station regulation. However, the congressional mandate to ensure that broadcasters serve "the public interest" lured the FRC, and later the FCC, into content areas.

In the early 1930s the FRC kicked Kansas broadcaster "Dr." John Brinkley off the air for prescribing sex-rejuvenation surgery and drugs of his own fabrication. Brinkley's "medical practice" was questionable, but especially so because he was not a licensed physician. About the same time, the FRC also jerked the Los Angeles license of the Reverend Dr. Robert Shuler, a fire-and-brimstone crusader who often broadcast personal attacks against his critics.

Orson Welles created a panic in October 1938 when he produced a live drama heard nationwide on CBS in which killer Martians invaded the United States. What made the broadcast so devastating was that the program reported the Martians' "progress" in news bulletin format. People tuning in late—and many did—thought they were hearing legitimate live news reports. Thousands panicked, thinking they faced vaporization by the Martian "heat ray." That fear turned to anger against CBS and Welles when listeners realized the broadcast was a Halloween prank. The FCC investigated but declined to take action, deciding Welles and CBS intended no overt deception. Even so, the FCC's "raised eyebrows" put broadcasters on notice they should be careful.

Government reaction to what many considered to be excesses continued in the 1990s, as evidenced by huge fines levied against stations airing "shock jock" Howard Stern's controversial program, as well as the development of the TV Parental Guidelines rating system (see Exhibit 9.d) and mandatory V-chip technology for blocking "bad" programs (see Section 9.4, p. 264).

2.9   RADIO IN WORLD WAR II, 1939–1945

During World War II, which the United States entered in 1941, radio escaped direct military censorship by complying voluntarily with codes. For example, broadcasters avoided live man-on-the-street interviews and weather reports; the former could risk transmission of a secret code and the latter would be useful to potential enemy bombers. Radio had a role to play internationally in psychological warfare, and in 1942 President Roosevelt appointed well-known newscaster Elmer Davis to head the newly created Office of War Information (OWI). The OWI mobilized an external broadcasting service that eventually became known as the Voice of America, which still broadcasts to foreign countries today.

Wartime restrictions on civilian manufacturing, imposed from 1942 to 1945, all but eliminated station construction and receiver production. Manufacturers of consumer goods devoted their capacity to military needs but continued to advertise their peacetime products to keep their names before the public. The government allowed them to write off these advertising costs as business expenses even though they had no products to sell.

Radio Drama
• • • • • • • •

Released from the normal competitive pressures to sell products by support of mass-appeal material, some advertisers invested in their public image by supporting high-quality dramatic programs of a type rarely heard on American radio, though such programs were common in Europe.

Radio developed playwrights such as Norman Corwin and Arch Oboler, who won their chief literary fame in broadcasting. CBS commissioned Corwin to celebrate the Allied victory in Europe with an hour-long radio play, *On a Note of Triumph*, in 1945. This emotional program climaxed an extraordinary flowering of radio art—original writing of high merit, produced with consummate skill, and always live, for the networks still shunned recordings. With the end of the war years and the artificial wartime support for culture, competitive selling resumed, and this brief, luminous period of radio drama and comedy creativity came to an end.

Wartime News

As the threat of European war loomed in the mid-1930s, NBC began developing European news operations to cover rapidly developing events. To counter NBC, Paley decided on a bold CBS stroke—a half-hour of news devoted to the 1938 Nazi incursion into Austria, originating live from such key cities as London, Paris, Rome, Berlin, and Vienna.

The networks' objections to the use of recorded material created tremendous problems of coordination and timing for the complex production. But in that historic half-hour, featuring Robert Trout, William Shirer, Edward R. Murrow (see Exhibit 2.g), and others, radio arrived as a full-fledged news medium. Thereafter, on-the-spot radio reporting from Europe became a daily news feature.

Later in 1938 came the Munich crisis, in which England and France abandoned Czechoslovakia to Germany's Adolf Hitler, climaxing 18 days of feverish diplomatic negotiations among the great powers. During these tense days and nights, pioneer commentator H. V. Kaltenborn achieved fame by extemporizing a remarkable string of 85 live broadcasts from New York, reporting and analyzing news of each diplomatic move in Europe as reports came in.

Thanks to CBS's early start, Paley's enthusiastic support, and a superlative staff of reporters and editors, CBS set a high standard for broadcast journalism during the war years, establishing a tradition of excellence that lasted into the 1980s. Then corporate takeovers, competition, and cutbacks (discussed in Section 3.6, p. 77) began to accelerate the erosion of news standards initiated in those pioneer days.

At this point in radio's story, television began a spectacular rise that was soon to affect the older medium profoundly. Radio's story picks up again in Section 2.13.

2.10 PRE-1948 TV DEVELOPMENT

Television experimentation took place for decades before it developed into a mass medium. Early television produced crude pictures without sound, interesting only as curiosities. Exhibit 2.h, on p. 42, shows the quality achieved in 1929, still far below

an acceptable level. Public acceptance awaited pictures with sufficient resolution (detail) and stability (absence of flicker) for comfortable viewing—a standard at least as good as that of the home movies then familiar to wealthier consumers.

Mechanical vs. Electronic Scanning

Inventors long sought a way to break down pictures rapidly into thousands of fragments for transmission bit by bit, and then to reassemble them for viewing. Two incompatible methods of accomplishing this developed simultaneously, one based on a mechanical approach and the other entirely electronic. The mechanical approach initially showed the most promise, but the equipment was large, required almost impossibly precise synchronization between transmitter and receiver, and lacked the picture resolution needed for public acceptance. In spite of these difficulties, John L. Baird of Great Britain doggedly pursued mechanical TV systems well into the 1930s. Baird's ultimate mechanical system, however, came out second best in 1936 when the British government compared it head-to-head with an all-electronic system developed by Marconi-EMI with assistance from RCA. Dr. H. E. Ives, Charles Jenkins, and others investigated mechanical approaches in America, but without Baird's zeal.

No single inventor can claim a breakthrough in electronic television, for television was a corporate rather than an individual achievement. However, two inventors are remembered for solving specific parts of the puzzle—Farnsworth and Zworykin. Philo T. Farnsworth, a self-taught American genius, devised an *image dissector*—a device for taking pictures apart electronically for bit-by-bit transmission. His patents required RCA, the corporate leader in American television development, to secure licenses from him to perfect its own system.

Vladimir K. Zworykin had immigrated from Russia in 1919 and worked as an engineer for Westinghouse. In 1923 he invented the *iconoscope*, the first electronic camera pickup tube suitable for studio operations (shown in Exhibit 4.h, p. 105).

As David Sarnoff focused RCA's resources into a drive to perfect television as early as 1930, he hired Zworykin to head a television research group that was based in Camden, New Jersey. Entrusted with the task of producing a marketable television system, Zworykin's "Camden team" tackled technical con-

EXHIBIT 2.g

Edward R. Murrow (1908–1965)

CBS news reporter Murrow, shown here during World War II, often broadcast from the BBC's Broadcasting House in downtown London. He and other American reporters used a tiny studio located in a subbasement. Once, when the building took a hit during a German bombing raid, Murrow continued his live report as stretcher-bearers carried dead and injured victims of the raid past the studio to the first-aid station.

First employed by CBS in 1935 as director of talks in Europe, Murrow came to the notice of a wider public through his memorable live reports from bomb-ravaged London in 1939–1941, and later from even more dangerous war-front vantage points. Unlike other reporters, he had a college degree in speech rather than newspaper or wire-service experience. The British appreciated his realistic and often moving word-and-sound pictures of their wartime experiences, and American listeners liked the way he radiated "truth and concern," as William Paley put it (1979: 151).

Widely admired by the time the war ended, Murrow became the core of the postwar CBS news organization. He served briefly as vice president for news but soon resigned the administrative post to resume daily newscasting. As an on-the-air personality, he survived the transition to television better than others, going on to appear in *See It Now* and in often highly controversial documentaries.

But Murrow's TV career was personally disappointing. Although he achieved almost mythic journalistic stature through his serious reporting, CBS also used him as host of *Person to Person*, a popular show that combined a Barbara Walters–style personality interview with a format resembling that of Robin Leach's *Lifestyles of the Rich and Famous*. Each week Murrow "tele-visited" the home of some celebrity, quizzing the famous person about his or her household furnishings and home life. Murrow found it professionally and personally embarrassing. Increasingly despondent about eroding values in broadcast journalism, he resigned from CBS in the early 1960s to direct the U.S. Information Agency under President John F. Kennedy.

Even after leaving CBS, Murrow continued to influence broadcast journalism at the network and elsewhere. In 1999, 34 years after his death, a panel of experts assembled by the New York University Department of Journalism and Mass Communication ranked Murrow's 1940 reports from London and his 1954 investigation of controversial senator Joseph McCarthy among the top ten examples of excellence in American journalism.

Source: Photo © Archive Photos.

EXHIBIT 2.h
• • • • • • • • • •
First U.S. Television Star

During early experiments with electronic television, RCA laboratories used, as a moving object to televise, a 12-inch papier-mâché model of a popular cartoon character, Felix the Cat, posed on a revolving turntable under hot lights. The image insert at left shows how Felix looked on television in 1929 when picture definition was still only 60 lines per frame. The blurred and jerky video images available to Internet users connected via a standard telephone modem are reminiscent of the picture quality at television's birth.

Source: Photos from NBC Studios.

cerns as well as the subjective problem of developing picture-quality standards sufficient to win full public acceptance. (See Exhibit 2.i on pp. 44 and 45 for more information about Zworykin and Farnsworth.)

TV Goes Public
• • • • • • • • • •

RCA and others had experimented with TV for many years, but it was not until 1939 that David Sarnoff was ready to display his company's product publicly. Sarnoff at the time was maneuvering to have the FCC adopt the RCA system and wanted maximum publicity. He got it by making RCA's TV debut during the New York World's Fair opening ceremonies. The demonstration was successful, and RCA began acting as though its system were already the national standard adopted for telecasting, an attitude that alienated competing manufacturers and members of the FCC.

Almost two years elapsed between the RCA World's Fair demonstration and the FCC adoption in July 1941 of technical standards for black-and-white television transmission. The 1941 standards adopted were largely those originally proposed by RCA. They remained the technical foundation for TV broadcasting for well over half a century, essentially unchanged until the FCC adopted digital TV standards in December 1996.

Within a year of the 1939 RCA demonstration, wartime needs halted production of most consumer electronics. Further development of civilian television was shelved for the duration. During World War II six experimental stations remained on the air—two in New York City and one each in Schenectady (New York), Philadelphia, Chicago, and Los Angeles. They devoted their few hours of airtime a week primarily to civilian defense programs. Perhaps 10,000 receivers were in use, half of them in New York City.

Postwar Pause

The end of the war in 1945 did not, as some expected, bring an immediate upsurge in television activity, despite a backlog of 158 pending station applications. Many experts believed that all-out development should await adoption of a color system. And potential investors wondered whether the public would buy receivers that cost many times the price of radios—or if advertisers would pay for expensive television time.

Meanwhile progress behind the scenes resumed. In 1945 a new *image orthicon* tube came into use, improving camera sensitivity while eliminating the uncomfortably high levels of studio light that older tubes had required. In 1946 AT&T opened its first intercity *coaxial cable* link between New York and Washington, DC, enabling the start of live interconnection.

2.11   GROWING PAINS: CHANNELS, COLOR, NETWORKS

By mid-1948 public demand and broadcaster interest inaugurated the long-anticipated television gold rush, the transition from mere experiments to true mass medium. Sixteen stations were on the air at the start of 1948—and nearly 100 two years later. The number of cities that television served grew rapidly. The audience expanded in one year by an astonishing 4,000 percent. Regular network service to a few eastern cities began. Major advertisers started experimenting with the new medium, and regular programs were launched.

Television's growing pains, however, still had not quite ended; the list of things yet to be resolved included increasing the number of channels available, adopting a color system, and establishing truly national networks.

The Freeze and *Sixth Report and Order*

In approving commercial television in 1941, the FCC had made available only 13 channels (later reduced to 12) to serve the entire United States. As more and more stations applied to go on the air, it became obvious that demand would soon exceed the supply of channels. And interference among the few stations on the air was growing because of a miscalculation of the distance needed to separate stations assigned to the same channel.

In September 1948 the FCC abruptly froze further processing of television-license applications pending solution of these problems. The freeze had no effect on applicants whose permits had already been approved. For the nearly four years of the freeze, 108 prefreeze stations enjoyed an enviable monopoly. The FCC held a lengthy series of hearings to settle the complicated engineering and policy questions that had precipitated the freeze. The much-anticipated decision, the basic charter of twentieth-century American broadcast television, came on April 14, 1952, in the FCC's historic *Sixth Report and Order* (FCCR, 1952).

The new rules relocated some of the original 12 VHF (very high frequency) channels to reduce interference and added 70 new channels (14–83) in the UHF (ultrahigh frequency) band. Experience with radar on these much higher frequencies during World War II suggested that UHF stations would suffer serious coverage limitations compared to those on VHF channels. Disagreements also flared over *intermixture* (assignment of both VHF and UHF channels in the same market), the distance needed between stations on the same channel to minimize interference, and the importance of adopting color standards. After more than three years of deliberation, the FCC combined four separate frequency bands to obtain sufficient spectrum space for TV. Exhibit 4.k (p. 110) summarizes their location in the spectrum and current TV channel numbers. The

EXHIBIT 2.i

Zworykin and Farnsworth: TV Pioneers

Vladimir K. Zworykin (photo on left) was born in Russia in 1889. In 1906 he enrolled at the St. Petersburg Institute of Technology, where he studied under Boris Rosing, a man whose experiments with cathode-ray tubes greatly influenced Zworykin's later work. He also studied at the College de France in Paris and later earned a Ph.D. in physics from the University of Pittsburgh. He began working on television while still employed at the Russian Wireless Telegraph and Telephone Company. When he moved to the United States in 1919, he brought his ideas about television with him.

A year later he began working for Westinghouse. Television was not a high priority at Westinghouse at the time, but the company did file for an initial patent on Zworykin's camera tube, which he called the *iconoscope*, in 1923. In 1930 RCA assumed control of GE's and Westinghouse's radio research, and Zworykin moved to RCA, where its leader and Zworykin's fellow Russian immigrant, David Sarnoff, was enthusiastically pushing television's technical development. Sarnoff was so impressed with Zworykin that within a few months he had his own lab at RCA headquarters in Camden, New Jersey. In 1929 Zworykin presented a paper to the Institute of Radio Engineers in which he described the *kinescope*, or picture tube. By 1933 he had dramatically improved his original iconoscope, which, along with the kinescope, formed a workable all-electronic television system.

Federal Communications Commission eliminated UHF TV channels 70 through 83 in 1970.

Channel-Allotment Plan

There was great pressure to leave the 108 prefreeze stations alone. The new UHF channels and the VHF channels not already in use allowed, at least in theory, the FCC to license more than 2,000 TV stations. The FCC increased the number of communities allotted a TV channel from 345 to almost 1,300, but about two-thirds of these were in the technically inferior UHF band. That was certainly the case for the 10 percent (later increased to 35 percent) of the channels reserved for noncommercial educational use. Of the more than 370 educational TV stations on the air in 1999, about 250 were on

Philo T. Farnsworth (photo on right) was born near Beaver, Utah, in 1906 but grew up on a ranch near Rigby, Idaho. While still a boy he began reading about electrical devices and radio, and he became especially fascinated with the photoelectric cell and cathode-ray tube. As early as 1922 he astonished his high school instructor by revealing a diagram for an all-electronic television system based on these devices. Four years later he lined up investors willing to support his work, and in 1927 he filed his first patent for his television system. The following year he demonstrated an all-electronic television system publicly in San Francisco. Development costs continued to mount, however, and in 1931 Farnsworth accepted the financial backing of the Philadelphia Storage Battery Company, better known as Philco. Philco was a leading manufacturer of radios and one of RCA's major competitors in the race to develop television. Farnsworth found it difficult to work within a large organization, however, and he left Philco in 1933.

In 1934 Farnsworth filed a patent interference suit against RCA, based on his original 1927 patents. He claimed that RCA's improved "image iconoscope" incorporated features covered in his patents, while Zworykin claimed that his 1923 disclosures provided him legal priority. Zworykin had visited Farnsworth in 1930 when Farnsworth was still in San Francisco, and the controversy continues regarding whether Zworykin's work was influenced by that visit and to what degree. In any case, Farnsworth's claim was upheld in 1934 and again in 1936 on appeal. RCA could not move forward with its inauguration of a commercially viable television technology without settling with Farnsworth over the image iconoscope issue. Accordingly, RCA agreed to a nonexclusive cross-licensing agreement with him. Farnsworth later won another fight involving RCA's use of other technology patented by him.

After 1940 both Zworykin and Farnsworth moved on to other projects, but with somewhat different degrees of success. Farnsworth bought an Indiana company and began manufacturing various types of entertainment equipment. By the late 1940s the company was facing bankruptcy. Farnsworth then conducted research on missile guidance systems, nuclear fusion, and other projects for International Telephone and Telegraph. In the late 1960s Farnsworth moved to Utah, established a small company, and continued his nuclear fusion research until he lost government funding and the company folded in 1970. Farnsworth had suffered ill health since the 1940s, and he died in 1971.

Zworykin, in contrast, continued his successful career with RCA. He worked on many important military projects during World War II, including the development of radar. He became an RCA vice president in 1947, continuing as an honorary vice president even after his retirement. He lectured widely, spent his winters in Miami, and generally lived the good life. He was honored with the National Medal of Science in 1967 and was inducted into the Inventors Hall of Fame a decade later. Zworykin basked in the honors his contribution to the technical development of television brought him, but he—like many other broadcast pioneers—was disappointed with the programming. When asked what he considered to be his greatest contribution to television, he is reported to have said the "off" switch. Zworykin died in 1982.

Sources: Zworykin photo (left) from Brown Brothers; Farnsworth photo (right) from Corbis-Bettmann.

UHF channels. Exhibit 2.j provides an example of how the allotment table separates stations on the same channel through-out the country, spacing them to avoid co-channel interference.

The number of on-air stations more than tripled in the first postfreeze year, as shown in Exhibit 2.k. Nevertheless, the new channel-allotment plan had serious defects. For one thing, there were still too few channels to give viewers in every market the same range of choices. Some had only one channel, some two or three, and so on. Viewers in any given market needed access to at least five channels to be able to choose among the three networks, an independent (non-network) station, and an educational outlet. For another, the FCC's decision to make both VHF and UHF allotments in many localities placed UHF stations in such markets at a serious disadvantage. A UHF station

EXHIBIT 2.j
. . . .
TV Channel-Allotment Plan

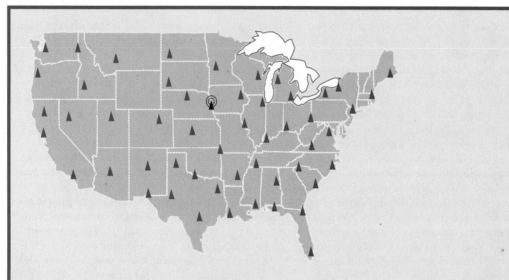

Using channel 7 as an example (a similar map could be made for any other VHF or UHF channel), this map shows the occupied channel 7 analog allotments (those actually on the air), except for three outside the contiguous 48 states. They are spread fairly evenly, separated from each other by a minimum of 170 miles. The list of analog channels available in Omaha is as follows:

 3—occupied by KMTV, a CBS affiliate
 6—occupied by WOWT, an NBC affiliate
 7—occupied by KETV, an ABC affiliate
15—occupied by KXVO, a WB affiliate
26—occupied by KYNE, a noncommercial educational station licensed to the Nebraska Educational Television Commission
42—occupied by KPTM, a Fox affiliate
48—reserved for noncommercial educational use; not on air
54—not on air

This market is also served by cable, multichannel multipoint distribution service (MMDS), and direct broadcast satellite (DBS). The FCC's digital TV channel-allotment plan also separates stations on the same channel geographically to minimize interference.

competing only with UHF stations faced a technologically level playing field. But a UHF station competing with one or more local VHF stations had a serious problem.

UHF Problem
. . . .

Stations assigned to UHF channels cannot cover as great an area as stations on VHF channels, given equal transmitter power and antenna systems. The FCC tried to ensure equal coverage for both VHF and UHF stations by authorizing UHF to use higher power. Yet even if added power could have helped, years passed before maximum-power UHF transmitters became available. And long after UHF television service began, set manufacturers still built VHF-only receivers, forcing viewers in markets with UHF stations to buy difficult-to-use UHF converters and often poorly performing UHF antennas.

EXHIBIT 2.k

Growth in Number of TV Stations, 1948–2000

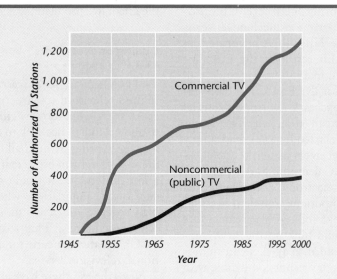

The modern TV era started in 1948, with 16 stations on the air. Only 108 stations had been authorized when the 1948–1952 freeze imposed a temporary ceiling. After that the number shot up, reaching 400 by 1955. Growth began to slow down at that point but has never actually stopped. Noncommercial stations developed more slowly, starting with the first two in 1954. Not included in this chart are the nearly 2,200 low-power TV stations or approximately 5,000 television translators.

Source: Adapted from *Stay Tuned: A Concise History of American Broadcasting* by Christopher H. Sterling and John M. Kittross. © 1990 by Wadsworth Publishing Company, Inc. Reprinted by permission of Wadsworth, Inc. 2000 figures from FCC.

Faced with overwhelming disadvantages, UHF television declined to a low of only 75 stations by 1960. The FCC continued to seek ways to assist failing UHF stations. Its most useful step came in 1962, when it persuaded Congress to require manufacturers to equip all receivers with built-in UHF tuners. Despite FCC efforts to improve UHF viability, the *Sixth Report and Order* saddled the television industry with long-term problems. The frustration of viewers who wanted more program options than the FCC's allotments made possible encouraged the eventual development of cable television and other substitutes for over-the-air reception.

By 1965 some of the FCC's UHF efforts had taken effect, and that part of the industry began a steady, though not spectacular, growth. FCC financial reports indicate that until 1974 UHF sta-tions as a group continued to lose money. Thereafter their average profit margin increased each year. By mid-1994 UHF commercial outlets on the air (594) had surpassed VHF (559). That trend continued, and by the end of the 1990s there were more than 680 commercial UHF stations on the air.

Concerns about UHF coverage escalated in the early 1990s when the FCC announced plans to migrate all existing TV stations to UHF channels as part of the transition from analog to digital transmission. The FCC later altered its plan, but many station owners discovered they would still lose their VHF channel for a UHF channel, adding to concerns abut the coverage potential of digital signals compared to traditional analog signals. See Section 5.7, p. 146, for more about digital TV transmission and reception.

Color TV

• • • • • • •

A battle began in the 1940s between CBS and NBC/RCA over a technical standard for color TV that came to a head during the freeze. CBS proposed a system based partly on mechanical scanning technology, while NBC/RCA advocated an all-electronic system. Interestingly, the FCC approved the CBS system in 1950, in the middle of the freeze. The next year five East Coast CBS stations hooked up to broadcast the first commercial colorcast, to a very small audience. The major equipment manufacturers, led by RCA, had simply refused to make TV receivers using the CBS system, in spite of the FCC decision. Finally, in 1953 all parties agreed to an electronic system patterned closely on RCA's. An important argument in its favor was that it had compatibility—black-and-white sets already in use could pick up the newly approved color signals and display them in monochrome. The CBS system was not compatible with monochrome sets already in consumers' hands.

However, color telecasts on a large scale developed slowly. Five years after the 1953 adoption of color standards, only NBC offered programs in color. Full network color production in prime time finally arrived in 1966. By 1972, nearly two decades after the adoption of color standards, only half of the country's homes had color TV sets. By the mid-1990s new black-and-white sets had become hard to find, except for very small "personal portable" models.

TV Relays and Recording

• • • • • • • • • • • • • • • • • •

Full-scale national network operations had to await completion of AT&T's coaxial cable and microwave relay links, which joined the East and West Coasts in 1951 and enabled the start of live network interconnection. Until the cable reached them, affiliates had to use kinescopes of network programs—films of television programs photographed as they were displayed on the face of a receiver tube—syndicated films, and their own live productions.

Engineers developed techniques for filming images on a kinescope—the technical name for the TV picture tube—in the late 1930s, but they became important only with the explosive growth of TV in 1948. In spite of improvements made over the years, the image quality of these "kinies" remained poor. In 1956 the Ampex Corporation demonstrated a successful videotape recorder, which saw its first practical use that fall on CBS. In a rare spirit of cooperation, competing manufacturers put aside rivalries and agreed on a compatible professional videotape standard. Videotape doomed the era of live production, as related in Section 2.12.

Ceiling on Networks

• • • • • • • • • • • • • • •

The FCC's 1952 *Sixth Report and Order* effectively limited television to three national networks because there were too few cities with four channels to allow a fourth network to compete nationally on an even footing. Even if all the independent commercial stations in every market combined to form a fourth network, it would reach, at best, 85 percent of the population, whereas ABC, CBS, and NBC each could reach more than 95 percent of television homes. This situation underlines the FCC's greater concern for local, rather than national, television service.

Nevertheless, there have always been pressures to create a fourth national television network. MBS, the fourth radio network, lacked the money to branch into television. From 1946 to 1955, however, a fourth chain did exist—Du Mont Television Network, founded by Allen B. Du Mont, a developer of the cathode-ray tube (CRT), the parent technology of the TV picture tube. The Du Mont network survived only as long as lack of interconnection facilities kept networking in check. Once interconnection relays became generally available, Du Mont could not sign up sufficient affiliates to compete with the larger, older, and richer networks.

It was not until the mid-1980s that Fox Broadcasting Company emerged to become the fourth network. In the mid-1990s the Warner Brothers Network ("The WB") and the United Paramount Network (UPN) followed Fox. In August 1998, TV station group owner Lowell "Bud" Paxson launched PAX TV, the seventh broadcast network. Each new network sought to capture a piece of the action dominated for decades by NBC, CBS, and ABC.

2.12 GOLDEN AGE OF TELEVISION, 1948–1957

Just as some people look back with nostalgia on radio's "golden era" of the 1930s and 1940s, some people reminisce about television's first decade as a mass medium, 1948 to 1957.

Programs as Motivators

In that decade, the networks put first priority on stimulating people to buy sets. Only attractive programs could do that:

> It was the only time in the history of the medium that program priorities superseded all others. If there was an abundance of original and quality drama at the time . . . it was in large part because those shows tended to appeal to a wealthier and better-educated part of the public, the part best able to afford a television set in those years when the price of receivers was high. (Brown, 1971: 154)

Most programs, local and network, were necessarily live—a throwback to the earliest days of radio. Videotape recording was uncommon, especially at the local level, until the late 1950s. Original television plays constituted the best artistic achievements of television's live decade. "Talent seemed to gush right out of the cement," pioneer *New York Times* critic Jack Gould observed years later.

It would be misleading, however, to suggest that all "golden age" programming achieved high artistic status. NBC's enormously popular *Texaco Star Theater*, for example, featured slapstick routines borrowed almost directly from vaudeville. Such programs appealed to less well-to-do, less elitist viewers, people for whom the purchase of a TV set was a major investment. Without such popular programs the cost of receivers would have remained high, and the growth of TV as a mass medium could have been slowed.

TV vs. Hollywood

Television programs could, of course, have been recorded from the very beginning by filming them (and many were—witness the continuing reruns of the early 1950s' *I Love Lucy*, a series shot using film cameras at an additional expense absorbed by the show's stars, Lucille Ball and Desi Arnaz). Economic, technical, and industrial barriers delayed widespread use of this solution. Hollywood's traditional single-camera method of production is slow and cumbersome, costing far too much for general television use; and it took time to adapt film pro-

duction to the physical limitations of television, with its lower resolution, smaller picture, and more restricted contrast range.

In addition, many critics opposed the use of film, expecting television to escape Hollywood's straitjacket in favor of a new, live, and less trite mass-entertainment genre. The two points of view were as far apart as their two production centers—television in New York and film in Los Angeles. But the economics of both industries drove them closer together. Inevitably, as technical barriers to producing filmed television programs fell, and as Hollywood accepted use of videotape, the production base for entertainment programs shifted to the West Coast by the late 1950s.

Hollywood's old-line feature-film producers saw television as a threat, much as publishers had feared that radio would undermine newspapers. Producers and distributors refused to release their most recent feature films to television for a dozen years. At first only pre-1948 films could be shown on television. Not until the early 1960s did Hollywood conclude that the television bane could also be a boon. Movie producers realized that audiences would not stop buying theater tickets just because they could see movies on television. And the networks were ready to pay big money to show "post-48" films—movies that in most cases had little or no residual value for theatrical release. Eventually, feature-film producers found they could profit even more from delayed release to cable television and the VCR market, as well as broadcasting.

Network Rivalries

In terms of program formats, many important ideas in television's first decade sprang from the fertile imagination of Sylvester "Pat" Weaver, who became NBC's vice president for television in 1949, leaving NBC six years later as board chair. Disregarding conventional wisdom about established viewing habits, Weaver occasionally replaced regular series with one-time *spectaculars*, 60 or 90 minutes long. Weaver also foresaw that the single-sponsor show (a hallmark of network radio) simply could not last in prime-time television, as costs soon exceeded what single advertisers could afford. Instead, NBC introduced *participating sponsorship*, which enabled several different advertisers to share the costs of a program. NBC's long-running *Today* and *Tonight* shows are further testimony to Weaver's lasting impact.

Despite Weaver's NBC innovations, CBS steadily gained in the audience ratings race. Much of CBS's success is attributable to William Paley. In the late 1940s Paley capitalized on a loophole in the tax laws that allowed him to entice talent from NBC and ABC by offering large-income performers deals they could hardly refuse. The other networks retaliated, but it was Paley who often seized the initiative.

In 1955 CBS achieved the number-one place in the ratings, a rank it would hold undisputed for 21 years. Meanwhile, the third network, ABC, found itself in somewhat the same position that CBS had occupied in the early days of network radio. Top advertisers and performers automatically chose CBS or NBC, turning to ABC only as a last resort.

A corporate breakup helped rescue ABC in 1953. When a Justice Department antitrust suit against the big Hollywood film studios forced them to sell off their extensive theater chains, one of the spun-off companies, Paramount Theaters, merged with ABC in 1953. Paramount injected much-needed funds and established a link with Hollywood that eventually paid off handsomely, helping ABC to achieve status equal to that of CBS and NBC.

In 1954 Walt Disney, the first of the major studio leaders to make a deal with television, agreed to produce a series of programs for ABC called *Disneyland* (1954–1957), later continued on NBC. The deal built ABC audiences and gave Disney free television promotion for his developing California theme park as well as for Disney feature films. Disney went on in later years to become a primary program source for broadcast and cable television. In keeping with the mergers and acquisitions characteristics of media industries in the 1990s, Disney's long association with ABC took a major turn when it purchased ABC in 1996 for around $19 billion. Much of the price was based on ABC's lucrative ESPN cable franchise rather than the broadcast network.

In search of a different image, ABC in the 1970s began to pay more attention to shows appealing to young, urban, adult viewers. Less concerned than older networks with traditional program balance, ABC leaned heavily on action, sex, and violence in much of its prime-time programming.

Like ABC, three of the new networks—Fox, the WB, and UPN—have tended to test the limits of taste in their effort to distinguish themselves from the well-established Big Three. They have been criticized for this, as well as for targeting and sometimes pandering to minority urban audiences. PAX has taken an entirely different approach, promoting its family-friendly programming as a safe haven in a sea of gratuitous violence and sex on the other networks.

2.13   RADIO'S RESPONSE

Television's "takeoff" year of 1948 marked both a high-water mark and the doom of full-service network radio. In that year radio networks grossed more revenue than ever before, excluding profits from their owned-and-operated stations. For more than 15 years, radio networks had dominated broadcasting, but television was about to end their rule.

Radio Network Decline

For the balanced news-and-entertainment network programming that prevailed before television, four radio networks seemed to offer as much as stations and audiences would use. The four supplied their affiliates a full schedule of varied programs, much as the major television networks do today.

Sponsors of major radio network programs and stars hastened the decline in radio network fortunes after 1948. As television captured its first audiences, it lured away major radio advertisers—and with them went star performers and their programs. By the early 1950s the radio networks were reeling. CBS's William Paley recalled that

> although [CBS's radio] daytime schedule was more than 90 percent sponsored, our prime-time evening shows were more than 80 percent sustaining. Even our greatest stars could not stop the rush to television. (Paley, 1979: 227)

The ultimate blow came when radio stations began to decline network affiliation—a startling change, considering that previously such an affiliation was a precious asset. But program commitments to the networks interfered with the freedom to put new, tailor-made, post-television radio formulas into effect. By the early 1960s only a third of all radio stations still held network affiliations. Networks scaled down service to brief hourly news bulletins, short information features, a few public-affairs programs, and occasional on-the-spot sports events.

Today's dozens of satellite-delivered audio networks are possible because there are more stations and cheaper satellite transmission, and because each modern service or network specializes in a specific kind of programming. But this kind of radio networking differs greatly from that of the 1930s—and perhaps foreshadows the future of television.

Rock to the Rescue

With the loss of network dramas, variety shows, and quiz games, radio programming shrank essentially to music and news/talk, with music occupying by far the majority of the time on most stations—just as it had in the 1920s. Only this time the music was off records rather than live, broadcasters' aversion to recorded programming having passed into history years before. Providentially for radio, this programming transition coincided with the rise of a new musical culture, one that found radio an ideally hospitable medium.

Southern stations such as WDIA in Memphis and WLAC in Nashville pioneered key elements of a strange new sound, and early in the 1950s Cleveland disc jockey (DJ) Alan Freed gained national recognition by playing music, which

> combined elements of gospel, harmony, rhythm, blues, and country. He called it "rock and roll." And people everywhere began to listen. . . . It transcended borders and races. . . . Rock and roll sang to the teenager; it charted his habits, his hobbies, his hang-ups. (Drake-Chenault Enterprises, 1978: 1)

Radio proved to be the perfect outlet for this new form of expression as stations in many markets converted to a "Top-40" format in the late 1950s. The format name referred to the practice of rigidly limiting DJs to a prescribed playlist of current best-selling popular recordings. Gordon McLendon, a colorful Dallas sportscaster and station owner, helped pioneer the format. Omaha-based Todd Storz played a key role in popularizing Top-40

when he adopted it for his string of group-owned stations. Such Top-40 specialists frequently reprogrammed bottom-ranked stations and lifted them to top ratings in a matter of months.

The announcer took on a much more important role in the Top-40 format. Before Top-40, announcers typically were wallpaper figures providing continuity between programs. After Top-40, listeners could hear the same songs on many stations, so an important distinguishing attraction became the personality of the DJ. Top DJs commanded high salaries, and even those in small markets sometimes achieved local cult hero status.

Paradoxically, Top-40's spectacular success came as much from its ability to repel listeners as from its skill in attracting them. Formula programmers relied on consistency above all else; they programmed relentlessly for their target audience, no matter what other audience segments might think. Initially, this often meant junking old programs attractive to older audiences. By the 1960s rock lyrics spread slogans of the disenchanted in a coded language, defying conventional, conservative standards that broadcasting had previously sought to maintain. In many communities there were protests by whites who feared rock music's close ties with black culture and its increasing popularity among white teens. Similar concerns were voiced in the 1990s when rap became popular with youthful listeners.

A second ingredient in Top-40's success was an equally single-minded dedication to constant promotion and advertising. Call letters and dial position were indelibly imprinted on the listener's mind, usually by endless repetition and promotional gimmicks, in an effort to ensure proper responses to rating surveys.

FM's Rise to Dominance

Radio's renewed focus on music brought FM to the fore. For its first quarter-century, broadcasting simply meant AM radio. In 1933 Edwin Howard Armstrong (see Exhibit 2.l) announced and patented a much-improved alternative method of audio transmission using frequency modulation (FM). Armstrong tirelessly promoted the superiority of FM, thinking he had the support of his friend David Sarnoff. But Sarnoff's RCA was

EXHIBIT 2.1

. . . .

Edwin Armstrong (1890–1954)

Edwin Howard Armstrong is perhaps the most tragic figure in broadcasting's technical history. Born on December 18, 1890, Armstrong grew up in Yonkers, New York, where the new science of wireless fascinated him even as a child. By age 14 he was experimenting with wireless gear in the attic of his home and erecting antennas higher than 100 feet. He later attended Columbia University, graduating with a degree in electrical engineering in 1913. Before graduating, Armstrong had struck on the idea of the feedback, or regenerative, circuit, the first of his four major inventions. The regenerative circuit dramatically increased the ability of Lee de Forest's Audion vacuum tube to amplify weak radio signals. In developing the circuit, Armstrong explored the Audion's operation thoroughly, and in 1914 he published a technical article on the Audion. In the article he presented a more cogent explanation of the tube's theory of operation than had its inventor. This article and his regenerative circuit immediately established Armstrong's reputation among scientists and engineers.

In 1917 Armstrong entered military service as a captain in the Army Signal Section. During this period he patented his second major invention, the superheterodyne receiver, a system superior to the earlier regenerative circuit. Returning to his civilian career, Armstrong in 1922 revealed his third major invention, the superregenerative circuit. By the mid-1920s, he was receiving royalties from receiver makers. Meanwhile, Lee de Forest had sued Armstrong over the regenerative circuit patent, claiming that he had invented the circuit years earlier. A vicious battle, both in the courts and beyond, raged between de Forest and Armstrong for more than 20 years. In 1934, to the astonishment of the engineering community, the U.S. Supreme Court ruled in favor of de Forest. Armstrong's patent for the superheterodyne circuit was also overturned because of a French inventor's prior claim.

While still fighting for his earlier patents, Armstrong began looking for a static-free broadcast technology, a search that led him to frequency modulation, or FM. Armstrong was contractually obligated to give RCA first right of refusal for any new invention. He demonstrated his FM system to his friend David Sarnoff and other RCA officials but received little encouragement. Sarnoff was simply stalling, convinced the next major development would be television, not a new radio service competing with AM. By 1936 Armstrong realized he must develop FM without RCA's help. He built an experimental FM station, W2XMN, in Alpine, New Jersey, where regular FM broadcasting began on July 18, 1939. He began licensing manufacturers to build FM receivers and created the Yankee Network of FM stations. Things were looking up until the United States entered World War II and consumer manufacturing ended. After the war the FCC moved FM broadcasting from 44 to 50 MHz to its present allocation, 88 to 108 MHz, instantly making half a million prewar receivers and about 50 stations obsolete.

Armstrong resumed his efforts to promote FM broadcasting and began receiving royalties from the makers of FM receivers, but not from RCA. In 1948 he sued RCA for infringing his FM patents. The years battling de Forest, his feelings of frustration and betrayal by David Sarnoff, and his disappointment in seeing FM languish while TV exploded, took their toll. In 1953 his wife of more than 30 years left him, and on February 1 of the next year, he jumped 13 floors to his death.

Interestingly, Armstrong seemed to get more attention after his death than before. In 1956 *Harper's* magazine ran an article outlining Armstrong's many contributions to modern electronics. Eighty-two-year-old Lee de Forest immediately fired off a four-page rebuttal, chastising *Harper's* for having credited Armstrong with inventions he claimed were his. Even death could not end the old rivalry. Today FM radio reigns supreme, and Edwin Armstrong is unquestionably its inventor.

Source: Photo from Brown Brothers.

making good money with AM, expected to make even more with TV, and devoted little attention to the development of FM.

About 50 stations made it on the air before the United States entered World War II. At the close of the war, however, the FCC shifted FM from its prewar channels to its present location at 88 to 108 MHz in the VHF band. This 1945 move made obsolete the half-million FM receivers that had been built to operate on lower frequencies.

Postwar interest in FM stations, mostly as minor partners in AM/FM combinations, peaked in 1948, with more than a thousand stations authorized. But as television's rapid climb to power began, FM faded into the background. Total station authorizations declined during most of the 1950s.

Faced with bleak prospects for FM, exhausted by continuing court struggles over his patents, and probably feeling betrayed by his onetime friend Sarnoff, Armstrong became deeply depressed. In 1954 he put on his overcoat and stepped out of the window of his 13th-floor New York apartment, plunging to his death.

In 1958 FM began to recover, not only because of greater audience interest in improved sound quality but also from the lack of spectrum space for additional AM stations.

A new pro-FM FCC policy also helped. For example, in 1961 the FCC approved technical stan-dards for FM stereophonic sound. In 1965–1967 the FCC adopted a nonduplication rule that required AM/FM owners, first in big cities and then in smaller markets, to program their FM stations independently of AM sister stations. These measures proved effective in moving FM out of AM's shadow and created FM as an independent service with its own formats. Many early FM receivers were manufactured as part of an AM/FM/phonograph (and often a television) appliance that was large, heavy, and expensive. Increased listener interest galvanized manufacturers into making FM receivers more widely available, more portable, and at a lower cost. By 1974 the majority of sets included FM, and two years later so did most car radios. By the 1980s a reverse trend had set in: Some radios were marketed with FM-only tuning, and 75 percent of all radio listening was to FM outlets. And ironically, when the Federal Communications Commission dropped the program nonduplication rules as a part of its *deregulation* in the 1980s, some FM stations allowed struggling AMs to duplicate their programming.

During the 1970s and beyond, television broadcasting, too, would find itself backed into a corner. The next chapter describes an era of new and increasing competition that threatened to undermine the entire broadcasting industry in its traditional form.

CHAPTER 3

Cable and Newer Media

Traditional broadcasting prospered into the 1980s, little changed from a decade or two earlier. But new multichannel delivery systems developed in the late 1970s, expanding so rapidly in the 1980s and 1990s that today only about one in five American homes now receive all their television programs directly from over-the-air stations.

Audience demand for still more viewing options, along with technical progress and important changes in regulatory thinking that favors marketplace solutions, have combined to create a fast-changing electronic media scene. Technologies introduced here are described in more detail in Chapter 4, *How Electronic Media Work*, and Chapter 5, *Relays, Recording, and the Digital Revolution*.

3.1 EMERGENCE OF CABLE

Broadcasting's greatest challenge emerged from attempts to improve TV station coverage. The FCC's 1952 television channel-allotment plan (the *Sixth Report and Order* described in Section 2.11, p. 43) left many "white areas"—places where people could not receive television service. Even in the biggest cities viewers might receive only five to eight channels. Most people received three or four, and virtually everyone wanted more.

Extending TV Coverage

At first low-power *repeater* transmitters met some of the demand by extending station coverage into fringe reception areas and into pockets that hills or mountains cut off from normal reception. The most common type, the *translator*, transmits on a different channel from the originating station so as not to interfere with the originating station's signal. Some stations operated several translators. Today, about 5,000 translators are in operation.

CATV

Another signal extender, *community antenna television* (CATV), originally worked something like a translator, repeating the signal of one or more nearby stations. Instead of transmitting signals over the air, however, a CATV system delivers them by wire to individual subscribers. A CATV system using *wide-band* cable can deliver signals from several stations, outdoing translators by giving eager subscribers a range of program choices.

CATV systems began appearing in rural areas of both Oregon and Pennsylvania in 1948, just as regular broadcast television began. And, like radio years earlier, CATV's original economic motivation was to sell receivers. Appliance dealers who built the first systems could market television sets to consumers who otherwise would not buy them because they could not pick up over-the-air signals. Community antennas on hilltops or on high towers, unaffordable by the average viewer, obtained unobstructed line-of-sight signal paths to originating stations with viewers sharing in their cost.

CATV systems remained a local concern in the 1950s. Regulation, if any, came from municipal governments that granted cable operators franchises, permitting them to string cables on utility poles or bury them along public rights of way. Early cable systems carried only five or six channels, all of them nearby television stations. Most systems served from a few dozen to a few hundred subscribers. Fewer than 1 percent of all television households had access to cable by 1960. A decade later (1970) the percentage of homes with cable was still under 10 percent (Exhibit 3.a).

Program Augmentation

CATV operators soon sought ways to enhance—or *augment*—their service by offering subscribers something new. This took several forms, as illustrated in Exhibit 3.b. Using microwave relays, some systems began delivering programs from more distant television stations. Others originated their own local programs. And some systems began to offer nonbroadcast programming—feature films, sports, and special events. Such augmentation, along with later additions, marked a fundamental change from CATV's original role.

Broadcaster Fears

Television stations welcomed CATV as long as it acted merely as a *redelivery service*, extending signals

EXHIBIT 3.a
• • • • • • • • •

Cable Growth Indicators, 1960–2000

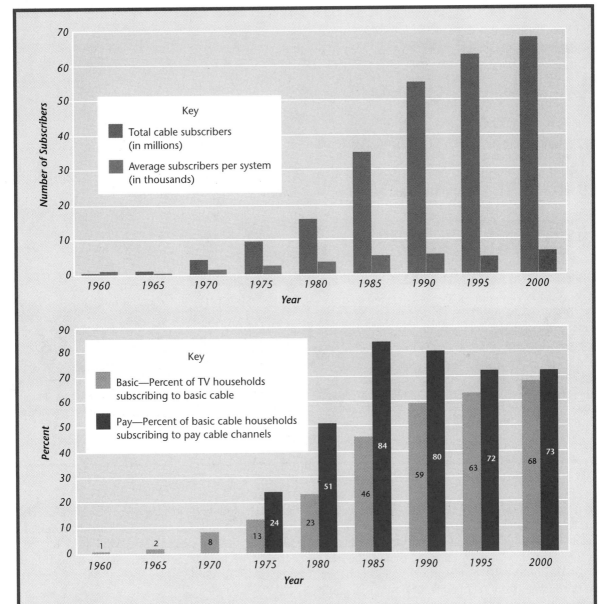

As cable grew, system size and channel capacity both increased. After 1975, pay cable spurred basic cable growth. The number of cable households continued to increase through the 1990s, but pay cable subscriptions as a percentage of cable households began declining in the mid-1980s. Recent figures show a relatively steady percentage of basic cable households subscribing to pay channels, perhaps signaling the end of a decade-long trend.

Sources: Adapted from *Stay Tuned: A Concise History of American Broadcasting* by Christopher H. Sterling and John M. Kittross. Wadsworth Publishing Company, Inc., 1990, pp. 660–661 for data through 1985; NTCA for data through 2000, citing various sources, including *Television & Cable Factbook*, A. C. Nielsen, and Paul Kagan Associates, Inc.

EXHIBIT 3.b
· · · · · · · · · ·

Changing Roles of Cable

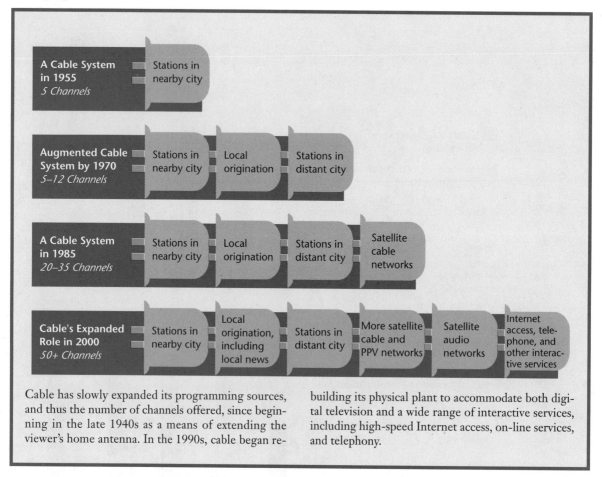

Cable has slowly expanded its programming sources, and thus the number of channels offered, since beginning in the late 1940s as a means of extending the viewer's home antenna. In the 1990s, cable began re-building its physical plant to accommodate both digital television and a wide range of interactive services, including high-speed Internet access, on-line services, and telephony.

to unserved areas, beefing up fringe reception, and overcoming local interference. Some stations eventually found their signals carried by dozens of cable systems that substantially enlarged their audiences.

By the mid-1960s, however, broadcasters began to see augmented CATV more as a potential competitor than as a neutral extender of their signals. The growing cable practice of importing signals from distant cities tended to fuzz normally fixed boundaries of broadcast markets. A network affiliate might find its programs duplicated in its own viewing area by a cable-delivered distant station affiliated with the same network. This undermined the concept of market exclusivity, under which broadcast programs had usually been sold. Program duplication added to CATV's tendency to

fragment audiences, leaving broadcast stations with lower ratings and thus less appeal for advertisers. In addition, CATV began to expand into big-city suburbs, no longer merely serving those unable to receive off-air service.

Regulation Limits Cable
· · · · · · · · · · · · · · · ·

The FCC initially adopted a hands-off policy regarding cable regulation, noting that cable operators did not transmit signals over the air. But as cable's economic impact grew, so did broadcaster complaints. In addition to complaining to the FCC, broadcasters appealed directly to Congress, pressuring politicians to encourage the FCC to take action. The FCC reversed its earlier position and began adopting protectionist rules. In 1962 it began to use microwave relay licenses as a justification for regulating cable systems that imported distant station signals. Four years later the Commission extended regulation to all cable systems, beginning a brief period of pervasive control of cable.

The FCC soon required cable to carry all television stations within each system's areas of coverage (the *must-carry rule*) and to refrain from duplicating network programs on the same day a network offered them. No new signals could be imported into any of the top 100 markets without a hearing on probable economic impact on existing television stations. In 1968 the Supreme Court upheld this FCC rule (US, 1968a).

As for cable-originated programs, broadcasters feared (long in advance of any real danger) that networks of cable systems might outbid them for transmission rights to popular programs such as major sports events. They spread scare stories about program *siphoning*—the draining away to cable of hitherto "free" broadcast programs. Broadcasters warned that fans might have to start paying to receive major league baseball and football games. They also claimed that network news and local public-service programs would suffer if cable cut into broadcaster advertising revenues.

Such tactics resulted in the FCC's so-called definitive cable regulations of 1972 (FCCR, 1972b). These rules severely restricted cable to being merely *ancillary* or secondary to over-the-air television. The new rules sharply limited the type and number of signals that cable could bring into the largest cities in competition with local stations. And cable operators had to provide, on request, access channels for local governments, educational institutions, and the public.

Later Deregulation

Only five years after the 1972 "definitive" rules appeared, however, an appeals court held that the FCC had "in no way justified its position that cable television must be a supplement to, rather than an equal of, broadcast television" (F, 1977: 36). This rebuff, plus a change in administration at the White House (and consequently in the FCC chair), led the FCC to reconsider. It began to remove itself step by step from cable regulation.

This *laissez-faire* spirit prevailed until the early 1990s. Prompted by complaints about high subscription fees, poor service, and the arrogance of cable system operators, Congress imposed new regulations with passage of the Cable Television Consumer Protection and Competition Act of 1992. Four years later, Congress again reversed course and passed the Telecommunications Act of 1996. See Chapter 12 for additional discussion of cable regulation.

3.2 CABLE BECOMES A MAJOR PLAYER

As program augmentation expanded and cable shed its regulatory constraints, CATV began to emerge as a full-fledged competitor for television viewers. As it matured, the term *CATV* faded out, replaced by *cable television*. The transformation of CATV into cable combined deregulation, use of new satellite relays, and increased original programming.

Domsats' Key Role

In the 1960s, communication satellites first came into use for international broadcast relays, bringing overseas events to home screens instantaneously. Demand for *domestic satellites (domsats)*, however, was delayed because of well-developed existing national systems of microwave and coaxial cable relays.

A 1972 FCC decision stimulated domsat demand. The commission adopted an "open-skies" policy, allowing any adequately financed and technically qualified business firm to launch satellites for domestic use (FCCR, 1972a).

Western Union's Westar I became the first American domsat in 1974. Such satellites operate as common carriers—point-to-point, rate-regulated communication services. They lease or sell satellite relay services, either to program distributors or to brokers who resell short-term access to users such as television news departments.

TVRO Antennas

A second satellite-related FCC decision, issued in 1979, deregulated *television receive-only* (TVRO)

EXHIBIT 3.c
. . . .

Ted Turner: Cable Empire Builder

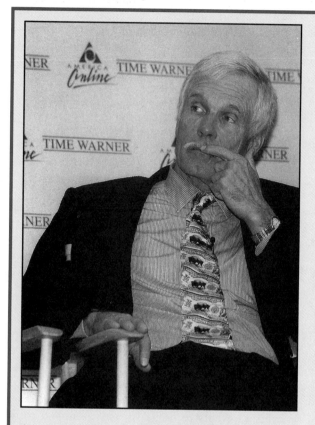

Ted Turner, former board chair and president of Turner Broadcasting System (TBS), is a creative man. He created the first superstation (WTBS, Atlanta), Cable News Network (CNN), CNN Headline News, CNN International, CNN Airport Network, Turner Network Television (TNT), Turner Classic Movies, The Cartoon Network, Turner Home Video, and Turner Program Services. He also likes acquiring things—things such as studio production and distribution companies (New Line Cinema, Castle Rock, Hanna-Barbera) and professional sports teams (Atlanta Braves and Atlanta Hawks). Ted Turner has become a symbol of the rough-and-tumble industry he helped create.

Referred to variously as "Terrible Ted," "Captain Outrageous," "The Mouth of the South," and "Man of the Year" (named by *Time* magazine in 1992), Turner is known for his outspoken opinions, his willingness to challenge the establishment, his aggressive and entrepreneurial spirit (signs reading "Lead, follow, or get out of the way" can be found on his desk, and the phrase is the title of the first Turner biography), his driving ambition, and his ego. Physically, he's the prototype of a southern gentleman: tall, lanky, with silver hair and mustache, a cigar in his mouth, and a Georgia drawl. He loves the movie *Gone With the Wind* so much he began his TNT network service with it and even named one of his five children Rhett.

Turner attended Brown University and was active in debate and the yachting club. College yachting served him well: In 1977 he won the prestigious America's Cup race. But Brown threw him out of college twice for poor grades and excessive pranks—and his fraternity dropped him for burning down its homecoming display.

Turner began his business career as a salesman with his father's advertising firm in Savannah, Georgia. In 1963 he became chief executive of various Turner companies, with headquarters in Atlanta. His interest in television began in the 1970s with his acquisition of Atlanta's channel 17, then a failing independent outlet. It was there, six years later, that he dreamed up the superstation concept.

On June 1, 1980, Turner launched Cable News Network (CNN), the first live, 24-hour, all-news

antennas, used to pick up programs relayed by satellite. This eliminated a cumbersome licensing process, opening the way to widespread satellite use for relaying program services to cable systems—and soon to broadcast stations as well. In 1977 fewer than 200 cable systems owned their own TVROs, but within a decade some 8,000 systems used them.

Turner and Superstations
. . . .

Domsats and TVROs gave cable systems a more efficient and less expensive way of obtaining new national program services with which to attract subscribers. However, the programs themselves still remained in short supply. Ted Turner, an innovative cable operation. Although it struggled for acceptance in its early years, CNN (including CNN Headline News, begun in 1981) now reaches virtually all of America's cable subscribers along with millions more in countries around the world. (In 1982 Turner also formed CNN Radio, a 24-hour all-news cable audio service.)

Turner does not always have the Midas touch. He stumbled in 1984 when he tried to compete with MTV. His Cable Music Channel lasted just over a month until MTV bought its assets for about $1 million plus free advertising on Turner's other cable services. His biggest battle—also an eventual failure—came in 1985 when he tried to take over CBS.

Within a year of the CBS debacle, Turner purchased the MGM studio's huge film library (including his beloved *Gone With the Wind*) for use on his TBS and, later, TNT cable networks. The purchase, however, pushed Turner deeply into debt, and he was forced to accept a consortium of large cable system operators as partners. They received positions on his board of directors in return for their investment.

TNT began in October 1988, promoted as the first cable network designed expressly to challenge the major television broadcast networks. Initially reaching some 17 million subscribers, the service offered mostly movies but began more original programming in the 1990s. It was the fastest-growing new cable network, reaching 60 million subscribers by 1994. Among other audience-building tactics, Turner backed the computerized "colorization" of classic black-and-white films from the MGM collection. And in 1991 Turner took over the Hanna-Barbera animation studio and its film vault, which became a feeder for The Cartoon Network begun in 1992.

By the early 1990s Turner had begun to leave most day-to-day control to managers, focusing instead on strategic planning and long-time interests—such as the environment—that he shared with his wife, Jane Fonda. Development of the Turner cable empire continued apace. TBS helped start the first independent television station in Moscow early in 1993.

Turner was mentioned as a possible bidder for CBS once again in 1995, but CBS sold to Westinghouse and, within weeks, a merger between Turner and Time Warner was announced. Turner came out of the deal as vice chairman of the merged company with stock worth an estimated $2 billion.

Media observers speculated whether Turner could adjust to being a subordinate executive in a media conglomerate. Turner associates deflected concerns about his new role. They noted that Ted Turner is the kind of guy who teaches cats to land on their feet.

Following the merger, Turner continued to manage his network properties agressively as a Time Warner executive. In 1997 *MediaWeek* reported that one in every three cable advertising dollars went to a Turner network (*MediaWeek*, 14 April 1997: 15). The American Marketing Association honored him with a 1998 Edison Achievement Award for his marketing vision and excellence. In 1999 the National Cable Television Center and Museum inducted Turner into its Hall of Fame.

Turner has also continued to make news as a celebrity apart from his role as a cable network executive. In 1997 he announced that he was donating $1 billion of his personal wealth to support United Nations causes around the world. In typical Turner fashion, he contrasted his generosity with that of his fellow billionaires and publicly chastised them for their lack of charity, calling Microsoft's Bill Gates a "philanthropic tightwad."

Source: Diane L. Cohen/SIPA PRESS.

entrepreneur with an almost mystical faith in cable's future, invented the superstation concept as one way to expand cable's program menu. His colorful career is outlined in Exhibit 3.c.

In 1970, Turner bought low-rated Atlanta station WTCG, operating on UHF channel 17 (renamed WTBS, for Turner Broadcasting System, in 1979). Its schedule soon heavy with movies and sports, channel 17 was uplinked for satellite distribution to cable systems throughout the country (see Exhibit 3.d). Turner initially charged nothing for his programs to entice cable-system operators to invest in TVROs and downlink WTBS for their subscribers. Cable systems paid 10 cents per subscriber

EXHIBIT 3.d
· · · · · · · · · · ·

Turner Pioneer Satellite Relay

When HBO and Ted Turner turned to domsats for distribution of their signals in the mid-1970s, they blazed a trail for more than 180 satellite-delivered cable networks. This exhibit shows how Turner initially distributed programming on his Atlanta-based UHF TV station to cable subscribers nationwide, making it the first "superstation." A changing cable market persuaded Turner to convert the service from a superstation to a cable network in 1998.

Source: Turner Broadcasting System, Inc.

per month for the relay service. Turner's eventual profit came from higher advertiser rates on WTBS, justified by the nationwide cable audience that it acquired as a superstation. By 1995 nearly all cable systems carried TBS (as the superstation was called to publicize Turner Broadcasting System), representing a potential audience of some 60 million homes.

Chicago's WGN and several other TV outlets also became superstations with schedules of films, syndicated fare, and sports. By the mid-1990s the superstation concept was losing some of its appeal. Turner concluded that economic factors placed his superstation at a competitive disadvantage with other cable program providers. Turner converted TBS to a cable network in 1998 but continues to use the term *superstation* for promotional purposes.

Superstation WGN suffered serious problems in the mid-1990s and also considered conversion. Late in 1996, WGN briefly lost rights to its national feed of Chicago Bulls basketball games and some cable system operators, notably, Tele-Communications, Inc. (later AT&T BIS), dropped the station in several markets. The National Basketball Association, however, restored rights to some games, and WGN won renewed consumer support by offering more than 140 major league baseball games in the 1998 season. Turner's conversion of TBS to a cable network also helped WGN by opening up a basic channel on many cable systems. WGN campaigned aggressively for these open channels, and by the fall of 1998, it was stronger than it had been only a few years earlier. WGN management accordingly postponed indefinitely plans to convert from superstation status but did not reject that option as a possibility in the future.

HBO and Pay Cable
• • • • • • • • • • • • •

Domsats played a key role in expanding another major programming innovation—*pay (or premium) cable*. Home Box Office (HBO) led this development in 1972. It introduced an advertising-free channel of (primarily) feature films for which subscribers paid an additional monthly fee over and above the charge for advertising-supported channels (later called the *basic tier*).

Lack of cost-effective distribution held back pay-cable progress until 1975, when HBO leased a satellite transponder—the receive-transmit unit on a satellite. HBO offered its programs to any system in the country able to buy or lease a TVRO antenna. Satellite delivery reduced distribution costs and enabled simultaneous reception throughout the country—an essential condition for national promotion of the service.

HBO's move from ground to space relays helped create cable networks. Although only a quarter of all cable systems carried a pay-cable channel in 1977, by the mid-1980s virtually all did (see Exhibit 3.a). By 1999, Time Warner-owned HBO and Cinemax, combined, enjoyed a subscriber audience of more than 34 million households. Today there are more than a dozen other pay-cable networks, though none has larger audiences.

Scrambled Signals
• • • • • • • • • • • • •

Pioneer cable systems controlled access to their program services by connecting and disconnecting individual subscribers. But distribution of programs to cable systems by domsats made it possible for individuals with TVROs to "pirate" signals directly off the satellites. Program providers began using several kinds of *encryption*—signal-*scrambling* techniques—to ensure that only those authorized to receive the programs could do so. The device used most widely to decode the encrypted signals leaving the satellites was the VideoCipher (see Exhibit 3.e). Today cable system operators increasingly use newer DigiCipher technology to decode the growing number of digital signals relayed by satellite. Cable system operators often used simple, relatively inexpensive traps at the subscriber's home to provide conditional access to channels for which the subscriber had paid. Traps worked reasonably well to control access to pay channels, but pay-per-view services (discussed in the next section) are typically impulse purchases that require a particular channel (or channels) to be scrambled during one hour but not the next. The answer is the addressable decoder, a set-top box (STB) that allows the system operator to scramble or unscramble specific channels in the homes of individual subscribers by sending an access code to a particular STB from the cable system's main office. Such "smart" STBs are also critical to the success of digital direct-to-home satellite service (discussed in Section 3.3) and to digital cable services.

Sophisticated STBs greatly enhance the ability of program providers to restrict access to proprietary services, but piracy remains a serious problem. Companies affected by pirates try to reduce losses by prosecuting unlicensed decoder dealers and those who try to bypass encryption systems. They also use social pressure by publicizing the names of those convicted. But a 1999 survey of cable subscribers by the Cable and Telecommunication Association for Marketing (CTAM) found that only about a third of those surveyed considered program piracy a serious criminal activity. Even fewer nonsubscribers expressed concern about people stealing cable service (CTAM, 1999).

EXHIBIT 3.e
• • • • • • • • • • •

Satellite-to-Home Viewing

(A)

Orbiting
satellite

Uplink   Downlink

Transmitting
dish

Descrambler

TVRO   TV

(B)

HBO/E   65   15

DIRECTV

(A) Satellite delivery of programs directly to viewers' homes, whether by DBS systems intended for such reception or by home pickup of relay signals designed primarily for cable system use, depends on a reliable means of scrambling the video and sound signals, which are then descrambled by devices supplied only to homes that pay monthly subscription charges for the service.

(B) The descrambling device, in this case a Video-Cipher along with its hand-held remote control unit, is installed between the viewer's TVRO and the television receiver.

(C)

(C) The smaller, less expensive antennas used by DBS and the wide range of program packages available from DBS service providers make these digital satellite-to-home systems popular with millions of viewers. The digital satellite system (DSS) receives and decodes the digital signal and also converts it to analog form for display on a conventional TV set or recording on a standard analog VCR. VCRs that record the unconverted digital signal from the satellite are also available.

Sources: Descrambler photo courtesy of Motorola, Inc.; DBS antenna photo courtesy of DIRECTV, Inc.

Rise of Cable Networks
● ● ● ● ● ● ● ● ● ● ● ● ● ● ● ●

Until about 1980, cable remained basically parasitic, feeding off existing broadcast programs and motion pictures. In the 1980s, however, cable began to flex its programming muscle. HBO led the way with cable-specific productions—programs made especially for cable. It taped special on-stage performances, obtained rights to sports events, and commissioned original programs. In 1996 HBO original programming won 14 Emmy awards, second only to NBC. In 1999, HBO won 23 Emmys, more than any other network.

Ted Turner made a major contribution in 1980 by introducing Cable News Network (CNN), a 24-hour schedule of news and news-related features that many broadcast critics dismissed as the "chicken noodle network" for its low budgets compared with broadcast network news departments. Turner spun off CNN Headline News 18 months later, another 24-hour service. But Turner was not alone. The cable industry in 1979 had collectively begun a full-time, public-affairs cable service, Cable-Satellite Public Affairs Network (C-SPAN), aimed at cultivating good public relations for the cable industry. Based in Washington, C-SPAN covers House of Representatives and (on a second channel begun seven years later) Senate sessions, congressional hearings, and other public-affairs events.

Other cable-specific program networks followed. MTV (Music Television) began in 1981 to provide 24 hours of hit recordings with matching video images, hosted by video disc jockeys. The Weather Channel, started a year later, offers 24-hour weather and environmental programming. Networks have appeared devoted to religion, country music, children's programs, home shopping, sports, ethnic programs, fine arts, and education. The growth of new cable networks continued in the 1990s, some of them competing with existing formats and others narrowly focused on subsets of older formats. A network restricted to sports seemed far-fetched when ESPN was launched in 1979, but its success was followed by ESPN2 and Fox Sports Américas in 1993. Classic Sports Network, launched in 1995, became ESPN Classic in 1997. In 1996 CNN/SI appeared and, along with many regional sports networks, further increased the options for sports programming on cable. The Golf Channel proved that a network devoted to a single sport could gain access, if not high viewership. The Golf Channel began in 1995 with a potential audience of only 140,000 subscribers, but by 1999 it could be seen in 26 million households. Indeed, by 1999, cable system operators had more than 180 national video satellite-distributed program services from which to select, with no end in sight. Data on some of these are found in Exhibit 6.e (p. 178).

Pay-per-View
● ● ● ● ● ● ● ● ●

A more flexible cable service requires subscribers to pay extra for individual programs. First appearing in 1984, *pay-per-view* (PPV) allows viewers to order individual programs from the cable company for a one-time fee. Movies, many of them "adult," account for a large number of PPV requests. "Events," predominantly boxing and wrestling, are important sources of PPV revenue. In the 1990s, boxing matches involving heavyweights Mike Tyson and Evander Holyfield were among the top PPV events. A 1996 bout between the two pugilists reportedly grossed more than $100 million, setting a PPV record. A return match in 1997 generated almost 2 million PPV requests and an estimated $95 million in revenue. Tyson received a year's suspension from boxing for biting Holyfield's ear during the fight, and some predicted an end to large PPV profits for a tainted sport. Boxing PPV did experience a decline in 1998, but by 1999 both Tyson and Holyfield were again attracting sizable PPV audiences. Although falling short of the excitement created by earlier Tyson–Holyfield fights, a 1999 sell-out match between Holyfield and Lennox Lewis still generated more than a million PPV orders.

As noted earlier, widespread use of PPV depends on *addressability*—an efficient means for the cable system office to receive program orders and address individual PPV decoders. Some PPV customers use a touchpad to communicate with a computer called a *video server* at the cable headend. The video server interacts with the customer's decoder and carries out record-keeping and billing operations. Other PPV subscribers must order PPV programming by calling the cable system operator or program provider. Eventually, all subscribers will be able to order PPV programs through their set-top boxes.

PPV developed rapidly in the late 1980s; by 1995 nearly a third of cable households had addressable decoders. In 1998 the National Cable Television Association (NCTA) reported there were almost twice as many national cable PPV networks as in 1997 (NCTA, 1999). PPV is not restricted to cable, however. Homes receiving direct broadcast satellite service also have access to PPV programming. DBS provides more PPV channels than are available to the typical cable subscriber.

3.3   NICHE SERVICES

Over the past two decades several alternative modes of program delivery have arisen to supplement mainstream broadcast and cable services. Each provides programming by subscription that is similar to or the same as that of existing cable channels (usually in areas lacking cable). A few offer new or different content.

Wireless Cable

The term *wireless cable* refers to systems that use terrestrial microwave radio signals to distribute cable programming to subscribers. Because the systems use radio waves rather than wires and provide the same programming as cable, the term is paradoxical but descriptive. The economic advantages of distributing programming without the expense of running cable to each subscriber's home are obvious, but wireless cable in its analog form has not been a serious competitor in areas served by cable.

Wireless cable appeared first as the *multichannel multipoint distribution service* (MMDS) operating on frequencies in the 2.5-gigahertz band. MMDS's location in the electromagnetic spectrum (discussed in Section 4.1, p. 88) limits its coverage to about 15 miles using analog transmission techniques. Home antennas must be within the line of sight of the transmitting antenna. The first MMDS system began operating in Washington, D.C., in 1985.

MMDS's head-on competition with cable made program suppliers (many owned by cable operators) reluctant to serve MMDS until FCC regulations required MMDS access to such programming in the early 1990s. This change brought new life to wireless cable. There were almost five times as many wireless cable systems operating in 1995 as there were in 1990. Wireless cable also received a boost in the mid-1990s when the telephone industry briefly seized on MMDS as a quick path to establishing a presence in the digital video distribution market. But the phone industry's interest in entering the video distribution business evaporated almost as fast as it developed. A few telephone companies, such as BellSouth and SBC Communications, continued to support and even expand their wireless cable systems. But by 1998 Heartland Wireless and other former industry leaders faced bankruptcy. By 1999 fewer than a million homes were receiving programming by MMDS.

In 1997 the FCC authorized another terrestrial microwave service somewhat like MMDS called *local multipoint distribution service* (LMDS). The term *local* is appropriate. LMDS uses much higher frequencies (in the 28-gigahertz band) than MMDS, which restricts coverage to about three miles. This means that LMDS must use many repeater transmitters covering small cells, much like cellular telephones. LMDS must also have receiving antennas that are in the line of sight of the transmitting antenna, and LMDS is even more susceptible to interference from anything in the path of the signal than is MMDS. Rain or even water vapor in the atmosphere can block LMDS signals, although a well-designed LMDS system can be made very reliable. LMDS has a huge bandwidth (more than 1 gigahertz), and because frequencies can be reused in the same market (like cellular telephone service), the effective bandwidth is even greater.

Yet another group of microwave frequencies in the 38-gigahertz band are available for wireless cable use. At these frequencies, however, signals take on lightlike characteristics, resulting in a coverage of less than three miles per cell and increasing the number of cells needed to provide reliable service. The need for a clear line-of-sight path also increases proportionally, but the bandwidth at 38-gigahertz is even greater than for LMDS, and frequency reuse in the same market is again possible.

Only marginally successful to date as a one-way delivery system for analog video programming, the wireless cable industry is experiencing a rebirth. In the wake of AT&T's 1998 acquisition of Tele-Communications, Inc. as a means of breaking into the local telephone market, other long-distance service providers began approaching wireless cable companies with attractive offers. The increasing demand for high-speed, one-way and two-way delivery of digital data (including telephony and video) triggered a reevaluation of this technology in the late 1990s. It seems unlikely, however, that wireless cable will challenge traditional cable or direct-to-home satellites for the distribution of television programming in the immediate future. At this writing many observers believe that terrestrial microwave systems once considered for wireless cable service may be better suited for providing high-speed Internet access and digital data delivery to businesses than for bringing TV programs into the home.

MATV/SMATV

• • • • • • • • •

A second niche service meets television service requirements of a building or group of buildings on a single piece of property. *Master antenna television* (MATV) and *satellite MATV* (SMATV) are special, limited-area types of cable television. An MATV installation combines a master television antenna on the roof of an apartment building with coaxial cable feeding programs to each unit. Modern multiple-dwelling buildings usually come with suitable cabling already installed.

Satellite MATV adds a TVRO antenna to the installation (Exhibit 3.f), enabling the system to pick up satellite-distributed services. Because SMATVs operate entirely within the boundaries of private property (usually an apartment or condominium complex), and thus do not cross public rights of way, they are exempt from most cable and broadcast regulations.

Cable system operators, who naturally want to sign up as many households as possible, oppose SMATV systems within their coverage areas. In fact, applicants for new cable franchises often press municipal authorities to ensure them equal access to SMATV-equipped buildings. There are several thousand MATV/SMATV operations (mainly in motels and large apartment buildings), though they serve only a small proportion of the total television audience.

Late in 1998 the FCC adopted rules on the use of satellite dishes and other over-the-air reception devices by renters that may profoundly affect MATV/SMATV. These rules allow individuals to install receiving antennas in areas where they have "exclusive use," such as balconies and patios. It is expected that most renters wanting this option will install DBS systems, although the rules allow the installation of conventional TV and wireless cable antennas as well. DBS systems have the advantage of offering a large menu of programming, and the signals arrive from angles well above the horizon. Both DBS and wireless cable requires a clear-line-of-sight path between the transmitting and receiving antennas, but it is much easier for renters on lower floors to get a clear shot at a satellite than a terrestrial wireless cable antenna several miles away. Most MATV/SMATV systems are in large multidwelling structures located in densely populated areas served by powerful local broadcast stations. Renters in these buildings may be able to supplement DBS service with indoor "rabbit ear" antennas for reception of local stations if they do not receive local broadcast stations as part of their DBS service.

MATV/SMATV operators may discourage renters' flight to individual DBS systems by placing DBS services on their systems. In 1999, DirecTV reported that its Master System Operator program was providing more than 200 multidwelling operators with technical, administrative, and marketing assistance. The company's lodging/SMATV business was providing free-to-guest programming to more than half a million rooms in hotels, motels, hospitals, dormitories, nursing homes, and office buildings. (DirecTV, 1999)

DBS: Round One

• • • • • • • • • •

There were approximately 12 million *direct broadcast satellite* (DBS) households in the United States in early 2000, and therefore, some would say DBS should no longer be considered a niche service. DBS's popularity is increasing, and most observers see it as a serious competitor to cable, as do cable industry leaders themselves. But at this writing, DBS serves only a fraction as many homes as does cable and it remains in cable's long shadow.

Widely touted when first announced in 1980, dismissed because of high costs and limited

EXHIBIT 3.f
• • • • • • • • • • •
MMDS and SMATV

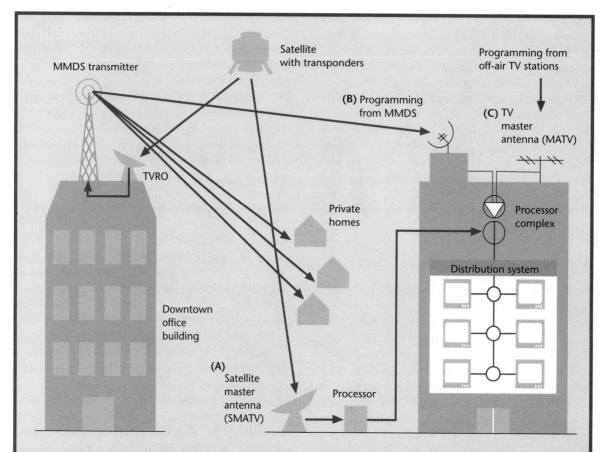

Viewers living in apartments or condominiums often receive their television signals by means of (A) a satellite master antenna television (SMATV) system, a satellite TVRO located near, as shown here, or on top of the building; (B) a multichannel multipoint distribution service (MMDS), a small terrestrial receiving antenna on the building roof; or (C) a master antenna television (MATV) system, also on the roof, picking up off-air signals. Each of these supplies a cable distribution service within the building. Individual homes also can subscribe to MMDS services as shown here. Cable system operators pressure local franchise authorities to eliminate options (A) and (C) when they can—or at least to allow cable equal access to such buildings. Viewer-owned DBS systems may make SMATV and MMDS less popular.

technology just a few years later, and becoming a reality only in the 1990s, DBS appears at first to render any land-based delivery system obsolete by delivering programs to consumers without going through an intervening broadcast station or cable system. A broadcast transmitter, relatively cumbersome and inflexible, reaches only a limited area—and hundreds must be linked to provide national network coverage. Cable systems must hard-wire each subscribing home. A single DBS transmitter, on the other hand, can serve the entire country (with special spot beams serving Hawaii and Alaska).

The FCC first received applications for DBS licenses in 1980. It authorized commercial operations in 1982, a remarkably fast regulatory go-ahead for a new service. Critics questioned whether DBS, primarily of interest to small and scattered audiences in uncabled rural areas (as elsewhere it must compete with cable and established program sources), could recoup its costs. The critics were right—the only American DBS operation (using a Canadian satellite) was short-lived because it attracted few subscribers.

This dim record resulted from a focus on technology at the expense of programming: No DBS operator specified how its service would differ from those already available from cable, broadcasting, and home VCRs. Even in rural areas, possible DBS interest was quashed by the growing use of TVROs for *C-band direct* reception from satellite channels intended as relays rather than for home reception (see p. 402). C-band satellites' requirement for larger and more expensive receiving antennas than higher-powered Ku-band DBS satellites did not discourage backyard-TVRO enthusiasts—some 3 million strong by the early 1990s. (For more about satellites, including a description of the C and Ku bands, see Section 5.2, p. 128.)

DBS: Round Two
• • • • • • • • • •

Despite the dismal first round, however, improving technology revived DBS interest less than a decade later. Once again the FCC received many applications for new services to begin in the 1990s. These new operators drew on DBS experience in Europe and the Far East, where the service faced fewer competing terrestrial channels than in the United States.

The very different second-round DBS projects planned eventually to utilize (1) high-power satellites and advanced-design TVROs; (2) video compression to allow multiple programs per channel (see Section 5.5, p. 138); and (3) both high-definition television and digital audio (see Section 5.7, p. 146, and Section 5.8, p. 153). But these new projects also incorporated a different DBS approach, including (1) a high priority on programming, with more specialized channels to compete with cable, and (2) projects backed by *consortia*, cooperating groups of companies, each focusing on a specific part of the larger picture. Huge DBS expenses were thus shared by several partners.

Both U.S. Satellite Broadcasting (USSB) and Hughes DirecTV began limited operations in 1994, sharing a satellite. Only a month after DirecTV and USSB began, a consortium of cable operators led by Tele-Communications Inc. introduced Primestar, a DBS service using a medium-power satellite requiring larger (three-foot-diameter) receiving dishes. In 1996 Echostar became a third high-power DBS service, and AlphaStar entered the market as a second medium-power service. Consumer response was good, and the new DBS providers signed up about 4 million households in the first two years, many of them former cable customers. DBS was attractive because it offered many more channels than cable and provided superior picture quality.

After a strong start, DBS growth began slowing. DirecTV expected to have 3 million subscribers by the end of 1996, but fell short by more than half a million. To attract subscribers, Echostar tried giving away receiving equipment in Boulder, Colorado, when Tele-Communications Inc. lost its local cable franchise. Months earlier the cable industry had initiated a nationwide anti-DBS advertising campaign emphasizing the high cost of connecting additional sets and the lack of local broadcast station carriage. The campaign may have had some effect because response to Echostar's offer was less favorable than the company had hoped. DBS growth continued, however, increasing to about 6 million subscribers in 1997.

Recognizing the enormous appeal of a DBS provider that could offer hundreds of cable channels plus local TV stations, Rupert Murdoch's News Corporation created ASkyB and proposed just such a service. Early in 1997 Murdoch announced an agreement to merge ASkyB with Echostar and outlined an ambitious plan of DBS development, sending shock waves through the DBS and cable industries. But things began falling apart almost immediately. Conflicts arose over operational control of the merged companies, and, as critics had predicted when he announced his bold DBS plans, implementation was much more difficult than Murdoch anticipated. There were also rumors that cable operators were threatening to drop News Corporation's programming (FX, fXM, Fox News Channel) if Murdoch joined forces with a rival DBS provider.

Only two months after Murdoch's announcement of an agreement with Echostar, he broke off the deal and Echostar filed a $5 billion lawsuit against him for breach of contract. Murdoch then attempted a merger with Primestar, but the U.S. Justice Department blocked the deal because Primestar was owned by a consortium of cable companies, a group considered unlikely to promote DBS as a competitor to cable. Murdoch returned to Echostar, which, in spite of previous disagreements, agreed to drop its suit and acquire Murdoch's U.S. satellite holdings, principally one of the last remaining high-power satellite slots serving the entire continental United States, for $1.25 billion. DirecTV announced plans to acquire USSB late in 1998, and then Primestar early in 1999.

In spite of the stormy deal making and breaking among DBS companies during 1998 and 1999, consumers continued to sign up for service. DirecTV ended 1998 with about 4.4 million subscribers, and Echostar had almost 2 million. DirecTV picked up another 2.2 million with its acquisition of Primestar. FCC figures show that DBS added about 2.2 million subscribers in 1998. During the same period, cable added only about 1.2 million subscribers.

Reducing the field of DBS providers to only two means the remaining companies are more likely to acquire the size necessary to move ahead aggressively. In mid-1999 both DirecTV and Echostar were adding at least 100,000 new subscribers each month. Some analysts believe that DBS will have around 15 million subscribers by 2002 and as many as 25 million by 2004. If these predictions materialize, DBS will still have less than 40 percent of cable's audience today, but much of that growth will be at cable's expense as cable subscribers defect to DBS.

Videotex/Teletext
• • • • • • • • • • • •

Videotex and teletext services were introduced in the United States during the 1970s, and Knight-Ridder and Times Mirror test-marketed them in the 1980s. The services failed to prosper in the United States at the time but were more successful in Britain and European countries, especially France.

Teletext is a one-way, broadcast-based information service delivering print and graphic materials to television screens in homes and offices. Teletext material goes out from conventional broadcast television transmitters, occupying an unused portion of the channel called the *vertical blanking interval* (VBI). Viewers use a keypad to select specific pages or screens of data from several hundred that are continuously transmitted and updated.

Videotex, often confused with teletext, involves two-way communication over telephone or cable lines. Not restricted by the limitations of the VBI, videotex offers unlimited numbers of information pages. Videotex delivery systems are often operated by common carriers, although the content is created by others.

The failure of these technologies in the United States during the 1980s sometimes is attributed to their use of an ordinary television set as the terminal device, combined with an excessive emphasis on graphics. By the early 1990s it was evident that digital computers with high-resolution monitors were far better adapted than TV screens for interactive information retrieval. In the mid-1990s efforts to use the television set as a terminal device resumed. Significant advancements in digital compression, sophisticated graphic design technology, and growing demand for interactive services made use of the TV screen as a terminal device more attractive. Skeptics, however, believe that most people accustomed to two-way interactivity on the Internet will continue using PCs or hybrid TV/PCs rather than conventional TV sets for anything other than the simplest interactive applications. Others say that consumers will use a variety of smaller, more portable, and less complicated Internet appliances to access specific Internet applications. The jury is still out.

Interacting with TV
• • • • • • • • • • • •

Cable visionaries in the 1970s foresaw interactivity as a logical next step in evolving home communications. Customers could talk with others on the cable—teachers could interact with students, and the public could more fully communicate with stores, banks, public utilities, and safety agencies—all without leaving home.

Warner Cable gave this type of interactive service a thorough trial beginning in 1977 on its Columbus, Ohio, system. Called Qube, it offered 10 channels over which viewers could respond by means of a touchpad. Though the experiment generated a lot of media coverage, Qube failed to attract sufficient subscribers to warrant its high cost, and after three years of mounting losses Warner closed it down.

Similar viewer apathy greeted later two-way experiments, including one operated by GTE in Cerritos, California, in the early 1990s. After a brief initial fascination, viewers either lost interest or were unwilling to pay the price necessary for the service.

The success of the Internet and the interactivity it provides, however, suggests that there was a dramatic shift in public interest during the middle and late 1990s. As the new century begins, there is a growing market for interactive services, available in the home, for sending and receiving text, audio, and video. Although some people will have little interest in particular features of the information superhighway, many are increasingly interested in doing more than simply viewing information on-screen. The Graphic, Visualization, and Usability Center at the Georgia Institute of Technology began conducting surveys of World Wide Web users in 1994. The first surveys revealed that most users considered on-line credit card purchases risky, but by 1998 about 82 percent said that on-line shopping and banking made their lives easier. Almost 71 percent reported having spent more than $100 on on-line purchases (Graphic, Visualization, and Usability Center). Stocks of "dot-com" Internet companies such as Amazon.com soared during the late 1990s, reflecting investor confidence in the future of electronic interactive commerce. The National Retail Federation/Forrester Online Retail Index reported that during January 2000 alone U.S. consumers spent $2.8 billion on-line, more than during all of 1997 (National Retail Federation/Forrester Online Retail Index).

Cable provides a large pipeline into millions of homes, and cable operators are rebuilding their systems to accommodate two-way transmission of digital information of all kinds. Conversion to provide digital video programming and increasing system channel capacity to compete with DBS are major goals, but delivery of interactive services is also extremely important.

That conversion is well under way, although progress has been slower than some had predicted in the early and middle 1990s, and than futurists such as Paul Allen (see p. 11) would like. In mid-1999 Paul Kagan Associates reported that fewer than 20 million homes were passed by cable offering high-speed Internet access. As this decade began, only about 10 percent of those homes were using cable modems to access the Internet (Paul Kagan Associates, May 1999). Industry analysts, however, predict that the availability of high speed Internet access, as well as the number of users, will increase rapidly in the early years of the century. As early as 1999, at least seven of the largest multiple-system cable operators were offering residential or commercial telephone service in more than 25 markets, including Los Angeles, Boston, Phoenix, and San Diego. AT&T's acquisition of Tele-Communications, Inc., and its proposed merger with MediaOne is expected to increase dramatically the number of homes using cable telephony in the future. Consumer demand for interactive services and cable's enthusiasm for services other than traditional TV programming appeared questionable at times during the 1990s, but now the pace of change is accelerating, and some, perhaps excessively optimistic, analysts believe that more than 80 percent of the nation's cable infrastructure will be capable of providing two-way interactive services by the end of 2000.

3.4 ELECTRONICS REVOLUTION

The sweeping changes taking place in electronic media today are made possible by a fundamental shift from analog to digital technology. Digital technology makes all forms of communication content—sound, images, and text—electrically equivalent. The original form of the information no longer matters to the equipment used to produce, distribute, or store it. This basic commonality of content is at the heart of the changes taking place in mass media and was made possible by the development of solid-state microelectronics.

Transistors

The "crystal and cat-whisker" detector employed in radios during broadcasting's infancy was the first solid-state electronic device, but it was not capable of amplification. Fleming's vacuum tube detector and de Forest's Audion eclipsed the

EXHIBIT 3.g

Changing Definitions of "Portable"

(A) The paper clip and pencil show the size of 1950s vacuum tubes in relation to 1960s transistors and current integrated-circuit chips. (B) The "father of FM radio," Edwin Armstrong, shows off an early "portable" radio that required either large batteries or household electric power. (C) Transistors began to replace vacuum tubes in the 1960s, leading to small, inexpensive portable radios. (D) Today integrated circuits provide two-way wireless communication of voice and data from almost any location.

Sources: Photo (B) Corbis–Bettmann; photo (C) courtesy of Zenith Electronics Corporation; photos (A) and (D) J&M Studios/Liaison International.

crystal detector and dominated electronics for almost 50 years. But vacuum tubes were hot, bulky, fragile, and power-hungry, and, worst of all, had a limited life. In 1947 a trio of Bell Laboratories engineers developed the transistor, a three-element, solid-state tube replacement that had none of these weaknesses. The engineers received the Nobel Prize in 1956 for their invention, a fitting distinction considering their transistor became the foundation on which all modern electronics is based. One of the first mass-produced transistor-based products on the consumer market, a tiny portable radio, appeared in 1954 and began the trend toward miniaturization depicted in Exhibit 3.g.

Transistors also made possible the development of that prime artifact of the late 20th century, the computer. Only computers could manipulate billions of information bits with the speed and intricacy needed for digital processing. In 1945 the first large experimental electronic computer used some 18,000 vacuum tubes plus miles of wire. The tubes generated so much heat that attendants had to stand by to replace them as they blew out. Transistors solved the heat and size problem while reducing costs and enormously increasing computer speed and memory capacity.

The Chip

Tiny computers and other devices remained impractical, however, as long as hundreds of separate transistors had to be meticulously wired together with hand-soldered connections. True miniaturization came with the invention of the integrated circuit—such as the computer chip shown in Exhibit 3.g.

Chips are made of crystalline wafers on which microscopically small electronic components are etched. A modern chip may contain many thousands of transistors and other components. This miniaturization makes it possible to crowd all the electrical circuits needed to construct entire radios and other electrical devices into a single chip. The integrated circuit chip is responsible for such dramatic developments as inexpensive multifunction digital watches, hand-held calculators, laptop computers, and cellular telephones. It has also made possible communications satellites and *electronic news gathering* (ENG) equipment, forever changing our concept of TV news. All forms of digital video delivered via wire or fiber-optic cable or over the air are the result of advancements in chip technology.

3.5 CONSUMER MEDIA

Solid-state technology made possible many new consumer products that widened viewer and listener options. Some have been dramatically successful.

VCRs: Giving the Viewer Control

Except for the TV set itself, no video device is more popular than the *video cassette recorder* (VCR). As early as 1967 CBS introduced a video recorder featuring a seven-inch tape cartridge that played automatically when the cartridge was inserted. Two years later Sony Corporation of America announced its first color VCR. Several years passed, however, before a consumer-oriented VCR became available.

Marketing of home videotape recorders began in 1975, when Sony introduced its Betamax VCR at an initial price of $1,300. Futurists predicted a new video revolution, now that consumers could choose not only *when* they would view something but also *what* they would view—broadcast or cable programs, prerecorded cassettes, or their own home-recorded videos.

Sony's monopoly ended in 1977 when Matsushita introduced its technically incompatible Video Home System (VHS). Various "bells and whistles" stimulated sales, and competition brought prices down below $1,000 in the late 1970s. By offering longer recording times and lower prices, VHS gradually monopolized consumer VCR markets. Today economy-model VHS machines sell for less than $100.

Sony had promoted Betamax's ability to record TV programs for later viewing, a practice called *time-shifting*, as its major feature. TV networks and major movie studios, whose products appeared on the networks, thought off-air recording was illegal pirating of their copyrighted products, not just a convenience to viewers. But in 1984 the Supreme Court ruled that viewers would not violate copyright law by recording programs broadcast over the air for their personal noncommercial use.

Hollywood distributors eventually recognized that the VCR provided a new market for their products. Renting and selling movies on cassette became a big business. Distributors soon discovered that consumers not only wanted recent releases but also would rent or buy old movies, providing additional unexpected profits. Box office blockbusters generate the greatest profits, but home video rentals and sales help producers reduce losses on theatrical duds. Today some movies rent for a few dollars and sell for $10 or less.

Consumer VCR picture and sound quality improved over the years with advances in tape and electronic circuit technology. In 1987, JVC introduced a new variation of the VHS format called Super-VHS (S-VHS), which provided improved image quality using a special metallic tape and S-VHS decks. In 1998, JVC announced Super-VHS Expansion Technology (S-VHS ET), a feature on some S-VHS machines that allows them to record S-VHS–quality images on conventional VHS tape.

During the 1990s, some TV stations adopted S-VHS for recording news events, but few consumers bought S-VHS systems for home use. Most people bought "standard" VCRs because people were more interested in the new convenience features such as wireless remote controls and easy on-screen programming rather than the enhanced picture quality provided by more expensive S-VHS systems. Constantly dropping prices also contributed to conventional VCR sales.

A new Digital-VHS (D-VHS) format also became available for home recording in 1998. The D-VHS format allows consumers to record DBS and digital broadcast signals in their original "raw" digital format, before they are decoded and converted to analog form for display. This means recorded programs will play back with the same excellent picture and audio quality as the original transmissions. D-VHS cassettes, because of their enormous 44-gigabyte capacity, are also seen as economical "bit buckets" for storing all kinds of digital information. D-VHS has the added advantage of being able to record and play back conventional analog VHS tapes. Prices for basic D-VHS machines at the time of their introduction started around $600.

Some observers believe a transition from tape to a more capable digital disc–based home video recording system is likely in the new century, although many people seem perfectly content with their inexpensive VHS systems.

Video camcorders are less common, although they are increasingly popular and prices continue to fall. Camcorders replaced 8-millimeter film cameras for shooting home movies. They produce pictures good enough that broadcast stations and networks encourage viewers to send in footage for possible use in newscasts, a practice that raises the ethical questions addressed in Section 9.8 (p. 276). Camcorders also contribute to entertainment programming. During much of the 1990s ABC filled an hour of prime time with *America's Funniest Home Videos*, featuring recordings of people doing goofy or embarrassing things. Today several tape formats are used in consumer camcorders. In addition to half-inch VHS, there are two other analog formats, 8-millimeter and Hi8. The trend is toward camcorders that record information digitally using tape that is only about one-quarter inch wide.

Audio Cassettes

Audio cassettes and the machines using them are small, making them well suited for use in cars and as personal portable playback devices. The audio cassette's popularity boomed following the introduction of Sony's Walkman in 1980. Consumers bought the Walkman and its imitators by the millions, and today some combination AM/FM/audio cassette players sell for less than $20.

Compared to an analog audio cassette, a digital audio cassette, or *digital audio tape* (DAT), provides superior sound quality. Unfortunately, DAT equipment is expensive. The music industry, fearing that DAT-to-DAT recordings would encourage flawless duplication of their products, successfully lobbied Congress to force DAT manufacturers to include technology preventing wholesale digital copying. Although DAT is widely used professionally, cost and limits on digital duplication have combined to keep consumer interest in it low.

Philips Electronics introduced a second digital audio cassette format, the *digital compact cassette* (DCC), in 1992. Philips reportedly sank about $55 million into developing DCC. Unlike DAT machines, DCC equipment uses cassettes the same size as ordinary audio cassettes, allowing DCC machines to play analog cassettes as well as record and play digital audio. Philips considered this a significant advantage, but DCC sales were disappointing. The expense of DAT and DCC, and competition between the two formats and from other technologies, made both products commercial flops.

CDs and the MiniDisc
• • • • • • • • • • • • • •

Introduced in 1983, the digital audio *compact disc* (CD) is one of the most successful products in consumer electronics history. The CD's audio quality, size, and durability make it ideal for use at home, in cars, and with personal portable players. The CD's success spurred the development of other disc-based products such as the CD-ROM. The audio cassette and CD practically killed the vinyl-based analog (LP) recording business. LPs remain popular among a devoted cadre of audiophiles, many of whom also prefer the sound produced by older tube-based audio amplifiers.

If the five-inch CD is a compact audio disc, Sony's MiniDisc (MD), at approximately half its size, is a subcompact. A recordable compact disc introduced in 1992, the MD gained some popularity because of its smaller size but failed to challenge seriously the larger CD, in spite of Sony's $100 million investment. Like DAT, Sony's MiniDisc is more popular in professional circles than among consumers.

MP3 Players
• • • • • • • • •

In the second half of the 1990s many computer owners began using an audio compression algorithm called MP3 to record music from CDs and other sources to their computer hard drives. MP3 is one of several compression standards developed by the Moving Picture Experts Group (MPEG), an international group of experts working to standardize compressed video and audio. MP3 can reduce the size of a CD audio file by a factor of twelve without significantly affecting the original audio quality. Because of its ability to shrink audio files, and thus speed up the transfer process, MP3 became quickly the most popular file format for distributing music on the Internet. One survey of almost 2,000 Internet-savvy college students found that more than 90 percent of them had downloaded an MP3 music file. In the late

1990s, manufacturers cashed in on the popularity of MP3 by offering portable MP3 players. MP3 players allow users to download MP3 files from a computer to a solid-state memory card. There are no moving parts in an MP3 player to wear out, and the player is immune to skipping caused by physical activity. Small size, light weight (less than 2 ounces in some models), and excellent audio performance make MP3 players logical successors to other portable audio devices. Early in 2000, however, portable MP3 players were selling for hundreds of dollars and typically offered only 30 minutes or an hour of CD-quality playing time, depending on the size of the memory chip. Extra memory cards for use in portables using removable storage media were also expensive, but prices of consumer electronics fall rapidly as large numbers of consumers begin buying. The entire MP3 phenomenon, and especially the appearance of portable MP3 players, sent shivers through the music industry concerned about protecting valuable copyrighted product. Copyright is discussed in detail in Section 13.10, p. 401.

Videodisc Players
• • • • • • • • • • • •

In the early 1980s both RCA and Magnavox marketed videodisc players for the home. Neither maker's machine could record video or play discs designed for the other. VCRs arrived slightly earlier, and they could, of course, both record and play back. The videodisc had limited appeal for most consumers, and in the mid-1980s RCA pulled out of the videodisc business.

Interest in the videodisc increased in the early 1990s. New machines appeared that could play five types of discs. Sales improved, but by the mid-1990s fewer than 1 percent of all households had videodisc players.

CD-ROM, CD-R, and CD-RW
• • • • • • • • • • • • • • • • • • • •

While the videodisc sputtered, consumers were buying *CD-ROM* (compact disc read-only memory) drives for their computers. CD-ROM provides the storage capacity needed to distribute large software programs and popular interactive games.

CD-R (compact disc recordable) allows consumers to record their own CDs. CD-R discs play on any CD-ROM drive or CD player. User-friendly, audio-only CD-R recorders appearing in the mid-1990s often provided better audio, but they were more expensive and less versatile than computer CD-R drives. They did not, however, require that consumers learn to deal with mice and menus.

Another CD development is CD-RW (compact disc rewritable). Introduced in 1996, a CD-RW computer drive can record, erase, and re-record blank CD-RW discs. CD-RW audio recorders are also available but, like CD-R machines, they require the use of special "music" CD recordable discs rather than the CD-R and CD-RW discs used by computers.

Digital Video Disc
• • • • • • • • • • • •

Often called the digital versatile disc, the *digital video disc* (DVD) provides up to 25 times as much storage space as a conventional audio CD (see Exhibit 5.n, p. 145). Because of its huge capacity, DVD is an ideal medium for distributing full-length movies in wide-screen high definition with surround sound audio.

In 1998 discount electronics merchandiser Circuit City introduced a version of DVD called *Divx*, or digital video express. Consumers could buy a movie on a Divx disc for $5 or less and play it as many times as they wished during a 48-hour period. Forty-eight hours after the first playing, the Divx disc ceased functioning. The consumer could then toss the disc, pay an additional fee of about $3 to reactivate it for another 48 hours, or purchase it for about $15 and play it indefinitely like any other DVD disc. Divx discs played only on specially equipped Divx-enhanced players costing more than standard DVD machines. Divx's appeal was that it provided far superior audio and video quality compared to movies on VHS tape and eliminated the hassle and late-return fees associated with tape rentals. But the concept did not catch on with consumers, Hollywood studios did not offer many movies using the Divx format, and other retail outlets showed little interest. In mid-1999 Circuit City withdrew the product and offered a $100 rebate to purchasers of Divx-enhanced players.

The first DVDs offered playback-only capability, but by early 1999 there were recordable DVD-RAM drives selling for about $500 available for use with computers. A DVD-RAM disc can hold about eight times as much data as a CD-RW disc, giving it a huge advantage. Early in 2000, Pioneer introduced the DVD-1000, a DVD recorder using the DVD-RW format, a format that is incompatible with DVD-RAM. The DVD-1000 was initially available only in Japan and cost about $2,400. Philips and several other major manufacturers, including Sony and Thomson, support DVD+RW, yet another DVD recording format. Supporters of DVD+RW say it is better than Panasonic-backed DVD-RAM because DVD+RW discs will play back on most DVD video players while DVD-RAM discs will not. Thomson plans to market a DVD+RW system late in 2000. Panasonic plans to launch a new generation of DVD players capable of playing DVD-RAM discs while simultaneously introducing DVD-RAM video recorders around the middle of 2000.

Streaming and Surfing
• • • • • • • • • • • • • •

At least 60 radio stations were "streaming" digitized audio programming in real time over the Internet by 1996, and others quickly developed an interest in it. Today thousands of web pages offer streamed audio. Many are provided by licensed broadcast stations or broadcast-related companies, but others are the products of organizations or individuals operating "virtual stations" that are available only via the Internet. Streaming's popularity is attributable in part to the fact that streaming software is available free to consumers and it is affordable for broadcasters and others wanting to distribute audio on the Internet. Software for streaming video is also available, and many web sites feature streamed audio and/or video.

The quality of streamed audio and video depends greatly on the bandwidth of the consumer's Internet connection, the traffic load on the network at the time, and other factors. Most consumers still connect to the Internet using conventional voice-quality telephone lines and modems rated at 56.8 kilobits per second or less. With this type of connection, audio quality can be good for voice and adequate for music, but hardly excellent by modern audio standards. Video is restricted to a small part of the screen, the images lack resolution, and the motion is jerky. Real-time delivery of

EXHIBIT 3.h

TV Web Surfing

WebTV Internet terminals and other set-top products (shown here integrated with Echostar's DBS receiving system) allow the traditional TV set to provide interactivity with newer media such as the Internet. Other "information appliances" providing less challenging user interfaces may open new technologies to a wider audience.

Source: © 2000 WebTV Networks, Inc.

high-quality audio and video via the Internet awaits a larger pipeline into the home than most users now enjoy.

Other consumer products, such as WebTV, are blurring the distinction between television and the PC. Microsoft-owned WebTV uses a set-top box that lets viewers surf the Internet using a conventional TV (see Exhibit 3.h). WebTV, America Online (AOL) TV, and similar interactive products may attract "technology-challenged" people wanting Internet access without having to purchase a computer and learn its operation. Microsoft also joined forces with DirecTV in 1997 to offer web site builders the opportunity to deliver audio and video to PCs using DBS signals.

Anything advancing the convergence of broadcasting with computer technology excites technophiles and futurists, but it is unclear how successful these products will be with consumers who lack their enthusiasm. WebTV, for example, was in only about 800,000 homes in 1999 after being on the market for several years, in spite of the awesome marketing power of its parent company.

3.6   BROADCASTING: CHANGING COURSE

Development of cable and other delivery and recording options gave people their first real electronic media choices beyond traditional broadcasting; the impact of those choices soon became evident. Dramatic changes in the fortunes of the Big Three broadcast television networks provide one means of assessing such an impact. In the 1980s all three changed hands, and new networks appeared.

Traditional Network Competition

In the 1970s ABC finally achieved competitive equality with its two older rivals. It gained affiliates (and thus audience), expanded its evening news to a half-hour (years after its competitors), and converted fully into color. Late in the decade ABC even managed to lead network ratings for several seasons in succession.

NBC sought the number-one slot by hiring away ABC's program chief, Fred Silverman, as president. Silverman's ABC success did not travel with him, however, and NBC remained in third place until former independent producer Grant Tinker took over the network presidency in 1980. Within five years he had piloted NBC into the top spot, where it remained for the rest of the decade.

In the 1970s CBS focused on finding William Paley's replacement. Paley restlessly hired and fired a string of potential successors. Paley finally retired in 1983, leaving Thomas Wyman as CBS chief executive officer, only the second one in the network's 55-year history. Wyman lasted only two years.

EXHIBIT 3.i
· · · · · · · · · ·

Rupert Murdoch: Media Magnate

In 1985 Australian publisher Rupert Murdoch became a U.S. citizen almost overnight to acquire Metromedia's six independent major-market television stations (WNEW-TV, New York; KTTV-TV, Los Angeles; WFLD-TV, Chicago; WTTG-TV, Washington, DC; KNBN-TV, Dallas; and KRIV-TV, Houston) and one network affiliate (WCVB-TV, Boston) for $2 billion. He resold WCVB-TV to the Hearst Corporation for $450 million but picked up another Boston station, WFXT-TV.

Having purchased half-ownership of Twentieth Century-Fox Film Corporation in 1984 for $250 million, Murdoch acquired the remaining half in 1985 (after the Metromedia deal) for $325 million. He thus gained complete control over a company with an extensive film library (including such hits as *Cocoon* and *Aliens*) and rights to numerous television series (*L.A. Law* and *M*A*S*H*, for example). In 1985 Murdoch announced plans to form a new national television network, Fox Broadcasting Company. A core of six *owned and operated* (O&O) stations (reaching about 20 percent of all U.S. television households) served as the network.

Fox premiered in October 1986 with *The Late Show*, starring comedienne Joan Rivers. Rivers had been the primary substitute host on Johnny Carson's *Tonight Show* on NBC. The Rivers show lasted only seven months. Fox tried several other programs before giving up on the late-night time period altogether in order to concentrate on other parts of the day.

In its first season, Fox averaged between 2 percent and 6 percent of the national audience. Fox programs typically languished at the bottom of the Nielsen list, although some fared better.

The quality of the Fox affiliate line-up remained a problem. More than 120 stations carried Fox pro-

gramming, but most were UHF, some only low-power outlets, and nearly all were the weakest stations in their markets. The network lost about $80 million in its first year of operation, but Murdoch marched on.

He concentrated both on programming and on improving his affiliate lineup. In 1994 Fox acquired New World Communications, which switched affiliation of its big-city stations to Fox (one from NBC, three from ABC, and eight from CBS). Before long Fox had become a major—and profitable—force in network competition, with some shows (*Married . . . With*

FCC Studies Networks—Again
· ·

Such boardroom and ratings battles masked more important underlying changes. In Washington the FCC, from 1978 to 1980, conducted its third investigation of broadcasting networks (FCC, 1980b). The Justice Department had previously signed agreements with all three networks requir-

ing them to loosen control over independent programmers and cease serving as sales representatives for their affiliates.

The latest FCC study concluded that FCC network rules developed during the previous two decades merely restricted legitimate business interaction between networks and their affiliates without actually protecting the stations from

Children and *The Simpsons*, for example) appearing on Nielsen's list of top-rated programs. With the help of 22 O&O stations reaching almost 35 percent of the nation's homes, Fox programming by 1997 often tied, and sometimes beat, at least one of the three older networks on several nights of the week. In 1996 Murdoch launched the Fox News Channel (FNC), a cable network intended to compete with newcomer MSNBC as well as the firmly established CNN. FNC made immediate waves by hiring 23-year ABC News veteran Brit Hume, a recognized star. Murdoch also paid cable system operators large up-front fees to ensure FNC's access to a national audience in excess of 17 million homes. The same year, Murdoch kicked off FOX Sports Net, a group of regional sports networks, initially serving more than 20 million cable TV homes.

Meanwhile, Murdoch in 1988 bought Triangle Publications (publisher of *TV Guide*) for $3 billion. He controlled 60 percent of metropolitan newspaper circulation in Australia, 36 percent of national circulation in Britain; he also had part-ownership in 10 book publishers and reached more than 13 million homes in 22 European countries with Sky Channel, a satellite broadcasting service for cable TV viewers, and still more millions with his StarTV satellite service in Asia.

Early in 1997 Murdoch announced plans to combine his U.S. direct broadcast satellite venture, ASkyB, with Echostar. The plan included offering local TV station carriage in addition to the full range of cable networks and pay-per-view services, but the ASkyB-Echostar merger failed to develop.

Murdoch tried to make a deal with another DBS company, cable-controlled Primestar, but government regulators intervened. He finally sold his U.S. satellite holdings to Echostar. Almost simultaneously, Murdoch spent about $2 billion to acquire shared ownership in International Family Entertainment (IFE) and announced plans to program IFE's Family Channel with kids' shows 12 hours daily, competing directly with Nickelodeon, The Cartoon Network, and the Disney Channel.

During the late 1990s News Corporation's subsidiaries produced some of the most popular entertainment available, including network TV programs such as *The Practice*, *The X-Files*, and *Ally McBeal*. Murdoch also merged *TV Guide's* assets with a subsidiary of Tele-Communications, Inc. (now AT&T BIS), a move designed to transform the property into a print, television, and Internet company (see Exhibit 8.f, p. 247). Somewhat belatedly, Murdoch realized that other major media firms had moved ahead of News Corporation in capitalizing on the Internet, and he began taking corrective steps. He also continued expansion of News Corporation's nonmedia holdings. In 1998 he acquired the Los Angeles Dodgers and, for a record $1 billion, Great Britain's largest soccer franchise.

The late 1990s were an exciting time for Murdoch on a personal level as well. In 1999 he divorced his wife of more than 30 years and, less than a month later, married 32-year-old Wendi Deng. Although well past the usual retirement age, Murdoch says his marriage to Deng has reenergized him and if anyone wants him to vacate his post as head of News Corporation he or she will have to either push him out or carry him out.

Source: Reuters/Russell Boyce/Archive Photos.

network dominance, or improving service to consumers. The study specifically blamed the FCC's own 1952 channel-allotment decision for the continued dominance of ABC, CBS, and NBC. So few markets had more than three VHF channel allotments that any fourth network faced a hard time in obtaining a competitive line-up of affiliates.

New Player: Fox

This barrier did not deter Rupert Murdoch, an international publishing tycoon with immense resources (see Exhibit 3.i). He launched his Fox network in 1985, the first serious attempt at a fourth national television network (since 1967, when a "United Network" failed within a month).

Initial limitations on the Fox network's size actually worked to its advantage. Fox escaped the FCC's financial-interest rule that restricted ABC, CBS, and NBC in producing their own entertainment programs (see Section 6.3, p. 171). This technicality gave Murdoch freedom to use his Twentieth Century-Fox movie studio as a source of programs for the network.

And he was aggressive in developing Fox into a serious competitor. In mid-1994 Murdoch made a $500 million investment in a group of stations, gaining a number of major market affiliates, eight taken from CBS alone. Station values began to climb as broadcast networks were forced to increase payments for station time, and a virtual bidding war for affiliates developed in larger cities.

Fox also made major investments in programming, especially sports programming. Fox successfully bid away rights to NFL football from CBS in 1993. During the last week of October 1996 Fox's weekly ratings topped the other networks' for the first time in new season, prime-time, head-to-head competition, and 30-second commercials in the Fox broadcast of the 1997 Super Bowl sold for around $1.3 million each.

Networks Change Hands
• • • • • • • • • • • • • • • •

The three established networks had already undergone traumatic corporate changes. After more than three decades of stable ownership, all three changed hands within a two-year period. This abrupt break with a 30-year tradition coincided with the retirement of long-time leaders, declining ratings in the face of cable competition, weak financial performance, and concern about increasing future multichannel competition.

ABC went first. In 1985 Capital Cities Communications (Cap Cities) announced that it would acquire American Broadcasting Companies, parent of the ABC network, in a friendly deal valued at more than $3.5 billion (the 1953 merger of ABC with Paramount Theaters had a $25 million value— one indication of how network importance had grown). The combined firm had to shed its cable interests and several TV stations to meet FCC ownership rules (discussed in Section 13.6, p. 387).

Within weeks after the Cap Cities/ABC announcement, Ted Turner revealed plans for an unfriendly takeover of CBS. The network managed to thwart Turner by repurchasing nearly a fifth of its own stock from the public. The battle left CBS deeply in debt and torn by internal dissension. Severe cutbacks followed, and many CBS employees seemed relieved when Laurence Tisch (chairperson of Loews, the entertainment and investment conglomerate), purchased nearly 25 percent of CBS's stock and became chief executive. In just two years, however, Tisch closed down CBS Laboratories and sold off virtually all of the network's nonbroadcast subsidiaries, including extensive publishing interests, culminating in 1988 with the sale of Columbia Records (the world's largest record company) to Sony for $2 billion. He focused all CBS's efforts on traditional broadcasting. Despite CBS's initial ratings success in the 1990s, however, Tisch seemed to tire of the business and began considering potential buyers.

NBC's turn came in December 1985. Its parent company, RCA, weakened by years of inept management and huge losses from failed computer and videodisc ventures, welcomed a friendly buyout offer from General Electric. RCA and its NBC subsidiary sold for $6.28 billion.

And Change Again
• • • • • • • • • • • •

For several years after the takeovers, news from all three networks went from bad to worse. Profits fell, audience shares continued to slip, and thousands of employees lost their jobs as the networks slimmed down to cut costs and compete more effectively. Competition from cable and independent stations lessened the Big Three networks' once dominant role.

By the 1990s, however, stabilized audience levels, growing advertising expenditures, and the approaching demise of fin/syn rules (see Section 8.2, p. 230) had restored to television networking much of its former luster. Two new national networks, The WB and UPN, premiered in January 1995 (see Section 6.3, p. 171). And later that year two of the Big Three disclosed that they would once again change ownership.

On July 31 the Walt Disney Company surprised both the broadcast industry and the financial community by announcing that it would acquire

Capital Cities/ABC in a deal valued at $19 billion—the second-largest takeover in history. Brief and confidential negotiations had produced the most powerful entertainment empire in the world, combining the most profitable TV network and its ESPN cable service with Disney's Hollywood film and television studios, its theme parks, and its Disney Channel.

The next day, after somewhat longer and certainly more public bargaining, CBS said it would abandon its nearly 70-year legacy of corporate independence by accepting a $5.4 billion buyout offer from the broadcasting and industrial conglomerate Westinghouse Electric Corporation. Together they would own 15 television stations and 39 radio stations, as well as program production and distribution facilities.

The original Big Three networks continued to lose prime-time audience share to cable, DBS, the newer broadcast networks, and the Internet during the second half of the 1990s. Questions arose about the continued viability of the newer broadcast networks, especially UPN. But these concerns did not prevent a seventh broadcast/cable network, PAX TV, from arriving in 1998 to further fragment the viewing audience and compete for network broadcast advertising dollars.

Rumors of sales or mergers continued, especially regarding General Electric's NBC (which purchased a third of PAX TV in 1999). But it was CBS that, in 1999, announced it would merge with Viacom in a deal estimated to be worth about $36 billion. The merger placed another major TV network and a Hollywood production studio under combined corporate control. It also dramatically enhanced CBS's involvement in cable through Viacom's MTV network (see Exhibit 6.n, p. 202). Questions remained regarding the status of Viacom-Paramount's interest in UPN and the FCC's rule preventing any company from owning TV stations reaching more than 35 percent of the national audience.

Radio
••••

Despite intense competition, the number of radio stations continued to climb after 1970, dividing listeners into ever-smaller segments. FM station numbers rose to match AM by the 1990's as FM's audience appeal overcame AM's former attraction.

FM's popularity with audiences grew, and advertisers persuaded the FCC to squeeze in more than 700 channel allotments to allow still more FM outlets.

Meanwhile, looser FCC technical oversight, combined with a growing number of stations on the air, had weakened AM's already-inferior signal quality. Under pressure from AM interests, the FCC adopted several technical and ownership rule-changes in an attempt to halt AM's decline. But when the FCC approved AM stereo in 1982, it declined to specify a technical standard, arguing that "the marketplace" could resolve that issue. Although five competing systems soon dwindled to two, the resulting confusion among manufacturers and consumers led to few stations adopting stereo by the early 1990s. Congress finally required the FCC to pick a standard, but this seemed a decade too late to help AM reverse its downward spiral.

The FCC also added 10 new channels to the top of the AM radio band. This *expanded AM band*, located between 1605 and 1705 kHz, should reduce AM interference by moving selected AM stations to the new frequencies. But the FCC had to revise its assignment plan several times, delaying implementation.

More important than technical improvements to radio in the 1990s were changes in the rules governing ownership. In the early 1990s many owners operating only one station—especially an AM station—were destined for bankruptcy. In late 1992 the FCC loosened its rules to allow a single owner to operate two AM and two FM stations in markets with at least 15 stations. The Telecommunications Act of 1996 further lifted radio's fortunes by eliminating the national cap on radio station ownership, although restrictions on the number owned in a single market remain. Interest in radio station ownership increased dramatically as a result. By 1997 several groups owned more than 100 stations. In 1999, Clear Channel Communications, Inc., announced plans to merge with AMFM, Inc., giving it control of more than 825 stations, a number unimaginable only a few years before.

Although much stronger and more profitable than in the early 1990s, the radio industry still faces challenges. There will be increased competition from audio services delivered by cable, DBS, the Internet, and new satellite digital audio radio services (see Section 5.8, p. 153) available to listeners nationwide at home and on the road. Radio also must eventually convert to digital transmission if it is to match the sound quality of its competitors and enhance revenues by distributing digital data. But economic efficiencies resulting from greater concentration of station ownership improve the chances of radio meeting these challenges successfully.

Television in Transition
● ● ● ● ● ● ● ● ● ● ● ● ● ● ● ●

TV broadcasters also face challenges, the most immediate being the shift from analog to digital signal transmission. In 1996 the FCC adopted technical standards for *digital television* (DTV) after eight years of discussion. The following year Congress passed legislation instructing the FCC to allow a station to retain its analog channel until 85 percent of the station's viewers can receive its digital signal over the air, by satellite, or via cable. The legislation otherwise requires the return of analog channels to the FCC by December 31, 2006. Rather than establish a single image display standard as it did for analog TV in the 1940s, the FCC allows broadcasters to use any DTV picture format. Most broadcasters applaud flexibility, but lack of a single government-mandated format also creates uncertainty.

The organization that developed the DTV standards approved by the FCC also proposed 18 image-scanning formats. So far the industry has adopted them as de facto standards, but broadcasters are still left to select which they will use. In 1998 the four largest broadcast networks announced their format choices for high-definition DTV transmission, but there was no consensus among the networks to provide a guide for the industry as a whole, and the networks promised to change their initial selections as marketplace demands dictate.

The networks' flexibility reflects their uncertainty about the future. At its best, DTV can provide wide-screen images with filmlike resolution. But the highest-resolution images require the use of the format demanding the maximum channel capacity.

This format provides truly beautiful pictures and superior sound, but there is little room left in the channel for much else. Lower-resolution formats

that qualify as "high-definition television" still leave room in the channel for simultaneous transmission of standard-definition programs. Many broadcasters believe they will reap larger profits by using lower-resolution formats and devoting the remaining channel capacity to broadcasting multiple programs (or digital services) simultaneously. The broadcasters also claim that even standard-definition digital television will provide much higher quality images than traditional analog methods. But industry critics say the government did not give broadcasters a new piece of the spectrum so they could fill it with multiple program streams that are not significantly better than those previously available. Ultimately, the marketplace will provide answers, but until then broadcasters must make educated guesses about audience demands and expectations while watching their competitors and their own bottom lines.

And the bottom line is a big concern. The cost of converting to digital is especially significant to those who own only one or a few stations or who own stations in smaller markets. It is likely that the cost of conversion contributed to the consolidation of TV station ownership during the late 1990s. Some owners found it advantageous to sell out, often at a nice profit, to a larger organization with the resources to absorb high expenses rather than assume the risks of conversion. Simply passing through high-definition network programming typically costs several million dollars per station, regardless of market size. Equipping a local station to originate wide-screen high-definition programming dramatically increases the costs. Many broadcasters doubt that local advertisers will pay more for ads presented in high-definition format, yet sponsors will probably complain if their messages look inferior when compared to the network programming in which the ads appear.

Broadcasters are also concerned about reception of their signals during and after the transition to digital. Pioneering DTV stations in large markets such as Chicago experienced problems getting permission to erect new DTV antennas. Environmentalists and people living near the proposed DTV transmitter sites raised concerns about safety. Broadcasters worry that interference levels in congested areas may effectively block DTV reception. Broadcasters are especially concerned about reception during the period when a great number of both analog and digital broadcast stations will be operating at the same time, before analog transmission is phased out. Carriage by cable, and possibly DBS, should ensure high-quality signals, but agreement over cable carriage of broadcast DTV remains a subject for negotiation.

Yet another concern is the dramatic growth of two-way communication services offered by competitors. Traditional one-way delivery of programming, the fundamental business of broadcasters, is a mature market, while two-way interactivity is viewed as a major growth area in the new century. Broadcasters control a wide lane on the information superhighway capable of handling heavy traffic, but it is one-way. To benefit from the growth in two-way interactive services, broadcasters must develop new business models that may include strategic partnerships with former competitors. See Section 5.7 (p. 146) for further discussion of DTV.

3.7 SORTING IT OUT

The development of new delivery systems and dramatic changes in older services has placed enormous pressures on manufacturers, broadcasters, competing systems, and policymakers. In the 1990s even sympathetic trade publications used terms such as "merger mania" and "feeding frenzy" to describe the wild scramble among media industries trying to position themselves in an unpredictable environment in which plans made one month often seemed foolish the next.

These struggles have produced some interesting contrasts. Today consumers have more media channels available than ever before. In the United States there are more than 12,000 radio stations, more than 3,500 TV stations, almost 11,000 cable systems, and satellite-delivered programming. There are more cable and broadcast networks. There are also wireless cable outlets in some communities. Although not comparable in quality to the media listed above, the Internet provides streamed audio approximating AM radio quality and video that is at least suggestive of TV programming. Americans have so many sources of

electronically delivered entertainment and information that they suffer from sensory overload. A burgeoning market exists in navigation devices that viewers use to help them keep the barrage at bay.

Looking a little deeper, however, reveals that many of these program sources are actually owned or controlled by a relatively few companies. Although there are more radio stations now than in 1990, there are fewer radio station owners. In 1999 the largest 25 cable system owners provided service to almost 61 million U.S. households, about 90 percent of all cable subscribers. The consolidation of media ownership, even as the diversity of media choices appears to increase, is an important trend to consider when attempting to understand what is happening.

Another factor to consider is the increasing involvement of the same owner in previously separate media. There has been cross-ownership of media since the beginning, but in the 1990s the barriers were lowered dramatically. Today the largest media firms are involved in practically every type of ownership arrangement allowed by law. Broadcasting and cable compete for the same audiences. Disney, Time Warner, and News Corporation, to name a few, are involved heavily in both broadcasting and cable. These companies delegate responsibility for the operation of different media to their subsidiaries, which may appear to compete as they maneuver for supremacy in the marketplace, but the type and amount of competition is decided at corporate headquarters.

Return of an Old Player

Yet another factor to watch in the new century is the evolving role of the telephone industry. Throughout the 1990s broadcast and cable leaders alternated between fear and relief as the telephone industry sent conflicting signals about its interest in becoming a competitor in the distribution of programming. AT&T bowed out of direct competition with broadcasters in 1926, although it retained its lucrative business networking stations. By the 1970s changing technology and competitive forces renewed telephone interest in entertainment/information media.

New telephone competitors, such as MCI and Sprint, charged that AT&T used its great size to hamper their efforts and renewed Justice Department concern over the Bell System monopoly. In 1982, after an eight-year court battle, AT&T ended the resulting antitrust suit by signing a consent decree. AT&T agreed to split off its 22 local operating companies, restructuring them into seven *Regional Bell Operating Companies* (RBOCs), or "Baby Bells." AT&T withdrew from local telephone service, focusing on increasingly competitive long-distance, manufacturing, and information services.

Meanwhile technology opened exciting new vistas for *telco* (telephone company) operations. Fiberoptic cable, for example, offers dramatic advantages over both copper cable and radio. As shown in Exhibit 5.b (p. 127), it is small and easily installed, yet it has enormous broadband channel capacity and freedom from environmental interference. Fiberoptic cable is a natural for heavy-traffic telephone routes and for main distribution trunks of cable systems. This new technology created new opportunities for telco operations in video delivery, but rules established when the Baby Bells split off prevented them from providing their own "information services." By the mid-1990s, court decisions had removed most of these restrictions, and the Telecommunications Act of 1996 (discussed in Chapter 12) opened the field further.

Both broadcasting and cable saw expansion of *plain old telephone service* (POTS) into information and entertainment as a threat, although cable foresaw how its own broadband capacity could carry new services in competition with telcos. In the mid-1990s it appeared the telcos would move aggressively into direct competition with cable by providing interactive television. The telcos found, however, that interactive television technology was extremely expensive, and, in spite of their enormous technical resources, they had problems getting it to work. They shifted away from true interactive programming and tried digital wireless cable as a way to get at least multiple video channels into the home. But they found digital signals on wireless cable frequencies did not penetrate obstacles such as trees as well as expected. Meanwhile the Internet shifted interactivity from the television to the computer, and competition for local and long-distance phone service intensified. The telcos scaled back their earlier plans significantly but continued to expand and improve their fiber-optic networks.

While the local telcos enjoyed continued growth, AT&T saw its long-distance business eroded by competitors such as Sprint and MCI Worldcom. Frustrated with its attempts to reenter the local telephone services market through alliances with local telcos, the company announced plans in 1998 to acquire Tele-Communications, Inc. (TCI), at the time the nation's second-largest cable operator. Then in 1999 AT&T announced that it would acquire the third-largest cable operator, MediaOne. This acquisition, if approved by government regulators, would make AT&T the largest cable system operator, with more than 16 million subscribers. AT&T's primary interest in TCI and MediaOne, however, was not to get into TV program distribution but to offer local phone and data services using cable's huge pipeline into the home. These announcements by AT&T underscore a trend for telcos to develop cooperative alliances and mergers with cable rather than to compete with it directly as a separate service in delivering TV programming to the home.

For Better or Worse?
• • • • • • • • • • • • • •

Some observers argue that consumers win when there are more choices. It is hard to argue that consumers have not enjoyed an ever-growing menu of channels during the 1990s or that it is not to consumers' advantage that they can now get news, sports, and weather 24 hours a day rather than having to wait until the local broadcast station gets around to it. It is much easier for individuals today to find programming that suits their particular interests than at any time in history. But critics say this bonanza of choices is not without its flaws.

In the scramble to attract viewers in a crowded marketplace, critics say programming standards have suffered. Critics acknowledge that competition, even controlled competition among divisions of the same corporation, is desirable if it improves the diversity of media content and its quality. But they suggest the opposite has occurred. They say that although there are exceptions such as The Weather Channel, MTV, and even ESPN, much of what the new channels offer is lacking in innovation. If technology and competition only produce 100 new ways to view the same episode of *Gilligan's Island*, what is accomplished? If they make possible a 24-hour Jerry Springer channel, is this progress? What about repeating the same lurid descriptions of President Bill Clinton's extramarital sexual behavior hour after hour simply because cable networks have plenty of time to fill and catering to prurient interests will attract a certain audience? The critics charge that competition for ratings in a fragmented media environment too often encourages producers to use violence, sex, and foul language to make their programming stand out in a crowded field. They point to the TV Parental Guidelines (see Exhibit 9.d, p. 267) and the government-mandated V-chip as evidence that many viewers find much of television content repulsive. Few Americans want to turn back the clock to the time before cable and the newer technologies arrived, but the changes have brought with them costs as well as benefits.

CHAPTER **4**

How Electronic Media Work

Technology largely determines both the potential and the limitations of electronic media. Where and how far signals will travel, how much information they can carry, their susceptibility to interference, and the need for technical regulation—all of these factors depend primarily on the signal's physical nature.

In spite of the digital revolution that promises to transform electronic communication, Americans still get most of their programming through analog media. This chapter focuses on how conventional (analog) broadcasting and cable operate and how digital television is similar in some ways but different in others. Digital television is also discussed in Chapter 5 along with other digital technologies. Some basic concepts are common to the way media work whether they use analog or digital technologies, and this chapter begins by exploring those concepts.

4.1   ELECTROMAGNETISM

A basic natural force, electromagnetism, makes possible a host of communication services. All forms of electromagnetic energy share three fundamental characteristics:

- They radiate outward from a source without benefit of any discernible physical vehicle.
- They travel at the same high velocity.
- They have the properties of waves.

Radio Waves

Visible waves that ruffle the water when a stone is dropped into a pond radiate outward from the point of disturbance caused by the stone. Unlike water waves, radio waves have the ability to travel through empty space, going in all directions without benefit of any conductor such as wire. This "wirelessness" gives broadcasting its most significant advantage over other ways of communicating. Radio waves can leap over oceans, span continents, penetrate buildings, pass through people, go to the moon and back.

Students learn in elementary school that the "speed of light" is 186,000 miles per second. Radio shares this characteristic with light energy, as do all forms of electromagnetic radiation. This means a radio signal can circle the Earth more than seven times in one second. Radio waves also share two other characteristics with other forms of electromagnetic energy, *frequency* and *wavelength*.

Wave frequency refers to the fact that all electromagnetic energy comes from an oscillating (vibrating or alternating) source. The number of separate waves produced each second determines a particular wave's frequency. Differences in frequency determine the varied forms that electromagnetic energy assumes.

Wavelength is measured as the distance from the origin of one wave to the origin of the next. Waves that radiate out from the point of disturbance—the stone in the water or the radio transmitter—travel at a measurable velocity and have measurable frequencies and lengths.

Frequency Spectrum

A large number of frequencies visualized in numerical order constitutes a spectrum. The keyboard of a piano represents a sound spectrum. Keys at the left produce low-pitch sounds (low frequencies). Frequency (heard as pitch) rises as the keys progress to the right end of the keyboard. Exhibit 4.a diagrams the electromagnetic spectrum and some of its uses.

As frequency goes up, the practical difficulty of using electromagnetic waves for communication also increases. Frequencies suitable for radio communication occur near the low end of the spectrum, the part with the lower frequencies and longer wavelengths.

Communication satellites employ the highest frequencies currently useful for broadcasting, mostly in the range of 3 gigahertz (3 *billion* oscillations per second) to 15 gigahertz. Some terrestrial microwave services used to distribute data and video, such as local multipoint distribution service (LMDS), use frequencies in the 28-gigahertz range. Even these frequencies are nowhere near as high as those of light.

As Exhibit 4.a illustrates, only a relatively small part of the electromagnetic spectrum can be used for radio communication with the current technology. This makes the threat of interference among stations a serious problem, especially in the crowded AM and FM radio broadcast bands. The

EXHIBIT 4.a

• • • • • • • • • •

Electromagnetic Spectrum Uses

This diagram is not drawn to a uniform scale—as the frequencies rise, the scale gets smaller (otherwise it would be impossible to show this much spectrum in a single diagram). The usable radio spectrum with which we are concerned (B) is but a small part of the larger electromagnetic spectrum (A). The higher the frequencies, above the usable radio spectrum, the more potentially dangerous the output of electromagnetic radiation can become (infrared and ultraviolet rays), though even these can be useful (X-rays).

The radio spectrum is divided into bands by international agreement. Low-frequency radio, a long-distance form of AM broadcasting, is used in Europe but not the United States. Though not shown here, virtually all usable spectrum space is assigned to some use (aviation, land mobile, ship-to-shore, CB radio, cellular telephone, satellite uplinks and downlinks, and military applications). There is no unallocated spectrum, meaning that new services have to move older services aside (and usually upward) in spectrum. See Exhibit 4.k later in the chapter for more detail on broadcast service allocations.

Changes in frequency nomenclature avoid awkwardly long numbers. Thus one kilohertz (kHz) = 1,000 Hz (hertz, or cycles per second); a megahertz (MHz) = 1,000 kHz; a gigahertz (GHz) = 1,000 MHz.

demand for more space for new services makes efficient spectrum management critical. New digital and compression technologies are, however, making possible more efficient use of the spectrum. The government also is turning over more of the usable spectrum to the private sector for commercial applications.

In 1993 Congress decided to reallocate for civilian use some 200 MHz of spectrum controlled by the military and other agencies. While allowing for a long period of transition (up to a decade or more), the law required that the government *auction* some of the reallocated spectrum, earning money for the U.S. Treasury. In 1994 the FCC initiated its first auctions—for personal communication and interactive video data services. Congressional pressure for more spectrum auctions remained a concern for broadcasters during the second half of the 1990s, as discussed in Chapter 12.

4.2   RADIO AND AUDIO WAVES

A burst of radio energy consists of a series of waves radiating from an antenna, much like waves radiate outward on the surface of a pond when a rock falls into the water. Exhibit 4.b depicts basic wave motion and wave characteristics. From a smooth surface (zero in illustration (A) in Exhibit 4.b), the wave of energy goes to a maximum positive peak, back down through the zero point, to a maximum negative peak, then back to zero. This complete process is a *cycle*. If it takes a second to complete the cycle, the wave has a *frequency* of one cycle per second, or one *hertz* (Hz), named in honor of Heinrich Hertz, who proved the existence of radio waves. If an electrical current in a wire goes through these cycles increasingly rapidly, the electrical energy eventually begins to leave the wire as electromagnetic radiation. If frequency increases enough, almost all of the electrical energy is radiated and the wire becomes an antenna.

Notice that the distance the wave travels, its *wavelength*, depends on how fast the wave moves and the number of new waves created each second. Because the speed of radio energy is constant (186,000 miles a second), *frequency* alone determines wavelength. The more waves created each second, the shorter the wavelength. Either wavelength or frequency can be used to distinguish one radio wave from another. Initially the common practice was to describe a radio wave in terms of its length, and many international broadcasting stations still use wavelength to describe their signals. Modern practice in the United States is to use frequency. Bands of frequencies have names such as MF (medium frequency), HF (high frequency), and VHF (very high frequency). Exhibit 4.a shows the various frequency bands and their associated designations.

Because radio frequencies involve very large numbers of hertz it is convenient to use abbreviations: (k)ilo instead of 1,000; (M/m)ega instead of 1,000,000; and (G)iga instead of 1,000,000,000. Thus a radio wave with a frequency of 650,000 Hz becomes 650 kHz (650 × 1,000).

Radio waves lose their energy as they radiate outward, a phenomenon known as *attenuation*.

How quickly a radio wave loses its energy depends on its frequency and the physical characteristics of obstacles encountered along the way. As a rule, high-frequency waves are more easily absorbed and reflected by objects in their path. This was one reason Marconi tried to use the lowest possible frequencies during the infancy of wireless communication. It also explains why VHF TV channels (2–13) are more desirable than higher-frequency UHF channels.

Scientists have known for years that X-rays and other electromagnetic waves with much higher frequencies than those used for radio communication can have hazardous effects on human tissue. Only recently has the FCC recognized the possible health effects of non-ionizing radiation, such as that produced by broadcast stations. The threat to public health created by broadcast stations is considered minimal, but stations must take precautions to protect employees whose duties require that they work near the transmitter and antenna system.

Radio Frequencies and Channels

Broadcasters know most people are not interested in terms like *kilohertz* and *megahertz* when they want to hear their favorite song or watch *Monday Night Football*. Still, people must know where to tune their receivers, and usually that means remembering a number. Most stations simplify the number as much as possible and associate it with a promotional phrase. An AM radio station featuring nostalgic songs of the 1940s may call itself "Mellow 66," for example. The number 66 really means 660 kHz, the station's frequency. The same is true for FM radio stations. Sometimes a radio station may call itself "Channel 82" (820 kHz) or something similar. This is correct in one sense because the government allots each AM and FM radio station a range of frequencies, or *channel*, in which to operate. The nominal frequency, such as 820 kHz, is the center of the frequency range in which the station can broadcast. Designation by channel rather than frequency is typical of TV stations and accurate because, for example, a station assigned to "Channel 6" operates within a band (a channel) of frequencies ranging from 82 to 88 MHz.

EXHIBIT 4.b
. . . .
Wave-Motion Concepts

(A) Starting at zero, energy moves up to a positive peak. This represents one-quarter of a full cycle and one-quarter of the final wavelength.

(B) Passing the peak, energy moves down to zero, the point at which the wave began. This represents one-half of a full cycle and one-half of the final wavelength.

(C) Having gone to a positive peak, energy falls to a maximum negative value. This represents three-quarters of a full cycle and three-quarters of the final wavelength.

(D) Starting at the negative peak, energy moves up to the zero point, the point of origin. This completes a full cycle, and the distance the wave travels is one full wavelength.

These illustrations show a wave completing a full cycle in one second, a frequency of one cycle (1 Hz). If the energy were traveling at the speed of light, the wavelength would be 186,000 miles.

(E) The effect of doubling the frequency to 2 Hz is illustrated. Note that the wave now completes two cycles (2 Hz) in a period of one second. Doubling the frequency has caused the wavelength to be half as long, "only" 93,000 miles.

(F) As frequency is doubled again (4 Hz), the wavelength is again halved, making it a quarter as long as when the frequency was 1 Hz.

Because electromagnetic energy travels at a constant speed, increasing frequency reduces wavelength proportionally. Wavelength, and therefore frequency, is critical to the way a radio wave behaves once it leaves an antenna.

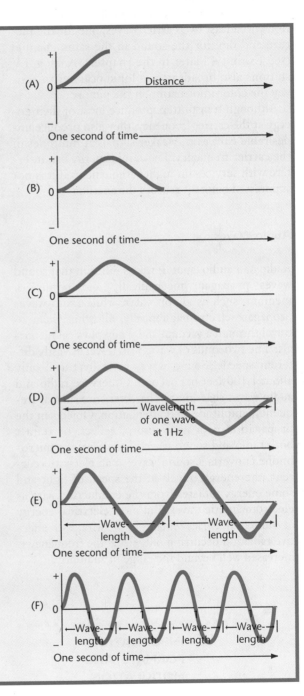

The center frequency of a radio station's channel is the *carrier wave*, so named because it conveys information (sound) superimposed on it, much as a road carries traffic. Conventional analog TV stations broadcast two carrier waves, one for the picture and one for the sound in the same channel (see Exhibit 4.j later in the chapter). Analog TV stations also broadcast a color subcarrier to convey the color information in the picture.

Although transmitters produce most of their energy at the carrier frequency, they also produce undesirable *harmonics*, weaker signals at multiples of the carrier frequency. Harmonic energy can interfere with services on higher frequencies if it is not suppressed before it reaches the antenna.

Audio Waves

Radio and audio (sound) waves differ in that sound waves propagate mechanically and require a medium, such as air or water, while radio waves propagate electromagnetically, allowing them to travel through a vacuum and many physical materials. The two kinds of waves also travel at vastly different speeds. Sound waves typically travel only about 1,100 feet per second. Otherwise radio and audio waves share many characteristics such as frequency, amplitude, and attenuation. Objects in the propagation path may absorb, reflect, or refract both radio and sound waves. However, if a microphone converts a sound wave to an electrical current, the energy travels at the speed of light and some energy radiates from the conducting wire as electromagnetic waves. But most electrical energy at audio frequencies remains in the wire, with effective radiation occurring only when the frequency is increased well beyond the range of audibility.

4.3 INFORMATION, FORMATS, AND MODULATION

The information transmitted by a station can be in either analog or digital format. Analog techniques have long been the standard in broadcasting, but American television is shifting to digital in the new century.

Analog and Digital

Analog techniques impose a pattern on the carrier wave that resembles (is *analogous* to) the wave pattern of the original information (sound or picture). When, for example, radio waves carry words in analog form, they consist of *amplitude patterns* (loudness) and *frequency patterns* (tones or pitches). When a person speaks, air molecules form pressure waves in response to the amplitude/frequency patterns of the words. A microphone responds to air-pressure patterns, transducing (converting) them into corresponding electrical patterns. The microphone outputs a sequence of waves with amplitude and frequency variations that approximate those of the sound-in-air patterns.

Ultimately, those *audio frequency* (AF) electrical variations are superimposed on a transmitter's *radio frequency* (RF) carrier, causing its oscillations to mimic the AF patterns. The same kinds of variations in color and brightness in an image can be superimposed on the video carrier wave produced by a TV transmitter. The result is a radio wave containing patterned variations analogous to the original sound or picture. To recover the information, it is necessary to detect the picture or sound by "stripping away" the RF carrier, leaving the wave pattern of the original picture or sound.

Digital techniques, in contrast, first encode the wave patterns of the original image and sound into a series of numerical symbols that represent the original wave patterns but are not analogous to them (see Exhibit 5.k, p. 141). Digital encoding reduces all forms of information to numbers for transmission; everything takes on the same form. As long as the transmission system can accommodate the large stream of numbers pushed through it, the original form of the information makes no difference. To recover the information, it is necessary to decode the numerical information conveyed, converting it back to what it originally represented, whether it be video, audio, or data.

It is important to remember that pictures and sound always begin as analog waves and must be converted back to analog form for humans to perceive the information. There must be an analog-to-digital (A/D) conversion at the source and a digital-to-analog (D/A) conversion at the receiver. Converting information to digital form prior to modulation is beneficial because of the advantages digital provides during storage and transmission.

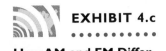

EXHIBIT 4.c
. . . .
How AM and FM Differ

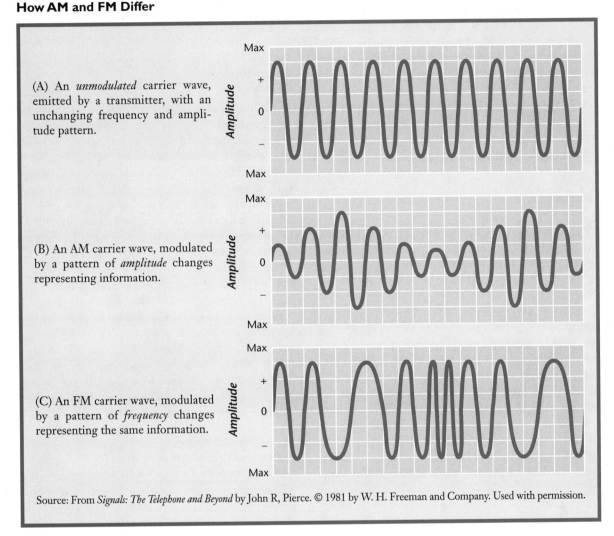

(A) An *unmodulated* carrier wave, emitted by a transmitter, with an unchanging frequency and amplitude pattern.

(B) An AM carrier wave, modulated by a pattern of *amplitude* changes representing information.

(C) An FM carrier wave, modulated by a pattern of *frequency* changes representing the same information.

Source: From *Signals: The Telephone and Beyond* by John R, Pierce. © 1981 by W. H. Freeman and Company. Used with permission.

AM vs. FM
. . . .

Modulation refers to ways of imposing meaningful variations on a transmitter's carrier wave to enable it to carry information.

Broadcasting transmitters use either amplitude or frequency modulation. Exhibit 4.c shows how they work. AM (amplitude-modulated) radio stations are called *standard* stations because the technology of amplitude modulation was developed before that of FM (frequency modulation). Conventional analog television uses both types, AM for picture signals and FM for sound. The digital television system adopted

by the FCC uses a complex form of AM called *8-VSB* (vestigial sideband modulation with eight discrete amplitude levels). See Section 5.7 (p. 146) for additional information about 8-VSB modulation.

Interference caused by lighting or electrical machinery has more effect on the amplitude of a radio wave than on its frequency. This is why such interference causes listeners to hear *static* on AM stations but not on FM stations. Overlapping signals from more distant stations on the same channel also easily distort AM radio signals. The circuits required for picking up FM signals cause receivers to put only the strongest signal being received rather than all signals coming in on the same channel. This characteristic of FM reception is called the *capture effect*. The picture of conventional TV, being AM, is much more susceptible to interference than the FM sound portion of the signal. Although DTV (digital television) uses AM, the information is in digital rather than analog format and contains error-correcting coding to reduce the effects of interference. A portion of the data stream used to broadcast DTV consists not of picture and sound information, but of special digital codes that help the receiving set minimize the effects of transmission errors and interference.

Sidebands

A specific radio frequency, not a channel or band of frequencies, identifies a station's carrier wave. For example, an AM station at 600 on the dial has a carrier wave of 600 kHz (600,000 cycles per second).

If the station transmits no audio, the only energy radiated is the 600 kHz carrier. When the station begins transmitting sound, energy radiates not only on the carrier frequency but also on *sideband* frequencies both above and below the carrier. For AM stations, the audio frequencies being broadcast determine the width of the sidebands. Here is where AM radio has a problem. Popular music contains audio frequencies well above 5,000 Hz, the limit at which the sidebands start spilling over into other AM stations' channels. AM stations have the unpleasant option of either broadcasting the music's full audio range and risking interference or filtering out the music's higher frequencies, making it sound muffled.

The sidebands above and below the carrier frequency are identical, and, technically, no "information" is lost if a station transmits only one sideband. However, this is not always an attractive solution because a *single-sideband* (SSB) signal requires a special, more expensive receiver.

But some services do economize on spectrum usage by suppressing one sideband. Conventional television needs such wide channels that transmitting both sidebands would use up too much spectrum space and reduce the number of usable channels. Therefore one television signal sideband is suppressed, leaving only a vestigial lower

sideband, as shown later in Exhibit 4.j, p. 108. DTV uses a similar technique plus compression technologies (see Section 5.5, p. 138) that make it much more spectrum-efficient than conventional analog TV.

Bandwidth

Just as the quantity of water a pipe delivers depends on the pipe's diameter, the channel's bandwidth limits the capacity of a communication system. The bandwidth needed depends on the application. A taxi dispatch service needs little bandwidth because simple intelligibility, not sound quality, is necessary. Video is an infamous "spectrum hog" requiring much more bandwidth than audio. Bandwidth becomes critical when it is necessary to transmit lots of information "right now," in real time.

Government policy usually dictates providing the minimum bandwidth required to achieve a particular goal at the time of allocation. Allocating AM stations 10 kHz channels made sense at the birth of broadcasting because the goal was to license the maximum number of stations, not broadcast high-fidelity audio. But today broadcasters face competing program distributors with much larger "pipes." Finding ways to push more product down the same channel is a critical concern.

Multiplexing and Compression

Multiplexing is a process that conserves resources by combining related functions into a single entity. Multiplex theaters, for example, conserve money by offering several movies simultaneously within the same building. Electronic media use several kinds of multiplexing.

One form of multiplexing involves modulating two or more separate carriers in the same channel. Broadcasters turned to multiplexing when it became necessary to add color information to the signal without exceeding the existing TV channel's bandwidth. The solution was to add a color subcarrier to the signal. Exhibit 4.j (p. 108) shows TV's analog color subcarrier between the video and audio carriers. FM radio stations also use multiplexing to broadcast in stereo and provide nonbroadcast auxiliary services.

In the 1990s *compression*, a technique popular in the computer industry, became extremely important to all communication industries struggling with bandwidth limitations. Rather than adding new information to the channel, compression seeks to eliminate all information not essential to the communication task being undertaken. Section 5.5 (p. 138) explains compression in greater detail.

4.4 WAVE PROPAGATION

Modulated signals are piped from a transmitter to an antenna, the physical structure from which signals radiate into surrounding space. The traveling of signals outward from the antenna is called *propagation*.

Coverage Contours

In theory, an omnidirectional (all-directional) antenna would propagate signals over a circular coverage area. In practice, coverage patterns often assume uneven shapes. Physical objects in the transmission path, interference from electrical machinery and other stations, and even the time of day can affect propagation distances and patterns. The sun also influences propagation. *Sunspots*, blotches on the sun's surface that increase and decrease over an 11-year cycle, dramatically influence some radio frequencies.

The higher the frequency of waves, the more the atmosphere absorbs their energy and therefore the shorter the distance they can travel. Objects wider than a wave's length tend to block propagation, causing "shadows" in coverage areas. Later in this chapter, Exhibit 4.m, p. 113, shows how large buildings cause "shadows" in TV-station coverage. VHF, UHF, and still higher frequency waves (used for LMDS, microwave relays, and satellites) have such short lengths that relatively small objects can interfere with their propagation. Even raindrops can block the shortest waves.

Because many variables affect wave propagation, broadcasters define their coverage areas in terms of reception probability. Ideally, engineers measure the signal strength at various distances from the transmitting antenna. They then plot these measurements on a map and draw lines connecting those having a minimal strength level, as defined by the FCC. The result is a map showing the station's *coverage contour.* More often engineers use FCC-approved formulas for calculating the coverage contour rather than determining it by taking measurements.

Two coverage contours usually are drawn on the same map, as illustrated in Exhibit 4.d. The flat Iowa terrain accounts for the almost perfect circularity of WHO-TV's coverage contours. The growth of cable has made coverage contour maps less important for conventional TV, although stations continue to include them in their sales and promotional literature.

Questions about coverage contours have become more common with the arrival of DTV. DTV brings with it a change in the way information is formatted and compressed for transmission and, for some broadcasters, a move from a VHF to a UHF channel. Based on field-testing and theory, the FCC established channel-allotment tables, transmitter power levels, and antenna configurations to ensure that most broadcasters achieve essentially the same coverage with DTV that they had with analog transmission.

Although the type of modulation can affect reception, as noted in the previous discussion of AM and FM interference, it typically has much less effect on propagation than does the frequency of the carrier wave when conventional analog techniques are employed. Conventional VHF and UHF TV stations use the same types of modulation, for example, and yet the lower-frequency VHF channels propagate much better.

Frequency-related propagation differences divide waves into three types: direct, ground, and sky waves. Each type has advantages and disadvantages that must be considered in matching frequency bands with service needs.

Direct Waves

At FM radio and television frequencies, in the VHF and UHF bands, waves follow a line-of-sight path. They are called *direct waves* because they travel directly from transmitter antenna to receiver antenna, reaching only about as far as the horizon, as shown

EXHIBIT 4.d

Predicting Station Coverage

SCALE OF MILES

0   50   100

This contour map of WHO-TV, channel 13, Des Moines, Iowa, shows Grade A and B contours. The Grade A contour includes the area within which satisfactory service is projected for 70 percent of the receiving locations 90 percent of the time; within the Grade B contour such service is projected at 50 percent of the receiving locations.

Source: From *Television & Cable Factbook*, 1996 edition, p. A-441. Used by permission of American Map Corporation and Television & Cable Factbook.

in Exhibit 4.e. Line-of-sight distance to the horizon from a 1,000-foot-high transmitting antenna propagating over a flat surface reaches about 32 miles. The signal does not cease abruptly, but after reaching the horizon fades rapidly.

Engineers locate direct-wave antennas as high as possible to extend the apparent horizon. Raising the height of the receiving antenna also extends the horizon limit.

Ground Waves

AM radio broadcast stations in the United States use medium-frequency (MF) waves. Waves in the MF band travel as *ground waves*, propagated through and along the surface of the Earth. They can follow the Earth's curvature beyond the horizon, as shown in Exhibit 4.e.

Ground waves can cover wider areas than can direct waves. In practice, however, a ground wave's useful coverage area depends on several variables, notably *soil conductivity*—the degree to which soil resists passage of radio waves—which varies according to dampness and soil composition. The area in which the ground wave provides a signal free of objectionable interference and fading defines an AM station's coverage contour.

Sky Waves

Most radio waves that radiate upward dissipate their energy in space. However, waves in the medium-frequency band (AM radio) and the high-frequency band (shortwave radio) when radiated upward tend to bend back at an angle toward the Earth when they encounter the ionosphere.

EXHIBIT 4.e

Ground-, Sky-, and Direct-Wave Propagation

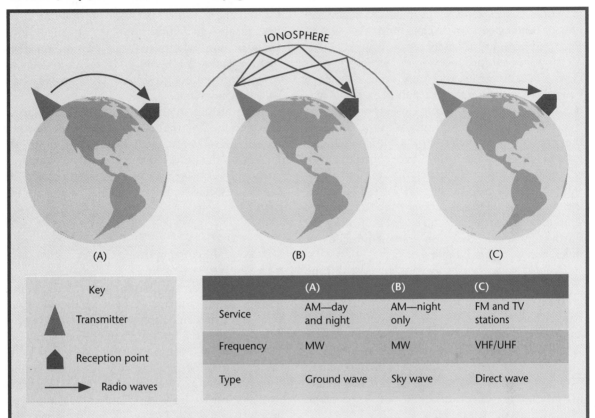

	(A)	(B)	(C)
Service	AM—day and night	AM—night only	FM and TV stations
Frequency	MW	MW	VHF/UHF
Type	Ground wave	Sky wave	Direct wave

Key

Transmitter

Reception point

Radio waves

Because of their location in the spectrum, radio and television services have different signal propagation paths.

AM radio is most complicated because its *medium-frequency* spectrum location gives it two different means of wave propagation: ground waves in the daytime (A), and both ground and sky waves at night. Ground waves travel along the surface of the Earth. Given good soil conductivity and sufficient power, ground waves outdistance direct waves, reaching well beyond the horizon. As discussed in the text, sky waves (B) bend back toward the Earth when they encounter the ionospheric layer of the atmosphere at night. Sky waves may bounce off the Earth and back to the ionosphere several times, and different frequencies are affected at different times of the day and night. In short, they are unpredictable and thus cause more interference than they provide useful service. Because of potential sky-wave interference, some AM stations leave the air at night. *Shortwave* international services make use of sky waves, changing frequencies to best match predicted ionospheric effects on sky waves.

FM radio and television stations, because of their *VHF* and *UHF* spectrum location, are subject only to direct line-of-sight wave propagation at any time of day or night (C). These waves are sent out from elements on top of antenna towers and directed downward toward the reception area, blocking off radiation that would otherwise scatter upward and out into space. The "horizon" limit, of course, depends heavily on local terrain.

The ionosphere consists of several atmospheric layers located from about 40 to 600 miles above the Earth's surface. Bombarded by high-energy radiation from the sun, these layers take on special electrical properties, causing *refraction* (a gradual type of reflection, or bending back) of AM and short-wave signals. Refracted waves are called *sky waves*.

Sky waves can produce some interesting situations. For example, listeners in Florida may hear a Chicago radio station that listeners in St. Louis cannot, in spite of St. Louis being closer to the station's transmitting antenna. St. Louis is in the *skip zone*, the area between where the Chicago station's ground wave ends and its sky wave returns to Earth.

Under the right frequency, power, and ionospheric conditions, returning sky waves bounce off the surface of the Earth, travel back to the ionosphere, bend back again, and so on. Following the Earth's curvature, they can travel thousands of miles, as depicted in Exhibit 4.e.

The ionosphere's effectiveness varies with time of day and with frequency. AM broadcast stations produce sky waves only at night. Sky waves can extend a station's coverage area, but often they have an opposite effect. Sky waves from stations on the same carrier frequency may arrive simultaneously, or sky waves from a distant station may compete with ground waves from a local station. The resulting interference effectively decreases the coverage of all stations on that frequency.

Shortwave stations, however, are not limited to using a single carrier frequency as are AM broadcasting stations. They can switch frequency several times throughout the day to take continuous advantage of the ionosphere's changing refractive abilities. These stations must be careful to tell listeners where to tune when they change frequencies. Shortwave broadcast stations estimate propagation conditions weeks or months in advance and publish the frequencies they plan to use. They also announce, on-air, the new frequency to which they are moving.

Antennas

Whichever type of propagation may be involved, antennas are needed, both to transmit signals and to receive them. Antenna size and location can have a critical influence on the efficiency of transmission and reception.

Small antennas built into receivers can pick up strong AM and FM signals. The higher the frequency, however, the more elusive the signal and the more essential an efficient antenna becomes—for radio perhaps a retractable whip antenna or, for best results, an outdoor antenna. In many locations, indoor "rabbit ears" and loop antennas suffice for analog VHF and UHF television signals, respectively, although some UHF signals require an outdoor antenna.

Transmitting antennas (see Exhibit 4.f) vary in size because for them to work efficiently, their length must be mathematically related to the length of the waves they radiate. For example, channel 2 (VHF) television transmitters radiate 20-foot waves, but channel 48 (UHF) waves are less than 2 feet long.

Directional Propagation

A flashlight or an auto headlight focuses light into a beam. A reflector creates the beam by redirecting light rays that would otherwise radiate backward or sideways, thus reinforcing the rays that go in a forward direction. Transmitting antennas can also be designed for *directional propagation*—beaming reinforced signals in a desired direction. Directional propagation has value both for increasing signal strength in a desired direction and for preventing interference between stations.

Concentrating radio waves by means of directional propagation increases their strength. This increase, called *antenna gain*, can be many-thousandfold. Gain, however, is not always easily achieved. Concentrating radio waves in a particular direction sometimes requires large or complex antenna designs. AM radio stations need physically large antennas because of their long carrier waves. A directional system requires at least two tall towers, often more, so a directional AM broadcast antenna requires space and is expensive. AM stations need directional antennas more than FM and TV stations, to reduce sky-wave interference.

Receiving antennas are also directional; this is easily demonstrated by turning a portable AM radio in different directions. Signals from a given station come in most strongly when the radio's built-in antenna points in the direction of that station's transmitter. Outdoor antennas used for TV (and FM radio) reception also display directional

EXHIBIT 4.f

Transmitting Antennas

(A)

(B)

Antennas differ depending on the service for which they are built.

(A) *AM antennas* use the entire steel tower as a radiating element. For efficient propagation, the height of this tower must equal a quarter the length of the waves it radiates. Propagation of ground waves depends on soil conductivity, among other things. The photo shows heavy copper ground cables being buried in trenches radiating from the tower base to ensure good ground contact. There are several antennas here for sending out a directional signal.

(C)

(B) *Shortwave antennas* are very complex. Each transmitting site has many antennas to feed different frequency signals, along with other suspended wires that act as reflectors to beam signals in specific directions to reach designated overseas target areas. The Voice of America uses scores of antennas like this one.

(C) FM and TV broadcasters often mount transmitting antennas atop tall buildings and mountains, providing the height needed for effective signal propagation without the expense of building supporting towers.

Sources: Photo (A) courtesy of Stainless, Inc., a division of Spectra-Site Broadcast Group; photo (B) courtesy of Voice of America, Washington, DC; photo (C) © Joel Gordon.

characteristics. Reception of signals arriving from several directions often requires outdoor antennas mounted on rotatable masts.

4.5   MUTUAL INTERFERENCE

Mutual interference between stations limits the number of stations that can be licensed to operate in any particular community.

Co-Channel Stations

The primary troublemaker, *co-channel interference*, comes from mutual interference between stations operating on the same frequency. Co-channel stations must be located far enough apart geographically to prevent their coverage contours from overlapping. Separation rules must take into account that signals too weak for useful reception by an audience may nevertheless cause co-channel interference. A station's *interference zone* therefore extends far beyond its service area.

In keeping with U.S. policy of allowing as many local communities as possible to have their own stations, the FCC makes co-channel separation rules as liberal as possible. Changing daytime versus nighttime coverage areas of AM stations, caused by sky-wave propagation at night, complicate the problem. The FCC seeks to prevent such sky-wave interference by requiring many AM stations to sign off or use lower power at night. Some stations have to reduce power and use a directional antenna system. Sometimes the authorized power level (measured in watts) is less than that of a bedside lamp. The resulting coverage area can be so small the station owner may find it more profitable simply to leave the air after dark. Relocating some stations in the expanded AM band (see Section 4.6) will help reduce these interference problems.

Adjacent-Channel Stations

Engineers design TV and radio receivers to be selective, to tune in one station and reject those operating on different frequencies or channels. But building highly selective receivers would be expensive, so set manufacturers balance *selectivity* against cost. The result is that receivers sometimes pass not only the desired signal but also those of strong stations on adjacent channels, those just above and below the desired station. To reduce *adjacent-channel interference* the FCC avoided placing analog stations on adjacent channels in the same market. But what about communities that have TV stations on channels 6 and 7 or 13 and 14? As seen later in Exhibit 4.q, p. 118, there is a big jump in frequency between these channels. This allows the assignment of both pairs of channels to the same market without any danger of adjacent-channel interference, in spite of their consecutive numbering.

Adjacent-channel signal strength falls off rapidly with increasing distance, so the separation needed between adjacent-channel TV stations is only about 60 miles. Some TV stations operating on the same channel (co-channel) need a separation of about 220 miles. Maximum-power FM stations using antennas high above the ground must be about 180 miles apart to avoid co-channel interference and 150 miles apart to prevent adjacent-channel interference. Because of sky-wave propagation, AM radio stations can suffer both co-channel and adjacent-channel interference from stations many hundreds of miles away.

Adjacent-channel interference is seldom a problem in the analog TV environment, but it is a concern as more DTV stations go on the air. During the transition to digital, many markets will have many more TV signals on the air than were allowed previously. Digital and analog signals differ greatly, so an analog TV tuned to an analog station should not notice a DTV station on a nearby channel. A strong digital signal on an adjacent channel, however, can swamp (overpower) the analog set's selectivity circuits and create interference. A strong analog station can also interfere with a DTV signal on an adjacent channel. The FCC has spaced digital and analog stations to avoid mutual interference, but many broadcasters expect some adjacent-channel analog/digital interference during the transition period.

4.6   AM STATIONS

As a broadcast station class name, AM is somewhat misleading because the term means simply *amplitude modulation*, and many services other than standard radio broadcasting use such modulation. The video component of the analog television signal, for example, is amplitude modulated.

Band and Channels

For years the government assigned AM broadcast stations to frequencies ranging from 535 to 1,605 kHz. This band provided 1,070 kHz of spectrum in which to assign stations. Each AM station received a channel 10 kHz wide, so there were 107 channels (1,070 divided by 10) on which AM stations could operate. In 1988 an international agreement expanded the U.S. AM broadcast band to 1,705 kHz. This expansion made available 10 new regional channels.

Rather than adding new stations in this expanded band, the FCC decided to use the space to reduce AM broadcast station interference by moving existing stations. Using a complex formula, the FCC in 1997 identified 88 AM stations for "migration" to the new frequencies. Stations allowed to move have five years in which to do so, and during that period, may transmit on both their old and new frequencies. This will help listeners follow these stations up the band. AM stations will still operate in a restrictive 10 kHz channel, but spreading them out should reduce co-channel and adjacent-channel interference.

Channel and Station Classes

The FCC has a rather complex method of categorizing AM stations according to the type of channel assigned and their class. There are three types of channels: clear, regional, and local. Clear channels provide stations assigned to them a considerable degree of freedom from interference, allowing them to cover large geographical areas. Regional channels cover smaller areas, and local channels typically serve only a single community.

There are four classes of AM stations, ranging from Class A to Class D. Class A stations have the most "privileges" in terms of maximum transmitting power and permissible hours of operation, while Class D stations have the least. There is considerable overlap, and the FCC rules devote several paragraphs to explaining the differences (see 47

CFR 73.21). Exhibit 4.g outlines the method of categorizing AM stations in greater detail.

Transmission

High AM radio transmitter power improves efficiency of both ground-wave and sky-wave propagation, providing greater coverage and overcoming interference. Daytime power authorizations for domestic U.S. AM broadcasting range from 250 watts to a maximum of 50,000 watts, as indicated in Exhibit 4.g. Cincinnati's WLW used "superpower" (500,000 watts) during the second half of the 1930s, but the experiment ended shortly after the beginning of World War II.

AM radio stations have wavelengths about 550 to 1,800 feet long, depending on their exact carrier frequency. Because antennas at least a quarter-wave long are necessary, AM stations must use vertical steel towers several hundred feet tall to radiate their signals.

In choosing sites for AM antennas, engineers look for good soil conductivity, freedom from surrounding sources of electrical interference, and distance from aircraft flight paths. Ground waves are critical to the coverage provided by AM stations, so the station's antenna system must include an effective ground system. Much of an AM radio station's antenna consists of many buried copper wires radiating from the base of the vertical steel tower. Engineers often have trouble convincing station owners that the expensive buried parts are no less important than the tower aboveground.

Low-power AM signals can be fed into building steam pipes or power lines, which serve as distribution grids. The signals radiate for a short distance into the space surrounding these conductors. Services using this propagation method are called *carrier-current* stations. Colleges and universities often use carrier-current stations, which require no licensing.

Unfortunately, carrier-current signals on power lines will not pass through standard transformers. Modified transformers will pass the signals, but "jumping" every transformer on campus can be difficult and expensive, often negating the advantages of unlicensed operation. *Travelers Information Service* (TIS) is a licensed application of carrier-current sometimes used to help motorists navigate in congested areas such as busy airports.

EXHIBIT 4.g

AM Radio Station Classification Structure

AM Radio Stations and Channel Classes

Station Class	Channel Class	Number Channels in Class	Power Range (kW)	Percentage in Each Channel Class
A, B, D	Clear	60	.25–50	35
B, D	Regional	51	.25–50	44
C	Local	6	.25–1	21

AM Radio Stations by Station Class

Station Class	Number of Stations *	Percentage of Total Stations[†]
A	77	1
B	1,730	35
C	1,023	21
D	2,072	42

*Data provided by FCC. The numbers vary as new stations sign on and others leave the air.
[†]Percentages do not add up to 100 percent because of rounding.

Class A stations operate unlimited schedules only on clear channels with powers ranging from 10 to 50 kW. Class A stations are protected from other stations on the same or adjacent channels.

Class B stations operate unlimited schedules on clear and regional channels with powers ranging from .25 to 50 kW. Class B stations in the extended AM band (1,605–1,705 kHz) operate with a power of 10 kW.

Class C stations operate only on local channels with powers ranging from .25 to 1 kW.

Class D stations operate either daytime, limited time, or unlimited time with nighttime power less than .25 kW. Daytime powers range from .25 to 50 kW. Nighttime operations of Class D stations are not protected from interference, and Class D stations must protect all Class A and Class B stations from nighttime interference.

Roadside signs tell drivers where to tune their AM radios to get helpful information.

AM Stereo

Because of its narrower channel, AM took longer to develop stereo than did FM. AM stereo uses a *matrix* mode that melds left and right channels in a way somewhat similar to the two channels of FM stereo but without the full high-fidelity benefit. The few AM stereo stations in existence in the early 1990s had mostly chosen the Motorola C-QUAM standard finally adopted by the FCC (see Section 12.10, p. 363). Early excitement regarding AM stereo had largely dissipated by the middle 1990s.

Few listeners were aware of its existence, and programming on many AM stations benefited little from being offered in stereo.

Shortwave AM

Shortwave broadcasting, which also uses AM, has been allocated parts of the high-frequency (HF) band between 6 and 25 MHz. The ionosphere refracts waves in this band both day and night, enabling round-the-clock coverage in target zones thousands of miles away from the originating transmitter.

The U.S. public makes little use of shortwave broadcasting. However, some foreign countries use short waves extensively for domestic services. A few privately operated American HF stations, mostly evangelistic religious outlets, broadcast to foreign audiences. Short waves are used extensively for international diplomacy.

4.7   FM STATIONS

Referring to AM as *standard broadcasting* has become sorely out-of-date because FM's inherently superior quality has enabled it to forge ahead of AM in numbers of both listeners and stations.

Band and Channels

U.S. frequency-modulation broadcasting occupies a 20 MHz block of frequencies running from 88 to 108 MHz in the VHF band. This 20 MHz block allows for 100 FM channels of 200 kHz (.2 MHz) width. Exhibit 4.k, p. 110, shows their spectrum location in relation to other broadcast channels.

The FCC numbers FM channels 201 to 300, but licensees prefer to identify their stations by their midchannel frequency (in megahertz) rather than by channel number (88.1 for channel 201, 88.3 for channel 202, and so on). The lowest 20 FM channels (88.1 to 91.9 MHz) are reserved for noncommercial use. The analog FM channel width of 200 kHz is generous—20 times the width of the AM channel.

Transmission

The coverage area of an FM station depends on the station's power, the height of the transmitting and receiving antennas above the surrounding terrain, and the extent to which obtrusive terrain features or buildings block wave paths. In any event, FM uses direct waves, which reach only to the horizon.

Once determined by these factors, an FM station's coverage remains stable, day and night. This stability is one reason the FCC had an easier time preventing FM interference than it did AM interference.

The FCC divides the country into geographic zones and FM stations into three groups according to coverage area: Classes A, B, and C, defined by power, antenna height, and zone. Classes B and C are further subdivided, but generally the maximum coverage areas for FM stations by class are as follows: Class A about 15 miles; Class B about 30 miles; Class C about 60 miles. The maximum power/height combination permits 100,000 watts (twice the maximum AM station power) and a 600-meter (almost 2,000-foot) antenna elevation.

In January 2000 the FCC, at the urging of chairman William Kennard, created two new classes of *low-power FM* (LPFM) stations. The FCC established a new LP100 class station operating with a maximum power of 100 watts and an antenna no higher than 30 meters (98.4 feet) to provide an approximate service radius of 3.5 miles. A second class, called LP10, provides for stations operating with 1 to 10 watts and antennas no higher than 30 meters to serve a one or two mile area around the station. LPFM stations must be noncommercial and, at least initially, must be owned by entities living or headquartered within 10 miles of the station. The FCC chose not to create a more powerful LP1000 (1,000 watt) service some LPFM advocates had proposed.

The FCC created LPFM in response to those claiming that liberalized ownership rules have concentrated radio station ownership in the hands of large commercial interests, leaving no opportunity for community-oriented broadcasting or a diversity of voices. Commercial FM broadcasters objected to LPFM, saying that the creation of an indefinite number of LPFM stations (possibly hundreds in the top fifty markets alone) will bring to the FM band the kind of interference that plagues AM. They also said that increasing FM band congestion will negatively affect ongoing efforts by the industry to bring in-band on-channel (IBOC) digital broadcasting to American listeners. (See Section 5.8, p. 153, for a discussion of digital radio.)

Frequencies used for FM broadcasting are much higher than those of AM stations, so they have proportionally shorter wavelengths and much smaller antennas. FM stations use tall towers only to place their small antennas high above the ground, not to actually radiate the signal, as do the towers of AM stations. FM stations take advantage of this by mounting their antennas on mountains or tall buildings. FM stations depend on direct, not ground, waves to provide primary coverage and do not need expensive buried radial systems. Finally, the small size of FM antennas allows engineers to design radiating systems that exhibit *antenna gain*. Using an FM antenna with gain increases the station's *effective radiated power* (ERP), as does boosting transmitter power, but over time it is much less expensive than a larger transmitter's monthly electric bill.

Reception

FM receivers with quality loudspeakers can reproduce the audio frequencies essential for high-fidelity sound. In addition, FM radio has greater *dynamic range* (the loudness difference between the weakest and strongest sounds) than does AM, making for more realistic reproduction of music.

FM's inherent immunity to static and interference gives FM radio a significant quality advantage over AM radio. FM's static-free reception is particularly noticeable in the southern part of the country, where subtropical storms cause much natural interference. FM also has an advantage in large cities, where concentrations of electrical machinery and appliances cause static. Large buildings in urban areas can, however, create reception problems for FM stations because they reflect FM radio's VHF signals.

FM Stereo

The left and right channel audio off stereo tapes, audio file servers, and CDs pass through the FM radio station's equipment as entirely separate signals until they reach the station's stereo generator, a subsystem of the transmitter. Here the two audio channels combine with supersonic (above the range of human hearing) pilot and subcarrier signals. The result is a composite signal that modulates the

station's carrier wave. FM receivers designed to reproduce stereo audio use the subcarrier and pilot signals to separate the two audio channels once again. Just as it is possible to watch color TV signals on a monochrome set, a monaural FM radio reproduces the audio in full fidelity, minus the separation between audio channels required for stereophonic effect.

Subsidiary Communications Service (SCS)

FM stations can also multiplex additional information to provide a nonbroadcast service received using special equipment. This multiplexed signal most frequently provides background ("elevator") music to stores and offices, or a reading service for the blind. SCS can accommodate a wide range of other applications, including paging, distribution of stock and bond prices, detailed weather reports, and traffic information. SCS may also provide station promotional information (such as music format, call letters, or location) in a visual format on special receivers, as discussed in Section 5.8 (p. 153.)

4.8   ELECTRONIC PICTURES

The basic structure of a TV station remains much as it was in the late 1940s (see Exhibit 4.l, later in the chapter). Cameras and microphones still convert sounds and pictures into electrical signals, which are fed to a transmitter, where they modulate RF carrier waves eventually radiated from an antenna. A DTV station is much the same. The major difference involves the way stations encode the information and modulate the transmitter. Most receivers still use a vacuum tube for display of the video image (see Exhibit 4.n, later in the chapter), and broadcasters continue to broadcast the analog signal much as they have for more than five decades. Transition to solid-state technology and microelectronics has, however, changed the way modern cameras operate.

Camera Technology

From the 1940s until the early 1990s tube-type cameras dominated the broadcasting industry.

EXHIBIT 4.h

The TV Camera

(A)

(B)

(A) As technology improved, tubes became smaller. Commercial broadcasters first used the *iconoscope*, the odd shaped tube at the top. Then *image orthicons*, with 3-inch and 4.5-inch faces, replaced iconoscopes. Smaller tubes such as the *vidicon* and *plumbicon* in turn replaced the image orthicon. (B) In the 1990s, smaller, more rugged and energy-efficient chips largely replaced tubes as imaging devices for cameras. (C) Light striking the surface of the CCD is converted to electrical charges. These charges are then collected in the storage area before being passed to the output area. From the output area the signal representing the light and dark parts of the televised image are clocked out to the camera's video amplifiers.

Sources: Photo (A) courtesy Frank Sauerwald, Temple University; photo (B) courtesy of Sony Electronics, Inc.

(C)

It was not until the 1980s that broadcasters began using cameras with a solid-state *charge-coupled device* (CCD) rather than tubes to convert light energy to an electrical signal. The video produced by these "chip" (CCD-equipped) cameras relegated them mainly to shooting news footage, where picture quality is less critical. Solid-state technology advanced dramatically in the 1990s, and cameras using CCDs are now commonplace. Some excellent studio cameras using tubes remain in service,

but broadcasters are replacing them with CCD cameras when they fail. CCDs offer the usual advantages of solid-state devices compared to tubes: less weight and bulk, longer service life, greater reliability, and reduced energy consumption.

Exhibit 4.h shows a photo and simplified functional diagram of the CCD used in modern cameras. CCDs use hundreds of thousands of tiny light-sensitive elements to convert (transduce) light to electricity. The electrical charge depends on the intensity of the light striking the light-sensitive element, so the CCD creates a pattern of electrical charges analogous to the bright and dark portions of the image. Each transducer produces a *picture element (pixel)* that combines with others to form the complete image. The electrical charges produced by the light-sensitive transducers are transferred and stored in another part of the CCD and, finally, read out in a line-by-line order.

Broadcast-quality cameras use three CCDs, one for each of the primary colors of light, which combine their output to produce all the colors we see on TV. Consumer-quality cameras typically have only a single CCD and deliver a less crisp color image, but one that is still amazingly good considering their price.

Scanning and Frame Frequency Standards

The technology of tube-based display devices requires that the video information be transmitted in a pattern that allows the receiver to "paint" images on-screen one line at a time using a moving electron beam. The U.S. analog TV system (standards differ from country to country) calls for a nominal *525 lines* of video information to be painted in horizontal rows from top to bottom to create the image on the TV screen. The number of lines making up the image helps determine the *resolution*—the clarity—of the image: the more lines, the greater the resolution. However, not all 525 lines of the U.S. analog standard are used to convey pictures. Some time—and thus some lines—must be devoted to the auxiliary signals used by the receiver to control the scanning. Transmission standards call for 483 active video lines in the 525-line analog system.

As illustrated in Exhibit 4.i, TV uses a method called *offset* or *interlaced scanning* in which all the even-numbered lines are scanned first, followed by the odd-numbered lines. Each of these scans constitutes an odd or even *field*. Two fields combine to make a complete *frame* of video. This odd-even scanning of lines was a way to prevent the image from appearing to flicker without having to increase the channel bandwidth required for TV transmission. The U.S. analog TV system paints the screen with 30 complete frames (60 fields) every second. The human eye and brain cannot distinguish individual images flashed in rapid succession. This *persistence of vision* causes viewers to see a smoothly moving image rather than a series of individual frames of video.

Progressive scanning display devices paint the lines on the screen one after the other, in numerical sequence. Computer monitors use progressive scanning, and the DTV transmission standards adopted in 1996 allow DTV display devices to use either interlaced or progressive scanning methods.

In analog broadcasting the scanning format and frame frequency standards are fixed. Stations must transmit using interlaced scanning and 30 frames per second, while all analog TV sets must display images using these same standards. A DTV station, in contrast, may use any picture format, although by convention the format used probably will be one of those recommended by the Advanced Television System Committee (ATSC). At the receiver, however, the displayed image format need not be the same as that broadcast. Many people will initially watch DTV transmissions on analog receivers by using a set-top box to convert the digital signal, in whatever format, to the only format the set can display: an interlaced 30-frames-per-second image. Viewers can watch a wide-screen, high-definition program on a conventional TV, but the displayed image cannot, of course, be any wider or exhibit higher definition than that which the set was designed to provide initially. The decision of the FCC not to link the receiver's scanning format and frame frequency to that being broadcast is an important difference between analog and DTV transmission standards.

Auxiliary Signals

Several types of *sync (synchronization) pulses* included with the analog video information are needed to ensure that each pixel in each line appears on the receiver screen in the same location it was in when leaving the camera. *Horizontal sync* pulses tell the receiver where to begin and end each horizontal line of video. *Vertical sync* pulses tell the receiver when an odd (or even) field is complete and the next field should begin. Additional *blanking pulses* turn off the flow of video information while the "paintbrush" (actually the electron beam in the receiver's picture tube) is repositioned to begin the next scan line or the next field. In the analog system the time required to position the beam to begin a new field is equivalent to that needed to paint 21 lines of video.

EXHIBIT 4.i

∙ ∙ ∙ ∙ ∙ ∙ ∙ ∙ ∙

Interlaced and Progressive Scanning

Field A—even lines

Frame
(Fields A+B)

Field B—odd lines

Progressive scanning—each
line is scanned in sequence,
top to bottom

Analog television uses interlaced scanning. Several of the DTV picture formats recommended by the Advanced Television Systems Committee (ATSC) also use interlaced scanning. In interlaced scanning a complete electronic picture, or frame, is made up of two fields, each of which makes up half the picture definition. One field (A) scans the even-numbered lines, and the other (B) scans the odd-numbered lines (shown here as broken). Together the lines of the two fields "interlace" to make up a complete frame (C).

At the end of each scanned line and each field, the electron beam has to fly back to start a new line or field. In analog TV transmission a blanking signal is sent to prevent the beam from activating any picture elements during these retrace times. When the beam reaches the bottom of the screen, another blanking signal turns off the beam while it is pulled back to the top to begin another field. This period is called the *vertical blanking interval* (VBI), and it is during this time that analog TV stations can transmit nonpicture information, such as closed captions and other data.

With progressive scanning formats (D) the electron beam traces one line at a time, top to bottom, in sequential order. Each scan produces one complete electronic picture, or frame. Computer screens use progressive scanning. Some engineers believe that interlaced scanning is better for motion, while progressive scanning provides better resolution for graphics, but the point is debatable. The computer industry supports progressive scanning formats for DTV, eliminating the need to convert an interlaced signal for display on the progressively scanned screen.

No picture (video) information flows during this relatively long *vertical blanking interval* (VBI), allowing broadcasters to insert other information.

The VBI is often used to provide *closed captioning* for hearing-impaired viewers, market reports, or on-screen *electronic program guides* (EPGs). Capturing and displaying information embedded in the VBI requires special decoder circuits either built into the TV receiver itself or in a separate set-top box designed for the purpose.

DTV also uses auxiliary signals embedded in the digital data stream to help receivers control the electron scanning beam (if the display device is a cathode-ray tube, see Section 4.11) to create an image. Although the scanning format, frame frequency, and other image characteristics (such as the shape of the displayed image) can be altered by the receiving equipment to accommodate the display device, receivers still need sync information to function properly.

TV Sound

∙ ∙ ∙ ∙ ∙ ∙ ∙

U.S. analog television uses FM sound. The sound carrier is located in the upper part of the television channel, as shown in Exhibit 4.j. Sound needs no synchronizing signals to keep it in step with the picture in the analog TV system. Sound and picture occur simultaneously and go out in real time.

The analog sound channel is designed to respond to a sound range from 10 to 15,000 Hz. It also has room for subcarriers that enable multiplexing stereophonic sound and alternative-language soundtracks within the channel.

With DTV, the sound, like the picture, is encoded in digital form and becomes part of the transmitted data stream. The audio standard adopted for U.S. DTV is called Dolby Digital. It is also sometimes called Dolby AC-3 or 5.1 channel surround-sound audio. Dolby Digital uses compression techniques to reduce the amount of

EXHIBIT 4.j
• • • • • • • • • •

The Analog TV Channel

The architecture of the 6 MHz television channel includes a vestigial (sort of an "electronic leftover") lower sideband that takes up the 1.25 MHz below the video carrier frequency. Note that .25 MHz is unused above the audio subcarrier. These seemingly "wasted" spaces reduce adjacent channel interference. When modulated by the black-and-white video (luminance) information, the main video carrier's upper sideband overlaps the sideband of the color subcarrier, but *interleaving* minimizes conflict. Uneven distribution of frequencies carrying the luminance signal makes such interleaving possible. If you visualize the frequencies of the monochrome information as the teeth of a comb, the color information occupies frequencies represented by the spaces between the teeth.

Note the audio carrier near the top of the TV channel. A multichannel television sound (MTS) station can use this area to accommodate the main audio signal, a stereo audio signal, a secondary audio program (SAP) signal, and a closed-circuit "professional channel" for transmitting telemetry signals or other data requiring only 3 kHz of bandwidth. The SAP signal may be used to provide a foreign-language version of the program being broadcast.

Source: Adapted from 47 CFR 73.699.

channel bandwidth required for audio transmission. The Dolby Digital standard allows for six separate audio channels: left, center, right, right rear (or side) surround, left rear (or side) surround, and a limited-bandwidth low-frequency (subwoofer) channel. The ".1" refers to the subwoofer channel that handles only very low frequency sounds. The center channel is for dialogue audio, and the remaining channels provide the surround-sound effect. Producing programs with surround sound is a complex and costly process. It is likely that most programs produced for transmission in wide-screen high-definition format will fully utilize the capabilities inherent in Dolby Digital audio. One of the advantages of the audio standard adopted is that it allows for multiple streams of audio, which

can be used, for example, to deliver a separate audio program, dialogue in several different languages, program-related commentary, or specially enhanced audio for the hearing impaired. Unlike analog TV sound, DTV sound requires synchronizing audio with its associated video in the data stream. Some early DTV transmissions featured sound that did not match the lip movement of people on-screen.

Color TV
• • • • • • •

The signal specifications for analog color television provide for *compatibility*—black-and-white (monochrome) receivers can reproduce a color signal without the aid of an adapter. Multiplexing permits adding a color component without enlarging the television channel. Exhibit 4.j shows the position of the color subcarrier in the conventional television channel.

RCA developed this color system, and, with National Television System Committee (NTSC) approval, the FCC adopted it in 1953. The NTSC color system has served the industry well for more than 45 years, but broadcast engineers are critical of it, saying NTSC really means "Never Twice the Same Color," a reference to the difficulties of getting colors on-screen to match exactly those seen in the studio.

Color information in a DTV transmission is, like the sound, digitally encoded and included in the data stream rather than being added to the channel by a separate color subcarrier. DTV will not suffer from image artifacts (abnormalities introduced by the transmission process itself) caused by adding color to an existing analog black-and-white TV system through multiplexing. However, DTV may sometimes exhibit image artifacts caused by excessive compression (see Section 5.5, p. 138).

Color cameras pick up images in the three primary colors of light: red, green, and blue (RGB). This is sometimes confusing because red, blue, and yellow are the primary colors familiar to artists who mix pigments that *reflect* light. TV pictures are created by mixing light itself. TV starts with a black screen, and the primary colors of light combine to create white; artists start with a white canvas, and the pigments mix to create black.

The camera divides light energy entering the lens into the three primary colors before it reaches the tubes or CCDs that convert light energy to electrical energy. The fidelity of *hues* (colors) and *saturation* (richness of each color) in the transmitted image depends on the camera's ability to process the three primary colors properly. Additionally, the color information has a brightness attribute called *luminance*. The luminance information in the signal provides most of the fine detail in the picture, and luminance alone suffices for monochrome receivers.

4.9 TV CHANNELS

U.S. broadcast standards call for a 6 MHz television channel—600 times the width of an AM radio broadcast channel. Indeed, *all* AM and FM broadcast channels together occupy less spectrum space than only four television channels. Exhibit 4.j shows how the 6 MHz of the channel are utilized when transmitting an analog TV signal.

Channel Width and Resolution

Even with such wide channels, conventional television achieves relatively low picture resolution by the standards of theatrical motion pictures and good-quality still photography. The average home receiver displays about 150,000 pixels per frame, but a projected 35-millimeter film frame contains about a million.

Television resolution standards represent a compromise. Higher quality has to be sacrificed to avoid using too much spectrum space. The same spectrum-saving goal motivates suppression of the lower sideband of the TV picture signal.

The resolution required varies with the application. Early PC users tried to use TV sets as computer monitors. They found that, at a viewing distance of 18 inches, the TV lacked adequate resolution for word processing. Such work required a more expensive, higher-resolution monitor. Yet most people happily watch rented movies on VHS tape—a delivery system providing less resolution than conventional broadcast TV—using their "inadequate" TV sets, but at typical viewing distances of 10 feet. This suggests that some programming may not require high resolution, with its associated high-bandwidth demands. Broadcasters struggle with this issue as they weigh the advantages of using their DTV channels for one or two programs in high definition or for many more lower-definition programs requiring the same channel bandwidth instead. Compression technology allows DTV to increase dramatically the resolution possible while occupying less spectrum space than standard analog television. But within the DTV realm, a higher-resolution image requires more spectrum space than a lower-resolution image.

Location in Spectrum

Exhibit 4.k shows the location of television channels in the spectrum relative to the location of channels other broadcast services use. An FCC table allots specific channels by number to specific towns and cities throughout the United States.

EXHIBIT 4.k
. . . .

Summary of Broadcast Channel Specifications

This table pulls together information scattered throughout the text to allow comparison of the major channel specifications of radio and TV services. Note especially the allocation of TV channels into four separate frequency allotments in two different bands. DTV will use only channels 2–51. The spectrum occupied by UHF channels 52–69 will be reassigned to other services.

Service	AM Radio	FM Radio	VHF Television	UHF Television
Frequencies	535–1705 kHz (MF band)	88–108 MHz (VHF band)	54–72 MHz (channels 2–4) 76–88 MHz (channels 5 and 6) 174–216 MHz (channels 7–13) (VHF band)	470–806 MHz (channels 14–69) UHF band
Total channels	117	100	12	56
Bandwidth (single station)	10 kHz	200 kHz (equivalent to 20 AM channels)	6,000 kHz (equivalent to 600 AM or 30 FM channels)	6,000 kHz
Classes of stations and power limits	Class A 10–50 kW Class B .25–50 kW Class C .25–1 kW Class D <.25–50 kW	A .1–6 kW about 15 miles B >6–50 kW about 30 miles C 25–100 kW about 60 miles	Channels 2–6 about 65 miles up to 100 kW* Channels 7–13 about 55 miles up to 316 kW*	Channels 14–40 about 40 miles up to 5,000 kW* Channels 41–69 about 30 miles up to 5,000 kW*
Educational allocation	None	88–92 MHz (20 channels)	137 specific channel allotments	559 specific channel allotments
Factors affecting station coverage and signal quality	Frequency Power Soil conductivity Day/night (sky wave) Thunderstorms Many directional antennas	Antenna height Power (to a degree) Unlimited time Few directional antennas High fidelity Little static Line-of-sight range	Antenna height Frequency Power (to a degree) Rigid spacing Line-of-sight range Unlimited time	Antenna height Frequency Power (to a degree) Rigid spacing More limited line-of-sight range Unlimited time

*DTV is limited to 1,000 kW and will eventually occupy only channels 2–51.

The FCC now has two TV channel-allotment tables, one for conventional analog stations and another for digital stations. The map in Exhibit 2.j (p. 46) gives an example, showing all the places to which one particular analog channel has been allotted.

Exceptions to the allotment tables are *low-power television* (LPTV) facilities. In 1982 the FCC gave final approval for LPTV service. LPTV stations operate on the same VHF and UHF channels and have the same technical characteristics as other analog TV stations, but they use greatly reduced power, from 10 to 1,000 watts. Coverage under ideal conditions is only about 10 miles, but in congested areas is often much less. LPTV stations were considered secondary services and, provided they did not cause interference to full-power TV stations in the area, could operate on any channel. LPTV's secondary status became a serious liability when the FCC began assigning channels for DTV. In announcing the DTV rules, the Commission made some concessions and pledged continuing efforts to accommodate LPTV, but the service's future in a digital world remained uncertain.

In 1997 the Community Broadcasters Association (CBA), on behalf of LPTV licensees, petitioned the FCC to create a new and more protected class of LPTV stations that originate programming. The FCC responded to the petition by issuing a Notice of Proposed Rulemaking in the fall of 1999. Meanwhile, the CBA also lobbied Congress, and in November 1999, President Clinton signed a law creating a new protected class of LPTV station. To be eligible for the new Class A status and the protection it offers, an LPTV station had to have been broadcasting at least 18 hours a day since September 1999, including a minimum of 3 hours of locally produced programming each week. In addition, no LPTV station could achieve Class A status if its signal posed an interference threat to an existing full-power station's planned DTV coverage area. In effect, the law protects LPTV stations against new high-power stations wanting to take their frequencies, but offers little protection against established full-power stations wanting to maximize their DTV service areas at the expense of LPTV. Some observers believe that only about 20 percent of the nation's almost 2,200 LPTV stations will qualify for Class A status. Prior to passage of the law, the FCC had estimated that

35 to 45 percent of all LPTV stations would have to either find another channel or cease operation entirely to protect the new DTV service.

4.10   ANALOG TV TRANSMISSION

This section deals principally with analog TV transmission, although some digital transmission characteristics are introduced. The development of DTV, along with other players in the digital revolution currently in progress, is discussed in more detail in Chapter 5.

Synchronization

The *sync generator* keeps the camera's output in step with the receiver display tube. Exhibit 4.l shows the sync generator in relation to other control-room equipment.

In addition to providing the horizontal and vertical sync pulses, the sync generator also inserts blanking signals during retrace periods and synchronizes inputs from external picture sources such as recordings and network feeds. Synchronizing and blanking cues become part of the composite picture signal.

Combining Picture and Sound

The picture (video) and sound (audio) components travel from the studio to the transmitter as independent signals. The transmitter site may be miles away from the studio so that it can be near the antenna. At the transmitter, video and audio signals modulate separate audio and video transmitters.

A television station's power is usually stated in terms of the effective radiated power (ERP) of its *video* signal. The video transmitter operates on higher power than does the audio transmitter because it has a much greater information load, requiring up to 20 times as much power.

Video and audio signals meet for the first time in the *diplexer*—a device that combines video and audio signals before feeding them as a composite signal to the antenna for propagation.

EXHIBIT 4.1
. . . .

Analog TV System Components and Signals

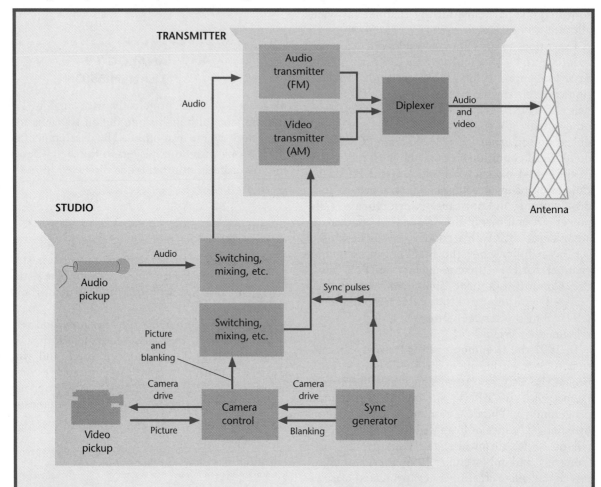

Each block stands for a function that in practice may involve many different pieces of equipment. The lower portion of the diagram represents the basic items and functions involved in studio operations; the upper portion represents those involved in transmission. Separate sets of equipment handle the video and audio components all the way to the diplexer, the de-

vice in transmitters that finally marries the two signals so that they can be transmitted as a composite. (Exhibit 4.j shows how the audio and video components are multiplexed in a single channel.)

Source: Based on Harold E. Ennes, *Principles and Practices of Telecasting Operations* (Indianapolis: Howard W. Sams, 1953).

Propagation
. . . .

Engineers elevate TV transmission antennas as high as possible—on mountain peaks, roofs of tall buildings, or the tops of tall steel towers. The antenna towers themselves do not radiate signals as do AM towers. As is true of FM stations, they only

support the radiating elements, which are relatively small, in keeping with the shortness of television's VHF and UHF waves. Television (and FM radio) station antennas provide *beam tilt*. Beam tilt concentrates the radiated energy downward so little of it skims over the line-of-sight horizon to be lost uselessly in space. Objects in

EXHIBIT 4.m
. . . .

TV Propagation Paths

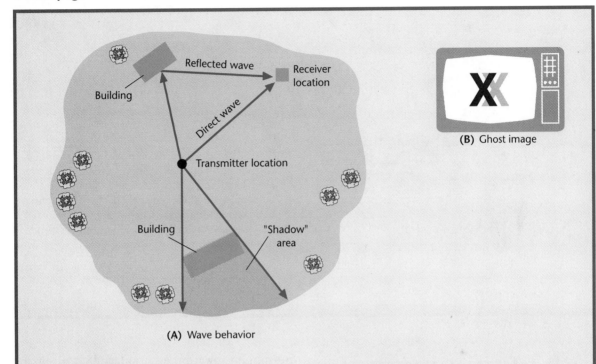

(B) Ghost image

(A) Wave behavior

(A) Simplified coverage pattern of a TV station, showing some characteristics of direct-wave propagation: The waves carry only to the horizon; some may encounter surfaces that reflect signals; some may encounter obstructions that cause "shadow" areas in the coverage pattern.

(B) When a receiver detects both a direct wave and a reflected wave, the reflected wave will have traveled over a longer path and will therefore arrive at the receiver slightly later than the direct wave. In the analog TV system, when the receiver displays the signal of the delayed reflected wave, the image lags slightly behind that of the direct wave, appearing as a "ghost."

In 1994 the FCC approved TV station use of *ghost-cancellation reference* (GCR) signals aimed at eliminating ghosts. GCR allows television receivers to identify and delete the second signal. As noted in Section 4.11 (p. 114) and Section 5.7 (p. 146), DTV also suffers from multipath signal interference caused by delayed reflected waves. If not corrected, multipath interference can result in the complete loss of DTV reception rather than the displaying of ghost images. DTV receivers have adaptive equalizer circuits that help to correct for errors caused by delayed reflected waves and help to prevent signal loss.

the propagation path block UHF waves more easily than they block VHF waves. UHF waves also attenuate more rapidly than VHF waves. The FCC therefore allows analog UHF stations to use very high power (up to 5 million watts ERP) to compensate to some extent for the inherent coverage limitations of UHF waves. DTV stations will operate with power ranging from 50,000 to 1 million watts ERP. In setting TV power limits,

the FCC uses a formula that takes antenna height into consideration.

A TV station's coverage distance and the shape of its coverage area thus depend on several factors: transmitting and receiving antenna height and efficiency, obstructive terrain features, transmitter frequency, and effective radiated power. Exhibit 4.m shows how obstructions can affect propagation.

EXHIBIT 4.n
•••••••••••
Color Kinescope Tube

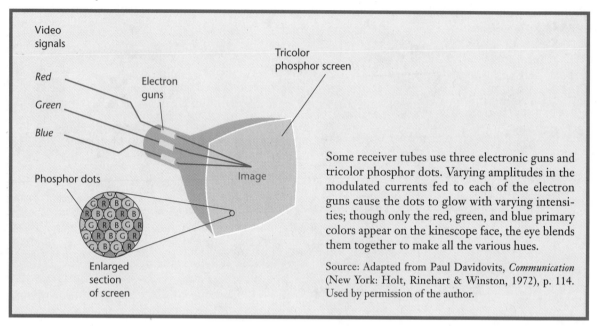

Video signals

Red

Green

Blue

Electron guns

Tricolor phosphor screen

Image

Phosphor dots

Enlarged section of screen

Some receiver tubes use three electronic guns and tricolor phosphor dots. Varying amplitudes in the modulated currents fed to each of the electron guns cause the dots to glow with varying intensities; though only the red, green, and blue primary colors appear on the kinescope face, the eye blends them together to make all the various hues.

Source: Adapted from Paul Davidovits, *Communication* (New York: Holt, Rinehart & Winston, 1972), p. 114. Used by permission of the author.

4.11   TV RECEPTION

Antennas and Contour Grades
•••••••••••••••••••••••

Simple indoor antennas often suffice if the viewer is within an analog station's Grade A contour. Some viewers had trouble receiving DTV signals using indoor antennas with early-generation DTV receivers, even if they lived within the Grade A contour, because of multipath signal interference. DTV set makers are continually improving the ability of receivers to cope with multipath problems, however. (See Section 5.7, p. 146) for additional discussion of DTV transmission and reception.) Viewers located in either an analog or DTV station's Grade B contour probably need an outdoor antenna for acceptable reception. Antennas operate at maximum efficiency on only one frequency, so antennas used for receiving many channels require a compromise. Even so, the short wavelengths used for TV allow receiving antennas to be highly efficient and directional. Like outdoor FM radio antennas, TV antennas must point toward the station, and if stations are at widely different points of the compass, the antenna must be rotatable.

Your TV Set
•••••••••

Television receivers process the video and audio parts of the signal separately. In most receivers, the video information goes to a *kinescope*, a type of *cathode-ray tube* (CRT) (see Exhibit 4.n). This is the tube the viewer looks at as the receiver "screen."

On the inside face of the kinescope, *pixels* consisting of phosphorescent particles glow when bombarded with electrons. Within the neck of the kinescope resides an electron gun. It shoots a beam of electrons toward the face of the tube. Guided by external deflection coils, the electron beam "paints" an image on the screen (actually the inner face of the tube).

The beam stimulates pixels that glow with varying intensities, depending on the beam's strength. It lays down an image, line by line, field by field (when using interlaced scanning), and frame by frame (when using progressive scanning). Synchronizing signals originated by an analog TV station's sync generator activate the deflection coils, keeping the receiver scanning in proper sequence. Codes in the digital data stream provide the necessary synchronizing signals in DTV transmission.

Color Reception

Phosphors that glow in the RGB primaries coat the inner face of the color CRT, arranged either as narrow parallel stripes or as triads of dots. Receiver circuits decode the television signal into components representing the energy levels of the three primaries in each pixel. Three streams of electrons, one for each primary, strike the tube face, using one or more electron guns.

Only primary colors appear on the kinescope. The eye perceives varying hues in accordance with varying intensities of the primaries as delivered by the electron beams. One can confirm that only the primaries appear by looking at part of a color picture on the face of a kinescope with a magnifying glass. Only dots or stripes in red, green, and blue will be visible.

4.12 SOLID-STATE RECEIVERS

Most television receivers still rely on the CRT (or kinescope tube) for picture display. However, researchers have developed several technologies for producing flat screens that can be hung on a wall.

Flat screens initially appeared in tiny, portable TV receivers and in laptop computers. Flat-screen displays are also replacing conventional CRTs as viewfinders in camcorders. Most flat-screen devices currently available use some form of *liquid crystal display* (LCD) technology. This technology employs a liquid crystal fluid, containing thousands of pixels, sandwiched between two plates of glass. When voltage is applied to the pixels, they twist on or off, blocking a backlight that passes through a polarized filter. The quantity of passed light varies, depending on how much the pixels have twisted or untwisted.

LCDs are of two types. *Passive-matrix* LCDs continuously supply low voltage to all pixels, increasing or decreasing the voltage when an element needs to be turned on or off. This process results in flickering, ghosting, and poor contrast.

LCD leaders such as Toshiba, Sharp, and Hitachi made advances in passive-matrix technology during the late 1990s that reduced problems with ghosting and lack of contrast, but these advances were incremental rather than revolutionary. Passive-matrix technology is most useful for digital watches, calculators, and other applications that display static or slow-moving images.

For moving video images, manufacturers use *active-matrix* LCDs, which turn each picture element on and off independently (see Exhibit 4.o). The sheets of glass between which the liquid crystal is trapped carry thin films of connected transistors etched onto their inner surfaces. Each tiny transistor is, in effect, a switch that can be turned rapidly on or off to polarize the liquid crystal at that point, and hence let light through. Three transistors are needed to create a single pixel capable of displaying a range of colors. Unlike passive-matrix LCDs, active-matrix LCDs are colorful, bright, and flicker-free, and they provide contrast comparable to that of a CRT. During the late 1990s, improvements in active-matrix technology and marketplace competition combined to reduce the cost of LCDs significantly. The small space requirements, reduced power consumption, and elimination of high-voltage electrical currents (with their associated electromagnetic radiation) make LCDs an attractive alternative to CRTs for many applications. CRTs, however, still enjoy an advantage in brightness, useful life, and relative cost to the consumer.

The *plasma display panel* (PDP) is an increasingly popular flat-screen technology. A PDP uses a grid of tiny channels (pixels) filled with gas, usually a mixture of neon and xenon. An electrical charge causes the gas to change to a plasma state, generating ultra violet (UV) light, which, in turn, reacts with red, green, and blue phosphors to produce visible light. PDPs provide bright images that do not wash out under most ambient light conditions. A traditional weakness of LCD displays is that they have required users to be positioned almost directly in front of them to see the image clearly. This is a minor inconvenience for an individual using a computer, but it becomes a larger problem when several people seated around a room try to watch a TV program. PDPs provide viewing angles of about 160 degrees, so even people sitting off to the side get a clear view. PDPs are relatively lightweight compared with other

EXHIBIT 4.0

Liquid Crystal Display

Liquid crystal display (LCD) screens vary in their construction and in the materials they use, but they all sandwich some form of light-emitting elements between two plates of glass.

Source: Adapted from "Flat-Panel Displays" by Steven W. Depp and Webster E. Howard, *Scientific American*, March 1993, p. 91. Used by permission of Michael Goodman.

displays of similar size, and they are also very thin, allowing them to be hung on walls. Early in 2000, ReVox introduced a PDP less than 3 inches deep while providing a 42-inch image (measured diagonally). However, the price of the ReVox PDP at the time of introduction was almost $19,000. PDPs from other manufacturers are similarly expensive, although some industry analysts believe PDP prices could fall to around $3,000 within a few years.

Other flat-screen technologies, such as field emission displays (FED) and digital micromirror devices (DMD), also are being developed. They may eventually challenge LCDs and PDPs in the marketplace.

Converging Technologies

Just as flat-screen displays first used in laptop computers migrated to TV sets, other computer-oriented technology is merging with TV receiver technology. Digital-analog incompatibility initially delayed progress, but $300 digital set-top boxes appeared in the mid-1990s, enabling viewers to use their TV sets to surf the Internet and view web pages. Concurrently, add-on circuit boards selling for around $150 became available for personal computers that effectively turned the computer into a TV/radio receiver. Some TV programmers began using the vertical blanking interval (VBI) to

EXHIBIT 4.p
• • • • • • • • • • •

Convergence

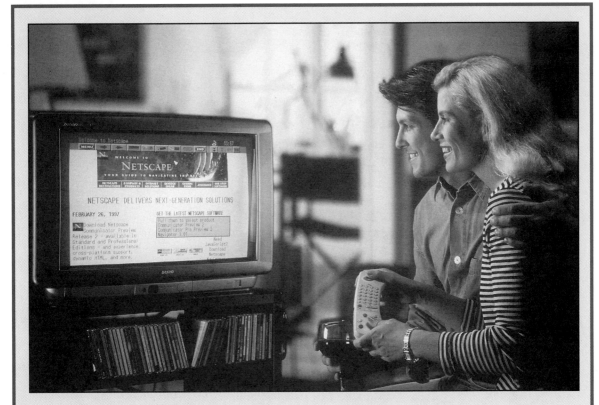

Advancing technologies increasingly blur the traditional distinctions between the TV set and the personal computer.

Source: © Mauritius/Index Stock Imagery.

intercast web pages and interactive data related to the program being broadcast. These developments reflect the evolutionary convergence of the TV receiver and the computer into a single device that can perform the functions of both (see Exhibit 4.p).

• •
4.13 CABLE SYSTEMS

As discussed in preceding sections, TV broadcasting suffers from spectrum crowding and interference inherent in any over-the-air communication. Cable systems avoid these problems by sending signals through the artificial, enclosed environment of *coaxial* and *fiber-optic cable* (see Exhibit 5.a, p. 126), from which the service derives its name.

Cable Advantages
• • • • • • • • • • • •

Properly installed and maintained, cable is a closed system preventing cable signals from radiating beyond the cable and external signals from over-the-air services from getting in. This signal isolation means that cable can use many frequencies denied to broadcasting and other over-the-air services that would otherwise cause interference (see Exhibit 4.q). This extremely wide range of available frequencies allows cable systems to offer viewers many channels, compared with an over-the-air station's single

EXHIBIT 4.q
· · · · · · · · · · · ·

Cable Spectrum Architecture

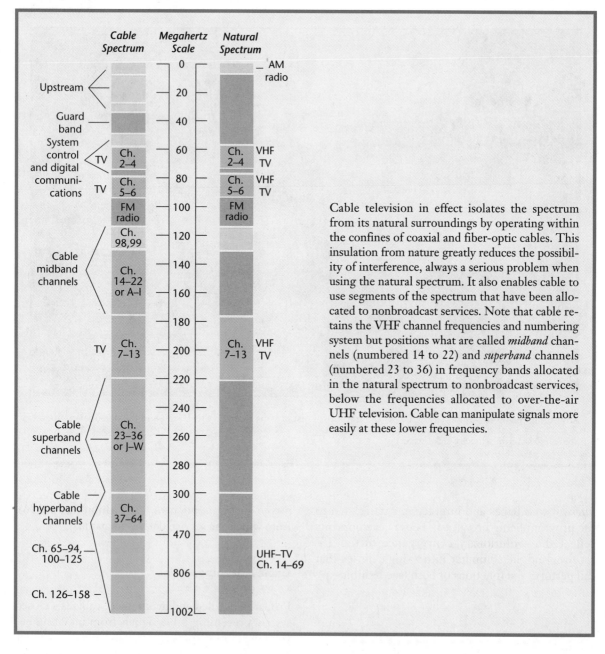

Cable television in effect isolates the spectrum from its natural surroundings by operating within the confines of coaxial and fiber-optic cables. This insulation from nature greatly reduces the possibility of interference, always a serious problem when using the natural spectrum. It also enables cable to use segments of the spectrum that have been allocated to nonbroadcast services. Note that cable retains the VHF channel frequencies and numbering system but positions what are called *midband* channels (numbered 14 to 22) and *superband* channels (numbered 23 to 36) in frequency bands allocated in the natural spectrum to nonbroadcast services, below the frequencies allocated to over-the-air UHF television. Cable can manipulate signals more easily at these lower frequencies.

channel in the same geographical area. Greater use of fiber-optic cable is increasing channel capacity even more, as discussed in Section 5.1 (p. 126) and Exhibit 5.b, p. 127.

Signals conveyed by a cable system have the same characteristics as broadcast signals but do not suffer from ghosting, fading, co-channel and adjacent-channel interference, or other problems caused by over-the-air propagation. Keeping cable a closed system requires constant vigilance, because damage by weather and rodents or inept in-house wiring by subscribers can compromise the system's integrity. Cable operators spend large sums annually to maintain their systems.

Another increasingly important advantage of cable is its physical connection to the home, which offers the possibility of two-way data flow. Though initially designed as a one-way delivery system for TV programming, cable technology inherently provides an easier transition to two-way communication than does broadcasting, DBS, or wireless cable. Cable can offer interactive services, including telephony and Internet access, over a single wire. Telephone and electrical power companies also have a physical (hard-wired) pipeline into the home and share cable's inherent two-way connectivity. But cable's pipeline currently enjoys a significant bandwidth advantage that allows it to offer more services simultaneously. Cable rebuilds during the 1990s have focused on increasing both bandwidth and two-way delivery capabilities by adding more fiber-optic cable to the system and moving fiber nodes closer to the individual subscriber (see Exhibit 4.s, p. 122).

Cable Drawbacks
• • • • • • • • • • •

Cable sacrifices the unique asset of radio communication—the ability to reach audiences without the aid of physical connections. Cable systems are also expensive to build—and to rebuild later in order to increase channel capacity.

Signals traveling through "wave guides" such as coaxial cable—other than those using optical fibers—attenuate rapidly, making it necessary to reamplify signals at frequent intervals, thus increasing costs. Where possible, firms mount cables on existing utility poles. But within cities, cables are often buried underground in conduits and tunnels at even greater expense.

Cable System Design
• • • • • • • • • • • • • •

At the system's *headend*, the system operator aggregates programs from various sources and delivers them via cable to subscriber homes. Besides reception facilities, a headend contains equipment for reprocessing incoming signals, equalizing and feeding them to a modulator for transmission over the system's delivery network, and assigning each program source to a specific cable channel.

Traditionally cable systems have used a *tree-and-branch pattern*. As Exhibit 4.r shows, *trunk cables* branch to lighter *feeder cables* that carry the signals to neighborhoods of homes, where still lighter *drop cables* connect the system to individual households.

This configuration allows a headend to feed programs over a radius of about five miles; covering wider areas requires subsidiary headends that receive the programs via special *supertrunk coaxial* (or fiber-optic) cables or microwave relays. Newer systems adopt a star pattern with a series of local hubs. This restructuring is especially useful in installing fiber trunk lines.

A higher performance system would bring fiber "to the curb" of each subscriber, reducing the amount of copper coaxial cable between the headend and the subscriber. Such a system, however, would be more costly, requiring a fiber-to-coaxial conversion device for each subscriber. Cable system operators must weigh the performance advantages of placing fiber nodes close to subscribers where relatively few homes connect to each node through a coaxial feeder line against the added cost of building additional fiber nodes. When cable operators started installing fiber-optic cable in the early 1990s, each node served about 10,000 homes. Today the typical node serves about 500 homes and provides dramatically improved bandwidth performance. It is likely that the number of homes per node will continue to drop, and some cable operators may begin exploring the possibility of eventually bringing fiber into the home, in spite of the enormous cost.

As suggested in Exhibit 4.r, programs for cable systems come to the headend from five sources: over-the-air reception of nearby TV stations, relays of more distant stations, locally produced or procured material, satellite-to-cable networks, and special audio and data services.

EXHIBIT 4.r
. . . .

Cable System Plan

The cable system headend contains its amplifiers and sometimes studios for locally originated programming, although such facilities may be separately located. The headend receives off-the-air TV station signals picked up by special antennas, and possibly also signals from more distant stations fed by microwave relay. The most important antennas are usually several TV receive-only (TVRO) antennas for picking up satellite signals relayed from a variety of program sources. Trunk and feeder distribution cables, shown mounted on poles in the sketch, are often run underground within urban areas. Newer cable systems employ a hub and star wiring configuration rather than this tree-and-branch pattern typical of older analog systems.

Tuning Cable Channels
. . . .

Cable television uses the same radio-frequency energy as do over-the-air transmitters. Exhibit 4.q shows the broad band of frequencies fed through a cable system. Early systems carrying only a few channels relied on the VHF tuners in subscribers' TV sets. Modern systems with more channels supply customers with an adapter unit, or *converter box*, that has its own tuning facility. It feeds into a receiver channel (usually 2, 3, or 4, whichever is *not* an over-the-air channel locally).

However, virtually all conventional TV receivers now come "cable ready" or "cable compatible," meaning that they can receive all analog channels offered on a cable system, including those used by cable but not by over-the-air broadcasters, without the need for a converter box.

Systems offering many channels typically have several encrypted (scrambled) channels that require set-top converters even with cable-ready receivers. An *addressable converter box* is necessary to receive pay-per-view (PPV) programming. A cable subscriber orders a PPV program, and the cable head-end sends a special signal allowing that subscriber's converter box to pass the program.

Smart Boxes and Interactive Services

The cable set-top box (STB) is evolving rapidly into a powerful digital computer capable of performing a multiplicity of functions. Some people see the STB as the focal point of a future home network connecting a wide variety of digital and analog communication devices and services. Some consumers may want STBs that do little more than what today's converter boxes do, but many will want sophisticated STBs with huge hard drives and software that allow them to enjoy many new features. A primary function will be providing access to the growing number of cable digital networks. The STBs of the future will offer near-video-on-demand (NVOD) and, ultimately, they will provide video-on-demand (VOD) so viewers can watch what they want when they want. Many viewers will enjoy STBs that allow them to pause a broadcast program in progress while they answer the phone, then resume viewing at a higher speed until they catch up with the program (zipping through or skipping commercials in the process). Or they may want to go back and review something they have just seen, perhaps magnifying the image to get a better view, without worrying about missing parts of the program. They may want the STB to automatically record programs of a particular type or featuring a particular person. They also may want an STB that serves as the connection between the cable system and telephone wiring within the home as more cable operators offer telephony. Internet users will want an STB that includes a modem allowing access via a high-speed cable system connection. Connection to the Internet will allow interactive software developers to download and upgrade software stored in the STB at any time, without requiring action by the consumer. Software upgrades can also be delivered during the VBI of analog programming or as part of the data stream during digital programming. These software upgrades will allow consumers to enjoy additional or improved features without having to buy expensive new boxes.

The interactivity provided by smart STBs will, some say, expand the dramatic late-1990s boom in e-commerce to the 80 percent of consumers receiving TV programming by cable and satellite. As we move into the new century, there is a battle developing among vendors to capture a part of the valuable real estate atop the TV set that will be the connection to the "electronic wallet." It is clear there cannot be a separate box for every feature available. The bundling of products into a single box is under way. In 1999 Microsoft announced plans to include its WebTV software in about 10 million cable STBs marketed during 2000. Other companies with existing interactive products, such as TiVo and ReplayTV, are also likely to bundle their products with the advanced digital STBs used for cable (and DBS) service. Wink Communications' interactive service, which displays text and simple graphics over TV programs, was supported by about 9 million advanced analog cable STBs in early 1999. Migration to digital cable (and DBS) STBs is clearly the path envisioned for the deployment of Wink's product. The number of available features and services is expected to increase as new generations of powerful STBs become available in retail stores, rather than being distributed solely through cable system operators.

Progress

The government and the cable industry pledged to develop open specifications that ensure that the new breed of STBs can connect to and work with other electronic devices and services. A major step toward those goals occurred in 1998 when Cable Television Laboratories (CableLabs) and the Consumer Electronics Manufacturers Association (CEMA) selected IEEE 1394 (popularly called Firewire) as the digital interface standard. Adoption of this standard ensures connectivity between new DTV receivers and STBs, as well as other

EXHIBIT 4.s

A Hybrid Fiber-Coaxial (HFC) Network

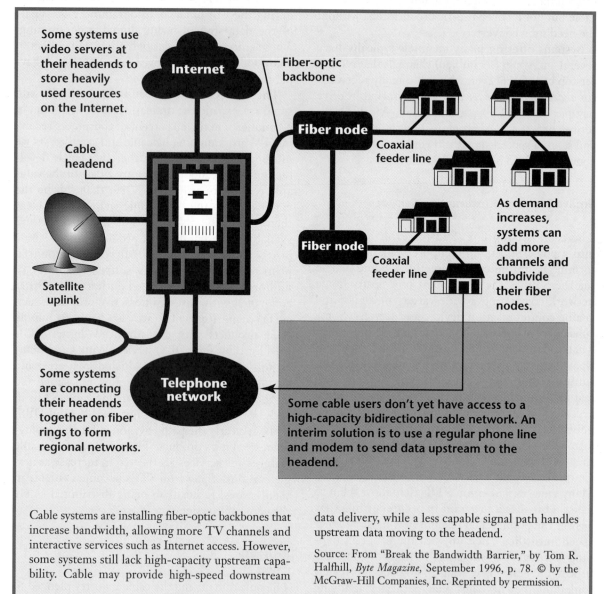

Some systems use video servers at their headends to store heavily used resources on the Internet.

Internet

Fiber-optic backbone

Cable headend

Fiber node

Coaxial feeder line

Fiber node

Coaxial feeder line

Satellite uplink

As demand increases, systems can add more channels and subdivide their fiber nodes.

Some systems are connecting their headends together on fiber rings to form regional networks.

Telephone network

Some cable users don't yet have access to a high-capacity bidirectional cable network. An interim solution is to use a regular phone line and modem to send data upstream to the headend.

Cable systems are installing fiber-optic backbones that increase bandwidth, allowing more TV channels and interactive services such as Internet access. However, some systems still lack high-capacity upstream capability. Cable may provide high-speed downstream data delivery, while a less capable signal path handles upstream data moving to the headend.

Source: From "Break the Bandwidth Barrier," by Tom R. Halfhill, *Byte Magazine*, September 1996, p. 78. © by the McGraw-Hill Companies, Inc. Reprinted by permission.

devices such as DVD players, digital camcorders, digital VCRs, and computers. Progress stalled, however, when the motion-picture industry and consumer product manufacturers disagreed over the technology that protects copyright holders against pirating of their products. Movie producers want total control over digital duplication, while consumer product manufacturers, who sell digital recording devices, want consumers to continue to have digital recording rights for purposes such as time-shifting programming. In 1999 the National Cable Television Association (NCTA) announced that the cable industry would begin marketing STBs late in 2000 with a copy protection technology called 5-C. The NCTA's action did not resolve the copyright protection conflict, but it did help advance progress toward the development of a copy protection technology standard for STBs.

Barriers to Adoption
• • • • • • • • • • • • • •

Delivery of the services promised by sophisticated STBs and other terminal devices in the home depend to a degree on the ability of the cable system to provide a two-way path between the cable headend and the consumer. Most interactive products currently available to consumers, such as WebTV, Wink, and TiVo, do not depend on cable for the upstream path between the consumer and the product provider. Note in Exhibit 4q that cable systems traditionally have devoted only a small portion of the cable spectrum (from about 5 to 40 MHz) to upstream signals flowing from the subscriber to the headend. Cable operators are upgrading their physical plants rapidly with fiber-optic cable to accommodate more video channels and new interactive services. But until higher-speed two-way capability becomes widely available, cable will provide, as does DBS, an alternative upstream communication channel, as illustrated in Exhibit 4.s. The alternative upstream channel is typically a conventional telephone line and modem with limited bandwidth, but the asymmetrical arrangement works in many interactive situations in which the amount of data moving downstream far exceeds that going upstream. Upstream data demand will increase as more people begin using interactive services, but it seems unlikely that upstream traffic will soon match that going downstream. The increased use of fiber and new system configurations has increased system capacity, as have advances in compression technology and STBs. But implementation of a flexible, efficient management plan for ensuring adequate upstream bandwidth, while the number of video channels and downstream data traffic expands simultaneously, remains a challenge for system operators.

Another challenge for cable operators is educating consumers about the new services. Although many people are excited about adding new interactive features to the television, the number of contented couch potatoes is huge. Operators are also learning that it is not enough simply to put a new STB in the subscriber's home, hand over the remote control, and walk away. Users will need considerable help in mastering the new technology, or the technology will need to be extremely user-friendly, before the wondrous world of interactivity becomes synonymous with the typical consumer's viewing experience.

CHAPTER 5

Relays, Recording, and the Digital Revolution

The most dramatic recent changes in electronic media technology have come in how programs are stored and moved from place to place. A useful way to relate the many technologies involved is to examine three basic functions:

- *relay* or distribution of programs from one or more central sending points to many receiving points (stations or systems)
- *recording* or storage of programs for repeated and shared use by networks and stations or systems
- *delivery* of programs to the home (by broadcast station, cable system, wireless cable, satellite, or the Internet), as described here and in Chapter 4.

5.1   TERRESTRIAL RELAYS

Electronic means of relaying media signals include those that use physical interconnections (wire and coaxial/fiber-optic cable) and those that operate over the air (satellite and microwave relay systems).

Wire Relays

Any point-to-point or point-to-multipoint linkage, whether by wire or radio, can function as a relay. A relay's channel bandwidth determines how much information the channel can handle in a given instant, which in turn determines which types of material the relay can distribute most efficiently.

Early broadcasters tried using telegraph lines in the 1920s when the telephone company refused to lease lines for networking (see Section 2.5, page 30). But telegraph lines lacked the needed bandwidth. Voice required lines designed for that purpose. In turn, lines designed for voice were inadequate for the demands of television's full-motion real-time video signal. Internet users face a similar bandwidth limitation when they use conventional modems and the "twisted pair" of the *plain old telephone system* (POTS) to send and receive video. Television requires a special form of transmission line known as *coaxial cable*, illustrated in Exhibit 5.a. Coaxial cable's outer conductor provides a barrier that both confines the signal within the cable and protects it against external electromagnetic interference. Attenuation still occurs, but repeater amplifiers inserted at intervals compensate for the loss. Coaxial

EXHIBIT 5.a

Coaxial Cable

The name *coaxial cable* derives from its two conductors with a common axis: a solid central conductor surrounded by a tubelike hollow conductor. Radio energy travels within the protected environment of the space between the two conductors. Cable television relies on this type of conductor, as do many terrestrial relay links that convey television signals, telephone calls, data, and other types of information. Fiber-optic cable is increasingly taking over trunk communications for these industries, as discussed in Exhibit 5.b.

Source: Adapted from Walter S. Baer, *Cable Television: A Handbook for Decisionmaking*, R-1133-NSF, Santa Monica, CA: RAND, 1973, p. 4. Used by permission.

cable's broad bandwidth is useful not only for television but also whenever it is desirable to relay lots of data from point to point rapidly.

Fiber-Optic Cable

The high capacity of fiber-optic conductors makes them ideal for handling digitally processed information.

In fiber-optic cable, hair-thin strands of extremely pure glass convey modulated light beams with a bandwidth in the thousands of megahertz (see Exhibit 5.b). A single glass filament has more

EXHIBIT 5.b
. . . .
Fiber-Optic Relays

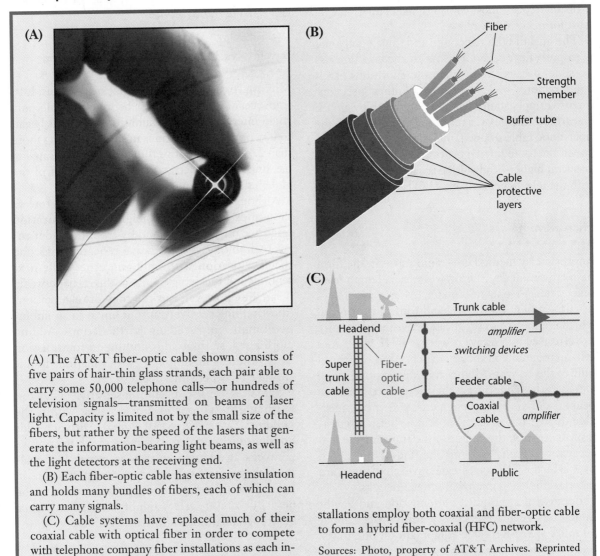

(A) The AT&T fiber-optic cable shown consists of five pairs of hair-thin glass strands, each pair able to carry some 50,000 telephone calls—or hundreds of television signals—transmitted on beams of laser light. Capacity is limited not by the small size of the fibers, but rather by the speed of the lasers that generate the information-bearing light beams, as well as the light detectors at the receiving end.

(B) Each fiber-optic cable has extensive insulation and holds many bundles of fibers, each of which can carry many signals.

(C) Cable systems have replaced much of their coaxial cable with optical fiber in order to compete with telephone company fiber installations as each industry provides a growing menu of broadband services to individual households. Most contemporary installations employ both coaxial and fiber-optic cable to form a hybrid fiber-coaxial (HFC) network.

Sources: Photo, property of AT&T Archives. Reprinted with permission of AT&T. Cable fiber diagram from National Cable Television Association.

than 600 times the information-carrying capacity of a coaxial cable.

Ordinary light will not travel efficiently through an optical fiber. Instead, *lasers* (light amplification by stimulated emission of radiation) or *light-emitting diodes* (LEDs) must be used to generate light consisting of a very narrow band of frequencies.

Fiber-optic cables have many advantages as relay links, especially for very heavy traffic. Little loss occurs from attenuation, reducing the number of repeater amplifiers required. Cables are small and lightweight. They neither radiate energy to interfere with other circuits nor receive interference from the outside.

Fiber-optic cables have been permanently installed on heavy-traffic telephone routes and are increasingly used for the main distribution lines of cable systems. Eventually, fiber-optic cables will probably replace most conventional copper wire, at least for major trunk relays.

The goal of both telephone and cable companies is to push fiber relays as close to the consumer as possible, because fiber offers the enormous bandwidth required to deliver video and other spectrum-intensive applications. In this contest, cable enjoyed an early advantage over telephone because, once the fiber-optic cable stopped, the critical "last mile" between the cable service provider and the consumer was with higher-bandwidth coaxial cable rather than with the telephone company's POTS "twisted pair."

Microwave Relays

Another relay method uses radio waves in the form of microwave links. Microwaves vary in length from one meter to one millimeter. Microwave relays usually employ the UHF band (only a portion of which is allocated to broadcast television channels).

Terrestrial microwave relays use UHF waves that attenuate rapidly in the atmosphere and can travel only to the horizon. Higher transmitter powers and antennas that concentrate the energy into a narrow beam (see Exhibit 5.c) compensate to a considerable degree for losses from atmospheric attenuation. Increasing the height of the transmitting and receiving antennas extends the line-of-sight path over which UHF waves can travel, but there are practical limits on microwave tower heights.

Because of atmospheric attenuation and height limitations, relay towers are typically spaced about 30 miles apart. Repeater equipment on one tower receives transmissions from the previous tower, amplifies the signals, and retransmits them to the next tower in the series. It takes more than a hundred towers to span the continental United States from coast to coast.

5.2   SATELLITE RELAYS

Like coaxial cable, satellites first served a relay function and only later began to deliver programs directly to consumers. Most satellite traffic still consists of relays, such as network programs relayed to affiliates and on-the-spot news events relayed to studios. But improvement in satellite technology now also accommodates another hybrid—direct broadcast satellite (DBS) services.

Satellite vs. Microwave Relay Distribution

Microwave relay networks cannot span oceans. Live transoceanic television first became possible only when international communication satellites began to function as relay stations in space. Orbiting more than 22,000 miles above the Earth, a single satellite has line-of-sight access to some 40 percent of the globe's surface, as suggested by Exhibit 5.d.

Although often likened to microwave towers thousands of miles in height, communication satellites differ fundamentally from the older relay technology. A microwave repeater links one specific location with only two others: the next sending and receiving points in the relay network. A satellite, however, links a group of relay stations (the satellite's receive/transmit units) to an unlimited number of receiving Earth stations. Adding more Earth stations adds nothing to transmission costs, whereas linking up new destinations in microwave relay networks does.

Satellites are also distance insensitive. They can reach Earth stations at any distance within the satellite's *footprint* (coverage area). Distance adds no transmission expense, as it does with microwave networks. In addition, microwave signals lose quality as they go through scores of reamplifications in being passed on from one repeater station to the next. But satellite relays amplify a signal only once before sending it down to Earth stations.

Geostationary Orbit

Early satellites moved across the sky like the sun and moon, requiring huge receiving antennas with costly tracking mechanisms to keep them pointing toward the moving signal source. But today's satellites can be positioned so that they remain stationary with respect to their target area on the Earth below. Once a receiving antenna has been adjusted to point in the right direction, it needs no further attention.

Satellites that appear to stay in one location above the Earth operate in *geostationary* (or *geosynchronous*)

EXHIBIT 5.c
• • • • • • • • •

Microwave Relay Antennas

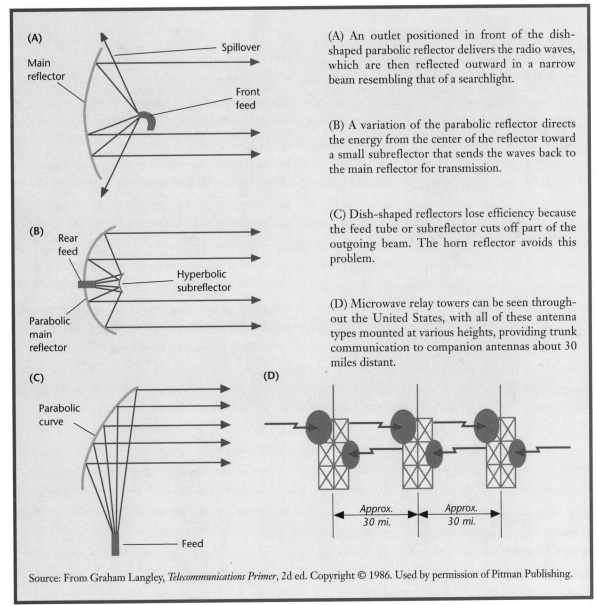

(A)
Main reflector
Spillover
Front feed

(B)
Rear feed
Hyperbolic subreflector
Parabolic main reflector

(C)
Parabolic curve
Feed

(D)
Approx. 30 mi.   Approx. 30 mi.

(A) An outlet positioned in front of the dish-shaped parabolic reflector delivers the radio waves, which are then reflected outward in a narrow beam resembling that of a searchlight.

(B) A variation of the parabolic reflector directs the energy from the center of the reflector toward a small subreflector that sends the waves back to the main reflector for transmission.

(C) Dish-shaped reflectors lose efficiency because the feed tube or subreflector cuts off part of the outgoing beam. The horn reflector avoids this problem.

(D) Microwave relay towers can be seen throughout the United States, with all of these antenna types mounted at various heights, providing trunk communication to companion antennas about 30 miles distant.

Source: From Graham Langley, *Telecommunications Primer*, 2d ed. Copyright © 1986. Used by permission of Pitman Publishing.

orbit—an orbital position directly above the equator at a height of about 22,300 miles. At that height, objects revolve around the Earth at the same rate that the Earth turns on its axis. Moreover, the centrifugal force tending to push a satellite outward into space cancels the gravitational force tending to pull it back to Earth, keeping it suspended in space for years.

The geostationary orbit consists of an imaginary circle in space. Satellites in that enormous orbit actually move through space at about 7,000 miles per hour. From the perspective of an observer on Earth, however, they seem to stay in one place, keeping in step with the Earth as it rotates. In practice, geosynchronous satellites tend to drift out of

EXHIBIT 5.d
· · · · · · · · · · ·

Geostationary Satellite Orbit

(A) Looking down from space over the North Pole, one can see how three satellites orbiting above the equator can "cover" most of the Earth ("most" because their signals fade at polar latitudes). The satellites appear to remain stationary with reference to Earth when orbited at an altitude of 22,300 miles—the geostationary (or geosynchronous) orbit (GSO). The INTELSAT international satellite system operates satellites over the Atlantic, Pacific, and Indian Oceans.

(B) American *domestic satellites (domsats)* orbit in the GSO above Colombia, each with an assigned position expressed in degrees west from the prime meridian at Greenwich, England. Ku-band, C-band, and DBS-band satellites are shown at different distances from the Earth for clarity, but they actually occupy the same orbit. Linkages between some satellites indicate they are *hybrids* carrying both Ku- and C-band transponders. Although some satellites occupy the same assigned positions, there is room for several "birds" without fear of collision. Satellites age and sometimes fail unexpectedly in midlife, as did *Telstar 401* on January 11, 1997, and must be replaced. New satellites are also placed in orbit at irregular intervals, and sometimes moved from one location to another. This diagram, therefore, is illustrative rather than definitive, and Ka-band satellites are not included.

Source: Based on diagram in *Broadcasting & Cable* (13 March 1995: 36–37); additional data provided by the American Distance Education Consortium.

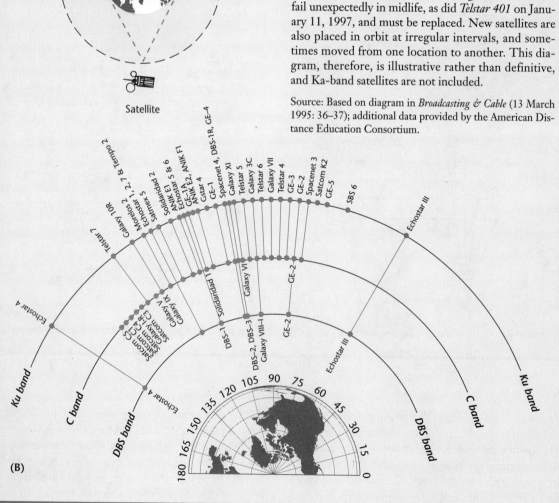

position, but ground controllers can activate small on-board thrusters to nudge a satellite back to its assigned orbital slot.

Through the International Telecommunication Union (ITU—see Section 14.3, page 411), the nations of the world have allotted each country one or more specific slots in the geosynchronous orbit for domestic satellites as illustrated in Exhibit 5.d. The ITU identifies positions in degrees of longitude, east or west of zero longitude, called the prime meridian. Zero longitude arbitrarily runs through Greenwich, England.

Allotments in the segments of the orbit suitable for "looking down" on areas of high traffic density, such as continental North America, are in high demand. This demand has created a potential slot scarcity. This problem is especially critical for those wanting occasional, short-term use of a relay satellite, because much of the total capacity available for domestic U.S. service is tied up by full-time users. Video compression (described in Section 5.5) provides a partial solution to this dilemma. The increased availability of terrestrial fiber-optic networks is also helping to reduce the capacity crunch, provided the need is primarily for point-to-point rather than point-to-multipoint relays.

Spectrum Allocations
• • • • • • • • • • • • • •

Like Earth-based transmitters, satellite transmitters need internationally allocated transmission channels. Satellites used in broadcasting transmit on microwave frequencies in the 3 to 7 GHz region (C band) and the 11 to 15 GHz region (Ku band). Most communication satellites use the C band. More powerful satellites, intended primarily for direct reception by small home antennas, use the higher-frequency Ku band.

In 1996 the FCC announced its plan for licensing a variety of communication services in the immense Ka band (17–30 GHz). The plan set aside frequencies for those wanting to offer terrestrial as well as satellite-based services. This additional band is reducing pressure on the increasingly cluttered C and Ku bands.

In areas of heavy terrestrial microwave usage, ground-based services often interfere with Earth stations receiving C-band signals. Ku- and Ka-band signals escape this drawback. But Ku- and Ka-band waves are so short that raindrops in heavy downpours can interfere with their propagation.

Each satellite needs two groups of frequencies, one for *uplinking* (on-board reception) and one for *downlinking* (on-board transmission). These frequency groups must be far enough apart in the spectrum to prevent interference between uplink and downlink signals. Thus satellite frequency allocations come in pairs—4/6 GHz, 12/14 GHz, and so on—with the lower frequencies used for downlinking.

The downlink frequency bands must be large enough to accommodate a number of different channels for simultaneous transmission by the satellite's *transponders*—combination receive/transmit units. Most relay satellites carry 24 transponders. Each can transmit 2 television channels (more with video compression), for a capacity of 48 television channels per satellite (or many more narrowband channels, such as those for telephone or radio transmissions).

Ku-band satellites usually carry fewer transponders, because their higher power means more weight per transponder. Some satellites combine both C- and Ku-band transponders for maximum flexibility.

Transmission/Reception
• • • • • • • • • • • • • •

Satellite transmitting antennas focus their output into beams to create footprints of varying size. Exhibit 5.e shows an example. The narrower the beam, the stronger the signal within the footprint, because directionality causes signal gain.

A satellite downlink beam is strongest at its center, growing progressively weaker at reception points farther out. Earth stations located near the margins of footprints therefore need larger-diameter antennas than those closer to the center. Besides being on the margin of the continental footprint, stations near the poles have an additional problem using geostationary satellites. The receiving antenna of an Earth station must be pointed toward the satellite in orbit over the equator. For stations near the poles, this requires that the antenna be pointed almost parallel to the Earth's surface. This, in turn, increases the likelihood that the Earth station will pick up interfering signals from land-based transmitters. Special spot beams are sometimes used to help boost signals in these areas.

EXHIBIT 5.e
. . . .

Satellite Footprints

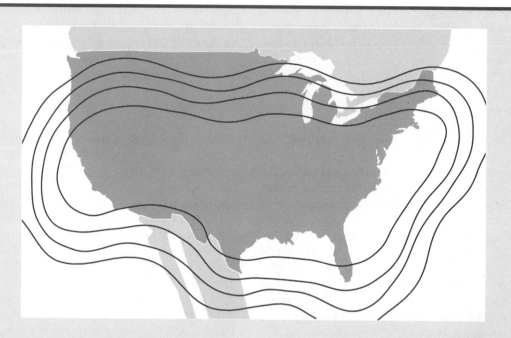

A satellite's effective coverage area, known as its *foot-print*, appears on this map in terms of levels of signal strength. The inner contour defines a satellite's *bore-sight*, its area of maximum signal strength allowing use of smaller receiving antennas. Outlying areas beyond the borders of the country receive only *beamedge* power, beyond which satisfactory reception cannot be expected even with large receiving antennas. This map shows the footprint of the same Satcom K2 satellite diagrammed in Exhibit 5.g, located at 81 degrees west longitude. The satellite actually orbits far to the south of the United States, a position above the equator near the Galápagos Islands. Not shown are smaller regional beams that can be directed toward Hawaii and Alaska to serve antennas in those areas.

Source: Mark Long, ed., *World Satellite Almanac: The Global Guide to Satellite Transmission and Technology*, 3rd ed. © 1993 Mark Long Enterprises, Inc., Ft. Lauderdale, FL. Used by permission.

The diameter of Earth antennas varies from more than 100 feet to less than 1 foot. The variously shaped antennas of receive-only satellite Earth stations, such as those shown in Exhibit 5.f, have become familiar sights. Round or square antennas about 18 inches in diameter have become the standard for direct broadcast reception in homes.

The signals captured by a satellite receiving antenna are extremely weak. They therefore need beefing up by a special high-quality amplifier—a *lownoise amplifier* (LNA). LNAs magnify the incoming signal strength by as much as a million times. Even this enormous amplification is inadequate if the antenna's efficiency is decreased by snow or ice collecting in the dish. Manufacturers design dishes to minimize the effects of ice and snow, sometimes including built-in heating elements, but Earth station dishes located in the nation's Snowbelt sometimes must be swept out. The LNA feeds the satellite signal to a *down-converter*, which translates the high satellite frequencies into the lower-frequency range that television receivers use. Some systems use a *lownoise block converter* (LNB), which combines the functions of amplification and frequency conversion in a single device.

EXHIBIT 5.f
. . . .

TVRO Earth Stations

Each of the relatively expensive *television receive-only* (TVRO) Earth stations (antennas) shown here concentrates the weak satellite signals into a narrow beam directed at a small second reflector mounted on the tripod. This secondary reflector beams the signal into a horn at the center of the TVRO dish, from which it is fed, still as a very weak signal, to a lownoise amplifier (LNA) or lownoise block converter (LNB).

Source: Photo courtesy Christopher H. Sterling.

Satellite Construction
. . . .

Communication satellites need five essential groups of hardware components (see Exhibit 5.g):

- *transponders*, the receive/transmit units that pick up programs, amplify them, and transmit them back to Earth
- *antennas* for receiving uplink signals and transmitting downlink signals (both program material and telemetering information)
- *power supplies*, consisting of arrays of solar cells and storage batteries
- *telemetering devices* for reporting the satellite's vital signs to, and receiving instructions from, the ground controllers
- *small thrusters* for moving the satellite, orienting it, and holding it in its assigned position, activated on command from ground controllers

Orientation is vital to a satellite because its antennas must always point in the target direction and its arrays of solar collectors, located on the satellite's body or on extended wings, must be positioned to receive direct rays from the sun. These solar collectors provide electricity to operate the satellite. They also charge on-board batteries that take over during periods when the Earth's shadow interrupts sunlight.

Satellites operate at low power relative to terrestrial relays. Power per transponder varies from about 5 watts to 240 watts (the higher power for Ku-band satellites designed for direct-broadcast reception). Most satellite transmitters use no more wattage than ordinary electric light bulbs, though their power is focused by directional antennas.

It seems paradoxical that, despite atmospheric absorption, satellites send signals such great distances with so little power. However, for most of their 22,300-mile journey, satellite signals travel through the near vacuum of space. When at last they encounter the Earth's relatively thin atmospheric envelope, they pass almost straight down through it, experiencing little attenuation. Terrestrial radio signals, in contrast, travel nearly parallel to the Earth, impeded by atmospheric absorption along their entire route.

Satellite Launching
. . . .

Most commercial U.S. communication satellites have been launched from Cape Canaveral, Florida. Launching involves two phases: A powerful rocket vehicle overcomes the initial drag of gravity and air resistance, carrying the satellite into low orbit. After being released from the rocket, the satellite's own, less powerful on-board rockets propel it into the high, geostationary orbit. Exhibit 5.h illustrates the sequence of events.

Most satellite launches are successful, but in the late 1990s the satellite industry experienced a string of failures. The most spectacular was in August 1998, when a Boeing rocket carrying *Galaxy X* exploded in the summer sky less than two minutes after its launch. Nine months later another Boeing rocket failed to reach proper orbit, making its satellite payload useless. These and several similar launch disasters caused a number of delays in planned launches.

EXHIBIT 5.g
• • • • • • • • • •

Satellite Components

(A) The largest part of most communication satellites is an array of solar panels (which power on-board batteries)—in this case two huge wings on either side of the electronic core of the satellite.

(B) The major parts of a satellite as defined in the text: transponders, antennas, power supplies (batteries and the "wing" solar panels, shown here folded up for launch), and thrusters for minor orbit adjustments.

General Electric built the Satcom K2 satellite, and the National Aeronautics and Space Administration (NASA) launched it into orbit. Launched in the mid-1980s, Satcom K2 used Ku-band frequencies for—among other services—distribution of NBC television signals to the network's more than 200 affiliates, satellite news gathering (SNG) transmissions, and Conus news cooperative delivery transmissions to some 120 affiliated stations in the United States and abroad. A number of nonbroadcast services also used its transponders.

Source: Mark Long, ed., *World Satellite Almanac: The Global Guide to Satellite Transmission and Technology*, 3rd ed. © 1993 Mark Long Enterprises, Inc., Ft. Lauderdale, FL. Used by permission.

EXHIBIT 5.h

Satellite Launches

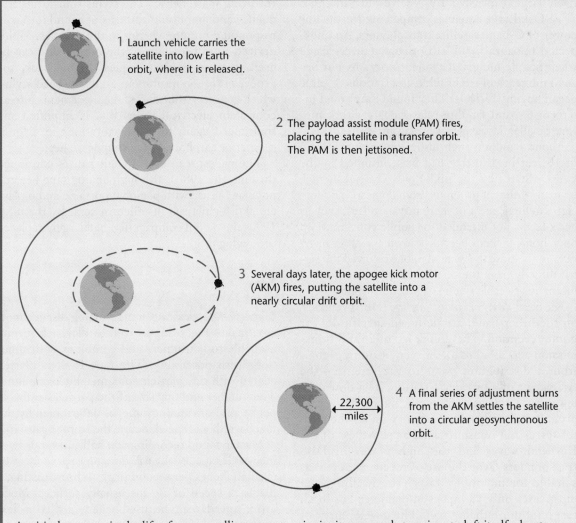

1 Launch vehicle carries the satellite into low Earth orbit, where it is released.

2 The payload assist module (PAM) fires, placing the satellite in a transfer orbit. The PAM is then jettisoned.

3 Several days later, the apogee kick motor (AKM) fires, putting the satellite into a nearly circular drift orbit.

22,300 miles

4 A final series of adjustment burns from the AKM settles the satellite into a circular geosynchronous orbit.

A critical moment in the life of every satellite occurs when it first leaves the Earth. An enormous, unmanned launch rocket is needed to (a) overcome gravity and atmospheric resistance, (b) attain escape velocity, and (c) place the satellite in an initial low orbit. The fragile satellite starts its journey as a mere passenger, protected within a nose cone. Only after it reaches an altitude of about 200 miles does a satellite ignite its own rocket engines to loft itself—by stages as shown in the diagram—into its 22,300-mile-high geosynchronous orbit.

Source: Mark Long, ed., *World Satellite Almanac: The Global Guide to Satellite Transmission and Technology*, 3rd ed. © 1993 Mark Long Enterprises, Inc., Ft. Lauderdale, FL. Used by permission.

Direct Broadcast Satellite Service

Cable television systems, broadcast stations, and other satellite-relay users can afford relatively large, expensive TV receive-only antennas (TVROs). Large antennas compensate for the low power of C-band satellite transponders. An unexpected bonanza for TVRO manufacturers came when people hungry for video programs but beyond the reach of either television stations or cable began buying TVROs. They found they could intercept C-band satellite-relay signals with somewhat smaller, less expensive backyard dishes.

About 4 million such dishes on the order of 6 to 10 feet in diameter had been installed in the United States by the mid-1990s. They can pick up many different programs from domestic and foreign satellites, most of them private relays, such as news feeds, not intended for public consumption. Such home pickups became known as *C-band direct reception* because the general public received the signal directly from the satellite rather than from an intermediary cable system or broadcast station.

Ku-band *direct broadcast satellite* (DBS) vehicles, on the other hand, are designed specifically for home reception. The uplink leg of a DBS transmission acts as a broadcast relay, sending signals from a distribution point to the satellite—the equivalent of a broadcast station. The downlink leg in turn acts as a broadcast signal, delivering programs directly to consumers without the need for terrestrial transmitters. The shortness of the Ku-band waves and the high power of DBS transponders favor the use of receiving antennas suitable for mounting on private dwellings—in most cases only 18 inches in diameter but in all cases not more than 3 feet. The development of DBS systems in the United States is discussed in Section 3.3 (page 66).

5.3   ANALOG SOUND RECORDING

We turn now from relays to recording technologies, beginning with the earliest to emerge—the art of sound recording. We deal first with *analog* recording and move next to digital.

Discs

In analog disc recording, a sound source causes a *stylus* to vibrate as it cuts a concentric groove in a revolving master disc. The stylus transforms frequency and amplitude patterns of sound into corresponding minute wiggles in the grooves. Molds derived from pressings of the master disc can be used to mass-produce copies. In playback, the grooves cause vibrations in a pickup-head stylus, which converts (transduces) the movements into an equivalent electrical signal that is amplified and transduced again to cause vibrations in a loudspeaker or earphones—the sounds we hear.

The widespread use of analog discs continued through the 1980s, although analog tape became popular for use with personal portable audio players and for players installed in cars. By the mid-1990s the digital compact disc had largely replaced the analog disc.

Tape

Magnetic-tape recording, unlike analog disc technology, makes recording as well as playback readily available to consumers and combines both functions with portability. The tape itself is plastic, coated with tiny particles of a metallic compound. The number and tiny size of the particles available per second of running time, as determined by the tape's width and speed, define the storage capacity.

Master sound recordings on half-, one-, or two-inch-wide tape usually call for a tape speed of 15 or even 30 inches per second (ips). In broadcasting, a playback speed of 7½ ips usually suffices. Much lower speeds can be used when quality is less crucial, as in office dictation and station output monitoring. Multitrack master recording and other specialized tasks call for tape stock that is wider than the standard quarter-inch.

Audio signals from a microphone or other source cause electrical variations in a recording head over which the tape passes. These variations create patterned arrangements of the metallic particles. During playback the tape passes over another electromagnetic head, where the tape's magnetic patterns create a modulated electric current for delivery to amplifiers and speakers. Running tape over a third electromagnet, an erase head, re-

EXHIBIT 5.i
. . . .

Analog Sound Recording Formats

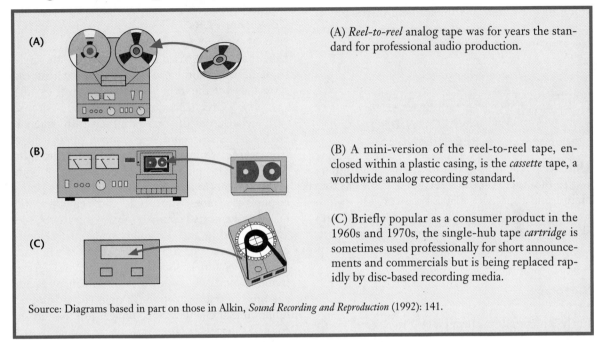

(A) *Reel-to-reel* analog tape was for years the standard for professional audio production.

(B) A mini-version of the reel-to-reel tape, enclosed within a plastic casing, is the *cassette* tape, a worldwide analog recording standard.

(C) Briefly popular as a consumer product in the 1960s and 1970s, the single-hub tape *cartridge* is sometimes used professionally for short announcements and commercials but is being replaced rapidly by disc-based recording media.

Source: Diagrams based in part on those in Alkin, *Sound Recording and Reproduction* (1992): 141.

arranges the particles to neutralize (erase) any stored magnetic patterns so that the same tape can be reused repeatedly.

Originally, all tape recorders—both professional and consumer—had a reel-to-reel configuration, with each reel separate and accessible. Now, however, enclosed cassettes or cartridges protect the tape and are more convenient to use than open reels.

Cassettes
. . . .

A cassette incorporates double hubs, one each for feed and take-up reels, in a single small housing (see Exhibit 5.i). After playing, the cassette must be rewound. Alternatively, in the case of half-width recordings, some equipment reverses the tape direction, or the cassette may be manually flipped over to play a second "side." The cassette lends itself equally to both analog and digital recording methods.

Cartridges
. . . .

A cartridge, often called a *cart*, has a single hub and contains an endless recordable tape loop that repeats itself (see Exhibit 5.i). Carts recognize the next recorded program or commercial on the tape and come to a stop until the operator manually presses the play button again. This self-cuing feature freed radio disc jockeys to spend more time answering the phone and performing other chores. Carts also encouraged automated operation of radio stations. Many carts were loaded into an automated player, with each cart containing a single program item such as a commercial, station ID, or promotional announcement. Musical selections often played off reel-to-reel tape machines built into the automation equipment because they provided audio quality not matched by carts. Inaudible cues on the various tapes instructed the on-board computer when to play a particular cart.

Cart-based automation systems were electro-mechanical monsters with dozens of motors, relays, and other gizmos. It was not unusual to hear two commercials running at the same time when something in the system malfunctioned. In the 1990s, carts rapidly lost ground to smaller, cheaper, and less error-prone tapeless automation systems.

5.4   ANALOG VIDEO RECORDING

Until 1956 the only way to record TV programs was to film them off the screen of a TV set as they aired using a specially adapted camera. These *kine-scope recordings* lost much of television's already skimpy detail. When used for rebroadcast, the filmed programs looked flat and hazy.

Videotape

Early videotape recorders (VTRs) were costly *quadruplex* studio recorders. VTR designers needed to increase greatly the speed at which tape passed over the recording and playback heads to capture the large amount of information contained in pictures-plus-sound. They solved the tape-speed problem by mounting *four* recording heads on a re-volving drum (hence the name *quadruplex*). The drum rapidly rotated *transversely* (across the width of tape) while the tape itself simultaneously moved longitudinally, as it does in sound recording (see Exhibit 5.j). The combined movements of heads and tape produced an effective head-to-tape speed of 1,500 inches per second.

Later, simpler and cheaper professional VTRs using one-inch and smaller tape stock came on the market. They retain the principle of combining head and tape movements but use fewer heads. Instead of laying down the track transversely, the heads cross the tape at an angle, producing a *slanted track*, as shown in Exhibit 5.j. The heads spin on a disc mounted inside a stationary drum, or capstan. The tape wraps around the drum in a spiral (helical) path—hence the name *helical* for slant-track recorders.

For portable equipment, professionals opted first for three-quarter-inch tape (called *U-matic*), but as

the technology improved, half-inch, 8-millimeter (about one-third of an inch), and even quarter-inch videotape evolved.

Consumer VCRs

Like many VTRs used by professionals, those de-signed for the home use tape enclosed within cas-settes—hence the term *videocassette recorder* (VCR).

Home VCRs depend on the user's television set for playback, but they contain their own tuners to enable users to record one channel off the air or from cable while watching a different program. VCRs can also play or record the output of home video cameras (*camcorders*), as well as display rented or purchased tapes. Most VCRs come equipped with many sophisticated computer-assisted features, such as slow motion and the ability to be programmed days in advance to record a sequence of shows on different channels. Some VCRs provide freeze-frame storage "windows" to monitor as many as nine channels at once on-screen, along with a "mosaic" function that changes video images into patterns of colored squares.

5.5   DIGITAL SIGNAL PROCESSING

Satellites rely heavily on digital signal processing. Broadcasting, however, has been based on an older method that creates signals analogous (similar) to original sounds and images. Analog processing imi-tates nature. For example, the continuously rotating hands of a clock imitate the movement of shadows on a sundial. A digital timepiece, however, tells time directly in numbers, jumping from one number to the next. Digital signals bear no resemblance to nat-ural forms because they consist entirely of digits—simple strings of numbers.

Sampling

Digital processing can be likened to cutting up a picture (the analog signal) into thousands of tiny pieces, selecting every other piece, assigning a num-ber to each of those pieces representing its ampli-

EXHIBIT 5.j
• • • • • • • • •

Analog Video Recording Formats

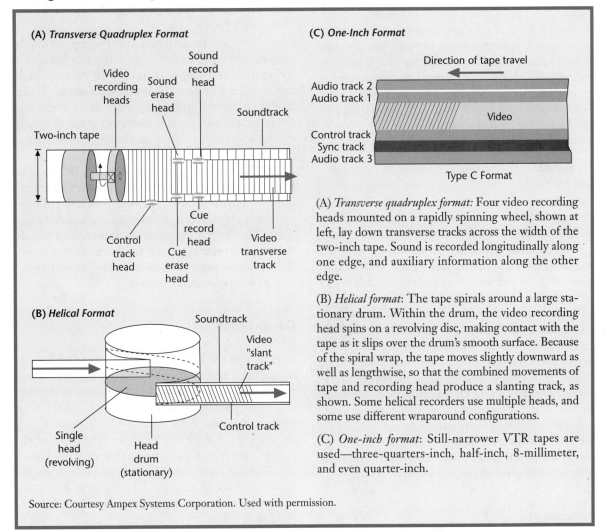

(A) *Transverse Quadruplex Format*

Sound record head

Video recording heads   Sound erase head   Sound record head

Soundtrack

Two-inch tape

Control track head   Cue erase head   Cue record head   Video transverse track

(B) *Helical Format*

Soundtrack

Video "slant track"

Single head (revolving)   Head drum (stationary)   Control track

(C) *One-Inch Format*

Direction of tape travel

Audio track 2
Audio track 1

Video

Control track
Sync track
Audio track 3

Type C Format

(A) *Transverse quadruplex format:* Four video recording heads mounted on a rapidly spinning wheel, shown at left, lay down transverse tracks across the width of the two-inch tape. Sound is recorded longitudinally along one edge, and auxiliary information along the other edge.

(B) *Helical format*: The tape spirals around a large stationary drum. Within the drum, the video recording head spins on a revolving disc, making contact with the tape as it slips over the drum's smooth surface. Because of the spiral wrap, the tape moves slightly downward as well as lengthwise, so that the combined movements of tape and recording head produce a slanting track, as shown. Some helical recorders use multiple heads, and some use different wraparound configurations.

(C) *One-inch format*: Still-narrower VTR tapes are used—three-quarters-inch, half-inch, 8-millimeter, and even quarter-inch.

Source: Courtesy Ampex Systems Corporation. Used with permission.

tude, transmitting the numbers, and then using the numbers to reassemble the picture with every other piece missing. If the pieces are tiny enough, the absence of the missing pieces will not be noticed.

Cutting up an original analog signal and leaving out some pieces is done by high-speed sampling. Each sample consists of a short pulse of energy, proportional in strength to the original signal's amplitude at that point. Each energy pulse is then *quantized*—labeled with a number representing the momentary amplitude of the analog signal.

Encoding
• • • • • • •

The numbers attached to each sample are encoded by converting them from the decimal number system (0, 1, 2, 3 . . . 9) into the binary (two-part) number system (0 and 1 only). Computers have made the binary system familiar, along with the term *bit*, which stands for *binary digit*.

Numbers expressed in binary digits consist of nothing more than strings composed of only two digits. They are conventionally expressed as "zero" and "one" but can also be regarded as equivalent to "on" and "off" or "high" and "low."

As an example, the output of a microphone consists of an analog signal, a continuously varying electrical amplitude (that is, voltage). A digital processor samples this continuous amplitude pattern, breaking it down into a series of small, discrete amplitude values. An encoder quantizes each value by assigning it a number in binary form representing its momentary amplitude. The digitized output consists of a string of "power-off" signals (zeros) and "power-on" signals (ones).

Bit Speed

Capacity of a digital channel is measured by *bit speed*—the number of bits per second that one channel can handle. Digitally processed signals inherently need wider channels than do the same signals in analog form. Sampling takes place many thousands of times per second. Those thousands are in turn multiplied by the number of binary digits it takes to quantize each one. Exhibit 5.k shows how expressing a three-digit decimal number digitally results in 12 binary digits. An analog telephone channel requires a bandwidth of only about 3 kHz. When converted to digital code, however, a telephone call requires a 32 kHz transmission channel to accommodate its bit speed of 64,000 bits per second.

Exhibit 5.k offers more detail. In essence, digital processing converts a continuous signal into a series of samples that are given numerical values encoded as binary numbers.

Advantages

The extreme simplicity of digitized signals protects them from many extraneous influences that distort analog signals. A digital signal cannot be distorted or misunderstood as long as the elementary difference between "off" and "on" or "high" and "low" can be discerned. In contrast, each new manipulation of analog signals lowers their quality. Recording, relaying, and other processing of analog information inevitably introduce noise, causing quality loss. However, each new digital copy of a digitally encoded sound or picture is in effect an original.

Because digitally processed information exists as binary digits, the stored signal can be taken apart

and reassembled at will in infinitely varied forms. The possibilities for manipulation are endless.

Spectrum Disadvantage

Digitally processed signals need wider channels than do the same signals in analog form because a string of binary digits is needed to identify each tiny sample. The first communication modes to use digital processing, therefore, were those with relatively simple signals that made no great demands on the spectrum—data processing and telephone calls, for example.

The need for high-capacity channels delayed the application of digital methods to broadcast transmission and reception. However, there were dramatic advances in compression technologies during the 1990s. Digital compression now enables broadcasters, cable operators, and DBS providers to transmit multiple programs in the same size channel required for one analog signal.

Compression

Digital signals' hunger for frequency bandwidth drives the ongoing development of *signal compression* (Exhibit 5.l). Compression involves determining those portions of the original digital data that are necessary for adequately reconstructing the analog wave. Often there are sequences of repeated information, and one or a few symbols saying "Do this *x* number of times" can substitute for many individual bits of data that do the same thing over and over. Since all of the original information is reconstructed during reception, this is called *lossless* compression. Some portions of an original audio or video analog wave contain information that humans do not consciously perceive. Compression techniques can delete this information without it being missed. This is *lossy* compression because the deleted parts of the original are not restored during reception. If lossy compression is taken too far, however, audio and video quality is degraded so much that consumers can begin to notice a difference. There is a trade-off between quality and conservation of channel bandwidth.

Some early experiences with compression proved less than "picture perfect." The process proved un-

EXHIBIT 5.k
· · · · · · · · · ·
More on Digital Signal Processing

Digital signal processing has become so pervasive in contemporary life that it's worth a little effort to learn how it works.

Actually, digital signal processing began with the first electrical communication system, the nineteenth-century telegraph. In that system, telegraph operators sent messages in Morse code by means of an on/off key that controlled electricity going down the telegraph wire. The code consists simply of varying lengths of "on" and "off," presented to the ear or eye as dots, dashes, and spaces, which in turn represent letters of the alphabet, punctuation marks, and numbers.

Modern digital signal processing also employs simple on/off signals. They represent the elements of a *binary code*, a two-digit number system that requires only two code symbols, conventionally written as 0 and 1. All communication content can be reduced to nothing more than strings of zeros and ones.

A system that communicates digitally needs to make only one elementary distinction. "On" and "off" differ so obviously that they leave little chance for ambiguity. That simplicity makes digital signals extremely "rugged"—able to withstand external interference and imperfections in transmission and copying systems.

The familiar ten-digit *decimal* system (0 through 9) is used in everyday life. In that familiar system, the values of digits depend on their *positions* relative to one another, counting from right to left. Each new position increases a digit's value by a multiple of 10.

Thus the number 11 means (counting from right to left) one 1 plus one 10 (1 + 10 = 11). The binary code also relies on position, but each digit's position (again counting from right to left) increases by a multiple of 2. Thus in binary code the decimal number 11 becomes 1011, which means (counting from right to

left) one 1 plus one 2 plus no 4 plus one 8 (1 + 2 + 0 + 8 = 11). Here's an example converting the three-digit number 463 to binary coded decimal (BCD) form:

Multipliers		8	4	2	1		
Binary numbers	0100	= 0	+ 4	+ 0	+ 0	=	4
	0110	= 0	+ 4	+ 2	+ 0	=	6
	0011	= 0	+ 0	+ 2	+ 1	=	3

As the examples indicate, it takes more digits to express a number in the binary system than in the decimal system. Thus, although the simplicity of digital transmissions makes them less subject to error, they need larger channels than analog transmissions.

Conversion of an analog signal to a digitized signal involves *quantizing* the analog signal—that is, turning it into a number sequence. Quantizing consists of rapidly sampling an analog waveform and assigning a binary numerical value to the amplitude of each momentary item in the sample.

A digitized waveform looks something like the diagram above. The bars indicate the points at which the wave's amplitude is sampled. The higher the sampling rate, the greater the fidelity of the digitized signal. There is an equation for calculating the sampling rate necessary to avoid distortion; it usually calls for sampling thousands of times per second.

able to handle some scenes that involved rapid motion. Basketballs passed from player to player, for example, looked like comets with long orange tails. But continuing dramatic improvements in compression technology are at the core of the digital revolution. Better compression techniques allow the transmission of constantly increasing amounts

of information through the same bandwidth channels and permit more data to be crowded into each unit of storage capacity.

For television broadcasters video compression offers the potential for multichannel delivery services and is central to squeezing HDTV's massive information load into an existing 6 MHz channel.

EXHIBIT 5.1

Video Compression

Multiple signals (only four are shown) are digitally encoded and compressed for transmission by satellite, optical fiber, or over-the-air services. The compressed signal is *de*compressed by an intermediary (such as a cable system headend decoder before passing the signal on to a household subscriber), by circuitry in digital home TV receivers, or by set-top boxes. Viewers will be able to select from dozens and eventually hundreds of signals more than they now receive.

5.6   DIGITAL RECORDING

The shift from analog to digital recording gained dramatic momentum in the 1990s. There are now many tape and tapeless recording formats available that provide the advantages of digital signal processing.

Digital Tape Recording

Digital and analog tape recording are similar in many ways, the primary difference being that digital recorders store audio and video as binary numbers rather than as analog waves. Tape, because it is a linear technology, often requires shuttling through many yards of tape to play a desired selection. Like analog tape, digital tape requires that the

tape make physical contact with the playback head(s) of the recorder, wearing away the coating that stores the information with each use. Eventually, the tape wears out and the recorded information is lost.

As noted in Section 3.5 (page 72), neither digital audio tape (DAT) nor Philips's digital compact cassette (DCC) is popular with consumers. Although many professionals like DAT, most people continue to use analog cassettes or a tapeless digital format for audio.

In the mid-1990s a small, consumer-oriented digital video (DV) format using quarter-inch tape reached the market, and professionals have more than half a dozen digital videotape formats from which to choose. All are cassette-based but otherwise differ significantly. Some handle the color and brightness components separately, each with its own channel. Others multiplex (mix) the color and brightness values in a single *composite* channel. The *component* formats have the potential of providing the best video quality. Formats that use little or no compression are superior to those that are heavily compressed, other factors being equal. Because of their electronic and physical differences, the formats generally are incompatible, although there are exceptions involving playback capability.

In 1997 both CBS and NBC adopted Panasonic's small (quarter-inch) DVCPRO digital format (also called D-7) for electronic news gathering (ENG) at their owned-and-operated stations. DVCPRO is an enhanced version of the consumer DV format that provides better performance. Like most of the broadcast industry, the CBS and NBC stations previously used a larger (half-inch) analog format. The switch was significant because it continued the trend from analog to digital and from larger formats to smaller ones. Acceptance of the Panasonic format by two industry leaders also signaled a possible end to the domination of Sony in the ENG equipment market.

Although Sony offers several competitive digital formats, and JVC has Digital-S (also called D-9), Panasonic's DVCPRO and an improved version, DVCPRO 50, are gaining widespread acceptance professionally. One 1999 industry survey revealed that about 60 percent of those responding planned to buy equipment using some version of the DVCPRO format (*Television Broadcast*, August 1999: 16). Panasonic also markets an even more advanced version of the format called DVCPRO HD, designed for recording high-definition programming.

Consumers also have several digital tape formats from which to choose. The most popular consumer digital format is *MiniDV*. MiniDV uses a matchbox-sized cassette tape that records an hour of digital video in standard mode, or ninety minutes in long-playing (LP) mode. Sony also offers *Digital8*, a format that uses the same size cassette as the 8-millimeter and Hi8 analog formats. The major advantage is that a Digital8 machine will play Hi8 tapes. Another digital tape format of interest to consumers is digital VHS (D-VHS), which allows consumers to record about five hours of digital DBS programming directly to tape before decompression and conversion to analog form. D-VHS and regular VHS cassettes are the same size, and a D-VHS VCR can play and record VHS tapes.

Tapeless Digital Recording

Digital recording on disc became much more popular during the 1990s. As noted in Section 5.3 (page 136), the compact disc (CD) has practically replaced analog LP discs for the distribution of audio. The audio CD and its variants provide random access to anything on the disc in microseconds. They also use a laser beam to read the information on the disc, which means the CD never wears out, regardless of how often it is played. The technology of the CD is illustrated in Exhibit 5.m. Laser technology may eventually displace most other types of information storage, including the magnetic floppy disks used in home computers.

A computer version of the CD, used to distribute software, interactive video games, and other types of information, is the CD-ROM (compact disc read-only memory). The CD-ROM is similar to the audio CD and theoretically shares the audio CD's approximately 680-megabyte capacity. CD-ROM manufacturers often use only the 550 megabytes nearest the center of the disc, however, to ensure reliability in critical applications. Once considered a luxurious computer peripheral, CD-ROM drives were in more than 40 percent of all home computers by 1996, and today practically every computer sold comes with CD-ROM.

Another CD family member is CD-R (compact disc recordable), which allows users to record on the disc only once but play back any number of times. In late 1996 CD-RW (compact disc rewritable) was introduced. CD-RW allows repeated recording and playback similar to magnetic tape, computer floppy disks, and hard drives, as well as Sony's MiniDisc, briefly discussed in Section 3.5.

Conflicts arising first over technical standards and later over objections from the entertainment industry regarding the copying of movies delayed the development of the *digital video (or versatile) disc* (DVD). In 1995 two competing and incompatible versions of the DVD were announced—one by Toshiba and Time Warner, the other by Sony and Philips. Observers saw in the conflict a replay of the 1970s VHS vs. Beta battle over videocassette formats. The competitors later compromised on DVD standards, which provide for disc capacities ranging from 4.7 to 17 gigabytes. A 17-gigabyte DVD stores 25 times as much data as a conventional CD, more than 11,000 times as much as a

EXHIBIT 5.m
· · · · · · · · · · ·

Compact Disc System

Sound is recorded on an optical disc in digital form. Microscopic pits alternate with a shiny reflective surface to create "on/off" binary digital signals encoding the original sound. A laser beam, guided by a super-sensitive tracking mechanism, strikes the pits, which reflect the beam back (as a modulated digital signal) to an optical sensor. The sensor feeds its digital output to a microcomputer that converts the digital information into analog form for delivery to speakers or earphones.

Source: IEEE Spectrum.

computer floppy disk. The DVD achieves this enormous capacity in part by using a shorter-wavelength laser beam than that used with audio CDs. This allows the pits and data tracks on the DVD to be compressed even more tightly than on earlier discs. The DVD also uses multilayer recording, with a second layer of information recorded, at a different depth, on top of the first. Finally, DVDs are two-sided devices, with data stored on both sides of the disc. Audio CDs and CD-ROMs store information on only one side. See Exhibit 5.n for a comparison of CD and DVD technology. The least expensive DVD players sold for about $500 in 1997, but prices dropped below $300 by 1999.

The Consumer Electronics Manufacturers Association (CEMA) reported sales of more than $171 million during the DVD's year of introduction. CEMA noted that these sales were better than either those for color TV or the CD format, products that later became extremely successful. About 4 million DVD players were sold in 1999. (CEMA, 1999)

Introduction of DVD-ROM drives spurred interest in DVD among computer users, because of DVD-ROM's larger capacity compared to CD-ROM. Interest in DVD among computer users increased with the introduction of DVD-RAM, a recordable/erasable version of DVD. DVD-RAM's 5.2-gigabyte storage capacity and the ability of DVD-RAM drives to read both DVD and

EXHIBIT 5.n

• • • • • • • • • •

Comparison of Conventional Audio CD and DVD

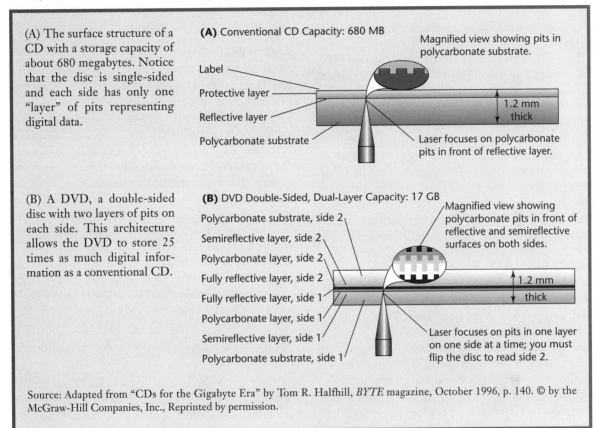

(A) The surface structure of a CD with a storage capacity of about 680 megabytes. Notice that the disc is single-sided and each side has only one "layer" of pits representing digital data.

(A) Conventional CD Capacity: 680 MB

Magnified view showing pits in polycarbonate substrate.

Label

Protective layer

Reflective layer

Polycarbonate substrate

1.2 mm thick

Laser focuses on polycarbonate pits in front of reflective layer.

(B) A DVD, a double-sided disc with two layers of pits on each side. This architecture allows the DVD to store 25 times as much digital information as a conventional CD.

(B) DVD Double-Sided, Dual-Layer Capacity: 17 GB

Polycarbonate substrate, side 2

Semireflective layer, side 2

Polycarbonate layer, side 2

Fully reflective layer, side 2

Fully reflective layer, side 1

Polycarbonate layer, side 1

Semireflective layer, side 1

Polycarbonate substrate, side 1

Magnified view showing polycarbonate pits in front of reflective and semireflective surfaces on both sides.

1.2 mm thick

Laser focuses on pits in one layer on one side at a time; you must flip the disc to read side 2.

Source: Adapted from "CDs for the Gigabyte Era" by Tom R. Halfhill, *BYTE* magazine, October 1996, p. 140. © by the McGraw-Hill Companies, Inc., Reprinted by permission.

CD formats make it a possible replacement for computer CD drives. DVD-RAM may face competition from several other recordable/erasable DVD formats, but DVD-RAM is establishing an early lead in the marketplace that may be difficult to overcome. Reports of laboratory experiments suggest the possibility of dramatically increasing the DVD's storage capacity by using even shorter wavelength laser technology.

In 1999, disc-based *personal video recorder* (PVR) products providing a form of video-on-demand entered the consumer market. TiVo and Replay-TV are set-top boxes (STBs) with sophisticated software and large digital hard-drive storage that record and manipulate programs in new ways. TiVo and ReplayTV were among the first advanced STBs (see Section 4.13, page 117) in what

will become a flood of "smart" STBs with massive digital storage capacity and interactive features. Both products allow viewers to pause a program in progress (for up to 30 minutes) and then resume viewing as the remainder of the program is being recorded. They also allow viewers to review something just seen, fast-forward through recorded material, or view recorded programming in slow motion or at faster than normal speed. Being disc-based, viewers have instant random access to anything on the disc for immediate viewing. ReplayTV features a 30-second fast-forward option, ideal for skipping over commercials. The systems connect by telephone modem to TiVo or ReplayTV's proprietary servers, which contain the program schedules for thousands of cable, DBS, and terrestrial broadcast systems, and can select programs to record based on the viewer's instructions or past viewing behaviors. This allows viewers to establish "virtual channels" devoted to, for example, Chicago Bulls basketball or any program in which Kelsey Grammer (star of *Frasier*) appears. Initial retail prices were high, starting at about $500 for TiVo and $700 for ReplayTV.

Interest in disc-based digital recording technology was strong among broadcast professionals as the new century began, but tape remains an amazingly strong competitor, especially for recording outside the production studio. A survey by *Television Broadcast* in 1996 indicated that 42 percent of respondents planned to buy a disc-based video camera before the end of the decade, but the publication's 1999 survey found that less than 1 percent had done so. Tape remained the major recording medium, and Sony's well-proven analog-format Betacam SP was still being used by more than 70 percent of the facilities responding. The 1999 survey showed a continuing commitment to disc-based video servers, however, with more than 40 percent of respondents saying that they would buy additional servers in the next year (*Television Broadcast*, August 1999: 16).

In the future, however, there will be more digital recording devices in the hands of consumers that use neither tape nor discs. Solid-state memory chips with constantly increasing storage capacity promise to replace moving media technology for many applications. Today's portable MP3 audio players with their memory cards are suggestive of the devices employing solid-state recording technology that many consumers will use in the new century.

5.7   DIGITAL TV TRANSMISSION

Digital technologies made great progress in the acquisition, editing, and storage of TV programming during the 1980s and early 1990s. In the distribution area, however, the calendar often seemed stuck at 1941, when American TV broadcast standards were developed. Movement toward more modern standards was slow. The major impetus for change emerged from a desire for TV pictures and sound rivaling that found in movie theaters. The initial goal was a *high-definition television* (HDTV) system pro-

viding a larger screen, vastly improved picture resolution, and multichannel sound. What has emerged is a new digital television (DTV) broadcasting environment in which HDTV as originally conceived is an important, but not an exclusive, component.

Analog Beginnings

Engineers in Japan, beginning in the 1960s, made the first significant progress with HDTV, developing a 1,125-line picture with a 16-to-9 aspect ratio. This analog system, called MUSE, required 36 MHz channels—six times wider than the U.S. 6 MHz standard. Japanese engineers later developed a broadcast MUSE version using a 9 MHz channel.

By late 1991 the Japanese were providing, eight hours a day, satellite delivery of MUSE service to selected viewing sites in Japan. However, the initial cost of receivers (about $18,000) drastically limited HDTV penetration. By the mid-1990s Japanese analog HDTV sets were still twice as expensive as ordinary sets, and fewer than 350,000 households had HDTV receivers.

America's Response

In 1988 the FCC began to consider U.S. HDTV standards to ensure U.S. terrestrial broadcasting's ability to keep up with Japanese and European efforts in the race toward improved television. Also, nonbroadcast services (domestic cable, VCR, and DBS), not being subject to the National Television System Committee (NTSC) 6 MHz channel constraint of broadcast television, could forge ahead with their own nonbroadcast versions of HDTV without waiting for a compatible broadcast version to emerge.

The FCC Advisory Committee on Advanced Television Service (ACATS), established to compare HDTV systems and recommend one to the Commission, sparked creative thinking among many domestic and foreign companies. The term "advanced," rather than "high-definition," in the committee's title was prophetic. The original goal was development of HDTV, but the focus changed in the early 1990s when several companies began concentrating on rapidly developing digital technologies. Rather than continuing to work as competitors in isolation, in May 1993 the developers formed a "Grand Alliance" and merged their ideas to accelerate the development and approval of a workable digital system. Later that year the Advisory Committee on Advanced Television Service selected a digital compression standard, agreed on the number of scanning lines (1,080), and adopted a Dolby CD-quality audio system for HDTV.

Although continuing to claim HDTV was the ultimate objective, broadcasters began urging Congress to allow them to use digital transmission technology for other services, such as interactive television, pay services, and high-speed data transmission. This shift in emphasis raised new questions and occasioned fierce debate regarding the transition from an analog to a digital broadcast service. It was not until mid-1996 that the FCC authorized three stations to begin broadcasting digital signals experimentally.

Just as a final transmission standard seemed ready for adoption, objections from leaders in the entertainment, computer, and cable industries once more delayed the decision. They claimed that some parts of the proposed standard interfered seriously with the smooth convergence of television and computer-based technologies. A compromise in late November 1996 allowed the FCC to adopt a digital television broadcast standard on Christmas Eve. Reflecting the compromise, the FCC chose not to impose a single picture format for all digital broadcasting. The Commission allows broadcasters to use any picture format they wish, but it is believed that only a few of the DTV formats recommended to the FCC by the Advanced Television Systems Committee (ATSC) will be adopted widely. The ATSC formats vary in their aspect ratio (relationship of image width to height), number of vertical lines scanned, number of picture elements (pixels) per horizontal line, type of scanning (interlaced or progressive), and number of frames transmitted each second. Exhibit 5.o summarizes the ATSC formats.

Originally, it was thought that only a format with a wide-screen aspect ratio of 16-to-9 and 1,080 vertical lines qualified as high-definition TV (HDTV). Today wide-screen formats with 720 vertical lines are also increasingly considered high definition. The other ATSC formats are considered *standard definition TV* (SDTV), even if they provide 16-to-9 aspect ratio. Exhibit 5.p compares HDTV formats to the NTSC analog standard.

Meanwhile, in 1994 the Japanese government announced an end to its support for and further development of analog HDTV, but then retreated under

pressure from Japanese manufacturers. The Japanese Ministry of Posts and Telecommunications later decided to promote a new digital system closely resembling that adopted by the United States, finally bowing to the inevitability of a digital world.

DTV Transmission

Exhibit 5.q, p. 150, shows the basic structure of a TV station transmitting a digital signal. A comparison with the structure of an analog station (see Exhibit 4.l, page 112) shows that there is a great deal of similarity before the audio and video signals reach the transmitter. This underscores the point that the FCC requires broadcasters to transmit a digitally encoded signal, but the FCC does not require that broadcasters provide a wide-screen, high-resolution image. As a practical matter, however, broadcasters will go digital well ahead of the transmitter. Earlier conversion to digital allows broadcasters to manipulate the audio and video with greater sophistication and present the transmitter a higher-quality signal. Indeed, analog TV stations routinely use digital techniques in various stages of the production process.

Much of the work necessary to transmit a digital signal occurs just ahead of and within the transmitter itself. First, the video and audio signals are encoded and compressed (using the FCC-mandated MPEG-2 and Dolby Digital/AC-3 compression algorithms), then divided into "packets" of information with unique identifying "tags," and integrated into the digital data stream. The transmitted data stream contains millions of data segments. A *data segment* is roughly equivalent to an analog line of video information. Within each data segment there is a *segment sync signal*, which performs approximately the same function as the horizontal sync pulse in an analog TV transmission. A series of 313 data segments make up a *data frame*, and a *data frame sync signal* marks the end of each frame. This means the data frame sync is roughly equivalent to the vertical sync pulse in analog TV. The segment and frame sync signals in the digital data stream are as critical to ensuring a properly received and displayed image in DTV as are the horizontal and vertical sync pulses of conventional TV. With DTV, transmission of auxiliary information, such as closed captioning, goes within the data stream itself rather than in the vertical blanking interval (VBI), as is done in analog transmission.

EXHIBIT 5.0
. . . .

ATSC Formats for DTV

Lines of Vertical Resolution	Pixels per Line	Aspect Ratio	Frames per Second	Type of Scanning
1,080	**1,920**	**16:9**	**24**	**Progressive**
1,080	**1,920**	**16:9**	**30**	**Interlaced**
1,080	**1,920**	**16:9**	**30**	**Progressive**
720	**1,280**	**16:9**	**24**	**Progressive**
720	**1,280**	**16:9**	**30**	**Progressive**
720	**1,280**	**16:9**	**60**	**Progressive**
480	704	16:9	24	Progressive
480	704	16:9	30	Progressive
480	704	16:9	30	Interlaced
480	704	16:9	60	Progressive
480	704	4:3	24	Progressive
480	704	4:3	30	Interlaced
480	704	4:3	30	Progressive
480	704	4:3	60	Progressive
480	640	4:3	24	Progressive
480	640	4:3	30	Progressive
480	640	4:3	30	Interlaced
480	640	4:3	60	Progressive

The formats in boldface are considered high-definition TV (HDTV) formats. The remainder are standard-definition TV (SDTV) formats. These formats have been proposed as standards by the ATSC, but the FCC allows broadcasters to use any picture format they wish. It is unlikely that formats other than those proposed by the ATSC will be employed widely. In 1999 an International Telecommunication Union (ITU) panel selected 1,080 P at 24 frames per second as a "common image format" for worldwide HDTV production purposes.

Source: Data adapted from Advanced Television Systems Committee Digital Television Standard Document A/53 (Annex A) Table 3.

EXHIBIT 5.p
. . . .

High-Definition Systems

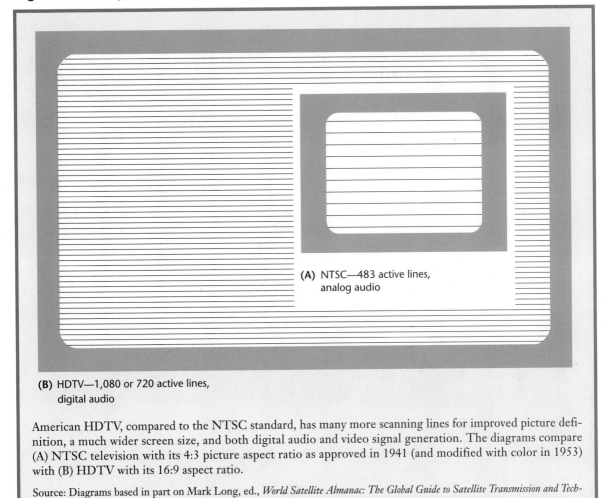

(A) NTSC—483 active lines,
analog audio

(B) HDTV—1,080 or 720 active lines,
digital audio

American HDTV, compared to the NTSC standard, has many more scanning lines for improved picture definition, a much wider screen size, and both digital audio and video signal generation. The diagrams compare (A) NTSC television with its 4:3 picture aspect ratio as approved in 1941 (and modified with color in 1953) with (B) HDTV with its 16:9 aspect ratio.

Source: Diagrams based in part on Mark Long, ed., *World Satellite Almanac: The Global Guide to Satellite Transmission and Technology*, 3rd ed. © 1993 Mark Long Enterprises, Inc., Ft. Lauderdale, FL. Used by permission.

As the data stream moves forward, additional information, as well as a pilot signal, is added to help the receiver reconstruct the original information in proper analog form and to compensate for errors caused by problems in transmission and propagation. The total data stream is then used to amplitude modulate a radio frequency signal using a modulation system called *8-VSB* (eight-level vestigial sideband). The modulated signal is then amplified and passed to the station's antenna system.

Exhibit 5.r shows the transmitted DTV signal as viewed on a spectrum analyzer. A comparison with the analog TV signal (see Exhibit 4.j, page 108) reveals both similarities and differences. Like analog TV, DTV uses the same 6 MHz channel. Unlike analog transmission, there are no separate carriers for video, audio, and color information. Note also the pilot signal located 310 kHz up from the lower edge of the channel, which is unique to DTV transmission.

The FCC allows DTV stations to use up to 1 million watts effective radiated power (ERP) to propagate their signals. The FCC assigns each station a power level and antenna height calculated to

EXHIBIT 5.q
. . . .

Digital TV System Components and Signals

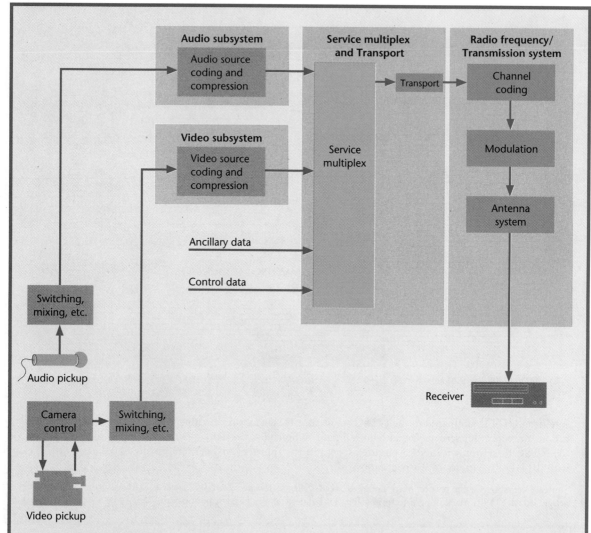

In some ways the DTV station is similar to the analog station illustrated in Exhibit 4.1 (page 112). In both analog and digital stations, audio and video are picked up by microphones and cameras, switched and mixed, combined with graphics, and otherwise manipulated in the production process. At this point in a DTV system, the audio and video are encoded and compressed in audio and video subsystems. The audio and video data stream is then divided into discrete "packets" with individual identifying data added. In the radio frequency transmission system additional information needed by the receiver is added. The final data stream then modulates the transmitter using the 8-VSB (eight discrete amplitude level vestigial sideband) technique mandated by the FCC. The signal is then transmitted to receivers, where the process is reversed to re-create the original analog sound and picture.

Source: Diagram adapted from Advanced Television Systems Committee Digital Television Standard Document A/53 Figure 5.1.

EXHIBIT 5.r
. . . .

The DTV Channel

The architecture of the DTV channel is similar to that of the analog channel (see Exhibit 4.j, page 108) in that they both occupy 6 MHz. The DTV system, however, uses a data stream containing audio, video, ancillary data, and control data to amplitude modulate a single carrier, rather than using separate carriers for the audio and video. During transmission a pilot signal is added 310 kHz above the lower edge of the channel. This signal helps the receiver lock onto the DTV signal. The carrier is suppressed, and only the upper sideband is transmitted, to conserve spectrum space much as with analog TV transmission.

The ATSC standard provides a transmitted digital bit rate payload of 19.28 megabits per second (Mbps) using the 6 MHz channel. Including required overhead data, DTV will transmit about 19.4 Mbps. High-definition formats require between 6 and 18 Mbps, depending on program content. Standard-definition formats require between 2 and 10 Mbps, leaving much of the channel capacity for transmission of other content.

Source: Diagram adapted from Advanced Television Systems Committee Digital Television Standard Document A/53 (Annex D) Figure 3.

duplicate, or almost duplicate, the station's analog coverage contour. Some broadcasters could not use their existing towers for DTV and experienced resistance when trying to rebuild or relocate their transmission facilities. Some stations in major markets such as New York, Chicago, and Denver had to delay installation of DTV antennas because of local concerns about safety and aesthetics.

DTV Reception
. . . .

DTV reception is the inverse of the transmission process. The receiver must tune to the frequency of the signal, isolate the digital data stream, and then decompress and decode it back into its original analog form. These functions can be performed by circuitry built into a DTV receiver or by a set-top box that then feeds the reconstructed analog signals to a separate display device.

Conversion to DTV transmission presents new propagation challenges. The picture quality of an analog signal degrades gradually as the signal loses strength, but digital signals exhibit a *cliff effect*. A digital signal provides perfect pictures up to the point the signal becomes too weak for the system's error-checking algorithms to compensate. At that point the picture is lost entirely, as if the system had fallen off a cliff. Conversion to DTV also requires that many VHF stations move to UHF channels, frequencies traditionally prone to propagation problems. The migration to UHF channels, combined with DTV's cliff effect, caused some broadcasters to fear that they might lose coverage with the conversion to digital broadcasting. In announcing the DTV channel allocations in 1997, the FCC acknowledged some stations might lose coverage but predicted that more than 93 percent of the new DTV stations will serve at least 95 percent of their analog service areas once conversion is complete. Many broadcasters remained doubtful.

Those doubts seemed justified when DTV stations began operating in the nation's largest cities and a new reception problem appeared. Broadcasters had initially worried most about viewers at the extreme edge of the coverage contour, where a fading signal could mean an abrupt and complete loss of audio and video. As more DTV stations began broadcasting in the late 1990s, concern shifted to what some people called reception *sinkholes*, congested urban areas where signals bounce off buildings, low-flying airplanes, and other objects, creating multiple signal paths. Viewers in such locations using small loop or "rabbit ear" indoor antennas often found it difficult to receive DTV signals. As illustrated in Exhibit 4.m, page 113, analog receivers display such multipath reception as ghost images, but early DTV receivers simply stopped functioning when faced with the task of sorting out the primary signal from the array of strong reflected signals. In 1999 the Sinclair Broadcast Group, owner of more than 50 TV stations, conducted tests of DTV reception in high multipath environments using typical indoor antennas. Sinclair's tests showed that just having someone walk through the room could cause complete loss of the DTV signal. (Sinclair Broadcast Group, 1999)

Part of the Sinclair tests involved comparing the FCC-approved 8-VSB modulation standard to *Coded Orthogonal Frequency Division Multiplexing* (COFDM), the modulation technique used in the terrestrial DTV system developed in Europe. Sinclair found that COFDM performed better with indoor antennas in multipath environments than did 8-VSB. Sinclair claimed that the ATSC used an unrealistic testing methodology (involving outdoor receiving antennas on 30-foot masts) when deciding to adopt 8-VSB rather than COFDM. The ATSC immediately defended its decision, noting that COFDM was tested against 8-VSB at least three times and rejected each time. The ATSC said that the problem was with the tuners in the first-generation DTV receivers, not the modulation standard adopted. DTV receiver manufacturers promised that future sets will offer better adaptive equalizers to improve multipath performance. Late in 1999, Sinclair, supported by several hundred stations and others concerned about the viability of 8-VSB, petitioned the FCC to allow stations to use either 8-VSB or COFDM. Others urged staying the course with 8-VSB only, noting that allowing each station to choose its modulation technique would further complicate and delay the implementation of DTV. In February 2000, the FCC re-

jected Sinclair's petition, expressing confidence in receiver manufacturers' ability to develop better technology that would solve the multipath problem. Those supporting Sinclair's petition remained skeptical.

Other problems surfaced as more DTV stations began operation in the late 1990s. For example, when WFAA-DT (operating on channel 9 in Dallas) signed on, the wireless heart monitors at Baylor University Medical Center malfunctioned. Previously there had been no station on channel 9 because another analog station in the market was on channel 8. Concern about new DTV stations on previously unassigned channels prompted the FCC and the Food and Drug Administration (FDA) to issue a joint statement in 1998 warning both broadcasters and the medical community to take precautions to avoid situations such as the one that occurred in Dallas. Broadcasters concerns about interference between old analog stations and new digital stations have been mentioned previously. Other unexpected problems may develop as the transition to digital continues.

Moving from NTSC to Digital

One of the most difficult decisions facing the FCC amid all this fast-moving technical change was how best to handle the transition from existing NTSC receivers and broadcast and cable production and transmission equipment to the selected digital standard.

Simultaneous with the development of transmission standards for digital broadcasting, the FCC wrestled with the issue of how best to move stations from their NTSC analog channels to their new digital channels. Knowing the transition would take years, the FCC needed to give stations new digital channels while allowing them to continue analog broadcasting on their old channels.

Early in 1992 the FCC decided on a 15-year transition period following adoption of a digital transmission standard. Broadcasters were to receive a second 6 MHz channel in the UHF band for digital broadcasting while continuing conventional analog transmission on their respective existing channels. Broadcasters would have to begin digital (at that time meaning high-definition) operations six years after the FCC's selection of a digital standard. They would have to simulcast half their programming in their second year of digital operation, and all of it by the third year. At the end of the transition, when presumably a substantial portion of television homes would have digital receivers, stations and networks would convert to digital operations exclusively. All TV broadcasting would be on UHF channels, the broadcasters having given up their old NTSC broadcast channels for reallocation to other services.

In 1996 the FCC revised its transitional plan dramatically by announcing a new channel-allotment scheme. Rather than move terrestrial broadcasters to all UHF digital channels, the Commission proposed packing most of the new digital licenses into a "core spectrum" comprising channels 7–51. Upon completion of the estimated 15-year transition to digital, 23 channels formerly used by broadcasters (channels 2–6 and 52–69) would be auctioned to other users providing other services. Broadcasters immediately countered that the entire range of TV channels (2–69) was necessary to avoid interference and maintain existing signal coverage patterns during the transition.

When the FCC finally assigned the new digital channels in 1997, the plan had changed again; the FCC made initial digital assignments on all available TV channels (2–69). Then in 1998 it announced that stations would eventually be assigned to a "core" DTV spectrum of channels 2–51. The FCC also sharply accelerated the phase-in period for DTV. Rather than allowing stations six years to begin digital broadcasting, the FCC required that at least three network-affiliated DTV stations be on the air in the top 30 markets by November 1, 1999. The mandate was designed to ensure DTV's availability to more than half the nation's TV households by the year 2000. More than two dozen stations failed to make the deadline, however, because of equipment delivery delays or problems acquiring access to usable antenna sites. The FCC also shortened the transition period to digital-only transmission from 15 to 10 years, until the end of 2006.

Later legislation allows a station to retain its analog channel until 85 percent of the station's viewers can receive the digital signal either over the air, by satellite, or by cable. The FCC also has scuttled earlier requirements that stations simulcast digital and analog programming during the transition, except during the last few years. The plan imposes no new public-interest requirements on digital broad-

casters, but the FCC expects to add some during the transition. Broadcasters were pleased that the plan does not require them to pay a spectrum fee for their new digital channels.

5.8 DIGITAL RADIO

Digital audio broadcasting (DAB) is developing more slowly than DTV. DAB ultimately promises (or threatens, depending on one's point of view) to replace analog radio broadcasting—both AM and FM—with a single new digital radio system.

If necessary, the FCC would likely set up a transition period during which stations would broadcast in analog AM and FM on their existing frequencies, while simulcasting DAB on new assignments (or, as appears increasingly likely, on their existing frequencies, thanks to new multiplexing technology). Then, when enough people have DAB receivers, stations will cease their analog transmissions.

By the early 1990s about 10 different technologies for achieving DAB existed, all of which fell into one of two categories. Some would require additional spectrum to achieve true digital sound quality, whereas others would operate in the present combined AM and FM bands, though likely with wider channels.

DAB Spectrum Needs

In 1992 the International Telecommunication Union (ITU) established frequencies (1.452 to 1.492 GHz) in the L-band for worldwide DAB despite objections from the United States. The United States used those frequencies for military and other purposes and did not want to reassign those services to accommodate DAB. Accordingly, later the same year the FCC announced that American DAB would use 50 MHz of S-band frequencies between 2.31 and 2.36 GHz. Both satellite-delivered and terrestrial DAB would use these frequencies if it were determined that digital and analog signals on the existing AM and FM bands could not coexist. By 1996 the FCC had reduced the range of frequencies from 50 MHz to 40 MHz because of potential interference with Canadian signals in the lower portion of the S-band. Meanwhile, Congress passed appropriation legislation mandating that many of the remaining frequencies be auctioned to a variety of wireless services. As a result, by 1997 only half (25 MHz) of the space originally reserved for DAB remained available. An industry group testing various DAB systems questioned whether 25 MHz could provide the quality and diversity of DAB services consumers expected.

DAB Progress

The National Association of Broadcasters (NAB) wants to offer DAB "in-band, on-channel" (IBOC), meaning by existing terrestrial stations using their currently assigned frequencies. The NAB argues that this approach will provide a smoother transition to DAB than reassigning stations to entirely different frequencies in a new band. The NAB also opposed satellite-delivered direct-to-home or direct-to-car DAB, which the FCC calls *digital audio radio service* (DARS).

In 1992 the FCC granted experimental authority for testing in-band DAB. In the mid-1990s the NASA Lewis Research Center lab tested seven systems (two of the systems operated in two modes each), five of them in-band systems. Four systems emerged for field-testing, but their proponents gradually withdrew their participation, claiming the testing procedure favored the European-based Eureka-147 DAB system. With the FCC's original S-band frequency allocation cut in half and proponents of in-band DAB unable to agree on test procedures that would lead to a standard, the status of terrestrial DAB in 1997 was little advanced from where it had been at the start of the decade. The FCC, however, went forward and auctioned two 12.5 MHz segments of S-band spectrum for satellite-delivered DARS. The winning bidders, Satellite CD Radio, Inc., (now Sirius Satellite Radio), and American Mobile Radio Corporation (now XM Satellite Radio), each paid more than $83 million for DARS spectrum.

Both Sirius Satellite Radio and XM Satellite Radio plan to launch their services late in the year 2000 or early 2001, but some industry analysts view those dates as being optimistic. Satellite-delivered DAB is targeted to listeners in cars, and both companies are moving forward with strategic alliances that will help ensure a successful launch. Sirius Satellite Radio has a deal with Ford, and XM Satellite Radio has one with General Motors, to put DARS receivers in some 2001-model trucks and cars. Both service providers also have agreements with various receiver manufacturers to make aftermarket DARS receivers for consumers with older vehicles or those with new vehicles made by other automakers. XM Satellite Radio and Sirius Satellite Radio say some DARS receivers may be available for as little as $200, and the subscription cost is expected to be about $10 a month. The companies are also forming alliances with content providers and building production studios in preparation for the initiation of services.

Receiving antennas for DARS have evolved during the years. An early design resembling a small silver-dollar dish has been abandoned for an antenna integrated into the trunk area that engineers hope will be almost invisible to consumers (and vandals). Although geostationary satellites will be the primary signal sources, DARS will need thousands of terrestrial repeaters to fill in areas where satellite signals may be blocked by natural and manmade obstructions.

FCC rules require that DARS radios be able to receive signals from both XM and Sirius, but the companies say that first-generation DARS receivers may lack this capability. It is expected that the companies will petition the FCC for a waiver of this requirement to ensure that DARS rolls out on schedule.

In the late 1990s the proponents of IBOC DAB continued their efforts to develop a workable system. By early 1999 the field of system proponents had narrowed to three organizations: Lucent Digital Radio (LDR), Digital Radio Express (DRE), and USA Digital Radio (USADR). DRE later formed an alliance with USADR, further reducing the field. The goal was to develop a system that could perform initially in hybrid mode and eventually in digital mode. A station operating in hybrid mode would broadcast both an analog AM or FM signal and a digital signal in the same channel without causing interference between the analog and digital signals or to other stations on adjacent channels. This would allow consumers to continue to use their conventional receivers but also provide programming to those buying new digital sets. Exhibit 5.s illustrates the hybrid system proposed by LDR. Making the transition to digital mode would be accomplished by simply turning off the station's analog transmitter

EXHIBIT 5.s

. . . .

In-Band, On-Channel (IBOC) Digital Audio Broadcasting

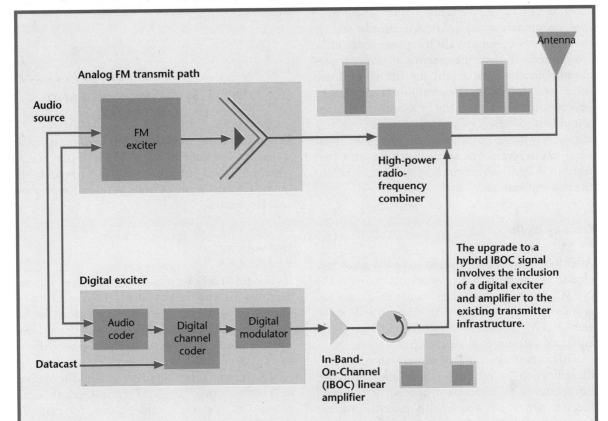

This diagram shows how Lucent Digital Radio proposes to provide an analog and a digital signal simultaneously within the same channel. A system providing DAB compatible with analog signals and operating within the existing AM and FM bands is favored by U.S. broadcasters. Other nations offer terrestrial DAB in a separate part of the spectrum.

once all consumers had digital receivers. The time required for this transition is uncertain.

The advantages of digital IBOC DAB are many. The IBOC systems proposed do not require allocating new spectrum space for AM and FM broadcasting, although they do require doubling the channel bandwidth (20 kHz for AM stations, 400 kHz for FM stations). Much like the transition to color television in the 1950s and 1960s, the transition to digital radio would occur without making existing analog receivers instantly obsolete. FM radio could deliver near-CD audio quality, and AM radio would move up to near-current-FM quality,

while eliminating or dramatically reducing problems with multipath FM reception and AM interference. Perhaps equally important, conversion to DAB would allow radio broadcasters to offer new or expanded information services, as discussed later in this section.

The FCC must approve one or more IBOC DAB standards before stations can begin the conversion. In October 1998 USADR petitioned the FCC to initiate a rule-making procedure for implementing DAB using USADR's IBOC technology. The following April the IBOC proponents agreed to conduct independent tests and submit their findings to the NAB's National Radio System's Committee (NRSC) by December 15, 1999. USADR, however, was the only competitor to meet the deadline. Meanwhile, on November 1, 1999, the FCC issued a Notice of Proposed Rule Making on terrestrial DAB, specifically stating that all systems, not just IBOC systems, would be considered. If the rule-making process moves ahead smoothly, a standard for IBOC (or some other type) DAB could be established early in the new century. But even if the FCC moves to adopt a standard (or several standards) quickly, it is unknown how many consumers will buy new digital receivers to receive the service. It is possible that terrestrial DAB will continue to flounder well into the new millennium.

Data by Radio

After long debate, both broadcasters and consumer electronics manufacturers, working through the NRSC, agreed on technical standards for a *radio broadcast data system* (RBDS), transmitted on an analog FM station's subcarrier for receivers equipped with a special microchip.

RBDS transmits a digital signal—an ad slogan, identification of the station's format, or other information—that can be displayed in a special receiver panel. The system also makes it possible for an RBDS-equipped car to tune stations by format, with the receiver jumping to a different but more powerful signal when the tuned signal becomes weak. And in an emergency situation—an approaching tornado, for example—RBDS can be used to turn on a receiver to transmit a warning message.

Conversion to an all-digital mode will dramatically increase radio's data transmission capacity. Bandwidth previously devoted to providing an analog signal will become available for additional purposes, such as paging, dispatching, e-mail delivery, computer communications, and networking. USADR claims that its IBOC technology will allow FM stations to accommodate data rates averaging about 48 kilobits per second (kbps). This approaches the fastest conventional telephone line connection that most people use for residential Internet access. AM stations would get only a fraction as much data-handling capability (about 1.2 kbps using the USADR system). This would allow AM

broadcasters to match the RDBS services offered by analog FM stations using a subcarrier, and is certainly better than having no data transmission capacity at all, as is the situation in the analog world.

5.9   NETWORKING AND SWITCHING

Whereas traditional broadcasting stations typically transmit all of their signals to all persons who have receivers, newer generations of multichannel media intend for only some persons to receive only some of their signals.

Networking

Radio and broadcast television use *networking* (see Sections 5.1 and 5.2) as relays to connect themselves to program services and to each other, not for delivery to the public. Early cable systems used networking in much the same way but also for delivery of all of their channels to all of their customers.

Most newer media, however, intend to deliver only some of their program services to only some consumers. A modern cable system, for example, intends to deliver its package of basic program services only to those who pay a monthly subscription fee and to deliver other programming (from HBO, Showtime, and others) only to those who pay an extra charge for those premium services.

Cable and DBS systems achieve selective audience delivery (sometimes called conditional access) through *scrambling* (or *encrypting*) some signals and unscrambling them for paying customers, or through the use of electronic *traps* in drop cables, which prevent delivery until they are removed.

Interactive media complicate the process even more. Only some signals from some sources are to reach only some receivers, which in turn are to send signals back *upstream* to the appropriate originating source. Two principal elements are required to accomplish this complicated task. First, sources must be linked with receivers, by terrestrial connection or otherwise, so that together they form a *network*. Then signals on that network must be routed, or *switched*, from source to intended receiver and, where interactivity exists, from receiver back to intended source.

The best examples of switched networks are those operated by the telephone companies. Few people appreciate the enormity and complexity of the *public switched telephone network* (PSTN). We simply push a few telephone buttons, and we are—almost miraculously—connected to precisely the person we want to reach.

The public telephone network resembles a cable system (see the cable layout in Exhibit 4.r, page 120). *Drop wires*, which connect individual homes to utility poles, join others to become a *distribution cable*, which in turn joins others to form a *feeder cable* that completes the connection from each home to a *central office*. Here the similarity ends, because unlike cable system headends, telephone central offices connect to each other by *trunk lines* and, via satellites and undersea cables, to their counterparts around the world.

Switching

• • • • • • •

This vast network interconnection must be combined with sophisticated *switching* capability to provide the one-to-one telephonic communication we have come to take for granted.

In its infancy, telephony accomplished its switches manually. A telephone operator asked what number you wished to reach and literally plugged a wire (called a *patch cord*) into a console so that your line would be linked to another (see Exhibit 5.t). This was an agonizingly slow process. The caller told the long-distance operator the number being called, then hung up and waited, perhaps an hour or more, for the various patches to be completed. Today switching is handled electronically by digital switches. The time needed to connect with a phone on the other side of the planet is little more than that needed to reach a friend across town.

As media converge, and as demand increases for voice, data, and video transmission, telephone companies and others that provide communication services are moving from circuit-switching to digital packet-switching technologies, such as *asynchronous transfer mode* (ATM). For most of its history, the telephone company, for example, usually connected only two people, and that connection continued unbroken until the two people finished

talking. If the two were only sighing every 10 seconds or so, the connection remained constant even between sighs. Nobody else could use the network resources needed for this conversation. This was inefficient. If the sighs and long silences were broken up, the circuits could be rapidly switched during the silent periods so that two other people could get a few words in between sighs. Four people rather than two could use the same resources simultaneously, and system efficiency would be doubled. If the switching was done at the proper times and quickly enough, nobody would realize that the network resources were being shared.

Advanced packet-switching technology takes this process further by digitizing the information, compressing it, and breaking it up into small pieces, or packets. Individual packets are switched through the network based on the resources available at a given moment. The packets representing the original information typically travel many different paths, but eventually they arrive at their destination and are reassembled to form the original message. The number of people using the same network resources, and the amount of information conveyed, are increased dramatically without an apparent loss in service quality, provided the packets move through the network rapidly enough for the application. Some applications, such as transmission of a text page, can tolerate considerable delays in packet transmission, but others, such as full-motion video, cannot.

The constant improvement in digital compression and switching technologies, along with broader-bandwidth interconnections, is increasing our ability to transmit even greater amounts of information in an instant. This greater capability is having a dramatic impact on what we can communicate and how we do it.

5.10   THE INTERNET

Some people believe that the Internet *is* the information superhighway. For them the Internet, and especially the World Wide Web, is the quintessential convergence technology combining text, audio, and video into a new medium that will, eventually, replace most traditional electronic media. Although many people think that is unlikely, the

EXHIBIT 5.t
. . . .

From Switchboard to Switch

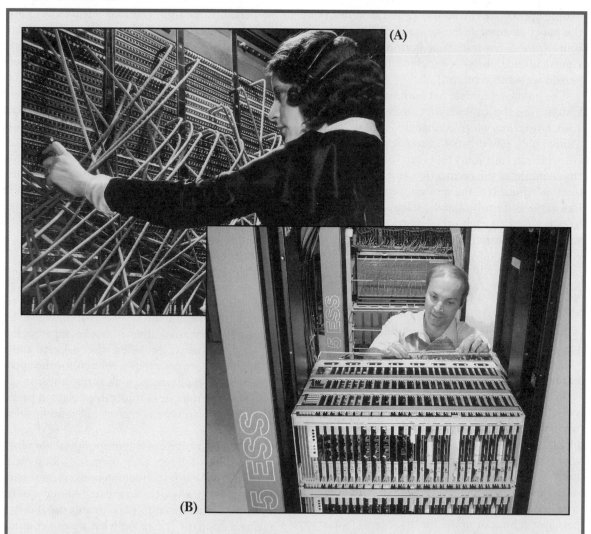

(A) Some people yearn for a return to the "good old days" when a telephone operator—a live human being—would ask each caller, "Number, please?" and then, using a manual switchboard with its many patch cords, make a physical connection of the caller's line to that of the intended receiver.

(B) But the volume and complexity of today's telephone traffic require sophisticated, though visually unarresting, digital switches, such as AT&T's 5 ESS system.

Source: (A) Corbis-Bettmann. (B) Property of AT&T Archives. Reprinted with permission of AT&T.

Internet was an important part of the digital revolution of the 1990s.

The Internet began in the late 1960s as a government-sponsored computing network called ARPANET. This network provided computer-to-computer communication among researchers working on defense-related projects at universities and research centers. Other research-oriented computer networks also developed. In the 1980s these networks became interconnected through the NSFnet (National Science Foundation network), a "network of networks." Interest in sharing computer data increased dramatically with the proliferation of computers in the hands of people outside the traditional research environment.

World Wide Web

The arcane protocols initially required limited use of the Internet to computer experts. There were attempts to make the Internet easier to use, such as providing a menu screen from which users simply selected the function they wanted performed rather than typing out complex instructions. In the late 1980s a Swiss nuclear research center (CERN) developed a new approach to linking computers, *hypertext transfer protocol* (HTTP). HTTP and hypertext markup language (HTML) have made possible the creation of the World Wide Web by allowing computer users to access files residing on other computers of any type in any location.

Use of the web increased dramatically following the 1993 introduction of *Mosaic*, a revolutionary software program providing network navigation aids and featuring the icons and point-and-click features familiar to users of Macintosh computers or PCs running Microsoft Windows. The Mosaic graphical user interface opened the door to computer-based communication for millions of people. Similar browser programs offering additional features and greater power appeared later, two of the most popular being *Netscape Navigator* and *Microsoft Internet Explorer.*

Netscape took an early lead in the browser wars. Microsoft began catching up rapidly when it bundled (included) Internet Explorer with the firm's ubiquitous Windows operating system software. This practice became one of the focal points during a Department of Justice investigation of Microsoft's business practices in the late 1990s. Netscape claimed that Microsoft used its quasi-monopolistic position in the operating system software market to unfairly leverage its web browser. America Online's merger with Netscape in 1999 weakened Netscape's argument that its value had been diminished by Microsoft's predatory practices. Nonetheless, late in 1999 a federal judge ruled that Microsoft had illegally used its monopolistic power to reduce competition in the software market. Microsoft's legal problems, however, appeared to have little effect on its browser business. By mid-1999 most estimates placed Netscape and Microsoft browser use about equal, with Netscape losing ground to Microsoft. The consumer's browser choice is important because browsers provide the gateway to the web. Individuals can change their "home" destination, but many start each web session by visiting the site programmed into the browser by its maker. Designers also try to develop consumer loyalty by allowing users to customize their browsers to bring up certain information automatically each time they log on. The race to remain competitive means that browsers constantly add new features that users can acquire by downloading improved versions on-line.

The number of sites on the web increased dramatically during the late 1990s, as did the number of users. A number of sites devoted to aggregating and searching the on-line universe became essential to finding information. Portals such as Yahoo and Lycos became prime cyberspace real estate because users go to them when trying to locate information on-line. They are highly valued by advertisers trying to get their messages to people using the web. And advertising is increasingly important on the web, as discussed in Section 6.8 (page 187). The Internet may have started in the noncommercial realm, but today web users are exposed to a plethora of banners, pop-up graphics, and jiggling icons promoting products and services. Those who used the Internet in its infancy often lament the increasing commercialization, much as the radio pioneers did in the early days of broadcasting. During the mid-1990s even the names of web sites began taking on great commercial value. In 1996 "tv.com" sold for a reported $15,000. The next year "business.com" went for $150,000. Businesses and organizations not interested in immediately launching a web site began registering domain names simply to protect their future options.

Internet Access

● ● ● ● ● ● ● ● ● ●

Most people access the Internet from their homes using a modem-equipped computer hooked to a conventional voice telephone line. The modem (*mo*dulator-*dem*odulator) is necessary because computers require digital data, while residential phone lines and their associated circuits are designed for analog audio. Internet access requires paying a fee to either an *Internet service provider* (ISP) or a commercial on-line service, such as America Online.

Subscribers connect to the Internet by calling their ISP or on-line service, where a computer answers. The user's modem and the host computer test the connection and then allow the user to log on. Once the user is identified as a subscriber, the host computer grants access to the services for which the user has paid.

During the mid-1990s businesses, schools, public libraries, and some individuals used high-capacity digital connections rather than modems and ordinary phone lines designed for analog voice communication. In the late 1990s higher-speed connections became more common.

Cable systems are offering high-speed Internet access in many communities. Cable modems can download data at speeds up to 10 megabits per second (Mbps) but usually the rate is less, and often much less. Upstream speed can reach several megabits per second but is also typically much slower. The speed depends on several factors, including the bandwidth allocated by the system operator to data transmission and the amount of noise or interference in the system. Cable capacity is a shared resource, so data speed drops as the number of subscribers simultaneously using the system for downloads increases. Some observers have suggested that being a shared network, cable may not provide subscribers as much security for their computer files as does a conventional phone line connection, but cable officials say that they are designing safeguards into their systems. System operators and cable modem manufacturers say that data delivery speeds may reach 30 Mbps or more as technology advances. One advantage of cable modems compared to conventional telephone access is that the cable modem is always on, and there is no need to go through the dialing/logging-on-process.

Telephone companies offer several alternatives to a conventional analog voice connection for In-

ternet users willing to pay more. *Integrated Services Digital Network* (ISDN) is one alternative, offering two 64-kilobits-per-second channels (128 kbps total) to carry the payload and an auxiliary 16 kbps channel for management and control functions. ISDN has not been popular with residential Internet users because of its cost and, sometimes, difficult installation and setup.

A newer technology called *digital subscriber line* (DSL) may be more successful. DSL is actually a family of technologies offering a range of performance. One promising version for residential consumers is *asymmetric digital subscriber line* (ADSL), which provides much greater downstream than upstream capacity. DSL technology is distance-sensitive, and the data rate drops as distance between the user and the nearest telephone switching office increases. Full-rate ADSL can provide downloads at about 6 to 8 Mbps and back-channel (upstream) rates of around 640 kbps at distances of about 18,000 feet (3.4 miles) under ideal conditions. As the new century began, however, the price for such performance was typically hundreds of dollars a month, assuming it was available at all. This placed ADSL at the top speeds beyond the reach of most residential Internet users. Slower-speed ADSL promising downloads at rates of about 700 kbps, was available from some telcos for about $50 per month, but prices and performance varied.

A variant of ADSL called *G.Lite* offers downstream speeds of about 1.5 Mbps and about 512 kbps upstream (over 18,000 feet) under ideal conditions. G.Lite was seen initially as an ideal service for residential Internet users because, unlike other types of DSL, G.Lite does not require a voice/data splitter at the subscriber's location. This significantly reduces equipment and installation costs. However, the later development of higher-rate ADSL technology that also dispenses with splitters may dampen the original enthusiasm that many people had for G.Lite.

The major advantage of DSL technology is that it uses the existing twisted-pair copper wires that already provide telephone service to most homes. Telephone companies can offer DSL service by simply installing new equipment in the subscriber's home and at the telco central office. Once installed, the same line used for normal voice communication can be used for data transmission simultaneously. This is an important consideration for heavy

Internet users who may be paying for two phone lines into the home, one for conventional Internet access and one for voice. DSL also provides the same "always on" convenience of cable modems. Unlike cable modems, DSL's bit rate does not decrease as more people use it, because DSL does not share capacity among users. Prices of DSL equipment and service are expected to fall as DSL becomes more widely available and popular with consumers. Consumer acceptance of new technologies is difficult to predict, but Allied Business Intelligence believes that more people will be using DSL than cable modems by the end of the year 2000. The firm estimates that there will be 9.4 million DSL subscribers, compared to 5.4 million cable modem subscribers, by 2005.

Satellites and wireless cable technologies also offer higher-speed Internet access, at least for downloads. DirecPC, for example, beams Internet data to PCs from geostationary satellites at a rate of about 400 kbps, and SPEEDUS.COM is using wireless technology in the local multipoint distribution system (LMDS) frequency spectrum to distribute Internet data around parts of New York City at download speeds exceeding 1.5 Mbps. The modems used by SPEEDUS.COM are capable of handling downloads at a blistering 48 Mbps. Both DirecPC and SPEEDUS.COM depend on conventional analog telephone lines for the back-channel (upstream) link from subscribers.

Several companies have placed, or are planning to place, dozens of satellites in low Earth orbit (LEO) to provide telephone, paging, and other personal communication services worldwide. Unlike communications satellites used for cable and broadcast network relays and direct-to-home broadcasting, LEO satellites are not in geostationary orbit (GEO). They are much closer to the Earth and do not hover over a fixed location on the ground. Several of the companies plan to offer Internet access as part of their service. The most ambitious plan is that of Teledesic, a venture conceived by cellular telephone pioneer Craig Mc-Caw and Microsoft founder Bill Gates. Teledesic wants to put hundreds of satellites in LEO so that a bird is available to cover every place on Earth at all times. Each satellite will not only communicate with terrestrial stations but also will be able to "hand off" information to other satellites in orbit, creating an "Internet in the sky." Teledesic says that most users will have two-way connections

providing up to 64 Mbps data rates downstream and up to 2 Mbps upstream. The system is designed for fixed-point Earth terminals but will accommodate mobile and portable users as well, including those aboard ships and planes. One advantage of the Teledesic plan, compared to GEO satellite systems, is that it will not suffer from delays associated with sending a signal 22,300 miles into space and waiting for a response. This delay, or *latency*, characteristic of GEO satellites is annoying for time-sensitive interactive communications, such as telephone service. The logistics and cost of building and maintaining such a huge system are awesome; initial cost estimates started at $9 billion. Teledesic says that the cost of the service to users will be competitive.

Some observers say that the Teledesic plan will never get off the ground. Others say that it will but that it will not provide the quality of service being promised, or that it will be too expensive. Teledesic does have significant support from major firms, including Boeing and Motorola. In July 1999 Teledesic announced a major launch contract with Lockheed Martin and another with Motorola to build its global, broadband Internet in the sky system. In mid-1999 Teledesic set 2004 as the target date for the start of its service. (In 1997 the target date was 2002.) If successful, Teledesic will bring high-speed Internet access to people and places where it has not been available previously.

Barriers to Growth
• • • • • • • • • • • •

The challenge of setting up a computer, making the necessary phone line or other connections, loading the browser software, and learning how to use it seems an enormous bother to some people. Manufacturers have responded by bundling hardware so everything needed is already installed. Software for the Internet also is coming already installed on the computer's hard drive. The graphical user interfaces of the browsers already discussed make using the Internet easier. Still there are those who find the idea of connecting to the Internet daunting.

Paradoxically, others are concerned that the Internet is too easy to use. Some parents would like having home access but hesitate for fear that their children will be exposed to pornographic content. This is possible because almost everything is available on the web, from a tour of the White House to directions for making poison gas. However, software is available that allows parents to block access to objectionable content.

The bother and fear of technology associated with using a full-featured PC for Internet access may fade as more products such as WebTV, smart set-top boxes with interactive features, and easy-to-use PC-like appliances that offer Internet services become commonplace. And, as noted in Chapter 1, young people do not suffer from technophobia to the same degree as do their elders. Today's well-educated under-thirty population grew up dealing with rapid technological change. Many like what the changing technology does for them and accept constant learning as part of their lives.

The lack of bandwidth provided by residential phone lines is also a barrier to the Internet's development. The bit speed of conventional modems increased dramatically during the 1990s. Compression technology also effectively increased the rate of delivery. But many dedicated Internet users find the rate inadequate for even motionless images, especially during peak use hours, when millions log on and begin accessing files simultaneously.

Cable modems, DSL, wireless cable, and both GEO and LEO satellites promise dramatically increased data delivery speeds. Availability and cost remain barriers to widespread adoption of these emerging technologies, and some of them may be commercial failures. But demand for broad-bandwidth "last mile" connections to the home is increasing as consumers want Internet services that require more bits per second. For some services, and in some locations, a wired system may prove most efficient. For other services, or for other locations, a wireless approach may be best. A combination of wired and wireless technologies will probably emerge to break the bandwidth barrier and satisfy consumer demands.

The Future

Estimates of Internet use vary wildly, but early in 2000 Nielsen NetRatings estimated that about 123 million Americans had Internet access in their homes. Add to this number millions more who access the Internet at work, at school, at libraries and other locations. Clearly, the Internet is an im-

portant new communications medium that many Americans use for a variety of purposes.

Currently, the Internet is not a serious rival to broadcasting and cable as a medium for the distribution of traditional programming. Video and audio streaming of programming is increasingly popular, but the quality is a pale shadow of what broadcasting and cable systems offer, especially for video. But broadcasters understand that the Internet offers both challenges and opportunities, and they are moving aggressively to use it for their own purposes. The National Association of Broadcasters estimated that more than 1,200 licensed commercial radio stations in the United States had web sites by mid-1999, many of them with streaming audio. It was a rare TV broadcast station or cable network that did not have a web presence by early 2000. However, most of those sites were designed to promote and complement the main programming service rather than compete with it directly.

Enthusiasts say that the Internet will become the predominant distribution medium for audio and video programming once consumers have a connection offering adequate bandwidth. That may be true, but not necessarily. The Internet is a packet-switched network that uses resources efficiently, but it may not be the best system for delivery of high-quality audio and video in real time. Streaming technologies currently compensate for the latency of the Internet with buffering. The user requests the audio or video program, but there is a delay before anything is heard or seen as the user's computer stores (buffers) the first part of the data stream. If the data stream slows down enough that the buffer runs dry, the program stops. Even with buffering, audio quality is lower than that of FM radio, and the video image is small, jerky, and badly out of synchronization with the audio. If a person had the option of experiencing a real-time event involving fast-paced action, such as a basketball game, on cable, DBS, or TV rather than on the Internet, the Internet generally remained a poor choice as the new century began.

The Internet may, however, be an effective means of delivering programming that is not presented in real time. The popularity of sharing MP3 audio files through the Internet demonstrates that this is a viable activity. And there is no technical reason that a TV show or movie can not be downloaded and stored for playback at normal speed at some later time, if the consumer has enough patience and storage capacity. This is only a variation on the time shifting many VCR owners do currently. The advantage to consumers is that the downloading is "scheduled" by them rather than their having to catch the show when the broadcaster or network decides to send it.

Even if the technological problems of merging the Internet with broadcasting are overcome, there may still be social and psychological restraints on blending them in the future. Internet surfing is different from watching TV or listening to the radio. The Internet requires a degree of active consumer participation, somewhat like reading a book or writing a letter. This interactive participation is one of the Net's most attractive features, but it does require a continuing series of decisions and actions. Many people like the fact that it does not take much mental or physical effort to watch TV or listen to radio. These people may wish to keep their viewing and listening separate from their time on the Internet. However, in 1999 Paul Kagan Associates found that about 20 percent of home Internet users watch TV while using the Internet (Paul Kagan Associates, 1999). This suggests that the Internet complements rather than replaces traditional media behaviors for many people. Broadcasters and advertisers, of course, hope people will blend viewing and listening with their use of the Internet by immediately purchasing on-line the products and services advertised in the programs.

The conversion from analog to digital TV will likely accelerate the convergence of the TV and the PC. Some observers say that computer manufacturers could build and sell 20 million to 50 million DTV-compatible computers by the time TV manufacturers build a million DTV receivers. If people use converged TV/PCs to receive DTV, the move to DTV may increase interest in the Internet and Internet-like services, because people will have greater access to the necessary tools. Conversely, merging TV into the PC may increase DTV viewing among those using the PC to access the Internet.

In summary the impact of the Internet on existing media distribution systems has been important, but it has not yet changed their essential function. As the century begins, cable and DBS, for example, are being used to enhance Internet delivery, but they are still viewed principally as vehicles for delivery of entertainment programming, with data (Internet-like) services a secondary function. However, it is clear that the development of the Internet during the last decade has had a profound influence on American society. It is equally clear that the Internet will continue to play an important role in the development of the electronic media in the future.

CHAPTER **6**

Commercial Operations

Commercial electronic mass media are businesses. Their primary motivation is to make money. And as in any business, the way to make money is to offer a salable product and to have income exceed expense. Advertising's dominant role in financing most broadcast media has had a profound impact on the types, number, and variety of program services offered. Cable and similar technologies rely mostly on subscriptions for their support, although advertising plays an increasing role in those media as well. Advertising on the Internet has grown explosively in the last few years and will probably continue to do so in the new century. This chapter focuses on commercial operations, leaving the special financial problems of public broadcasting to the next chapter.

6.1   THE BASICS

Exhibit 6.a summarizes the dimensions of the commercial and noncommercial broadcasting and cable industries in relation to total audience potential. The individual outlet—more than 16,000 stations and nearly 11,000 cable systems—forms the basic economic unit of these industries.

These 27,000 units function as local retailers, delivering programs directly to consumers. People sometimes use the terms *local station* and *local cable system*, but in fact *all* stations and systems are local in the sense that each is licensed or franchised to serve a specific local community. An exception to the local outlet model, *direct broadcast satellite* (DBS) systems, market their services directly to consumers on a nationwide basis. Similarly, a growing number of "virtual" radio stations are using the Internet to reach national and international audiences.

Despite the localism of licenses and franchises, economic efficiency favors centralization of station/system ownership and program production. Still more efficiency comes from vertical integration—common ownership of production, distribution, and delivery facilities (see Section 6.4). The large organizations that result increasingly dominate the media economy.

6.2   BROADCAST STATIONS

Definition

In the United States, the traditional commercial broadcast station can be defined as an entity (individual, partnership, corporation, or nonfederal governmental authority) that

- holds a license from the federal government to organize and schedule programs for a *specific community* in accordance with an approved plan
- transmits those programs *over the air*, using designated radio frequencies in accordance with specified technical standards
- carries commercial messages that promote the products or services of profit-making organizations, for which the station receives compensation

Within limits, an individual owner may legally control more than one station, but each outlet must be licensed separately to serve a specific community. Moreover, each license encompasses both transmission and programming functions. A station therefore normally combines three groups of facilities: business offices, studio facilities, and transmitter (including an antenna and its tower). Usually, all facilities come under common ownership, although in a few cases stations lease some or all of them.

Station Functions

Exhibit 6.b offers examples of *tables of organization* (sometimes called *organization charts*) for both a television and a radio station. These vary widely from facility to facility, but all outline the station's personnel structure and indicate who reports to whom.

All commercial stations need to perform four basic functions: general and administrative, technical, programming (including promotion and news), and sales.

General and administrative functions include the services that any business needs to create an appropriate working environment—services such as payroll, accounting, housekeeping, and purchasing. Services of a specialized nature peculiar to broadcasting usually come from external organizations, such as engineering consulting firms, audience re-

EXHIBIT 6.a
• • • • • • • • • •

Electronic Media Dimensions

Total U.S. Television Households (TVHH)	99.4 million
Commercial Radio Stations	
AM	4,784
FM	+5,720
	10,504
Noncommercial FM Stations	+2,040
Total Radio Stations	**12,544**
Commercial TV Stations	
VHF	561
UHF	+661
	1,222
Noncommercial TV Stations	
VHF	125
UHF	+245
	370
Low-Power TV Stations	
VHF	563
UHF	+1,594
	2,157
Total TV Stations	**3,749**
Total Broadcast Stations	**16,293**
Cable Systems	**10,700**

These data change almost daily but are representative for 2000.

Source: *Broadcasting & Cable*, February 7, 2000, p. 58. Copyright © 2000. Reprinted by permission of Cahners Publishing Company.

search companies, and program syndicators. For a network affiliate, the main such external contract is with its network.

Technical functions, usually supervised by the station's chief engineer, center on transmitter operations, which must follow strict FCC rules, and the maintenance and operation of studio equipment.

Programming functions involve planning and implementation. Major program planning decisions usually evolve from interplay among the programming, sales, and management heads. Because most stations produce few non-news programs locally, the program department's main job is to select and schedule prerecorded programs. A program manager typically serves as the department head in this area at a TV station (an operations manager in radio). Many stations have now dispensed with that title, leaving program decisions to the general manager or combining the function with that of the promotion department.

Promotion (increasingly called *creative services*) includes making its potential audience aware of a station's programs through advertising, on-air announcements, newspaper listings, and even T-shirts and bumper stickers.

News, though a form of programming, usually constitutes a separate department headed by a news director who reports directly to top management. This separation of news from entertainment makes sense because of the timely nature of news and the unique responsibilities news broadcasting imposes on management. At some TV stations news operations have taken on great importance by becoming major profit centers. At most radio stations, however, except for all-news stations, local news appears to be declining in importance.

Sales functions divide into local and national aspects. Station sales departments have their own staff members to sell time to local advertisers. To reach regional and national advertisers, however, a station usually contracts with a national sales representative firm that acts for the station in out-of-state business centers. A network affiliate benefits from a third sales force, that of its network. Exhibit 6.c, p. 170, suggests some of the tasks routinely performed by a television sales manager and other executives.

One job straddles the programming and sales functions—that of the traffic department. It coordinates sales with program operations, preparing the daily program *pre-log* (usually called simply the *log*), which schedules facilities, personnel, programs, and announcements. Traffic personnel ensure fulfillment of advertising contracts and arrange for *make-goods*—the rescheduling of missed or technically inadequate commercials when a TV program fails to

EXHIBIT 6.b
. . . .

Broadcast Station Tables of Organization

EXHIBIT 6.b
• • • • • • • • • • •

Broadcast Station Tables of Organization (continued)

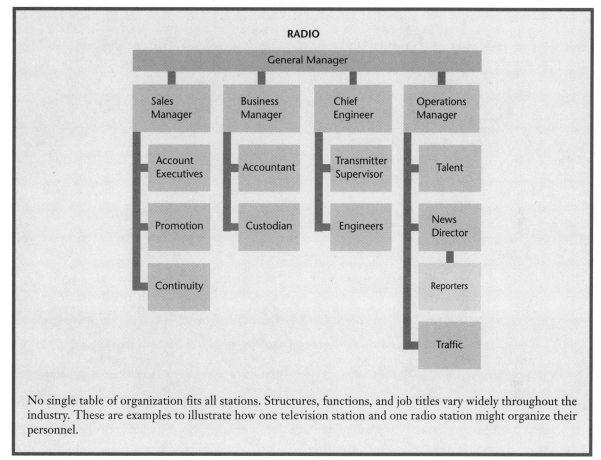

No single table of organization fits all stations. Structures, functions, and job titles vary widely throughout the industry. These are examples to illustrate how one television station and one radio station might organize their personnel.

deliver promised ratings. Traffic maintains a list of *availabilities* (or *avails*), keeping sales personnel up-to-date on commercial openings in the schedule. Traffic usually fills unsold openings with public-service or promotional announcements. The traffic department also works with the accounting department to make sure advertisers are billed for spots aired. At many stations computers handle much of the complex work of the traffic department.

Station Groups
• • • • • • • • • • •

Like other business enterprises, broadcasting can benefit from economies of scale. A company owning several stations can buy programs, supplies, and equipment in bulk and spread the cost of consultants and attorneys across several stations, sharing experiences and ideas.

For reasons of public policy, the FCC places limits on the number of radio and television stations one owner may control (see Section 13.6, page 387, for details). Still, within those limitations, the trend is toward greater concentration of control. Congress assisted that trend by passing the Telecommunications Act of 1996, which removed all national limits on the number of stations any one company may own, although there are still limits on the percentage of television households (35 percent) that any one TV group owner may cover. Also, there are limits on the number of stations one entity can own per local market.

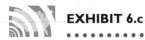

EXHIBIT 6.c
• • • • • • • • • •

A Day in the Lives of TV Executives

	General Manager	Program Director	General Sales Manager	News Director
8:30	Open mail; dictate letters and memos.	Check discrepancy reports for program and equipment problems; take appropriate action.	Check discrepancy reports for missed commercials; plan make-goods.	Meet with assignment editor and producer; plan the day.
9:30	Discuss financial statements with business manager.	Call *TV Guide* with program updates.	Local sales meeting; discuss accounts and quotas.	Meet with union shop steward; discuss termination of reporter.
10:00	Call group headquarters regarding financial status.	Prepare weekly program schedule.	Accompany local account executive on sales calls.	Read mail; screen tape of last night's newscast.
10:30	Meet with civic group angry about upcoming network program.	Select film titles for Saturday and Sunday late movies.	More sales calls.	Discuss noon news rundown with show producer.
11:00	Call network; ask for preview of questionable show.	Meet with promotion manager regarding *TV Guide* ad for local shows.	Call collection agency; discuss delinquent sales accounts.	Meet with chief engineer regarding satellite news gathering (SNG) failure.
12:00	Lunch with major advertiser.	Lunch with syndicated program saleswoman.	Lunch with major advertiser.	Monitor noon news; lunch at desk.
2:00	Department heads meeting.	Department heads meeting.	Department heads meeting.	Department heads meeting.
4:00	Meet with chief engineer regarding new computer system in master control.	Meet with producer/director to plan local holiday special.	Prepare speech for next week's Rotary Club meeting.	Meet with producer and director; plan rundown of 6:00 P.M. newscast.
6:00	Dinner with promotion director job candidate.	Attend National Academy of Television Arts and Sciences annual local banquet.	To airport; catch flight to New York for meeting with national sales rep.	Monitor 6:00 P.M. news.

Not all television station executives work 10- or 12-hour days, but many do, especially as competition from cable and the other new media increases.

Consequently, there was a flurry of activity in 1996 and thereafter as group owners sought to increase their holdings. Within a year of the act's passage several radio groups owned more than 100 stations, and the largest groups increased their holdings over the next two years to more than 300 stations. At the end of 1999 two radio group owners, Clear Channel Communications and AMFM, merged, creating the largest radio group owner in the United States, with 850 stations (see Exhibit 6.n, p. 202). The largest television group owner at the turn of the century, Paxson Communications, owned 90 stations. Interestingly, at the same time Congress allowed for such increased concentration, it also ordered the FCC to find new ways to promote the ownership of media by small businesses, presumably in an effort to diversify electronic media ownership.

All stations owned by a group do not necessarily affiliate with the same network—or, for that matter, with any network. Stations owned by the networks themselves, known as network *owned-and-operated* (O&O) stations, are the exceptions. Each national television network's O&O group reaches between 24 and 35 percent of the nation's television households, ensuring ABC, CBS, NBC, and Fox their own prestigious outlets in major markets.

6.3   BROADCAST TV NETWORKS

Definition

Although the term *television network* means one of the major national commercial networks to most people, in fact there are many television networks in operation. Some are national, some regional, and some exist only for the purpose of distributing one or two programs. The rules dealing with affiliation agreements between networks and their affiliate stations define network, or "chain," broadcasting as "simultaneous broadcasting of an identical program by two or more connected stations" (47 U.S.C. 153[p]). Until recently there were special rules dealing with program syndication and financial interests in independently produced programming that applied only to the Big Three national networks, but those rules have been rescinded for the most part.

For years only three major national commercial television networks existed—ABC, CBS, and NBC. Fox premiered in 1986, long after the Big Three networks were well established, and slowly began to increase its programming and to improve its affiliate line-up. In May 1994 Fox made enormous strides toward parity with the Big Three when it entered into an agreement with New World Communications Group whereby New World would switch affiliations of its major-market stations to Fox. The move created a domino effect as the major networks scrambled to line up new affiliates. Fox moved to near parity with the established networks by gaining the network rights to NFL Football (outbidding long-time rights holder CBS) and by developing a network news service, Fox NewsEdge. Each of the Big Three plus Fox now have about 200 affiliates, through which they can reach virtually all U.S. households.

Other national networks include The WB, the Warner Bros. entry into the television network business. The WB premiered on January 11, 1995, and was followed five days later by United Paramount Network (UPN). A seventh broadcast network focusing on family programming, PAX TV, debuted in the fall of 1998. Like Fox before them, these networks began with limited hours of programming but had aspirations of becoming fully competitive national broadcast television networks. PAX TV received a definite boost in that regard when NBC purchased a 32 percent stake in the network. In addition, more than a hundred regional and part-time networks are in operation.

Like stations, networks vary in their organizational structure, yet each must fulfill the same four basic functions of administration, programming, technical engineering, and sales. Networks, however, enjoy the luxury of a much higher degree of specialization than do stations. NBC, for example, has separate units to handle operations, entertainment, news, sports, the television network, and NBC owned-and-operated stations.

Affiliation

More than half of all full-power commercial television stations affiliate with one of the major networks. Most function as a *primary affiliate*—the only affiliate of a given network in a given market. *Secondary affiliates*, typically those in markets with only two stations, share affiliation with more than one network. A few markets have only a single station. In Glendive, Montana, for example, KXGN-TV has the unusual privilege of picking and choosing programs from CBS, NBC, and Fox.

Affiliation does *not* mean that a network owns or operates the affiliated stations. The networks' parent companies do, of course, own some television stations. But they contract with hundreds of other affiliates, agreeing with each to offer it the network's programs before offering them to any other station in the same market.

The station in turn agrees to *clear* time for all or portions of the network schedule. However, it has the right to decline to carry any specific program, or it may offer to carry the program at a time other than that of network origination (a *delayed broadcast*, or DB); the network may or may not agree to this last option.

In effect, traditional television networks use affiliates' time to deliver station audiences to network advertisers. In exchange, a broadcasting network offers its affiliates

- *monetary compensation* (in some cases) based on audience size and composition
- a structured schedule of *network programs*, usually at no direct cost
- *simultaneous program distribution* so that affiliates can receive the service at the same time
- an *advertising environment* that appeals to the affiliates' clients
- a *sales organization* that finds national clients to purchase advertisements that occupy the network's portion of the affiliates' commercial time

Compensation
• • • • • • • • • •

Networks and their affiliates formalize the link between them through an *affiliation contract*. A clause at the heart of the network-affiliate contract defines the terms on which the network will make payments—called both *station compensation* and *network comp*—to the station for the right to use the station's time.

Each television network uses a different plan for calculating compensation, and the plans have changed significantly since the mid-1990s (see "Changing Network-Affiliate Relations" on page 173). The amounts paid to affiliates vary from station to station, reflecting differences in market size, station popularity, and other factors. Rates in top markets such as New York and Los Angeles are as high as thousands of dollars per hour. In smaller markets they are as low as $50, and some affiliates in the smallest markets receive no compensation at all.

Network compensation represents a surprisingly small percentage of the gross revenues of network-affiliated stations—on the average, less than 5 percent. But stations measure the value of affiliation less in terms of compensation than in terms of the audiences that network programs attract.

Affiliates profit from the sale of spots in the 90 seconds or so that the network leaves open for affiliate station breaks in each prime-time hour of network programming and the seven or eight minutes made available at other times of the day. Moreover, the stations' own programs (whether locally produced or purchased from syndicators) benefit from association with popular and widely promoted network programs.

Clearance
• • • • • • •

Networks and their affiliates experience a somewhat uneasy sharing of power, complicated by political and economic factors too subtle for contracts to define. In one sense the networks have the upper hand. Affiliation plays a vitally important role in an affiliate station's success. However, without the voluntary compliance of affiliates, a television network amounts to nothing but a group owner of a few stations rather than the main source of programming for some 200 stations.

The complex relationships between networks and their affiliates hinge on the act of *clearance*—an affiliate's contractual agreement to keep clear in its program schedule the times the network needs to run its programs. An affiliate might fail to clear time or might preempt already cleared time for several reasons. Network public-affairs and other nonentertainment offerings usually get low ratings and therefore most often fail to get clearance. Sometimes stations skip low-rated network programs in favor of syndicated shows simply to keep audiences from flowing to the competition. Often a station wants to increase the amount of commercial time available. It can run more commercials than the network allows by substituting a syndicated program or a movie.

Networks rely on affiliates not only to carry their programs but also to carry them *as scheduled*. Delayed network broadcasts erode national ratings. Networks also need simultaneous coverage to get the maximum benefit from promotion and advertising.

In practice, affiliates accept about 90 percent of all programs offered by their networks as scheduled, most of them on faith. Stations can request advance screening of questionable programs but usually feel no need to do so, even though, as licensees, stations rather than networks have the ultimate legal responsibility. Because most television programs come in series, affiliates know their general tone, so the acceptability of future episodes can usually be taken for granted.

Thus affiliates have little or no direct influence over the day-to-day programming decisions of their networks. In the long run, however, they have powerful leverage. Network programming strategists take serious note of the feedback that comes from their affiliates.

Changing Network-Affiliate Relations

Starting in the late 1980s, when rising costs and increasing competition combined to weaken ABC, CBS, and NBC, the traditional network-affiliate relationship began to crack under the strain. Affiliates felt that network compensation failed to reflect the true value of their time to their networks. The networks increased the amount of precious advertising time available to themselves in their schedules, a move affiliates resented. Preemptions reached all-time highs, costing the networks millions of dollars in lost revenues.

In the mid-1990s, after a flurry of affiliation switches, the networks experienced a renewed appreciation for their station "partners." In many cases they raised compensation rates. And whereas historically they had limited affiliation contracts typically to 2 years' duration, they began writing agreements with their most desired affiliates that ran as long as 10 years. With the longer contracts, however, came network expectations—and increased pressure—for more complete program clearances by their affiliates.

By the end of the decade, changes in the business led to strained relations between the networks and affiliates. When The WB began operations in 1995, it broke with tradition in several important ways. First, the network affiliated with cable television systems in markets where no broadcast affiliate was available. Second, instead of paying compensation, The WB follows a "reverse compensation" plan under which affiliates, rather than receiving payments, pay the network 25 percent of all incremental profits they derive from their association with it. Following this lead, several other networks "floated" the idea—much to the dislike of the affiliates—of ending compensation and requiring affiliates to pay for network programming. Fox succeeded in convincing its affiliates of the needed change, and the network estimates that it will receive more than $200 million dollars from its affiliates over a three-year period. The trend continued in 2000 when Granite Broadcasting Company agreed to pay NBC $36 million per year so that its station, KNTV, could be NBC's new affiliate in the San Francisco market.

Other issues increased the strain between the networks and their affiliates. The increased cost of network programming and the networks' declining revenues forced them to seek new arrangements. Fox, for example, convinced its affiliates to pay a share of the network's royalties to the National Football League. It then irked its affiliates by announcing that it would take back several minutes of local advertising time during prime-time programming. Other networks clashed with affiliates over program "repurposing"—the reuse of network programming on other media, such as cable networks. Affiliates argue that this repurposing undermines the importance of the broadcast outlet. Tensions ran so high between ABC and its affiliates that the parties simply stopped talking to each other for several months in 1999.

But affiliates are not helpless in this struggle with the networks. Technology has freed many affiliates from their former, almost total dependence on networks for nonlocal news. Using satellites and minicams, stations can now cover not only local but also national and even international news. In fact, the roles are partially reversed, making the networks dependent on affiliates for coverage of local stories of national significance. An especially dramatic example of this dependence occurred in June 1994 when the networks relied on local station helicopters for pictures of O. J. Simpson as his car traveled Los Angeles freeways on the way to his home, where he was arrested for the murders of his former wife and her friend. Other examples of this role reversal occurred when the networks received much of their information and video of the April 1995 Oklahoma City bombing incident and the April 1999 Littleton, Colorado, high school shootings from their local affiliates.

Competition from Independents

Approximately 400 stations, most of them UHF, are known as *independents*. They have no full-service network affiliation, although they may be affiliated with Fox, UPN, The WB, PAX TV, or one of the regional networks. As Fox continued to enlarge its presence, however, calling a Fox affiliate an independent became at best a legal technicality.

The FCC's *prime-time access rule* (PTAR) long played a major role in turning some independents into profit makers. Starting in 1971 the PTAR gave independents their first chance to counterprogram effectively against network affiliates in the 7:00 to 8:00 P.M. (Eastern and Pacific time) period. They gained the advantage of rerunning network series during that hour, a program option that the PTAR denied to affiliates in the major markets. Citing a changing media environment, the FCC voted to rescind the PTAR effective in 1996.

Other factors favoring independent stations included their coverage of live sports events, their success in outbidding affiliates for popular syndicated program series, their aggressive promotional campaigns, and their establishment of the Association of Independent Television Stations (INTV) in 1972. Recognizing that most of its members had some form of network affiliation, the association changed its name in 1996 to the Association of Local Television Stations (ALTV). Cable also helped by making the signal quality of UHF independent stations equal (for cable viewers) to that of VHF affiliates.

The fact that so many formerly independent stations have now affiliated with one of the emerging networks demonstrates that affiliation with a national network remains one of the most valuable assets a television station can have.

Network Regulation

Unable to control the networks directly (because it does not license them), the FCC regulates them indirectly through rules governing the stations they own and the contracts affiliates make with them.

The FCC's network regulations were originally intended to ensure station autonomy and to pre-

vent the networks from becoming the overwhelmingly dominant force in broadcasting. For example, regulations prevent network contracts from forcing stations to clear time for network programs. However, as network power waned with the increasing strength of competing media, the FCC became less concerned with network dominance.

Nevertheless, the FCC still prohibits the national broadcast television networks from preventing an affiliate from accepting programs from other networks, nor may an affiliate prevent its network from offering rejected programs to other stations in the affiliate's market.

 6.4 CABLE

Cable/Broadcast Comparison

The economic organization of cable television systems and networks differs substantially from that of broadcast stations and networks.

Cable systems depend primarily on subscriber fees, not advertising, for their revenue. A cable system therefore owes allegiance only to the television households in its franchise area that choose to subscribe (in 2000 as many as 91 percent of the households in Palm Springs, California, for example, but averaging about 68 percent of all television households in the country).

A commercial broadcaster, in contrast, depends almost entirely on advertisers and has a legal obligation to serve the total audience in its market area, which is usually much larger than that of a cable system. Large cities often divide their municipal areas into several different cable franchises. Cable systems outnumber commercial television stations about 9 to 1. Whereas most viewers can tune in a number of local radio and television stations, almost without exception they can subscribe to only a single cable system.

System Organization

By the year 2000 more than 10,700 *cable systems* (the basic units of the industry) served about 68 million subscribers in the United States. More than 95 percent of cable subscribers had access to

EXHIBIT 6.d

Cable System Table of Organization

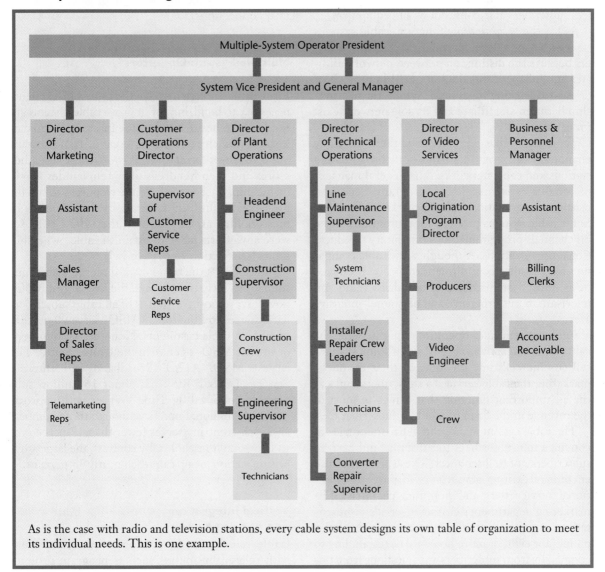

As is the case with radio and television stations, every cable system designs its own table of organization to meet its individual needs. This is one example.

30 or more channels; about 61 percent could see 54 or more; some could view more than 100. Exhibit 6.d shows how a typical system structures its organization. Whether small, with a few hundred subscribers, or large, with thousands, each cable system performs the same four basic functions as do broadcast stations, as described in Section 6.2.

However, a cable system's *technical* functions differ from those of a broadcast station. Broadcast technicians' jobs end when the signal leaves the transmitter—listeners/viewers are on their own when it comes to arranging for reception. Cable technicians must be concerned with the integrity of both the sending and the receiving aspects of the system. Often this division of responsibility results in a distinction between *inside-* and *outside-*plant personnel.

The inside group operates and maintains studio facilities, as well as the complex array of equipment at the *headend*, which receives programs by various means from program suppliers and processes the signals for delivery via coaxial and fiber-optic cables to subscribers. The outside group installs and services subscriber cable connections and equipment.

As for the *programming* function, cable systems start with a more level playing field than broadcasters because the distinction between network affiliates and independents does not exist in cable. Nor does broadcasting's elaborate symbiotic relationship between an affiliate and its sole network exist for cable.

Typically, cable systems fill their multiple channels with programs from both broadcasting stations and cable networks. The general manager of smaller systems makes program decisions, usually in consultation with the marketing director. Some cable systems produce local programs, often on channels programmed by community or educational organizations. Although a few cable companies have made a commitment to locally produced news, most systems defer to broadcast radio and television in this expensive and personnel-intensive area.

Cable system *sales* operations might better be called *sales and marketing*. Although some systems sell commercial time on some of their channels, marketing the cable service to subscribers ranks as the all-important function that brings in most of the revenue in the form of monthly subscriber fees.

The cable system marketing department tries to convince nonsubscribers to subscribe and current subscribers not to disconnect. These personnel also encourage existing subscribers to purchase additional service tiers and premium channels. The marketing department's customer service representatives interface with the public, answering telephones for eight or more hours a day, responding to complaints from subscribers and questions from potential customers. A system's ability to handle these contacts promptly and skillfully can have a profound effect on its financial success and, in extreme cases, on whether it keeps or loses its franchise.

System Interconnection
• • • • • • • • • • • • • • • •

Often several cable systems in a large market *interconnect* so that commercials (or programs) can be seen on all of them, rather than only one, providing combined advertising coverage equivalent in reach to that of the area's television stations.

System interconnection may involve physically linking the systems by cable, microwave, or satellite (*hard* or *true* interconnections), or it may depend simply on the exchange of videotapes (*soft* interconnections).

Multiple-System Operators
• • • • • • • • • • • • • • • • • •

Firms large enough to make the capital investments necessary to build, buy, or improve cable systems are not likely to be attracted by the limited potential of a single franchise. Thus emerged a trend toward *multiple-system operators* (MSOs)—firms that gather scores and even hundreds of systems under single ownership. Such firms have the resources to bid for high-cost, politically intricate cable franchises in metropolitan areas. Not until the 1992 Cable Act were any efforts made to establish cable ownership limits (see Section 13.6, page 387).

In the 1990s the trend toward mergers and consolidations increased. In fact, the five largest MSOs now serve almost two-thirds of all cable subscribers nationwide, and the top 25 MSOs serve 90 percent of all U.S. cable customers. Nonetheless, hundreds of smaller MSOs operate in the United States. The largest MSO, AT&T Broadband and Internet Services (AT&T BIS), has about 16 million subscribers, followed by Time Warner with approximately 13 million. Impressive though those numbers are, each company serves fewer than 17 percent of all television households. By contrast, the largest television station group can reach nearly 35 percent.

Vertical Integration
• • • • • • • • • • • • •

Cable television often involves ownership links with related businesses such as program production, distribution, and delivery. This type of linkage is called *vertical integration* and the cable industry resorts to it increasingly. Time Warner, by virtue of its acquisition of the Turner Broadcasting System, offers an excellent example. Time Warner operates the nation's second-largest MSO, serving 13 million subscribers. It also owns 69 percent of The WB television network, several film and television production companies, and nearly 20 cable television networks, including CNN, Home Box Office (HBO), TNT, and Turner Classic Movies. Not all cable networks are owned by MSOs, however. With the proliferation of new networks in the late

1990s, the percentage of cable networks owned by MSOs is actually declining.

Enter the Telcos
● ● ● ● ● ● ● ● ● ●

For much of the history of the cable television industry, telephone companies were prohibited from providing cable television services in the same area they provided local telephone services. The Bell telephone companies were prohibited from providing cable services altogether. This changed dramatically with the passage of the 1996 Telecommunications Act, by which Congress, attempting to stimulate competition in the cable television industry, allowed telephone companies to begin delivering cable services—both within and outside their local telephone service areas. According to the legislation, telephone companies can offer video programming services in a number of different ways: over the air as do wireless cable operators; as common carriers; by obtaining franchises and operating as a traditional cable system; or through a hybrid operation known as open video systems, which combines aspects of common carriage and the cable TV model of operation (see Section 12.9, page 362).

By the beginning of 1997, several telephone companies were taking advantage of their newfound freedom. Some companies began investing in wireless cable systems, and others chose to operate as traditional cable systems. Bell Atlantic operated a system in Dover Township, Delaware, and entered into a marketing agreement with DBS Operator DirecTV; Ameritech was supplying service to subscribers in Columbus, Ohio, and suburban Chicago and Detroit; Bellsouth initiated service in Chamblee, Georgia; GTE built a system in Clearwater, Florida; and US West was active in Omaha, Nebraska. Although many local telephone companies were more interested in entering the newly opened long-distance telephone market, some telcos were ready to compete head-to-head with the cable television industry. By the end of the decade, local telephone companies provided cable services to roughly half a million subscribers. Industry analysts expect that number to grow to more than 3 million by 2008.

Though not a local telephone service provider, AT&T entered the cable television business in a big way when it acquired the nation's largest MSO, Telecommunications, Inc. (TCI), plus several other large MSOs in 1999. In addition to providing traditional cable services, AT&T BIS intends to use its cable infrastructure to deliver local telephone and high-speed Internet services to its subscribers. By "bundling" several services together into one package, AT&T believes that it can offer its subscribers lower costs overall. Other long-distance carriers, notably MCI Worldcom and Sprint, have countered AT&T's move by acquiring wireless cable systems throughout the country, thus guaranteeing them direct local access to their customers.

● ●
6.5 CABLE PROGRAM SERVICES

Cable systems typically carry most broadcast television stations whose signals cover their franchise areas. In addition, some systems produce limited amounts of their own programming and offer *access channels* for programs produced locally by others. Beyond that, they draw on three main types of centralized program providers: basic-cable networks, superstations, and pay-cable/pay-per-view networks. Despite ever-increasing channel capacity, most cable systems are unable to carry all of the program services available to them. As the major cable networks began programming multiple channels rather than only one, competition for carriage intensified. Exhibit 6.e gives the data for a representative selection of cable program providers.

Basic-Cable Networks
● ● ● ● ● ● ● ● ● ● ● ● ● ● ●

The 1992 Cable Act requires cable operators to offer subscribers a *basic service tier* (or level) of program sources, including, at a minimum, local television stations and all public, educational, and government access (PEG) channels. (See Section 12.8, page 359, for TV stations' *must-carry* and *retransmission consent* options.) A cable operator may add other channels of programming to this minimum, or *lifeline*, service. Subscribers pay a monthly fee to receive the basic tier.

The 1992 Cable Act defines *cable programming service* as all video programming provided over a cable system except that provided on the basic service tier. This includes programming from such *cable networks* as USA, ESPN, and MTV, nearly all of which

EXHIBIT 6.e
• • • • • • • • • •

Examples of Satellite-Distributed Cable Program Services

Advertiser- and Industry-Supported Cable Program Services		
Network **(launch date/owner)**	**Homes Reached** **(millions)**	**Content**
ESPN (1979/ABC; The Hearst Group)	77.1	Sports events, news, and information
CNN (1980/Time Warner, Inc.)	77.0	News, sports, weather, business
USA (1980/Seagrams; Liberty Media)	77.2	Series, movies, specials, sports
C-SPAN (1979/Cable industry)	77.0	House of Representatives, public-affairs programs
TBS Superstation (1976/Time Warner, Inc.)	78.0	Movies, sports, specials, series
MTV (1981/Viacom)	73.2	Music videos, series, pop culture
QVC (1986/Comcast–AT&T BIS)	72.2	Homeshopping

Premium (Pay) Services		
Network **(launch date/owner)**	**Homes Reached** **(millions)**	**Content**
HBO (1972; 1975 via satellite/ Time Warner)	26.7	Movies, sports, specials
Showtime (1976; 1978 via satellite/ Viacom)	13.6	Movies, sports, specials
Viewer's Choice (PPV) (1985/Cable consortium)	27.0	Movies, sports, specials

Source: *Cable Television Development*, Spring/Summer 2000, National Cable Television Association (Washington, D.C.). *Television and Cable Factbook 2000*, Services Volume. Warren Communications News, Washington, D.C., 2000.

are *advertiser supported*. Some cable systems include such networks in the basic service tier, others on what they call an *expanded basic tier*. Some systems have several tiers, each for a separate fee.

For virtually all basic-cable network programs (home shopping networks and Video Jukebox—now named simply "The Box"—are exceptions), a cable system pays a fee directly to each network that it carries. Fees range from just a few cents up to 50 cents or more per subscriber per month. As competition for channel space increased in the late 1990s, some new cable networks, such as Animal Planet and Home & Garden Television, offered cable operators up-front cash payments of up to $8 per subscriber to secure carriage on their systems. This practice may well continue as the number of new cable networks increases. A basic-cable network sells commercial spots to its own advertising clients and usually leaves about two minutes of advertising time each hour for local sale by cable systems.

Advertising-supported cable networks have much smaller staffs than the Big Four broadcast networks, and they reach smaller audiences. They maintain commercial sales departments but also must devote much attention to selling *themselves* to cable systems. Unlike the major broadcast networks, some have difficulty finding affiliates because so many cable networks compete for outlets.

Superstations

The *superstation* is a paradoxical hybrid of broadcasting and cable television—paradoxical because, although the FCC licenses each broadcast station to serve only one specific local market, superstations also reach hundreds of other markets throughout the country by means of satellite distribution to cable systems (see Exhibit 3.d, page 62).

Cable operators often include superstations in one of their basic packages of channels, paying a few cents per subscriber per month for the service. The station gets most of its revenue through higher advertising rates, justified by the cumulative size of the audiences it reaches through cable systems.

Before superstations altered the rules, copyright holders selling syndicated programs to individual stations had based their licensing charges on the assumption that each station reached a limited, fixed market. For example, syndicators formerly licensed programs to WTBS to reach only the Atlanta audience. But once WTBS achieved superstation status, its programs reached audiences in hundreds of other markets, in many of which television stations may have paid copyright fees for the right to broadcast the very same programs.

Responding to complaints about the inequities created by duplicate television program distribution on superstations, the FCC reimposed its *syndicated exclusivity* (syndex) rule in 1988. Syndex requires each cable system, at a local television station's request, to delete from its schedule any superstation programs that duplicate programs for which the station holds exclusive rights.

The rule spawned a new industry: companies that supply programs cable systems can insert on a superstation channel to cover deleted programs. WTBS, meanwhile, worked out a *blackout-proof* schedule, free of syndicated programs subject to exclusivity clauses.

In 1996, the role of the superstation began to change. WWOR, a popular superstation from New York, stopped its nationwide distribution, and Ted Turner announced that WTBS would cease operations as a superstation. Turner replaced WTBS with a similar cable network called TBS Superstation. These events leave WGN in Chicago as the most prominent and popular superstation still in operation.

Some cable systems also carry *radio superstations*, notably Chicago's Beethoven Satellite Network (WFMT-FM), a classical music station. Listeners pay an extra fee to the cable system operator to receive such radio stations, as well as other audio services, which go through a special cable connection to their home high-fidelity stereo systems. Subscribership to premium audio services has never exceeded 3 percent of cable households.

Pay-Cable Networks

Subscribers pay their cable systems an additional monthly fee to receive *pay-cable*, or *premium*, networks. In exchange, the subscriber gets programs (recent movies, sporting events, music concerts, etc.) without commercial interruption. Pay-cable subscription fees average about $10 per month for each of the more popular premium channels. The cable operator negotiates the fee with the program supplier, usually splitting proceeds 50/50.

Included under pay cable are *pay-per-view* (PPV) networks. Instead of paying a monthly fee, PPV subscribers pay a separate charge for each program they watch, much as they would at a movie theater. Although their programs resemble those of regular pay-cable networks, PPV services usually offer recent theatrical releases before they are available on other channels, as well as special events that regular pay-cable networks decline to carry because of the high price tag. Professional wrestling and boxing matches typically draw the largest PPV audiences. Like other program services, PPV networks take advantage of multiplexing, offering their fare on several channels. This allows them to offer more programs and to schedule them more times each day.

6.6 ADVERTISING BASICS

Unlike cable television, broadcasting in America operates primarily as an advertising medium. Dependence on advertising colors every aspect of broadcast operations. As Exhibit 6.f indicates, except for direct mail, broadcast television ranks as the largest *national* advertising medium, though newspapers surpass television in total advertising volume. Radio comes in fourth in *total* advertising dollars, after television but ahead of magazines and the Yellow Pages. Advertising on cable and the Internet, although rapidly increasing, remains small relative to advertising on other electronic media.

Broadcast Ads

Broadcast advertising falls into three categories, defined by coverage area and image: *local*, *network*, and *national-spot*.

Local advertising comes mostly from fast-food restaurants, auto dealers, department and furniture stores, banks, food stores, and movie theaters. When such local firms act as retail outlets for nationally distributed products, the cost of local advertising may be shared by the local retailer (an appliance dealer, for example) and the national manufacturer (a maker of refrigerators). This type of cost sharing, known as *cooperative advertising*, or just *co-op*, supplies radio with a major source of its revenue.

When a station connects to a national network, it instantly becomes a medium of national advertising. For advertisers of nationally distributed products, broadcast *network advertising* has significant advantages:

- The advertiser can place messages on more than 200 stations of known quality, strategically located to cover the entire country.
- Advertising can be placed on all of those stations with a single transaction.
- The advertiser has centralized control over commercial messages and assurance that they will be delivered at the chosen times and in a preferred program environment.
- The advertiser benefits from sophisticated network market research.
- The advertiser gains prestige from the very fact of being on a national network.

Despite those advantages, some national advertisers find networks too costly or too inflexible. They have the option of using *national-spot advertising*. Advertisers using national-spot advertising work through their advertising agencies' and stations' national sales representatives to assemble ad hoc collections of nonconnected stations. The commercial announcements go out to the chosen stations by mail or by satellite. As Exhibit 6.g shows, television depends heavily on national-spot sales.

National-spot advertisers can choose from several program vehicles: station-break spots between network programs, participating spots in local or syndicated programs, and sponsorship of either local or syndicated programs. National-spot advertising thus enables advertisers to capitalize on audience interest in local programs, something the network advertiser cannot do. Exhibit 6.h shows that the largest national advertisers use spot and network advertising in combination to achieve better coverage than either could yield on its own.

Pros and Cons

As an advertising medium, broadcasting has unrivaled access to all family members under the changing circumstances of daily living. In addition, car and portable radios allow broadcasting to compete with magazines and newspapers as an out-of-home medium. Television is the only medium that can vividly demonstrate how a product works. Above all, the constant availability of broadcasting as a companionable source of entertainment

EXHIBIT 6.f
. . . .

Advertising Volume of Major Media: Local vs. National

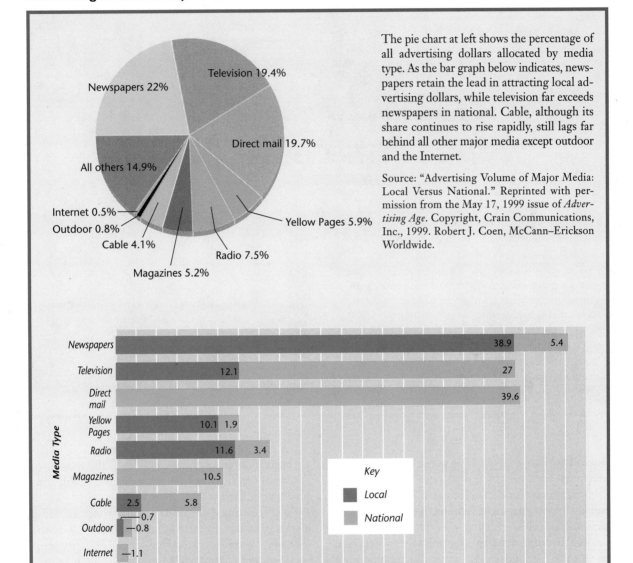

The pie chart at left shows the percentage of all advertising dollars allocated by media type. As the bar graph below indicates, newspapers retain the lead in attracting local advertising dollars, while television far exceeds newspapers in national. Cable, although its share continues to rise rapidly, still lags far behind all other major media except outdoor and the Internet.

Source: "Advertising Volume of Major Media: Local Versus National." Reprinted with permission from the May 17, 1999 issue of *Advertising Age*. Copyright, Crain Communications, Inc., 1999. Robert J. Coen, McCann–Erickson Worldwide.

EXHIBIT 6.g

Broadcast Station and Cable System Revenue Sources

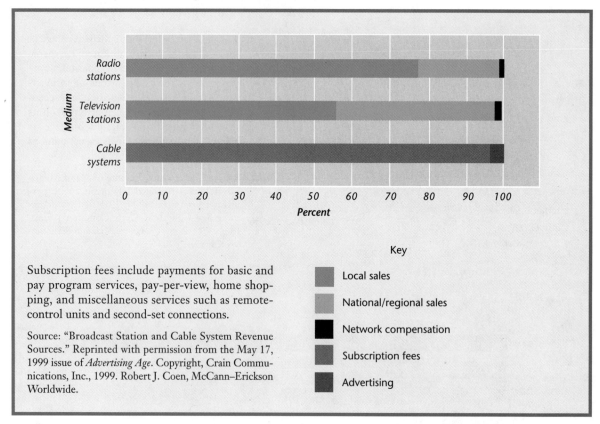

Subscription fees include payments for basic and pay program services, pay-per-view, home shopping, and miscellaneous services such as remote-control units and second-set connections.

Source: "Broadcast Station and Cable System Revenue Sources." Reprinted with permission from the May 17, 1999 issue of *Advertising Age*. Copyright, Crain Communications, Inc., 1999. Robert J. Coen, McCann–Erickson Worldwide.

Key
- Local sales
- National/regional sales
- Network compensation
- Subscription fees
- Advertising

and information gives it a great psychological advantage.

However, commercials must make their point quickly. Normally, most radio commercials last 60 seconds, and most television commercials last 30 seconds, though there is a growing trend toward 10- and 15-second spots. Even the longest spot cannot duplicate the detail of a large printed department-store ad or a supermarket ad with clip-out coupons. Listeners or viewers cannot clip out broadcast ads to consult later. Nor can broadcasting compete effectively with the classified sections of newspapers, despite attempts such as want-ad and home-shopping programs.

A broadcasting station has a limited commercial *inventory* of openings for commercials in its program schedule. Its single channel allows it only 24 hours of "space" each day. Moreover, only so many commercials can be crammed into the schedule

without alienating the audience. Print media, in contrast, can expand advertising space at will simply by adding pages.

Broadcasting also suffers the drawback of being unavailable to advertise some products. Congress forbade cigarette advertising in 1971, and most broadcasters voluntarily ban products such as hard liquor and "adult" movies.

Integrating Ads into Broadcasts

Most broadcast advertising takes the form of *announcements*, or *spots*. Advertisers usually prefer *scatter buying*—spreading their spots over several programs. That avoids risking too much on any one program and gains exposure to varied audiences.

Some programs have natural breaks where spots can be inserted without interrupting the flow—

EXHIBIT 6.h

How Top Advertisers Allocated Their Budgets in 1998

Rank	1	2	3	4	5
Advertiser	General Motors	Procter & Gamble	Phillip Morris	Daimler-Chrysler	Sears, Roebuck
Total estimated ad expenditures (in billions)	$2.9	$2.7	$2.05	$1.65	$1.58
Percentage of expenditures allocated to:					
Network TV	26%	25%	20%	19%	14%
Spot TV	12	6	9	31	4
Syndicated TV	4	6	3	*	4
Network radio	*	*	*	*	*
Spot radio	*	*	*	*	*
Cable networks	6	10	6	5	3
Newspaper	6	*	*	4	15
Magazines	16	15	19	21	4
Outdoor	*	*	3	*	*

* Less than 1 percent

All these premier advertisers chose broadcast television for their major expenditures. Even Phillip Morris, prohibited by law from advertising its tobacco products on electronic media, sells other products in sufficient quantity to warrant allocating almost one-third of its budget to electronic media. This chart does not include direct mail and cooperative advertising expenses.

Source: "How Top Advertisers Allocated Their Budgets in 1998." Reprinted with permission from the September 27, 1999 issue of *Advertising Age*. Copyright, Crain Communications, Inc., 1999.

between rounds of a boxing match or between musical selections on a radio show, for example. In other cases the break must be artificially contrived. The art of writing half-hour situation comedies includes building the plot to successive break-off points for the insertion of commercials. Viewers often complain about arbitrary interruptions of theatrical feature films, whose scripts do not, of course, provide seemingly natural climaxes every 10 minutes to accommodate commercial breaks. Some viewers also object to television "official" time-outs during sporting events, taken for the convenience not of the players but of the advertisers.

As a matter of both law and custom, stations make *station breaks*—interruptions in programming for the insertion of *identification announcements* (IDs). Commercials inserted during these interruptions are called *station-break announcements*.

Networks interrupt their program feeds periodically to allow affiliates time to insert mandatory IDs. They also leave time for affiliates to insert commercial announcements sold by the stations to local or national-spot advertisers. Because of their close association with popular network programs, commercials inserted in these *network adjacencies* are especially valuable.

Two quasi-commercial types of announcements also occur where commercials normally appear. *Promotional announcements* (promos) call attention to future programs of networks and stations. Most broadcasters consider on-air promotion their most effective and cost-efficient audience-building tool. *Public-service announcements* (PSAs), though they resemble commercials, are broadcast without charge because they promote noncommercial organizations and causes. PSAs give stations a way of fulfilling public-service obligations and, along with promos, serve as fillers for unsold commercial openings.

The time devoted to clusters of commercial and other announcements at station breaks, once limited to a minute or two, has constantly edged upward. Advertisers as well as audiences often complain of this proliferation, which is referred to as *commercial clutter*. And the problem continues to worsen. In March 2000 the American Association of Advertising Agencies (AAAA) and the Association of National Advertisers (ANA) reported that the Big Four networks averaged 16 minutes and 43 seconds of advertising during prime-time programming in November 1999. This represents a 21.8 percent increase from November 1991. UPN and The WB each aired more than 17 minutes of commercial clutter during prime time in 1999. Daytime commercial clutter for ABC, CBS, and NBC averaged just over 22 minutes per hour. (AAAA and ANA, 2000)

Cable Advertising
• • • • • • • • • • • •

Cable system operators face more complexities than do broadcasters when inserting commercials in their programs. They have to deal with many channels, each with different requirements for advertising insertions. Modern systems use computer programs triggered by electronic signals originated by cable networks to insert locally sold commercials automatically into cable network programming. Advertiser-supported cable networks and superstations include commercials within programs, in the same way broadcast networks do.

Perhaps more than anything else, lack of adequate statistical information about cable audiences has impeded the growth of cable advertising. Cable's many program choices so fractionalize its audience that few channels attract sufficient viewers to yield statistically valid measurements. Although in the past rating services supplied audience information for only the most popular national cable networks, a variety of local cable measurement services are available in larger cable markets (see Section 10.7, page 310).

6.7   ADVERTISING RATES

Two sets of variables affect the price of broadcast advertising time: one set is relatively stable; the other is relatively dynamic. Station managers have little or no control over *market size, station facilities* (frequency, power, antenna location), and *network affiliation*. Station managers *do* control *programming, promotion,* and *sales*. Good management can lure demographically desirable audiences away from competitors by offering popular programs supported by effective promotion. An efficient sales department can lure advertisers away from competitors with persuasive arguments, good prices, and careful follow-up.

Most large advertisers make their electronic media buys based on ratings, the statistical measurement of audiences exposed to a product's commercials (see Chapter 10). In particular, they order spot schedules designed to achieve a predetermined number of *gross rating points* (GRPs)—that is, to reach an overall number of viewers—or *target rating points* (TRPs), to reach audience subgroups such as teens or women 18 to 49 years of age.

Cost per Thousand
• • • • • • • • • • • •

No standard formula for setting appropriate broadcast rates exists. Market forces, however, eventually tend to bring prices into line. The industry uses *cost per thousand* (CPM) to compare relative advertising costs among competing media. CPM is the cost of reaching 1,000 (represented by the Roman numeral *M*) households or other defined targets.

CPM is calculated by dividing the cost of a commercial by the number of homes (in thousands) that it reached. Advertisers typically pay an average CPM of about $10 for a prime-time spot on one of the major television networks and about $6 on cable networks.

CPM calculations enable advertisers to compare the cost of advertising on one medium with the cost on another, one station with another, and one program with another. These and similar measurements also enable a station or network to guarantee a specific audience in advance. Advertisers whose commercials fail to reach the promised viewer level usually receive additional commercials—*make-goods*—

at no cost (almost never does a broadcaster give refunds), a practice called selling *on the come*. When the 1994 CBS miniseries *Scarlett*, for example, averaged an 18.5 rating, the network had to deliver a number of make-goods to advertisers who had paid as much as $325,000 for a 30-second commercial in the program, based on CBS guarantees of a 24 rating. Similarly, CBS reportedly aired more than $25 million in make-goods due to lower than expected viewership of the 1998 Olympic Games. Advertisers were promised an average rating of 19 or better, but the actual ratings rarely topped 17.

Price Variables

Broadcast advertising depends for its effectiveness on cumulative effect. A buyer, therefore, normally contracts for spots in groups (a *spot schedule* or a specially priced *spot package*). Prices vary according to the number of spots purchased (*quantity discounts*) and other variables, such as the following:

- *Time classes*. Typically, stations divide their time into specific dayparts, and even subclasses of dayparts, with different prices for each.
- *Spot position*. For a guaranteed place in the schedule, advertisers are willing to pay the premium rate charged for *fixed-position* spots. Less expensive *run-of-schedule* (ROS) spots may be scheduled by a station anywhere within the time period designated in the sales contract. Stations often *rotate* spots, both *horizontally* (over different days) and *vertically* (through different time periods) to give advertisers the benefit of varying exposures for their commercials.
- *Preemptibility*. Stations charge less for preemptible spots, which advertisers buy on the understanding that their contracts will be canceled, or their ads played at another time, if a higher-paying customer turns up.
- *Package plans*. Stations offer at discount a variety of packages, which may include several advertising spots scheduled at various times and on various days.

A television station may offer more than a hundred different prices for spots. Most publish *rate cards*, often in *grid* form (see Exhibit 6.i), although some limit their distribution to station personnel. Rate levels can be defined quite arbitrarily, enabling the station to quote different prices for the same spot. Such a grid gives sales personnel great flexibility in negotiating deals without having to resort to under-the-table rate cutting.

The rate for the same spot position in the same program may change over the course of a season if the audience for the program rises or falls significantly.

In the late 1990s the cost of a 30-second network spot in a regular prime-time television program averaged about $125,000. Highly rated programs and special events may cost much more: A 30-second spot on *Seinfeld* cost $550,000 in 1996, and Fox sold its 1999 Super Bowl spots for up to $1.5 million per half-minute.

Prices for spots in syndicated programming can vary widely based on the program's popularity, its age, and, for off-network product, its performance during its network run.

National 30-second spots on top off-network syndicated programs, *The Drew Carey Show*, *Friends*, and *Seinfeld*, for example, can cost from $100,000 to $145,000 each. Top-rated first-run syndication programs such as *Entertainment Tonight*, *Wheel of Fortune*, and *Jeopardy!* command rates between $60,000 and $85,000 per spot.

Popular daytime syndicated shows such as *Oprah* and the *Jerry Springer Show* fetch between $13,000 and $50,000 for a 30-second advertisement.

Radio network ad rates range widely, influenced by daypart and audience reach. In the 1990s the average spot on a major radio network ran about $1,000. In general, ad rates cover time charges only, not the cost of producing the commercials.

Cable Rates

Cable advertising rates vary according to three levels of audience potential: a single cable system, several interconnected systems, and national program services.

Cable systems offer many advertising opportunities in both locally originated programs and national program services. Ad rates vary accordingly, from as little as $2 for a listing in a system's classified-ad channel to $400 and sometimes higher for a 30-second local commercial inserted into one of the

EXHIBIT 6.i
. . . .

TV Station Rate Card

Rate Card No. C41b, Effective January 12

F = Fixed
P = Preemptible on two weeks' notice
Q = Preemptible on one week's notice
I = Immediately preemptible

All spots rotate horizontally and vertically

60-second = double the 30-second rate
20-second = 30-second rate
10-second = 50 percent of 30-second rate

Announcements between time blocks take the rate of the higher block.

		30-Second			
		F	**P**	**Q**	**I**
Daytime and Fringe					
Mon–Fri 7:00 A.M.–9:00 A.M.	Wake Up Call	$ 60	$ 50	$ 40	$ 30
9:00 A.M.–10:00 A.M.	*The Rosie O'Donnell Show*	70	60	50	40
10:00 A.M.–12:00 P.M.	Morning Block	70	60	50	40
12:30 P.M.–4:00 P.M.	Afternoon Block	80	70	60	50
4:00 P.M.–5:00 P.M.	*Judge Judy*	100	90	80	70
5:00 P.M.–6:00 P.M.	News	140	130	120	100
6:00 P.M.–6:30 P.M.	News	200	180	170	160
7:00 P.M.–7:30 P.M.	*Wheel of Fortune*	240	220	210	200
7:30 P.M.–8:00 P.M.	*Jeopardy!*	240	220	210	200
11:00 P.M.–11:30 P.M.	News	140	130	120	100
11:30 P.M.–1:30 A.M.	Movie	70	60	50	40
Weekend					
Saturday 7:00 A.M.–1:00 P.M.	Kid's Rotation	80	70	60	50
1:00 P.M.–2:00 P.M.	Wrestling	90	80	70	60
2:00 P.M.–5:30 P.M.	Various	80	70	60	50

This represents only a portion of a hypothetical small- to medium-market station's sales rate card. Many stations publish no card at all; others use cards only for internal operations and do not make them publicly available. As advertiser demand increases or decreases, sales managers direct their account executives to charge customers higher or lower rates for each program or time period.

more popular cable networks. Local cable operators are fond of marketing their advertising opportunities with the line "all the advantages of TV advertising but at radio rates."

System interconnects differ from cable networks. A network may deliver its programs to thousands of systems throughout the nation, but interconnects involve mostly the placement of commercials on a number of systems within a geographic region. The area covered by such a group of systems may represent anywhere from as few as 13,000 subscribers, as in the case of Rock Springs/Green River, Wyoming, to more than 3.5 million in the metropolitan New York City interconnect called WNYI. This wide variation in interconnect configurations results in rates that may run as low as $20 to as high as several thousand dollars for a single 30-second spot.

Cable networks and superstations, though national in scope, do not necessarily command the highest advertising prices. Again, audience size and composition determine price. In the 1990s, on average, fewer than 3 percent of U.S. television households tuned in to any advertising-supported cable program service at any given moment. By the end of the 1990s daytime rates varied from the low hundreds to the thousands of dollars per commercial. In prime time A&E, for example, charged as little as $4,000 per 30-second spot, while CNN and TBS charged as much as $15,000.

Alternative Ad Buys

Some types of ad-buying deals, used by both broadcasters and cable operators, fall outside normal rate practices.

- In a *trade deal* (also called a *tradeout*), a station or cable system exchanges commercial time for an advertiser's goods or services.
- In *barter deals*, stations exchange advertising time for programs. These are discussed in detail in Section 8.2 (page 230).
- *Time brokerage* refers to the practice of selling time in blocks to brokers, who then resell it at a markup.
- *Per-inquiry* (PI) buys commit the advertiser to pay not for advertising time but for the number of inquiries or the number of items sold in direct response to PI commercials.

6.8 SALE OF ADVERTISING

As noted in Section 6.7, the effectiveness of the sales operation largely determines the success of any commercial enterprise.

Network Sales

Each broadcast television network maintains its own sales department, typically headed by a vice president, usually with offices in New York and Los Angeles and sometimes in other cities such as Chicago and Dallas. They tend to organize into specialties such as prime time, late night, news, sports, and so on, and they deal almost exclusively with advertising agencies.

Ad-supported cable networks and superstations also have sales departments. Advertisers who wish to place orders with several networks often do so through time-buying organizations, which negotiate buys on their behalf for spots on a combination of cable networks.

Local Sales

Most television station sales departments employ a general sales manager, a local (and sometimes also a national) sales manager, account executives, and support staff. The number of account executives (a fancy name for salespersons) varies from station to station. About six usually suffice for a medium-market outlet. At some stations a sales assistant or even a secretary does all the support work; some stations include research specialists and commercial writers on the sales staff.

Sales managers hire, fire, and train salespersons, assigning them a list of specific advertisers and ad agencies as contacts. Beginning salespeople sometimes start without benefit of such an account list (other than, by industry tradition, the Yellow Pages). They must develop their own accounts, making *cold calls* on potential new advertisers.

Salespeople usually work on a commission basis, keeping a percentage of all advertising dollars they bring into the station. This arrangement gives them incentive to make more sales to raise their income.

Spot Sales

Stations and cable systems gain access to national advertising business through *national sales representative* firms (*reps* for short). Some stations also have regional reps for nonnational sales outside the station's service area. A rep contracts with a string of stations, acting as an extension of the stations' own sales staffs in national and regional markets.

Reps perform many services other than sales. Their national perspective provides client stations or systems with a broader view than that of local markets. Reps often advise clients on programming, conduct research for them, and act as all-around consultants. In return for their services, reps collect a commission of 8 to 15 percent on spot sales they make for their clients.

So-called *unwired networks* offer an alternative way of selling national-spot advertising. Under this concept, companies buy commercial time, usually in bulk and at a discount, from television stations throughout the United States. Then they sell it, at a markup, to national advertisers. One such organization, for example, offers a package of spots in prime-time movies on independent stations around the country, selling time on many stations in a one-invoice transaction. Traditional sales reps strongly oppose the unwired-network concept, viewing it as a threat to their exclusive national representation of client stations.

Cable Sales

Advertising-supported cable networks deliver relatively small audiences compared with the major broadcast networks. However, some advertisers target the specialized audiences attracted by cable's dedicated channels (sports fans who watch ESPN, for example).

Because of cable networks' limited audience reach, advertisers are likely to place commercials on several different networks, which can be a bothersome task. Time-buying organizations, such as Cable One, solve this problem by negotiating deals on behalf of advertisers for spots on any combination of cable networks.

Advertising on the Internet

One of the hottest areas of advertising growth at the end of the 1990s involved the Internet, specifically the World Wide Web. Advertisers increasingly place ads on the web, principally in association with browser home pages (such as Netscape), search engines (such as Yahoo), and web sites that attract a desired demographic. Midway through 1999, nearly all of the leading 100 national advertisers in the United States had begun advertising on the web. In 1995, ad expenditures on the web totaled only $35 million, but that figure ballooned to more than $2 billion in 1998. Some industry analysts predict that web advertising revenues will more than triple in the next few years. If this is true, the Internet will outrank cable television and magazines in advertising revenues and will rival radio as an advertising medium.

Advertising Agencies

All regional and national advertisers and most large local advertisers deal with media through advertising agencies. Agencies conduct research; design advertising campaigns; create commercials; buy time from cable systems, broadcast stations, and cable and broadcast networks; supervise implementation of campaigns and evaluate their effectiveness; and, finally, pay media on behalf of the advertisers they represent.

Agencies become intimately familiar with each client's business problems, sometimes even assisting in the development of new products or the redesigning and repackaging of old ones. Ad agencies that handle large national accounts determine the best *media mix* for their clients, allocating budgets among the major media, as indicated in Exhibit 6.h.

For decades ad agencies traditionally received a 15 percent commission on *billings*—the amount the advertising media charged. That is, an agency would bill its client the full amount of advertising time charges, pay the medium 85 percent, and keep 15 percent as payment for its services. By the year 2000, however, fewer than half of all ad-agency arrangements used this scheme. Some agencies now accept less than 15 percent or charge fees in addition to commission, some work on a straight-fee basis, and some work on a "cost-plus" basis.

In any event, that media allow a discount on business brought to them by agencies creates an odd relationship: The agency works for its client, the advertiser, but gets paid by the medium in the form of a discount on time charges. The travel business operates similarly. A travel agency works, at least theoretically, for the traveler but gets paid by the hotel or airline in discounts on charges.

Proof of Performance

After commercials have aired, advertisers and their agencies need evidence to show that contracts have been carried out. Broadcast stations log the time, length, and source of each commercial when broadcast. These logs provide documentary proof of contract fulfillment. Broadcast business offices rely on logs when preparing proof-of-performance affidavits to accompany billing statements. At many stations, computers do the logging automatically. Some stations also make slow-speed audio recordings of everything they air as backup evidence in the event of a dispute, retaining the tapes for three to six months. Cable systems rely mostly on equipment that not only inserts commercials into various cable programs but also provides verification of proper performance.

Advertisers and agencies can get independent confirmation of contract fulfillment by subscribing to the services of companies such as Competitive Media Reporting (CMR), a joint venture of Arbitron and VNU, which conducts systematic studies of radio and television commercial performance. Its Broadcast Verification service uses a computerized electronic pattern-recognition system to monitor television announcements on the eight broadcast networks, on 40 cable networks, and on stations in 100 markets. CMR's InterWatch service tracks Internet advertising on more than 335 Internet magazines and ad-supported web sites.

6.9 ADVERTISING STANDARDS

Advertising raises touchy issues of taste, legality, and social responsibility. Both legal and voluntary self-regulatory standards influence what may be advertised and what methods may be used.

Government Regulation

Section 317 of the Communications Act requires reasonably recognizable differences between radio/television commercials and programs. A station must disclose the source of anything it puts on the air for which it receives payment, whether in money or some other "valuable consideration."

This *sponsor identification rule* attempts to prevent deception by disguised propaganda from unidentified sources. Of course, anonymity is the last thing commercial advertisers desire. But propagandists who use *editorial advertising* (sometimes called *advertorials*) may not always be so eager to reveal their true identities, nor do those who make under-the-table payments to disc jockeys or others for on-the-air favors wish to be identified. Outright deception in advertising comes under Federal Trade Commission (FTC) jurisdiction.

Contrary to popular opinion, neither Congress nor the FCC has set a maximum number of commercial minutes per hour of programming—by broadcast or on cable. Limits have been set on advertising content in children's shows, however (see Section 13.9, page 399).

Program-Length Commercials

The FCC once prohibited program-length commercials—productions that interweave program and commercial material so closely that the program as a whole promotes the sponsor's product or service. The FCC lifted the ban on program-length commercials in 1981 for radio and in 1984 for television. They are still illegal when directed toward children (see Section 13.9).

A flood of program-length commercials, known also as *infomercials*, ensued, often touting merchandise of questionable value. Ads for nostrums, kitchen gadgets, astrology charts, and the like overran both broadcast and cable. Despite the aura of sleaze surrounding infomercials, some began to achieve respectability. General Motors, for example, used the form to introduce its new line of Saturn cars. And Bell Atlantic experimented in 1992 with what it called a *sitcommercial*, an infomercial masquerading as a situation comedy. By the mid-1990s, some low-power television (LPTV) stations devoted their entire program schedules to infomercials.

Taboo Products

Some perfectly legal products and services that appear in print and on billboards never appear in electronic mass media. This double standard confirms that special restraints are imposed on broadcasting and, to a lesser degree, on cable because they come directly to the home, accessible to all.

Nevertheless, canons of acceptability constantly evolve. Not until the 1980s, for example, did formerly unthinkable ads—such as those for contraceptives and those showing brassieres worn by live models—begin to appear.

Congress banned the broadcasting of cigarette ads in 1971 and later extended the ban to include all tobacco products. Up until 1996 the most conspicuous example of self-imposed advertising abstinence involved the industry's refusal to accept hard liquor ads. Broadcasters were concerned that carriage of such ads would provoke opponents of all alcoholic beverages to begin an assault on beer and wine ads. In 1996, however, several television stations began accepting advertising from Seagrams for two of that company's hard liquor products. By the end of the decade more than 300 radio stations and more than 100 television stations and cable networks were airing Seagram's commercials. As feared by many in the industry, critics of such advertising, both inside and outside the government, called for the ads to be banned, much as cigarette ads had been 25 years earlier. Congress, the FTC, and the FCC began considering the issue, at least informally. The FCC subsequently declined to initiate a formal inquiry, leaving that task to the FTC.

Prior to 1999 federal law prohibited broadcast advertising of gambling casinos. Responding to a challenge brought by a group of television broadcasters, the Supreme Court unanimously ruled the statute unconstitutional. According to the Court, it is the job of advertisers and broadcasters, not the government, to assess the value of truthful information about lawful conduct (U.S., 1999).

Self-Regulation

The ban on liquor ads is an example of the voluntary self-regulation once codified by the National Association of Broadcasters (NAB). Its radio and television codes, though full of exceptions and qualifications, set nominal limits on commercial time. For example, the NAB allowed radio 18 minutes per hour, and it allowed network television affiliates 9.5 minutes in prime time, 16 minutes at other times.

In 1984 the Justice Department charged that the NAB standards, even though voluntary, violated antitrust laws by urging limits that reduced compe-

tition. The NAB promptly disbanded its Code Office, apparently relieved to be rid of a thankless task.

Meanwhile, the television networks, and some group broadcasters, had begun dismantling their individual program and advertising codes. The networks had separate departments, variously called Continuity Acceptance, Broadcast Standards, and Program Practices. In the late 1980s ABC and CBS sharply cut back the number of employees assigned to standards and practices departments. Advertisers, agencies, and others expressed fears that these moves might result in an overall lowering of standards. Adverse public reaction, they felt, could be followed by attempts at governmental intervention.

Unethical Practices

Aside from issues of advertising length and content, four specific types of unethical advertising practices in broadcasting have proved particularly troublesome. In the past they triggered both FCC and congressional action.

A conflict of interest occurs when a station or one of its employees uses or promotes on the air something in which the station or employee has an undisclosed financial interest. Called *plugola*, this practice usually results in an indirect payoff. A disc jockey who gives unpaid publicity to her or his personal sideline business is an example.

Direct payments to the person responsible for inserting plugs usually constitute *payola*. It typically takes the form of under-the-table payoffs by recording-company representatives to disc jockeys and others responsible for putting music on the air.

Local cooperative advertising sometimes tempts stations into *double billing*. Manufacturers who share with their local dealers the cost of local advertising of their products must rely on those dealers to handle cooperative advertising. Dealer and station may conspire to send the manufacturer a bill for advertising higher than the one the dealer actually paid. Station and dealer then split the excess payment.

Clipping occurs when affiliates cut away from network or barter-syndication programs prematurely, usually to insert commercials of their own. Clipping constitutes fraud because the networks compensate affiliates for carrying programs in their entirety with all commercials intact.

In 1986, in keeping with deregulation policy then in vogue, the FCC redefined billing fraud as a civil or criminal matter, not an FCC violation. It left the networks to solve clipping problems on their own by bringing suit against offending stations—a cumbersome procedure at best. The Commission did say, however, that it would consider false billing charges when judging a licensee's character during licensing proceedings.

A relatively new practice that some consider unethical involves the digital manipulation of images, tricking viewers into thinking they are seeing something that is not really there. Networks now routinely inject digital advertising messages into sports programs, creating billboards and display ads that show up on the television screen but do not exist at the ballpark or racetrack. A particularly striking example of the use of digital manipulation aired in an ad for a financial services firm during the 2000 Super Bowl. In the ad, paralyzed actor Christopher Reeve appeared to walk to a podium. In fact, Reeve is confined to a wheelchair; the action in the ad was the product of digital animation.

6.10   SUBSCRIPTION-FEE REVENUE

In sharp contrast to broadcasting, cable television, wireless cable, and DBS systems rely on subscription fees for most of their revenues. In the 1990s such systems averaged about $35 to $40 total monthly revenue per subscriber. Of that amount, only about $2 per subscriber came from advertising.

Cable-Fee Regulation

If a graph could chart the history of cable television rate regulation, it would resemble a very wild roller coaster. Regulation of cable rates was at the discretion of local franchising authorities during cable's early history. When the FCC established its comprehensive cable regulatory scheme in 1972, it *required* local governments to regulate basic-cable rates, only to rescind that requirement several years later. At the height of the deregulatory movement in the mid-1980s, Congress effectively deregulated cable rates for the vast majority of systems. This led to a period of rapidly rising prices and increasing consumer complaints.

In 1992 Congress enacted legislation that again required the FCC to assume supervisory responsibility over most cable rates. The Commission responded in early 1993 and again in 1994 with a series of rulings that not only limited most cable subscriber fees but also required some cable systems to reduce their prices and even make refunds. By 1996 the regulatory pendulum had begun to reverse direction, and Congress moved to end most cable television rate regulation, effective in 1999. (See Section 12.8, page 359, for rate regulation details.)

Cable Service Tiers

Some cable systems charge a single monthly rate for their service. But most divide their product into several levels of program service, called *tiers*, with a separate fee for each level.

Most modern systems offer a *basic service* that includes local television stations, one or more distant superstations, and some advertiser-supported cable networks. The monthly fee for this basic package can vary from a few dollars to $25 or more. Some systems break their basic service into two or more tiers. For example, they may pull several of the more popular ad-supported networks (such as MTV and ESPN) out of the basic package and offer them separately as an *extended* or *expanded* basic service at extra cost.

At first in anticipation of, and later in reaction to, a provision in the 1992 Cable Act calling for governmental rate regulation of the basic program tier, some systems restructured their service. Many now offer a *broadcast basic* or *lifeline* tier limited to local stations and public access, educational, and government (PEG) channels. They then offer ad-supported networks and superstations only on one or more expanded tiers. Some systems also offer *a la carte* program options, charging subscribers a per-channel fee for individual ad-supported program services, or for groups of services that are *bundled* together.

The next level of service includes *pay-cable*, or *premium*, channels, such as HBO, Showtime, and Cinemax. By the end of the 1990s more than 47 million homes—almost three-fourths of all cable households—subscribed to pay cable. Usually, subscribers pay a separate fee for each pay service they select. Such fees range from about $2 to $20 or more per service per month, averaging about $10.

Some cable operators require that their customers subscribe not only to the basic but also to an expanded-basic tier before they can buy any premium services. The 1992 Cable Act's *anti-buythrough* provision prohibits this practice, although systems were given 10 years to comply with the new rule.

Pay-per-View Programs

Cable systems with *addressable* converters dedicate one or more of their channels to *pay-per-view* (PPV) programs. A one-time PPV charge allows viewers to see a single program, either a movie or a special event such as a boxing match or a rock concert. By 1999 more than 32 million homes had access to PPV programs, for which they typically paid about $4 per movie and as much as $50 for each special event.

PPV programs come in one of two ways. First, individual cable systems—so-called *standalones*—negotiate directly with producers for PPV movies and events. Much more commonly, national program services, such as Viewer's Choice, acquire PPV rights to programs and distribute them to cable systems under an arrangement that splits revenues between the program service and the cable system.

As video compression produces greater cable channel capacity, more and more systems are devoting multiple channels to PPV. Some also are beginning to experiment with two-way, interactive *video-on-demand* (VOD), a PPV technology that will compete with neighborhood VCR rental operations by permitting subscribers to order their choice of movies at the touch of a button.

Satellite-to-Home Services

In 1994 two high-power *direct broadcast satellite* (DBS) services, Hughes DirecTV and United States Satellite Broadcasting (USSB), began active marketing to potential subscribers. Both DBS providers used the same high-power satellite to deliver their services to subscribers, although their program packages were different. Both companies offered an exclusive selection of cable television networks for a fee that ranged from $8 to $40 per month, depending on the package chosen. Pay channels and sports events packages cost extra—from $8 per month for movie channels and regional sports networks to $149 for a season of NBA basketball games. As many industry analysts predicted, the two services merged in 1999 when DirecTV acquired USSB.

RCA and Sony dealers market the equipment necessary to receive DirecTV, calling the package a digital satellite system (DSS). Cost to viewers for an 18-inch TVRO, set-top converter, and remote control initially ranged between $700 and $900 (plus optional professional installation charges). Pressure from new DBS competitors, notably Echostar's Dish Network, soon forced retailers to lower the cost of a basic system to as low as $100 in some markets.

Another satellite-to-home service, Primestar, had for years been one of several companies that supplied programs to homeowners who paid $2,000 or so for their own large-diameter TVRO. Primestar charged about $150 for a decoder and $30 a month to view a variety of program channels. In 1994 Primestar increased its satellite power, thus reducing required TVRO size to 39 inches. In 1999 DirecTV purchased Primestar, adding 2.5 million subscribers to its 4.8 million DBS customers. DirecTV plans to remove the older Primestar dishes and replace them with DSS equipment during a period of several years.

Because they were nationally distributed, none of these DBS systems could initially include signals of local TV stations. Customers had to either switch between TVRO and rooftop antenna or subscribe to basic cable to fill this void. This limitation put DBS at a disadvantage compared to cable systems, which carry both cable networks and local broadcast stations. To overcome this competitive barrier, Echostar, after acquiring additional satellite channel capacity in 1998, began beaming selected local TV signals into several local markets—a service that became known as *local into local*. Local broadcast stations and cable system operators objected, arguing that the service should have to deliver *all* broadcast signals in each market. Congress resolved the issue in 1999 by passing legislation that permits DBS operators to provide some local-into-local service immediately but requires operators to carry all broadcast stations in the markets served by 2002.

EXHIBIT 6.j

Broadcast and Cable Employment

Medium	Full-Time Employees (thousands)	Percent Women	Percent Minorities	Top Four Job Categories (thousands)	Percent Women	Percent Minorities
Broadcast	150	41.0	20.2	131	34.9	18.2
Cable	134	41.7	30.9	54	28.9	21.7

The top four job categories include officials and managers, professionals, technicians, and sales workers. These statistics represent broadcast employment in 1997 and cable employment in 1998, based on reports from only those non-network units that have five or more full-time employees. The FCC, which collected these data, does not have figures for the DBS and wireless cable businesses. The Commission stopped collecting broadcast employment data after 1997 because a court declared its broadcast equal employment opportunity program illegal. The Commission continues to collect cable television employment data. Total broadcast employment declined from the late 1980s (total employment in 1987 was 176,000) to a low of 144,500 in 1993. Employment levels have increased since then. Cable television employment levels also continue to increase, up from 84,000 in 1987.

Source: FCC, *1997 Broadcast and Cable Employment Report* and *1998 Cable Employment Report* (Washington, DC, 1998, 1999).

6.11 PERSONNEL

The number of people employed in an industry usually gives some indication of its importance. According to that yardstick, electronic media have relatively little importance. However, their social significance lends them far greater weight than their small workforce suggests.

Employment Levels

Exhibit 6.j provides employment data for the non-network units of the broadcast and cable industries. Overall, when networks and other related operations are included, these industries employ some 600,000 people full-time. By way of comparison, one lone manufacturing corporation, General Motors, employs nearly 650,000 people.

Many specialized creative firms support media, producing materials ranging from station-identification jingles to prime-time entertainment series. Such firms offer more opportunities for creative work than do the media themselves—performing, writing, directing, designing, and so on. Other media-related jobs are found in nonprogram areas. These include jobs in advertising agencies, sales representative firms, program-syndicating organizations, news agencies, common-carrier companies, and audience-research organizations.

Aside from the major television networks, most broadcast and cable organizations have small staffs. The number of full-time employees at radio stations ranges from fewer than 5 for the smallest markets to about 60 for the largest, with the average being 15. Television stations have between 20 and 300 employees. A typical network affiliate employs about 90 full-time people, and an independent station about 60.

Cable systems average about 30 full-time employees but range from family-run systems in small communities (with perhaps 5 or 6 employees) to large-city systems with staffs of more than 100. Cable MSO headquarters units average about 55 full-time employees.

EXHIBIT 6.k
• • • • • • • • • •

Salary Levels

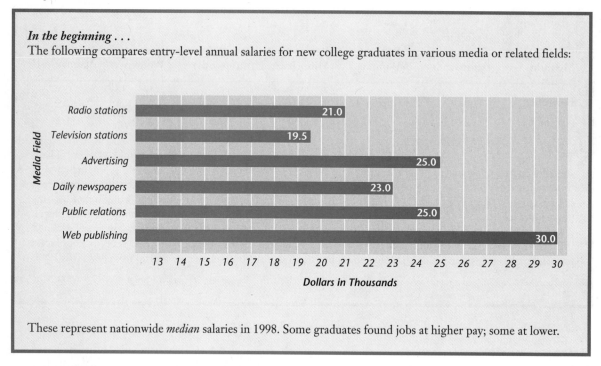

In the beginning . . .
The following compares entry-level annual salaries for new college graduates in various media or related fields:

These represent nationwide *median* salaries in 1998. Some graduates found jobs at higher pay; some at lower.

Salary Levels
• • • • • • • • •

The huge salaries reported in *People* magazine and on *Entertainment Tonight* go to top performing and creative talent and executives who work mostly at network headquarters and the production centers of New York and Hollywood.

Average salaries for jobs at most stations and cable systems rank as moderate at best. Typically, those working in sales earn the highest incomes at broadcasting stations. They are the ones who, quite simply, bring revenue directly to the enterprise so that bills—and salaries—can be paid. At the department-head level, general sales managers usually make the most money and traffic managers the least. Program directors fall somewhere in between. Exhibit 6.k offers a more detailed salary analysis.

Typically, television stations pay higher salaries than radio stations or cable systems for comparable positions, although in the late 1990s the salaries of radio general managers matched and then exceeded the salaries of their TV counterparts. Presumably, this is a result of local radio consolidation; local ra-

dio managers may be running up to eight stations per market. Fewer general managers are doing more work and are getting paid handsomely. Basic laws of supply and demand come into play throughout the compensation structure; because fewer people want—and are technically qualified—to be TV weathercasters than want to be sportscasters, for example, salaries for the former group tend to exceed those for the latter. Salaries for radio talent reflect the relative importance of various dayparts.

Employment of Women
• • • • • • • • • • • • • • • • •

As in all businesses, women in media often run up against corporate "glass walls" that limit their lateral movement to jobs that will provide the experience necessary for promotion to higher positions. Even when they have the appropriate qualifications, women frequently hit a "glass ceiling"—an invisible barrier that prevents their movement to the top. This may be changing, at least in television (see Exhibit 6.l). A 1999 report by the NAB

EXHIBIT 6.k
· · · · · · · · · · ·
Salary Levels (continued)

And on the average . . .
As they moved from entry-level to top-level positions, electronic media employees, *on the average*, earned the following annual salaries along the way. (Some salaries do not appear here because broadcast stations and cable systems use somewhat different job titles.)

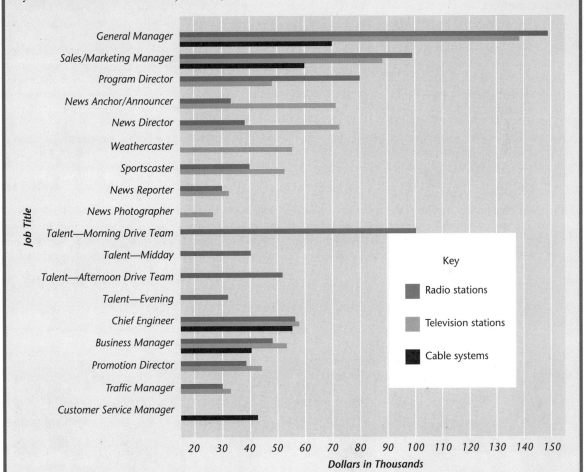

These are only *averages*. Salaries paid by organizations in large cities tend to run higher than those in smaller communities. And because averages typically rise each year, the above data prove most useful in comparing compensation levels for the various positions rather than in studying specific salaries themselves.

Sources: Based on data in *Annual Survey of Journalism & Mass Communications Graduates*, James M. Cox, Jr. Center for International Mass Communication Training and Research, University of Georgia (Athens, GA, 1999); *Television Employee Compensation and Fringe Benefit Report*, National Association of Broadcasters/Broadcast Financial Management Association (Washington, DC, 1998; *Radio Station Salaries*, National Association of Broadcasters (Washington, DC, 1999); and Pay Equity Initiative, Women in Cable and Telecommunications Foundation (Chicago, 1998).

EXHIBIT 6.1
• • • • • • • • • •

A Woman's Place Is in the Newsroom . . .

The path for women to equal job opportunity in broadcasting has been bumpy and long, but some progress has been made.

In 1948 ABC hired Pauline Frederick as the first woman network news correspondent. She remained the sole female hard-news network reporter for the next 12 years. In her early years, when assigned to interview the wives of presidential candidates at national political conventions, she was also required to apply their on-camera makeup.

Later she became famous for her coverage of the United Nations, first as a correspondent for ABC, then with NBC. In 1976 Frederick became the first woman to moderate a presidential debate, between Gerald Ford and Jimmy Carter. She died in 1990 at the age of 84.

In 1976 ABC hired Barbara Walters as the first woman anchor on a weekday evening network newscast. She shared the anchor desk with Harry Reasoner. Her salary: $1 million a year ($500,000 for anchoring and $500,000 for producing and hosting four entertainment specials a year), plus perquisites (a private office decorated to her taste, a private secretary, a makeup consultant, and a wardrobe person). Walters had established her reputation as co-host of NBC's *Today* show and as a successful, if sometimes controversial, interviewer of famous personalities as diverse as Princess Grace of Monaco, Fred Astaire, Ingrid Bergman, and Fidel Castro.

Her tenure as anchor ended in 1977, and she moved on to more celebrity interviews and to ABC's news magazine *20/20*. In 1999 Walters began co-hosting *The View*, a daytime talk show for women on ABC.

Following her graduation from Indiana University, Jane Pauley began her broadcast career in 1972 at WISH-TV in Indianapolis as a reporter and anchor. She then moved to WMAQ-TV in Chicago, where she was the first woman regularly to co-anchor a weeknight news broadcast in that city. In 1976 Pauley joined the NBC network as co-anchor of the *Today* show, a position she held for 13 years. Since 1992 Pauley has co-anchored the newsmagazine *Dateline NBC*. In addition to these duties, Pauley hosts *Time and Again* on the cable network MSNBC, a daily program that chronicles past events using actual NBC news footage.

Pauline Frederick

Barbara Walters

Jane Pauley

... and in the Boardroom

Kay Koplovitz created the first advertiser-supported cable television network, USA Network, in 1977 and became the first woman television network president as a result. Koplovitz was the first cable network executive to negotiate cable rights with the NHL, NBA, and Major League Baseball. Under her direction, USA became the largest provider of original programming among the basic-cable networks. In 1992 Koplovitz launched another popular cable network, the Sci-Fi Channel. After Koplovitz left USA Network in early 1998, President Bill Clinton appointed her chair of the National Women's Business Council. In addition to this post, Koplovitz serves on numerous corporate and philanthropic boards. Recently, she received the Champion of Liberty Award from the Anti-Defamation League for her role in promoting the USA Network "Erase the Hate" campaign, which is dedicated to combating hate and racism. Koplovitz now serves as chair of Broadway Television Network, a new pay-per-view service that will bring live Broadway productions to home viewers.

Following graduation from Harvard, Susanne Daniels began her broadcasting career as an assistant to Lorne Michaels at *Saturday Night Live*. She then joined ABC and spent two years as director of variety, reality, and special programming. After moving to Fox, where she was director of comedy programming, Daniels went on to become The WB's number two programming executive. In December 1998 she was named president of WB Entertainment—the person ultimately responsible for deciding which programs air on the network.

Pat Fili-Krushel's first job in broadcasting following her graduation from St. John's University in 1975 was with ABC as a secretary. After rising in the network to program controller of ABC Sports (and earning an MBA from Fordham), she moved over to HBO, where she worked for nine years. Fili-Krushel then programmed Lifetime for five years before returning to ABC, where she was head of daytime programming. In July 1998 she was promoted to president of the ABC Television Network—the top programming job at ABC. Ever on the move, in April 2000 Fili-Krushel left ABC to head the consumer division of Healtheon/WebMD, a new media company providing health information via web sites, TV programs, and a cable network.

Sources: Photos from AP/Wide World Photos (Frederick); Copyright 2000 ABC, Inc. (Walters); Corbis-Bettmann (Pauley); AP/Wide World Photos (Koplovitz); Photo by James Sorenson. Photo © 2000 The WB Network. All Rights Reserved. Used by permission. (Daniels); Steve Fenn/Copyright 2000 ABC, Inc. (Fili-Krushel).

Kay Koplovitz

Susanne Daniels

Pat Fili-Krushel

indicates that the number of women general managers in television increased 39 percent between 1998 and 1999. Still, only 15 percent of the approximately 1,200 general managers in the country are female. (NAB, 1999)

In its annual employment study of all broadcast stations with five or more employees, the FCC reported that in 1997 women occupied about 41 percent of all jobs, up from 32 percent in 1979. Perhaps more important, women represented 35 percent of employees classified as officials and managers (up from 23 percent in 1979) and more than 53 percent of those classified as sales workers (up from 31 percent in 1979).

For cable systems with five or more employees in 1998, the FCC reported that women constituted nearly 42 percent of all employees and about 29 percent of officials and managers. The potentially lucrative area of sales saw major improvement: Women held more than 47 percent of those jobs, up from only 30 percent in 1983.

Minority Employment

The FCC also reports on the electronic media's efforts to upgrade employment opportunities of minority-group members—defined by the FCC for this purpose as Aleutians, American Indians, Asians, Blacks, Hispanics, and Pacific Islanders. Although their overall progress has been slower than that of women, many minority groups have made outstanding contributions to broadcasting and cable. Exhibit 6.m discusses some outstanding African Americans in the field.

The most recent (1997) FCC figures indicate that minorities represented 20.2 percent of all broadcast employees, an improvement of only 6 percent over 1979. The percentage of minority members holding sales jobs increased from just over 8 percent in 1979 to just over 28 percent in 1997. The report for officials and managers showed a slower rate of growth: Minorities held more than 15 percent of the top jobs, up from just under 8 percent in 1979.

Cable has outpaced broadcasting in terms of minority employment. Of all cable system employees in 1998, minorities made up about 31 percent, more than double their 1979 proportion. Minorities held more than 21 percent of top jobs (officials and managers) and nearly 30 percent of those in sales.

EXHIBIT 6.m
. . . .

Outstanding African Americans in Electronic Media

Although African Americans have not yet gained full equality in the media industry, many have risen to positions of prominence in broadcasting and cable. The following are but a few.

Karen Barnes—vice president of Fox Children's Network

Don Cornwell—chairman, president, and controlling stockholder of Granite Broadcasting, television station group owner

Ragan Henry—chairman of U.S. Radio, a black-owned radio group

Catherine Hughes and Alfred Liggins—controlling owners of Radio One, Inc., the nation's largest minority-owned radio group

Robert Johnson—founder and president of the Black Entertainment Television cable network

Michael Moye—creator and executive producer of *Married . . . With Children*

Richard Parsons—president of media giant Time Warner

Bernard Shaw—news anchor, CNN

Carol Simpson—news anchor, ABC News

Source: Reprinted with permission of *Broadcasting & Cable*, September 9, 1991. © 1991 by Cahners Publishing Company. With updates by authors.

Labor Unions
. . . .

At networks, national production centers, and most large-market network television affiliates, unionization generally prevails. Not so at smaller stations and cable systems. The fragmentation of the industry into many units, each with a small staff, makes unionization impractical.

In small operations staff members often have to handle two or more jobs that in a union shop might come under different jurisdictions. For example, a small radio station cannot afford to assign two employees to record an interview, paying one as a technician to operate equipment and another as a performer, when the job can be done by one person.

Electronic mass media unions draw on types of personnel first unionized in older industries—

electrical work, music, motion pictures, stage, and newspapers. Thus the American Federation of Musicians represents every kind of professional musician, from pianists in bars to drummers on television.

People who work in broadcasting and cable can be grouped into two broad categories, the creative/performing group and the craft/technical group. Unions divide along similar lines; those representing the former usually avoid the word *union*, calling themselves *guilds*, *associations*, or *federations*.

The first pure broadcasting union, the American Federation of Television and Radio Artists (AFTRA), began (originally as AFRA) in 1937, representing that universal radio performer, the announcer. Most creative/performing unions, however, began with stage and motion-picture workers. Examples include the Writers Guild of America (WGA), the American Guild of Variety Artists (AGVA), and the Screen Actors Guild (SAG). A technical union, the International Alliance of Theatrical Stage Employees and Moving Picture Machine Operators of the United States and Canada (IATSE), expanded into TV from the motion-picture industry.

In 1953 NBC technicians formed an association of their own that ultimately became the National Association of Broadcast Engineers and Technicians (NABET). Later the union changed the word *Engineers* to *Employees* to broaden its scope. Recently, NABET has been negotiating with the broadcast networks about the status of new jobs in the digital era. NABET argues that the new jobs should be union positions; the networks counter that they need the flexibility to hire nonunion employees for certain positions.

Job Opportunities

Surveys of students enrolled in college electronic media programs indicate that most want to work in the creative/performing area, especially in on-camera or on-mike positions. The oversupply of candidates makes these the least accessible jobs for beginners. Broadcast stations and cable systems usually delegate creative work to outside production companies, which means that such work is concentrated in a few major centers, where newcomers face fierce competition and unions control entry.

News, the one field in which local production still flourishes, offers an exception to the scarcity of creative jobs at stations and systems. Nearly all television stations and some cable systems employ news specialists, and although there are typically more applicants than openings for on-air jobs such as news reporters and anchors, job notices for qualified news producers often go unanswered.

Sales offers another employment area likely to expand. All commercial networks and stations, as well as a growing number of cable systems, employ salespeople. Ambitious job seekers should also note that top managerial positions have historically been filled by sales personnel, although in recent years more general managers have come from news.

Advertising agencies and national sales representative firms offer entry-level employment opportunities. Nevertheless, personnel directors frequently complain that college-educated job applicants fail to understand or appreciate the financial basics of the industry.

As technologies proliferate and competition intensifies, so does the need for creative and effective promotion. Stations, systems, and networks have to promote themselves and their products to potential audiences and advertisers. Cable networks use promotion extensively to persuade cable systems to buy their offerings.

About two-thirds of all jobs in cable are technical, but marketing, market research, and advertising also have high priority. The need for creative people will also increase as more cable systems provide public-access and local-origination programming.

Corporate Video

Applicants who look only to broadcasting and cable for entry-level jobs unnecessarily limit their employment options. Virtually every large organization that has contact with the public uses electronic mass media in one form or another. Opportunities for production and writing jobs exist in business; religious, educational, and health organizations and foundations; government agencies; the armed services; and specialized production companies.

Many such organizations make extensive in-house use of satellite-fed, closed-circuit television. By the year 2000 about 80 private business television (BTV) networks were sending programs to more than 30,000 downlink sites in the United

States. Most BTV organizations direct their productions to their own employees for product introductions, training, and corporate communications. The investment firm Merrill Lynch, for example, transmits three or more product-driven programs each week to its sales force in 480 offices around the country.

This rapidly growing field of corporate video applies broadcast techniques to job skills training, management development, sales presentations, and public relations. Such nonbroadcast uses of television require trained personnel for production, direction, writing, studio operations, program planning, and similar functions. They also offer the following advantages over traditional broadcast positions:

• Many broadcast stations are reducing staff size at the same time that corporate video organizations are enlarging theirs.
• Jobs often are available for students right out of school, even in major markets.
• Beginning salaries, benefits packages, and working hours are better.
• Employees enjoy greater job stability.
• Job security does not depend on ratings.

6.12 INVESTMENT, PROFIT, AND LOSS

The electronic media industries, though not as capital-intensive as many manufacturing giants such as the automobile business, nonetheless require high capital investments to construct new facilities or acquire and maintain existing ones.

Capital Investments

Radio station construction costs range from $50,000 for a simple, small-market AM or FM outlet to several million dollars for a sophisticated station in a major market. Full-power, major-market television facilities may cost tens of millions of dollars. Upgrading operations with new technologies such as satellite news gathering (SNG) vans, helicopters, and newsroom computer systems calls for additional capital outlays in the hundreds of thousands.

On a per-home-served basis, a cable system costs even more to construct because each home must be physically connected to the headend. Owners now pay anywhere from $700 to $1,250 per home when building new systems. To *upgrade* a system—that is, to increase its channel capacity or convert from coaxial cable to optical fiber—costs between $10,000 and $15,000 per mile.

In terms of wireless delivery of multichannel video, wireless cable and DBS systems demonstrate the range of costs to begin operations. Industry experts estimate that start-up costs for DBS systems begin at $500 million. By contrast, a 33-channel terrestrial wireless cable system costs between $500,000 and $2.5 million, depending on the power of the transmitter.

HDTV

HDTV's sharper, wider pictures will require eventual replacement of most broadcast and cable video equipment, including the home television set. Industry experts can make only rough estimates of what all this will cost. The first local television stations to convert to HDTV paid $1.5 million to $2 million just for the equipment necessary to retransmit a network signal. To produce and transmit local programming costs another $10 million to $12 million. Some broadcasters who oppose the new technology, and thus may not be completely objective, project that the industry may pay as much as $10 *billion* for a technology that, in all likelihood, will neither increase a station's or a network's audience nor expand its advertising revenue.

Unlike other media, electronic mass media rely on the public to supply the largest part of the industry's basic capital equipment: broadcast receivers. For consumers to be able to watch HDTV in their homes on the new, wide-screen television sets, whether the programs come over the air or through cable or from a VCR, they will collectively pay between a conservatively estimated total of $80 *billion* and a high of $200 *billion* or more.

Ownership Turnover

Broadcasting and to a large extent cable have reached the point where investment in new facilities has come to a halt. Now the more likely route to ownership is through the purchase of existing facilities.

Factors such as market size and location, format (in the case of radio), network affiliation (in the case of television), and competitive position in the market all influence asking/offering prices for stations. Through the first half of the 1990s, television stations, FM stations, and AM/FM combinations typically sold for about 2½ times the station's annual gross revenue, or 8 to 10 times its *cash flow* (operating revenue minus operating expenses). For a standalone AM station, the price was closer to 1½ times gross revenue or 5 times cash flow.

The Telecommunications Act of 1996 changed all that. Because the national ownership limits for radio, and to a lesser extent television, were rescinded, station trading in 1996 exceeded the previous year's sales by almost 205 percent. By the end of the decade deals for radio groups and television stations brought prices of up to 20 times cash flow. Exhibit 6.n details the money spent on broadcast acquisitions since 1995.

Cable systems typically sell for 16 to 20 times cash flow. Another common rule of thumb for estimating the price of a cable system assigns a dollar value to each subscribing household. In the early 1990s that figure averaged between $1,700 and $2,000 per subscriber, but in 1999 that figure ballooned to nearly $5,000 for some systems. Like the other formulas, this per-subscriber method yields only very rough estimates. It ignores factors such as number of homes passed by the system (which may include potential subscribers), age and channel capacity of the system, and franchise requirements.

Profits and Losses

As they moved through the 1980s and into the early 1990s, most electronic media experienced increasing financial difficulties. The nation's overall economic problems, combined with growing and ever more diverse media players, typically resulted at best in lower profits and at worst in losses or even bankruptcies.

As the national economy improved, so did the economic fortunes of the TV broadcast networks. By 1993, after cutting costs and stabilizing their audience shares, the Big Four networks were in the

black. With the demise of the of fin/syn rules (see Section 8.2, page 230) and as advertising expenditures increased, the networks' outlook appeared positive. But by the end of the 1990s high broadcast rights fees for sporting events and popular prime-time programs significantly decreased the profit margins of most of the networks. In 1998 The WB, UPN, ABC, and CBS reported losses in their network operations. In 1999 only NBC and CBS reported profits from their TV networks. The network-owned station groups, on the other hand, showed significant profits.

At the television station level, profits continue to rise as advertising revenues increase yearly. Successful stations post 50 percent profit margins. Affiliates of the major television networks averaged pretax profits of more than $6 million at the end of the 1990s, and independent stations averaged nearly $4 million. Remember that these are average figures; some major-market network affiliates earn upwards of $90 million per year, while other stations may lose money.

On the whole, radio fortunes look positive. In 1994 radio advertising revenues exceeded $10 billion for the first time, and that figure increased to $15 billion in 1998. In the last half of the 1990s, radio advertising revenues increased between 7 and 9 percent annually, and most industry analysts expect that trend to continue in the new millenium. Networks, on the whole, benefited from increased advertising expenditures, as did major-market stations and groups. Nationwide most FM stations and AM/FM combos operate in the black. The outlook for standalone AMs, however, especially the few remaining daytime-only operations, is grim.

Cable presented a mixed profit picture in the 1990s. By the mid-1990s the cable audience had plateaued at about two-thirds of U.S. television households, and new competitors—notably DBS systems and the telephone companies—were beginning to siphon subscribers away from the cable operators. FCC-imposed cuts in subscriber fees in 1994 reduced cable system revenues and profits, and Wall Street lost some of its confidence in the industry, as evidenced by the fact that cable stocks lost 10 to 40 percent of their value in 1996. The industry recognized its problems and began addressing them by decreasing its debt, streamlining its administration, and consolidating its holdings. The 1996 Telecommunications Act cut back on

EXHIBIT 6.n

Dollars Flow with Dereg . . .

Note that "deals" in the chart may include more than one station in the sale.

Station Type	1995	1996	1997	1998	1999
TV/Radio			$1.85 billion/ 7 deals	$6.54 billion/ 7 deals	
TV	$4.74 billion/ 112 deals	$10.488 billion/ 99 deals	$6.4 billion/ 108 deals	$7.12 billion/ 90 deals	$4.72 billion/ 86 deals
AM/FM combos	$2.79 billion/ 213 deals	$12.034 billion/ 345 deals	$12.73 billion/ 322 deals	$7.54 billion/ 264 deals	$26.88 billion/ 196 deals
FM	$685.68 million/ 329 deals	$2.628 billion/ 417 deals	$2.1 billion/ 395 deals	$1 billion/ 341 deals	$1.48 billion/ 169 deals
AM	$106.76 million/ 195 deals	$212.82 million/ 254 deals	$361.6 million/ 235 deals	$596.2 million/ 248 deals	$238 million/ 213 deals
Total	**$8.32 billion/ 849 deals**	**$25.362 billion/ 1,115 deals**	**$23.44 billion/ 1,067 deals**	**$22.8 billion/ 950 deals**	**$33.32 billion/ 774 deals**

Source: *Broadcasting & Cable*

cable rate regulation to some degree, cable advertising revenues were increasing at a rate of more than 10 percent per year, and cable network ratings showed steady gains. In the late 1990s, even as cable subscribership slowed because of increased competition, optimistic revenue projections for new cable services such as Internet access and local telephony led to the increased valuation of cable systems and unprecedented system sales. Many of the largest MSOs were betting that a newly invigorated cable industry would see substantial successes—and profits—in the new century.

The outlook for the DBS and wireless cable businesses appears mixed. By the beginning of 2000, total DBS subscribership had reached 12 million nationwide, subscriptions were increasing an average of 250,000 per month, and consolidation was decreasing the intra-industry competition. Wireless cable, once seen as a formidable competitor to cable systems, experienced a loss of confidence among investors, notably the telephone companies. By the end of 1999 there were fewer than 1 million wireless cable subscribers nationwide. On a brighter note, wireless cable owners saw increased attention from firms interested in using multichannel multipoint distribution service (MMDS) frequencies to distribute voice and data services.

The Internet

The rise of the Internet as a commercial communications medium has created one of the fastest-growing and largest industries in the country. Analysts indicate that the Internet industry generated $301 billion in 1998, rivaling the automotive ($350 billion) and telecommunications ($270 billion) industries. With more than 100 million adults in the United States on-line, many of whom use the Internet to purchase goods and services, e-commerce reached $102 billion in 1998 and is expected to double annually to reach $1.3 trillion

. . . and Consolidation Increases

The 1996 Telecommunications Act's deregulatory policies unleashed a torrent of electronic media consolidation. Media mergers create new companies with broad and varied business interests. Long-distance telephone leader AT&T's takeover of TCI, the largest cable television operator, and AOL's acquisition of Time Warner are but two examples. This exhibit highlights two mergers, showing the combined media assets of the companies. If big is better, these companies are among the best.

The merger of radio giants AMFM and Clear Channel Communications, valued at $23.5 billion, created a company with the following holdings:	**The merger of Viacom and CBS, valued at $36 billion, created a company with the following holdings:**
850 U.S. radio stations	35 U.S. TV stations
240 international radio stations	2 TV Networks (CBS and UPN)
19 U.S. TV stations	3 TV film production companies
467,500 outdoor displays in 35 U.S. and 20 overseas markets	3 TV syndication companies
AMFM radio networks	163 radio stations
29% Hispanic Broadcasting Corp.	Westwood One radio network
30% Lamar Advertising Co.	25 U.S. and international cable networks
Katz Media Rep firm	2 home video chains
Chancellor Marketing Group	2 book publishing companies
SFX Entertainment	4 theme parks
	Interests in 12 commercial web sites

in 2003. By the end of the 1990s almost 2,000 secure web sites capable of conducting electronic transactions were being added to the Internet every month. The various Internet businesses employ a workforce of more than 1.2 million people. (Center for Research on Electronic Commerce, 1999)

6.13   CRITIQUE: BOTTOM-LINE MENTALITY

From the public-interest standpoint, stations, cable systems, and networks need to earn profits. When they operate at a loss, their public-service programs tend to suffer first. Moreover, the lowering of standards by money-losing firms can be contagious: Their rivals tend to match lower standards to compete in the marketplace.

But obsessive concern for short-term profits has its dangers, too. Federal deregulation and permissive interpretation of antitrust laws have encouraged media acquisitions, mergers, takeovers, and consolidations. These transactions focus so single-mindedly on profits that they produce a *bottom-line mentality*—executive preoccupation with profit-and-loss statements to the exclusion of all else.

Cable television, not explicitly required to operate in the public interest, has made little or no effort to modify profit-driven goals beyond such services as C-SPAN and Cable in the Classroom. Broadcasting, which does have a public-interest mandate and which FCC rules formerly kept mostly in the hands of professional broadcasters, fell increasingly under the control of conglomerate officials with no broadcasting background.

Trafficking in Stations

The FCC's antitrafficking rule was designed to prevent station trading at the expense of public service. It required a broadcast station licensee to operate for a minimum of three years before selling it. Deletion of this rule in 1982 enabled new licensees to make quick entrances into and exits out of the broadcasting business for the sake of virtually overnight profits.

First-time broadcast buyers began to specialize in *leveraged buyouts* of television stations. These deals involve buying stock to gain a controlling interest in target firms. To do so, buyers incur huge debts that have to be repaid out of profits, leaving little money for quality programming and often requiring personnel layoffs. Increasingly, stations turned to quick and easy sources of income—program-length commercials, titillating shows exploiting sex and violence, tabloid pseudo-news shows, and the like.

Network Changes

The mighty networks were not immune. When new management took over at ABC, CBS, and NBC, cutbacks in operating budgets caused the layoff of thousands of employees and the early retirement of others. This new austerity took its heaviest toll in network news as veteran reporters were fired and entire news bureaus closed.

Making an unprecedented public criticism of his own network, Dan Rather, *CBS Evening News* anchor and managing editor, wrote a piece for the *New York Times* headlined "From Murrow to Mediocrity?" Rather pointed out that CBS chairman Laurence Tisch "told us when he arrived that he wanted us to be the best. We want nothing more than to fulfill that mandate. Ironically, he has now made the task seem something between difficult and impossible." Rather added,

> News is a business, but it is also a public trust. . . . We have been asked to cut costs and work more efficiently and we have accepted that challenge. What we cannot accept is the notion that the bottom line counts more than meeting our responsibilities to the public. Anyone who says network news cannot be profitable doesn't know what he is talking about. But anyone who says it must always make money is misguided and irresponsible. (Rather, 1987)

Four years later, as network belt-tightening continued, Rather's CBS colleague Mike Wallace observed that Tisch had disappointed broadcast veter-

ans, "probably mostly by the manner in which it was done. It was done too coolly." Wallace continued,

> When you've worked in an outfit 25 to 35 years and you've seen guys actually lay down their lives on the line in the course of doing their job and they are tossed aside, you are led to believe that "thus far and no further will be these cuts," and suddenly it's a lot further. You begin to feel a sense of disappointment and disillusionment. (Bier, 1991)

Toward the end of the 1990s network news personnel seemed to have come to terms with the new reality. In a 1999 interview, Mike Wallace was asked whether news had suffered in an era when it had to toe the bottom line. He replied,

> I suppose to some degree, yes. But you have to be realistic about it. You don't have the same kind of money. Therefore, bureaus close. The kind of on-the-spot coverage you used to get is too expensive. The marketplace today is no longer what it used to be. (McClellan, 1999)

Whoever owns the networks, the owners expect the news divisions to be profitable while maintaining high journalistic standards. News personnel are learning to do more with less and still get the story to the public.

Meeting the Competition

For the electronic media companies discussed in this chapter to succeed in the new millennium, they will have to respond aggressively to changes in technology and new competition. By the late 1990s most television broadcasters realized that the movement to digital broadcasting, while involving some costs, promised some advantages as well. In addition to broadcasting in high definition, television stations might well broadcast several different services at once, finally becoming multichannel operators. Broadcasters are also exploring high-speed data delivery and the integration of traditional television and computer services, leading to what some have termed "interactive TV." Radio broadcasters will be able to compete with the sound quality of CDs as digital audio broadcasting becomes a reality. On the other hand, FCC proposals to create a whole new class of low power radio stations could add hundreds of new competitors.

Traditional cable operators face a host of new competitors, most notably DBS systems and the telephone companies. The cable industry has responded by aggressively rebuilding its physical facilities to accommodate more channels and digital services. The development of cable modems allows cable subscribers to access the Internet at speeds far greater than those available via telephone lines. Some cable systems have decided to meet the telco competition head-on by offering interactive voice and data services. Wireless cable operators are planning a move to digital delivery to increase channel capacity, and DBS operators will be offering local broadcast signals as part of their service packages. Mega-media companies, created through acquisitions and mergers, may be competing in all these arenas.

One thing is certain: The technology, the structure, and the business of the electronic mass media will continue to change rapidly. Only by keeping up with such change and adapting to it can existing media companies hope to survive.

CHAPTER 7

Noncommercial Services

Most of this book deals with commercial electronic media. But many Americans believe that profit motives alone cannot be counted on to fulfill all the national cultural, educational, and informational needs that electronic media ideally could serve. Hence arose the concept of *noncommercial broadcasting*—motivated by public-service goals rather than by profit.

7.1   FROM EDUCATIONAL RADIO TO "PUBLIC" BROADCASTING

Today's *public broadcasting* began as a more narrowly defined service devoted explicitly to education. When radio boomed in the 1920s, educational institutions joined in the rush. Most of these pioneer not-for-profit stations operated on a shoestring for only a few hours a week.

AM Radio

With broadcasting's growing financial success in the late 1920s, commercial interests began to covet AM channels tied up by educational licenses. Some schools surrendered their licenses in return for promises of airtime for educational programs on commercial stations—promises that faded with the rising value of commercial time.

Many stations that held on found themselves confined to low-power, inconvenient hours (often daytime only, of limited value for adult education), and constantly changing frequency assignments. In 1927, 98 noncommercial AM stations were in operation, but by 1945 they had dwindled to about 25.

Reserved Channels

The decline in educational AM stations confirmed what critics had said from the first: Educational interests should not be expected to compete with commercial interests for broadcast channels, and the government should have set aside a certain number of AM channels exclusively for educational use.

When the Federal Communications Commission came into being in 1934, it reviewed the proposal to reserve AM channels. But the Commission accepted commercial owners' assurances that they would give ample free time for education. As stations that gave up their licenses for promises of time had already discovered, such promises never seemed to work out in practice.

The FCC finally endorsed the *reserved-channels principle* when it had an opportunity to allocate a brand-new group of channels in 1941—those for FM radio. In its final FM-allotment plan in 1945, the Commission reserved the lower portion of the FM band exclusively for educational use—20 channels (of the 100 total) running from 88 to 92 MHz.

The first noncommercial educational FM stations depended almost totally on local programs. They had neither a well-developed system of program exchange nor a network to help fill out station schedules. Their experience, of course, repeated that of commercial stations in the early 1920s—local resources alone simply could not provide adequate programming. Networking was as essential for noncommercial as for commercial broadcasting.

The arrival of television in the late 1940s stimulated a new campaign for reserved channels. It was considerably more intense than previous radio campaigns, because television attracted more attention. Educators and foundations foresaw far greater possibilities for educational television than they had for educational radio.

During the FCC's 1948-to-1952 freeze on new television station applications, some commercial interests made a concerted effort to block the educators' campaign. But a wide spectrum of educational and cultural groups combined into a counterlobby: the Joint Committee on Educational Television (JCET). As part of its pro-reservations exhibits, JCET prepared a content analysis of existing television programming (fewer than a hundred commercial television stations were then on the air). Analysis showed conclusively that commercial stations and networks did little to put television at the disposal of education and culture. JCET sounded a warning: If the forthcoming channel allotments failed to reserve channels for educational use, a once-in-a-lifetime opportunity would be lost.

The FCC took notice. Its *Sixth Report and Order*, which ended the freeze in 1952, provided for 242 reserved educational television (ETV) channel allotments, 80 VHF and 162 UHF. Many more were added in later years, bringing the total allocation to about 600, of which some 240 remain unused.

EXHIBIT 7.a

Growth in Number of Noncommercial Broadcasting Stations (1925–2000)

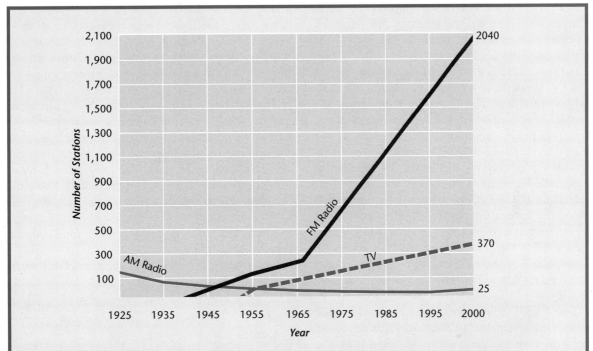

After a steady decline in the number of educational stations until 1945, noncommercial radio growth moved to the new FM service and later to television. AM figures are estimates because no official count exists.

Source: Data from FCC.

ETV Constituency

Educational radio leaders had kept the faith over many lean years, largely ignored in the seats of power. During the fight to achieve ETV channel reservations, however, the noncommercial broadcasting constituency grew larger and more diversified. Television's glamour attracted national educational, cultural, and consumer groups that previously had taken little notice of educational radio broadcasters.

Various enthusiasts developed conflicting views about the form ETV should take. Some held "educational television" to mean a broadly inclusive cultural and information service. Others took it to mean a new and improved audio-visual device, primarily important to schools and formal adult education. Some, following the model of commercial broadcasting, favored a strong national network and a concern for audience building. Others, expressly rejecting the commercial model, focused on localism and service for more limited, specialized audiences. Some wanted to stress high culture and intellectual stimulation. Others wanted to emphasize programs of interest to ethnic minorities, children, and the poor. These differences would plague public television when it appeared in the late 1960s.

A New Beginning

For its first dozen years, ETV developed slowly, long on promise but short on performance (see Exhibit 7.a). The few stations on the air depended on underfunded local productions and filmed programs of limited quality and interest. Their broadcast day lasted about half the length of commercial station schedules.

In 1959 many stations formed a program cooperative, National Educational Television (NET). It provided a few hours a week of shared programs, delivered to the cooperating stations by mail. Though a step in the right direction, NET still fell far short of meeting the enormous program needs of the noncommercial service.

In the mid-1960s a nonprofit foundation stepped into this bleak picture, hoping to transform it with a dynamic vision. The Carnegie Foundation felt that ETV needed well-articulated national goals, topnotch public relations, and leadership at the federal level. To generate highly visible recommendations for achieving these goals, the foundation set up the Carnegie Commission on Educational Television (CCET). The commission comprised nationally known figures from higher education, media, business, politics, and the arts. In its watershed 1967 report, *Public Television: A Program for Action*, the Carnegie Commission proposed that Congress establish a national "corporation for public television."

The CCET deliberately used the word *public* rather than *educational* to disassociate its proposals from what many had come to regard as the "somber and static image" projected by the existing ETV services. It also chose the word *public* to emphasize its recommendation for an inclusive service embracing not only formal instruction and classroom television but also a broad cultural/informational service intended for the public.

Six months after the 1967 Carnegie Commission report came out, its basic recommendations were enacted into law as part of President Lyndon Johnson's Great Society legislative program. The system it created survives to this day.

7.2 NATIONAL ORGANIZATIONS

The transformation of ETV into a national public television system succeeded in bringing it to public attention. In solving some problems, however, the Carnegie Commission created others. Public television stations were often highly individualistic. They made uneasy, sometimes even acrimonious, partners with their federal allies in Washington.

Corporation for Public Broadcasting

Congress set the Carnegie Commission's keystone recommendation in place with the Public Broadcasting Act of 1967 (*broadcasting* because Congress added radio at the last minute). The new law became part of the 1934 Communications Act. It created the Corporation for Public Broadcasting (CPB), a quasi-governmental organization whose key role is to funnel money into projects supportive of the nation's public broadcasters.

The Carnegie Commission had recommended that the president of the United States appoint only half of the CPB board. Congress, however, gave all the appointive power to the president. Congress also declined to legislate the long-term federal funding that the Carnegie Commission had recommended. These two departures from the plan left CPB at the mercy of presidential and congressional politics. Unfortunate consequences followed, as detailed in Exhibit 7.b.

The Public Broadcasting Act gives CPB such tasks as the following:

- making grants and contracts for obtaining and producing high-quality programs from diverse sources
- helping to set up network interconnection
- encouraging the development of new public broadcasting stations
- conducting research and training

CPB itself may not own or operate stations. Congress emphasized that the stations should retain local autonomy, remaining free to select and schedule programs according to local needs. The last thing Congress wanted was to create a centralized, federal broadcaster.

The act directs CPB to do its job "in ways that will most effectively assure the maximum freedom . . . from interference with or control of program content or other activities." Nevertheless, some stations regarded the very existence of a federal agency dispensing federal funds as a threat. The issue of local versus federal control has continually stirred up controversy between CPB and the stations.

Public Broadcasting Service

CPB launched its network in 1969–1970, calling it the Public Broadcasting Service (PBS). PBS was deliberately fashioned to differ considerably from the commercial network model. Over the years,

EXHIBIT 7.b
• • • • • • • • • • •
Political Manipulation of CPB

Incidents during three conservative presidential administrations provide textbook illustrations of the difficulty of insulating a broadcast service from politics when it depends on government for substantial economic support. Section 398 of the Communications Act, added by the 1967 Public Broadcasting Act, tries to prevent political influence by expressly forbidding any "direction, supervision, or control" over noncommercial broadcasting by officials of the U.S. government. This legal detail did not stop the Nixon administration in the 1970s or the Reagan and Bush administrations a decade later from manipulating CPB for their own ends.

When the Public Broadcasting Service (PBS) network began beefing up its news and public-affairs programming around 1970 by hiring ex-commercial-network personnel, the White House became concerned. Regarding PBS as far too liberal, the administration objected to public television's concern with national affairs, when, according to the administration's interpretation, it should be focusing on *local* needs. In 1973 the administration sent a more direct message when President Nixon vetoed a two-year funding measure for CPB. Long-range federal funding legislation did not pass until 1975.

By 1981, when the Reagan administration took office, the CPB board, never high on the priority list of political appointees, had become fertile ground for political gamesmanship. Reagan appointed several hard-right conservatives to the board. One Reagan CPB appointee advocated making a content study to determine whether public television programming leaned too far to the left. The study never materialized, but the legally mandated political impartiality of CPB had been seriously undermined.

That was evident during Bush's term (1989–1993), when CPB found itself under the combined pressure of an administration trying to trim federal funding and a Congress reviewing specific broadcasts while considering funding legislation. Late in 1991 funding for CPB for 1994–1996 was delayed for months in a controversy over both the amount of money proposed and two documentaries: *Tongues Untied*, about gay black life (1991), and *Color Adjustment*, about the role of blacks on television (1992). In the end, Congress did pass a record authorization for the three-year period, and President Bush, in the midst of an election campaign, signed it. But all sides agreed that the constant tension over funding and programming was anything but over.

Indeed, soon after the Republican Party took control of Congress in 1994, House Speaker Newt Gingrich (R-Ga.) issued a call to "zero out" federal funding for CPB. He accused the corporation of "eating taxpayers' money" and attacked its officials as "rich upper-class people" (*New York Times*, 20 January 1995:C3). Gingrich later backed off somewhat, saying that Republicans were committed to good television but that some things might have to change at the Corporation for Public Broadcasting.

Though not an example of manipulation, additional fodder for congressional critics of public broadcasting was delivered by public stations themselves in 1999 when word spread that several prominent PBS stations made a practice of trading their donor lists with political organizations. When the story broke, it appeared that the stations had traded lists only with Democratic groups, thus angering congressional Republicans. Later reports indicated that many stations used list brokers, who made the lists available to political groups of all stripes, conservative and liberal. When the dust settled, Congress made such list exchanges illegal, and any chance of public broadcasting receiving increased funding disappeared. Before this controversy arose, Congress had already approved CPB appropriations for the years 2000, 2001, and 2002 at $300 million, $340 million, and $350 million, respectively.

however, these differences have diminished. PBS operates the interconnection facilities and, until very recently, increased its domination of the all-important task of program selection.

A weak national network had in fact been a Carnegie Commission recommendation. Though the report stressed the vital importance of network interconnection, it also warned against tight centralization of program control. Public broadcasting was expected to have "a strong component of local and regional programming" to "provide the opportunity and the means for local choice to be exercised upon the programs made available from central programming sources" (CCET, 1967: 33).

As time proved, it was unrealistic to limit PBS's programming role, even though doing so was desirable to avoid the overcentralization typical of commercial networks. The Carnegie Commission failed to anticipate the practical problems of asking PBS to lay out a smorgasbord of programs from which affiliates would pick and choose at will. (Public broadcasters tend to avoid the term *affiliate* because of its commercial connotations—*member station* is preferred—but it is used here for convenience.)

Commercial experience had shown that a network needs the strong identity that can come only from a uniform national program schedule—identical key network programs available to everybody in the nation at the same time of day. But station licensees—zealous to protect their local identity, operating under varied ownership and funding patterns, and profoundly suspicious of political interference from Washington—resisted that type of network. They sided with "their" organization, PBS, against CPB in the battles over program philosophy that followed.

As a means of continuing service while the philosophical debate continued, PBS program selection from 1974 until 1990 occurred through a complicated mechanism called the *Station Program Cooperative* (SPC). Member stations (whose executives actually voted for which programs they would carry and help finance) praised SPC for its democratic participation. But SPC was also criticized for limiting program innovation, as stations nearly always voted and paid for programs successful in the past, rarely investing scarce dollars in new ideas.

Late in 1989, under pressure from a Congress concerned about SPC's decentralized and argument-ridden decision making, PBS stations concluded that the democratic SPC system had outlived its usefulness. They dissolved SPC and centralized all programming decisions under one PBS officer—much like the traditional commercial network model.

When former FCC commissioner Ervin Duggan became PBS president in 1994, he designed a new three-division management organization. The new structure returned more choice to individual stations, allowing them to buy less from the national program service and to select more single programs or series from a new unit, PBS Syndication Services. The revised plan was intended in part to address what cost-conscious members of Congress and others called the "overlap" problem—several PBS stations that serve the same metropolitan area spending money on essentially the same program.

PBS Operations

PBS differs from the commercial networks in that it carries only programs that others produce. Member stations (affiliates) contract with PBS, agreeing to pay dues determined by their respective budget and market sizes. Rather than the network paying for use of their time (long the case with commercial television network affiliates), public stations pay PBS for programs—an arrangement closer to a cable system's relationship with the many cable networks, or to program syndication in commercial broadcasting.

At present, PBS serves virtually all public television stations on the air. Its 35-member board consists of professional station managers and members of the public drawn from station boards. The PBS staff totals about 300 in Alexandria, Virginia (near Washington, DC), New York, and Los Angeles.

Satellite Interconnection

PBS pioneered the use of satellites for network relays. In 1966 the Ford Foundation proposed using satellites instead of AT&T's wire and microwave facilities for relaying public television programs. A decade later, with government support, PBS announced a plan to interconnect its member stations using transponders on a domestic communications satellite.

Benefits claimed for the system included cost savings, better-quality reception, ability to relay signals both east and west as well as variously within given regions, and simultaneous transmission of several signals, allowing stations to choose among more program options. Satellite facilities also offered fund-raising possibilities through the lease to others of excess transponder capacity.

Each station had to contribute about $25,000 (a hefty sum for most) toward the cost of TVROs. During 1978 public television stations abandoned terrestrial links and began using satellite interconnection; indeed, PBS was the first national broadcast service to do so (even before most cable networks had begun operation).

National Public Radio

In 1970 CPB created National Public Radio (NPR) both to interconnect stations (like PBS) and to produce some programs (unlike PBS). NPR provides its member stations, which do not have to carry any set amount of network programming, with about 22 percent of their daily schedules. NPR also coordinates satellite interconnection with 24 uplinks and more than 300 downlinks. Satellite distribution (the world's first satellite-delivered radio network) began in 1980 with 4 audio channels, which has now increased to 24.

CPB radio program funds go directly to NPR member stations, which support the network by paying a flat fee and then subscribing to morning and afternoon news services, NPR's musical and cultural programs, and Public Radio International programs. Stations pay anywhere from $25,000 to more than $300,000 annually, depending on market size, for NPR's program service, providing more than half of NPR's operating income.

7.3 PUBLIC STATIONS

Both public radio and public television stations vary enormously in size, resources, goals, and philosophy. Though all are licensed by the FCC as "noncommercial educational" stations, some flirt with commercialism and many play no discernible formal educational role.

Television

In general, four classes of owners hold public television station licenses. The four groups differ and even conflict in their concepts of noncommercial broadcasting.

States and municipalities hold about 40 percent of the licenses. Many belong to state educational networks, usually programmed by the station located at the state capital or other key origination points. Many southern states in particular chose this means of capitalizing on educational television.

Colleges and universities own about 25 percent of the stations. They often complement educational radio stations of long standing and sometimes closely tie to college curricula.

Public-school boards hold licenses for about 3 percent. Their stations naturally focus on in-school instructional programs, many produced by and for the local school system. The low number of such stations reflects the failure of public television to live up to its early promise as a major adjunct of formal education. In fact, as school budgets became tighter, several school stations left the air or transferred to other licensees.

Community foundations control a third of the outlets. The licensees are nonprofit foundations created especially to operate noncommercial stations. They recruit support from all sectors—schools, colleges, art and cultural organizations, foundations, businesses, and the public. Usually free of obligations to local tax sources, they tend to be politically more independent than stations that depend for revenue on local or state tax dollars. Among the best-known community stations, because they produce much of what appears on PBS, are WETA in Washington, DC, KQED in San Francisco, KCET in Los Angeles, and WGBH in Boston.

Radio

CPB built its national noncommercial radio network, NPR, around a cadre of professionally competent, full-service stations (referred to as "CPB-qualified"). These stations had to meet certain standards involving factors such as power, facilities, budgets, and number of employees. Today's standards for what are now called "CPB-supported" stations are much less rigorous: minimum hours of operation (12 or 18, depending on the level of financial support sought) and program schedules that do not advocate a religious or political philosophy or consist primarily of in-school or in-service educational instruction.

Stations that carry NPR programming do not necessarily have to be CPB-supported, although nearly all of them are. By the year 2000 more than 600 noncommercial stations were NPR members (like PBS, NPR does not use the term *affiliate*). A much larger group of more than 1,000 noncommercial stations either do not qualify for NPR membership or choose not to join the network. In 2000 the FCC authorized a new class of low-power noncommercial FCC station. Once implemented, this action could give hundreds of community groups and nonprofit organizations acces to the airwaves (see Section 4.7, p. 103).

7.4 ECONOMICS

Ask any noncommercial broadcaster to name his or her most serious problem, and the answer will almost always be "not enough money."

Compared with those in many other countries, America's noncommercial system, on a per-person basis, is woefully underfunded. For every tax dollar allocated to public broadcasting in America, the

EXHIBIT 7.c

Public Television Revenue by Source

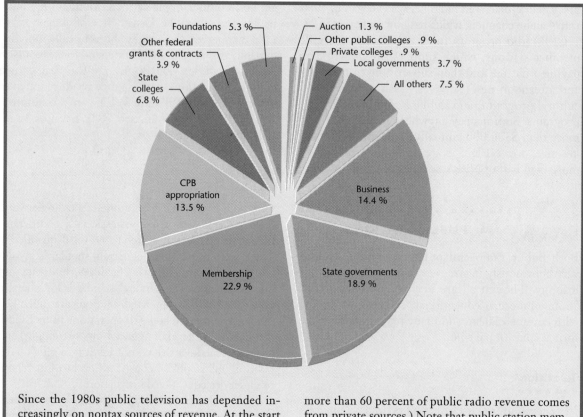

Foundations 5.3 %

Other federal grants & contracts 3.9 %

State colleges 6.8 %

Auction 1.3 %

Other public colleges .9 %

Private colleges .9 %

Local governments 3.7 %

All others 7.5 %

CPB appropriation 13.5 %

Business 14.4 %

Membership 22.9 %

State governments 18.9 %

Since the 1980s public television has depended increasingly on nontax sources of revenue. At the start of that decade, tax-based sources provided more than two-thirds of all revenue; in the 1990s tax support represented slightly less than half and has remained constant at that level over recent years. (By contrast, more than 60 percent of public radio revenue comes from private sources.) Note that public station members, businesses, and state governments provide the largest segments of public broadcasting support.

Source: Association of Public Television Stations, 1999.

governments of Britain and Canada fund their BBC and CBC services, respectively, with about $25, and Japan's NHK public network receives about $15 in government funding per person served.

Exhibit 7.c shows the diversity of public television's funding sources. But diversity has a price: Each source brings different obligations with its funding, and each has its biases.

Government Support

Local and state governments supported educational FM from its inception in the 1940s, and many states expanded into educational television over the ensuing decades. By the mid-1960s state tax funds provided about half of all public broadcasting income. In the face of rising budget shortages, however, both state and local support had declined substantially by the 1990s.

At the outset the federal government gave no financial assistance at all. Federal legislative recognition came first with the Educational Television Facilities Act of 1962. It granted up to a million dollars of federal money to stations in each state, subject to their matching federal dollars with money from other sources. Extended and revised, this act continues to support equipment and facilities purchases for public broadcasting. This program is administered by the National Telecommunications and Information Agency (NTIA), an arm of the U.S. Department of Commerce. As with CPB, federal funding of NTIA also is under attack, and there is considerable argument over whether this funding should continue.

Long-Range Funding

Attempts to fund CPB for longer than the usual federal one-year budget cycle failed until 1975 when Congress finally authorized funds for a three-year period. Although similar multiyear appropriations have followed since, Congress has sometimes rescinded appropriations to which it already agreed, and there have even been attempts to halt federal funding entirely (see Exhibit 7.b). This uncertain commitment by Congress—and several state governments as well—has made it difficult for public broadcasting to plan, let alone pay, for the future.

Foundation Grants

Foundations provide substantial support for noncommercial broadcasting. Without the backing of the Ford Foundation, early educational television might not have survived its first decade. Carnegie Foundation money paid for two important studies that helped shape public broadcasting.

In 1981 the largest single gift ever made to public broadcasting came from Walter Annenberg, then publisher of *TV Guide*. He donated $150 million, spread over 15 years, to fund a project to create innovative college-level courses and programs. The Annenberg/CPB project shifted focus in the early 1990s, in part because of changes in tax laws, to support programming for primary and secondary schools rather than university-level audiences.

Corporate Underwriting

Through a limited form of sponsorship called *underwriting*, businesses provide program producers funds to cover at least some production costs for specific projects. FCC regulations allow compa-

nies brief identifying announcements at the beginning and end of such programs. After 1981 the FCC also allowed corporations to display logos or trademarks. Some critics worry that excessive corporate underwriting may influence the content of noncommercial programming, a fear similarly raised regarding advertisers and commercial broadcasting.

Commercial Experiment

Seeking alternative or additional revenue sources for public broadcasting, Congress in 1981 established the Temporary Commission on Alternative Financing for Public Telecommunications (TCAF) to supervise an experiment with commercials as a means of support.

The TCAF found that most public television stations would not carry ads because of labor union contract requirements, local economic considerations, or concerns about advertising's impact on the character of and other financial support for public broadcasting (TCAF, 1983). The experiment ended in mid-1983, and outright commercials remained taboo.

Creeping Commercialism?

Influenced by TCAF recommendations, the FCC in 1984 authorized *enhanced underwriting*, leading stations to carry what former PBS president Bruce Christensen called "almost commercials" (Smith, 1985). The FCC tried to balance station and public needs: Stations may sell announcements up to 30 seconds long and mention specific consumer products but may sell no more than two and a half minutes between individual programs.

The line between "enhanced underwriting" and commercial spots has now become increasingly blurred. Critics deplore public television's increasing commercialization. After all, public broadcasting's primary justification is that it furnishes *alternatives*— programs not likely to appeal to advertisers but nevertheless useful and desirable. To the extent that pleasing advertisers motivates *both* services, the argument for public broadcasting is weakened. Still, the notion resurfaces from time to time. When a Republican-dominated Congress threatened in 1995 to eliminate CPB funding, full commercialization of public broadcasting was suggested as a possible replacement funding source.

Local Contributions

Federal appropriations (and often national foundation grants) trickle down only when stations match federal grants. The desperate scramble to match federal money forces local fund-raisers to push membership/subscription drives to the saturation point. Polls indicate that many viewers detest marathon fund-raising drives (derisively called *begathons*). On-air auctions of donated goods and services (most of them from commercial sources) promote the givers so blatantly as to amount to program-length commercials. However, as Exhibit 7.c demonstrates, these efforts are vital to the survival of noncommercial stations: Nearly 25 percent of public broadcasting's revenue comes from individual viewers and listeners.

Other devices used (in some cases merely proposed) to raise funds locally, and to some extent nationally, include the following:

- selling commercial rights to market merchandise associated with programs (The Children's Television Workshop defrays a large part of *Sesame Street*'s production costs from such sources.)
- selling newly produced programs to commercial television or pay cable for initial showing before release to public television (proposed)
- renting station facilities (usually studios) to commercial producers
- selling books, videotapes, and other items to viewers
- trading a reserved educational channel for a less desirable commercial channel for a price (Such channel reclassification requires FCC approval.)
- acquiring assured tax-based revenue by charging commercial broadcasters a tax on profits, a receiver licensing fee, or a spectrum-use fee (proposed)
- selling access to the FM subcarrier or TV vertical blanking interval for commercial use
- selling excess channel capacity after public stations convert to digital broadcasting (proposed)

In sum, public broadcasting executives have to serve too many masters and spend an inordinate amount of time on fund-raising. The year-to-year uncertainties of congressional appropriations, corporate underwriting, and local membership drives make it impossible for public broadcasters to plan rationally for long-term development. As a result, in the mid- and late 1990s some prominent legislators began calling for the establishment of a public broadcasting trust fund, the proceeds of which could provide predictable long-term funding. Interestingly, this is exactly what the Carnegie Commission suggested in its initial report in 1967. As of mid-2000 these proposals had not been formally addressed by Congress.

Public television's funding uncertainties become especially salient as the deadline for conversion to digital broadcasting approaches. The Association of Public Television Stations estimates that the conversion costs for its members collectively will exceed $1.7 billion (a sum public television stations do not have on hand). Public broadcasters requested $770 million in federal aid over five years, but in 1999 Congress appropriated only $15 million for the effort. As the new century dawns, public broadcasting advocates continue to seek additional conversion funds from the federal government.

7.5   TV PROGRAM SOURCES

Although noncommercial broadcasting's basic purpose is to supply an alternative to commercial services, program overlap does occur. Noncommercial stations often show feature films and syndicated series obtained from the same distributors that commercial stations use. On the other hand, noncommercial outlets do more local production and experimental programming than do their commercial counterparts.

Stations as Network Producers

A handful of major-market producer stations act as primary producers for PBS. In the 1990s they provided more than a third of national programming, with another 10 percent or so coming from lesser public television stations. Major producer stations, such as WGBH-Boston, have long histories of creative innovation in public television. They tend to develop specializations: WETA-Washington, DC, for news and public affairs, for example, and WGBH for science, documentary, and drama presentations.

Foreign Sources

By the turn of the century nearly a fifth of all PBS series contained some overseas material, while others came straight from sources such as British commercial and BBC television series. Best known are the long-running *Masterpiece Theatre* and *Mystery!* series, both co-produced by WGBH.

Independent and Syndicated Production

Under pressure from producers who claimed PBS largely ignored their output and abilities, Congress in 1988 required CPB to fund an Independent Production Service, separate from CPB, to encourage more independent program sources.

Easily the best known of these is the nonprofit Children's Television Workshop. It won international fame for *Sesame Street* and subsequent children's series. The *Sesame Street* series began in 1969 with initial funding from government and foundation grants. By the late 1980s ancillary commercial ventures, such as merchandising items using the program name and characters, defrayed well over half of its budget.

Local Production

The typical noncommercial television station produces and uses more local programming than do most commercial stations. Locally produced programs consist mainly of news and public affairs, along with some educational/instructional material telecast during daytime hours for in-school use.

However, local production is often one of the first things to go when budgets are cut, as was common during the economically difficult early 1990s. Accordingly, many stations cut local production by half from its mid-1970s level, now averaging slightly more than 5 percent of the total program schedule.

Cable Alternatives

Just as cable has increasingly taken on many functions of traditional broadcasting in entertainment, so has it done in education and public affairs. C-SPAN, in addition to its extensive public-affairs programming (see Section 3.2, p. 59; Section 9.5, p. 266; and Section 11.5, p. 332), offers C-SPAN in the Classroom, a free educational service that allows teachers to tape material for instructional use.

By the end of the 1990s about 75 percent of all schools in the country had been supplied with free cable service, most by local cable systems as a part of their franchises. Cable in the Classroom, a free service of the commercial cable industry, provides some national coordination. The service provides about 500 hours of commercial-free programming per month, supplied largely by 31 cable-delivered networks. A 1996 survey of some 800 elementary- and secondary-school teachers found that 70 percent used Cable in the Classroom programming at least once during the last school year, and 25 percent indicated they used it up to four times per week.

Cable television also provides noncommercial alternatives through its access channels—channels set aside for public, educational, and governmental use. Many communities have vital cable access operations, providing one or more channels of local entertainment, educational, and government programming. Public access channels provide a forum for residents to showcase their artistic (though in some cases that is certainly open to argument) talents on television, a proposition too expensive for most people to pursue on a broadcast station. Some cities use their government access channels to transmit council meetings, school board meetings, and various other local government activities, providing their citizens with greater access to local decision making.

DBS and the Internet

In the Cable Act of 1992, Congress imposed a requirement on all DBS operators that they reserve from 4 to 7 percent of their channel capacity, as determined by the FCC, "exclusively for noncommercial programming of an educational or informational nature." Following lengthy litigation over the new law, a federal court in 1996 upheld the regulation.

In 1998 the FCC ruled that DBS operators must set aside at least 4 percent (the statutory minimum) of their channel capacities for noncommercial educational and informational programming. Although some groups argued that a special board be created to decide what noncommercial programming the DBS systems should carry, the FCC decided to let DBS operators choose for themselves. Many programmers have applied to the

DBS operators for carriage of their channels. According to the FCC's rules, DBS operators may charge the educational programmers half of the direct costs they incur in carrying the channels. The regulations also limit carriage to only one channel per programmer.

Increasingly, the Internet is providing significant amounts of educational "programming" to a global audience. Many universities offer courses and even entire degree programs online. With the push to reach nontraditional students through "distance education," the Internet is unique in its ability to deliver interactive instructional programming to widely dispersed audiences.

7.6   NONCOMMERCIAL TV PROGRAMS

PBS began in 1970 with several programs either still on the air (*Washington Week in Review*) or well remembered (*The French Chef*, which featured Julia Child demonstrating that good French cooking is within everyone's reach). Many initial programs focused on news and public affairs.

News and Public Affairs

In contrast to commercial networks, PBS found its affiliated stations receptive to long-format (more than half-hour) newscasts.

The NewsHour with Jim Lehrer (and its immediate predecessor, *The MacNeil/Lehrer NewsHour*) has consistently been acclaimed as one of TV's best news and information programs, featuring in-depth interviews and analyses of current news stories. *NewsHour* provides a significant alternative to the largely news headline newscasts of the commercial TV networks and lacks the acrimonious exchanges that characterize some long-format cable news programs.

PBS began the series in 1976 as a half-hour news program, expanding it to an hour in 1983. In 1994 Robert MacNeil announced that he would leave the program in October 1995, with Lehrer continuing as sole anchor. Despite critical praise, *News Hour* has failed to build a large following, even by noncommercial standards. Still, its loyal audience has kept the program on the air for more than 20 years. In 1996 PBS introduced *NewsHour Online*, a continuously updated news service accessible on the PBS web page.

PBS provides a steady diet of other public-affairs programs as well, including *Washington Week in Review*, *Nightly Business Report*, *Wall Street Week*, and *Frontline*. Bill Moyers specials are irregularly scheduled but get both wide coverage and high praise for their in-depth analysis of and sensitivity to social problems.

When in 1992 the commercial networks abandoned full-time coverage of the quadrennial political conventions, PBS (and cable services) stepped into the breach, offering full prime-time coverage—combining Robert MacNeil and Jim Lehrer with NBC's top correspondents, including anchor Tom Brokaw when NBC was not airing convention news itself. Such coverage was repeated during subsequent conventions.

Another aspect of PBS news and public-affairs programs appeared in 1992 as well: PBS-distributed programs were now legally *required* (by CPB long-range funding legislation) to offer objective and balanced content. If an expert review panel finds questions of balance or fairness in a program, CPB may support the costs of producing balancing programs. As a part of this "Open to the Public" process, CPB announced more ways for its audience to "talk back" on fairness and other issues—a toll-free telephone number and quarterly meetings of the CPB board with public groups away from Washington.

In anticipation of the 1996 election, PBS introduced the "Democracy Project," which it described as a "laboratory for citizen-oriented news and public affairs programs." The Democracy Project featured a series of programs on elections, government, and public participation. A classroom component of the Democracy Project—activities and special events for elementary schools—focused on engaging youths in civic education and the democratic process. The project continued through the 2000 campaign.

History

Occasionally, documentaries attract viewers who do not normally watch the noncommercial service. This category was exemplified by *The Civil War*, an 11-hour 1991 series that won widespread praise and viewership—and helped sell thousands of copies of a handsome companion book and videotape copies

EXHIBIT 7.d
• • • • • • • • • • •

Ken Burns and *The Civil War*

For five consecutive nights in September 1991, some 14 million viewers watched *The Civil War* unfold, making it the most-watched series in the history of public television. It earned an average national rating of 9 and was seen by four times the normal PBS audience.

The fruit of a five-year effort and some $3.2 million in support from CPB, Washington, D.C.'s WETA, General Electric (the only corporate underwriter), and the National Endowment for the Humanities, *The Civil War* was the creation of filmmaker Ken Burns. Burns had done earlier documentary programs for PBS, but this was different—an 11-hour series that brought the 1861–1865 struggle to life as most viewers had never seen or thought about it before. Combining the creative use of period music, live shots of battlefields, quotes from experts, and old photos over which the cameras seemed to roam lovingly, *The Civil War* took a neutral view of the war's causes. (Viewers in both North and South later com-

plained about bias, suggesting that the neutral approach had been successful.)

Syndicated newspaper columnist George Will summed up the series and its maker: "Our *Iliad* has found its Homer: He has made accessible and vivid for everyone the pain and poetry and meaning of the event that is the hinge of our history." Concluded producer Burns:

> There's no way that *The Civil War* could have been produced anywhere but on public television. Public television not only tolerates but encourages the kind of experimentation and risk-taking that we take as independent producers. And on public television I don't lose the viewer's attention every six to eight minutes for a commercial. The ability to hold that attention is critical to me as an artist. Public television is also the only kind of television where the creator actually retains control. (CPB, 1991: 44–45)

The series proved financially valuable, too. Viewer contributions to local stations rose sharply, thanks to money pleas during breaks. The companion book sold more than 750,000 copies, and related video and audio recordings also sold well. Burns followed his success with a 90-minute program on early radio called *Empire of the Air*, which traces the lives of Lee de Forest, Edwin Armstrong, and David Sarnoff; a multipart documentary series tracing the history of baseball; and, with co-producer Stephen Ives, an eight-part series descriptively titled *The West*.

Noting that we often come to terms with our common past through the lives of other people, Burns has increasingly turned his attention to biography. He says that "the beginning of my interest in history, the beginning of anyone's interest in history, is in some ways about other people I want to know." This interest in biography has led Burns to produce films for PBS on the lives of Thomas Jefferson, Lewis and Clark, and Frank Lloyd Wright.

Source: Pam Tubridy Baucom, Florentine Films. Photo courtesy of WETA-TV, Washington, D.C.

of the series, proceeds from which helped meet the program's production costs (see Exhibit 7.d). A similar, though smaller, reaction greeted *The West*, an eight-part documentary that aired in 1996.

Public television's provision of such programming goes back more than two decades. The multipart *Civilization*, a review of world history and

cultural development featuring Sir Kenneth Clark, ran in PBS's first season and helped to introduce the companion book marketing idea.

Vietnam: A Television History, co-produced by Boston's WGBH and French television, became the highest-rated documentary of 1983 and won several awards. The documentary's less-than-positive assessment of America's role brought howls of protest from conservative groups—and a controversial decision by PBS to allow several critics to air their differences during an hour or so of airtime with the program's producers.

The American Experience has served as host vehicle for a series of historical documentaries on American history and key historical figures—with such varied programs as those on the building of the Brooklyn Bridge, the impact of rigged television quiz shows, immigration, and several studies of American presidents, including a well-received two-part study of *The Kennedys*.

Part adventure, part history lesson, and part treasure hunt, *Antiques Roadshow* debuted in 1997 and has become one of PBS's most popular series. Hosted by antiques expert Chris Jussel, the show brings antiques specialists together with citizens seeking information about their possessions. Television viewers experience the excitement as participants in the show learn the true value of their antiques. The show's experts also offer tips on finding, preserving, and protecting valuable collectibles. Each program is taped in a different city, and host Jussel takes the opportunity to show the viewers important historical and cultural sites in the area. As with many other popular PBS offerings, *Antiques Roadshow* is produced by WGBH.

Performing Arts

All of the performing arts—drama, dance, music—find a home in public television. Among public television evening programs, *Masterpiece Theatre* is probably best known. The Sunday evening hourlong staple provides British-made drama, ranging from short series built around historical events, famous novels, or contemporary views of the human condition, to such lengthy costume favorites as *Upstairs, Downstairs*, on life in a fine Edwardian house. Another popular program since its first broadcast in 1980 is *Mystery!* hosted first by Vincent Price and then by Diana Rigg. Viewers of this program watch as their favorite sleuths—Miss Marple, Hercule Poirot, Inspector Morse, Sherlock Holmes, Rumpole of the Bailey—solve crimes and other mysteries in often humorous and intellectually stimulating ways.

Music and dance appear regularly on three PBS programs that began in the mid-1970s and are still going strong: *Great Performances* (1974), *Austin City Limits* (1975), and *Live from Lincoln Center* (1976).

A more recent series, *Evening at Pops*, features the Boston Pops Orchestra.

Science and Nature

One of the first programs dealing with science and nature aired in the 1970s. *The Ascent of Man* featured philosopher Jacob Bronowski's view of developing mankind in a multipart series and yet another useful companion volume.

Since then PBS has offered the eight-part series *The Brain*, as well as the multipart *Cosmos*, with famous astronomer Carl Sagan introducing many Americans to the wonders of astronomy. The long-running series *Nature* and some of the most popular public television documentaries, such as the *National Geographic Specials* and *National Audubon Society Specials*, proved that Americans could also produce and enjoy topnotch nature films.

Over the last two decades, *NOVA*, the longest-running nationally broadcast science series, has presented on PBS more than 400 different hourlong science programs, each devoted to a single topic. Each week the program draws some 8 million viewers, the largest audience of any PBS weekly series.

Children

Public broadcasting has a clear mandate to provide constructive, imaginative children's programs. Accordingly, many public television stations fill a large portion of their daytime hours with in-school programs (paid for by school districts) and schedule additional special programs for children both early in the morning and again in the afternoon.

Since 1968 Pittsburgh station WQED's *Mister Rogers' Neighborhood* (see Exhibit 7.e) has taught social skills to very young preschoolers, focusing on their values, feelings, and fears with gentle conversation and songs tailored to their cognitive needs. It is second in viewing popularity only to the famous *Sesame Street*.

The Children's Television Workshop (CTW) brought all the technical resources of television—as well as all the capabilities of educational research—to bear with its 1969 premier of *Sesame Street*. A wonderfully original group of large-scale puppets, the Muppets, became the program's hallmark. Research showed that children who watched *Sesame Street* learned to read more quickly than did other children.

EXHIBIT 7.e

· · · · · · · · · · ·

Fred McFeely Rogers: American Hero

"Won't you please, won't you please, please won't you be my neighbor?" Since 1968 millions of young children have heard and sung along to these soothing words from Fred Rogers as he begins each episode—more than 850 of them—of *Mister Rogers' Neighborhood*, the longest running program on PBS.

After graduating from Rollins College in 1951, Fred Rogers began his television career at NBC but soon joined the staff of WQED in Pittsburgh, the country's first community-supported television station. He was co-producer, puppeteer, and musician of the first children's show for the station, *The Children's Corner*, in 1954, and in 1955 the show won the Sylva-

nia Award for the best locally produced children's program in the country. While working at the station, Rogers continued his education at the Pittsburgh Theological Seminary, and received his Bachelor of Divinity degree in 1962. He was ordained a Presbyterian minister, with a special charge of serving children and families through television.

Over the years Rogers took to the air himself and began to develop the gentle and affirming on-air style that has become his trademark. With the help of grants from the Sears, Roebuck Foundation and National Educational Television, WQED began producing *Mister Rogers' Neighborhood* in 1968 for national distribution. The program's format remains constant, an attribute that helps kids feel comfortable. Part of the time is spent in Mister Rogers' house, where he talks with his viewers, visits with guests, sings songs, and brings something to show. The other part of the program takes place in the Neighborhood of Make-Believe, where human friends interact with puppets in stories that help children deal with their everyday concerns.

Fred Rogers is CEO of Family Communications, Inc., which he formed in 1971. The nonprofit corporation produces *Mister Rogers' Neighborhood* as well as other audiovisual educational materials. Rogers has written several books for children and parents, and has recorded four albums of children's music. For his efforts, Rogers has won numerous honors, including two Peabody awards and four Emmys. In 1998 he received the Emmy Lifetime Achievement Award, and in 1999 he was inducted into the Television Hall of Fame.

Fred Rogers, who is married (since 1952) and has two sons, turned 70 in 1998, and still produces several weeks of new *Neighborhood* programs each year. For more than 30 years the program has entertained and educated kids and has made them feel good about themselves. Thanks for being our neighbor, Mister Rogers.

Source: © 2000 Family Communications, Inc.

No series on either commercial or public television has ever been given as much scheduled airtime as has *Sesame Street*, which now has taken on classic status as the preeminent program for preschoolers.

Building on *Sesame Street*'s success, CTW later branched out with *The Electric Company* for older children. In 1980 CTW began a daily science program, *3-2-1 Contact*, and later *Square One TV*, a math series for young children. CTW also began *Encyclopedia*, a children's series for HBO, and in 1991 launched a $20 million multimedia literacy project, *Ghostwriter*. The following year it joined with UNICEF and Mexico's Televisa to create a Spanish-language version of *Sesame Street*.

At the beginning of 2000, the national programming feed of PBS distributed 15 different children's television shows to PBS affiliates across the country. These programs target children from preschool age through their teens and include such popular favorites as *Teletubbies*, *Barney and Friends*, *Arthur*, and *Zoom*.

Education and Instruction

Since 1977 CPB has funded more than 20 instructional television (ITV) series for use in classrooms, providing texts and other support for each series. Since 1981 PBS has offered the Adult Learning Service (ALS), a cooperative effort with local stations that provides college-credit television courses for viewers at home. Many of these courses reach colleges directly by means of the Adult Learning Satellite Service, which began in 1988 and serves some 2,000 institutions of higher learning. In parallel fashion, PBS coordinates the distribution of programs for elementary and secondary schools.

As the new century began, PBS was planning a 24-hour adult learning channel, hoping that DBS operators would choose to carry it as part of their mandated noncommercial carriage requirement (see Section 7.5). Dubbed PBS-U, the channel will start with a four-hour block of programming from the ALS. Ultimately, PBS will program the channel with two 6-hour segments of instructional programming during the day, with each segment repeated at night.

In carrying out their classroom instruction mission, public television stations draw on several large libraries of instructional materials from companies that act as syndicators. Notable examples include the Agency for Instructional Television (AIT)

in Bloomington, Indiana, innovative producers of interactive videodisc programs for children; the Great Plains National Instructional Television Library (GPN) of Lincoln, Nebraska, producers of well-known classroom series for children; and the Annenberg/CPB Project, producers of several prime-time adult learning series.

Not all of the educational and information programming from PBS is intended solely for the classroom. Many of the most popular PBS series feature how-to themes. In the late 1990s programs such as *Baking with Julia* (Child), *This Old House*, *The New Yankee Workshop*, and *The Victory Garden* delivered entertaining instructional content to do-it-yourselfers across the country.

7.7 NONCOMMERCIAL RADIO PROGRAMS

Typical public radio stations affiliated with National Public Radio (NPR) and/or Public Radio International (PRI) fill about two-fifths of their weekly total airtime with programs from these networks and other syndicated sources.

National Public Radio

NPR's first continuing success was its weekday late-afternoon news-and-feature program, *All Things Considered*, which began in 1971. The program quickly won every radio program award in sight and became a listening addiction for its loyal fans. NPR took the same approach to news on *Morning Edition*, which premiered in 1979, and *Weekend Edition*, which debuted in 1985. In 1993–1994, NPR first offered around-the-clock hourly newscasts, continuing an 18-hour-a-day schedule of newscasts on weekends. NPR also provides continually updated news reports on its web site.

In addition to news programs, NPR distributes a series of talk, cultural, and informational programs to its affiliates. One particularly popular talk show is the Peabody Award–winning *Fresh Air* with Terry Gross. This weekday magazine of contemporary arts and issues features in-depth interviews with prominent cultural and entertainment figures plus criticism and comment. *Performance Today* takes listeners behind the scenes to learn more about composers and performers through interviews and features. Perhaps the most unlikely NPR offering—though one of the most popular and financially successful for stations that carry it—is *Car Talk*, featuring the Magliozzi brothers, whose answers to called-in car-maintenance questions are spiced with jokes, word play, and a general sense of fun.

Public Radio International

A second national public radio service, Public Radio International (PRI), began as American Public Radio (APR) in 1983. Formed by Minnesota Public Radio, it first served chiefly as a distribution channel for the popular *A Prairie Home Companion*, featuring Garrison Keillor (see Exhibit 7.f). By the early 1990s it was the largest distributor of public radio programming. Effective July 1994, it changed its name to reflect its increasing global emphasis on news and other programming. PRI now offers roughly 400 hours of weekly news, classical music, and cultural programming to more than 600 affiliated stations.

Unlike NPR, PRI provides programs to only one station per market. Also unlike NPR, it does not produce programs, but rather acquires them from member stations. In addition to classical and other music programs, PRI offers *Monitor Radio*, a news-and-feature program of the *Christian Science Monitor* newspaper, and the half-hour weeknight *Marketplace* business news program. It also brings BBC World Service programming to American listeners.

Formats

The primary appeal of public stations is that they offer content usually heard nowhere else—jazz, opera, local public affairs, folk music, and the like. Recently, a growing number of stations, especially in larger radio markets with multiple noncommercial licensees, have shifted their formats from music to all-talk. The single largest format is religious/gospel programming, reflecting the large number of noncommercial stations licensed to religious organizations.

Public stations tend to differ most from their commercial counterparts in that many vary their content considerably each day rather than consistently hewing to a single format. They come close to the "middle-of-the-road" or "something for everybody" formats that traditional radio stations provided in the 1940s and 1950s.

Several hundred noncommercial FM stations and a growing number of low-power AM stations (and a few cable-only FM services) provide one or more types of community radio. (College stations operating on student fees could be considered a type of community radio.) In the 1960s, reacting against rigidly formatted radio and dissenting from establishment values, some small stations began the *underground* or *free-form* radio format.

Such radio is now lumped under the *progressive* format, which mixes live and recorded music and talk, usually at the whim of the presenter. Community stations often feature, if not progressive music, a classical, jazz, or diversified format. Small, low-salaried staffs supplemented by volunteers mean low overhead, enabling such outlets to present formats that would not be commercially viable. (See Exhibit 9.i, p. 281, for a closer look at noncommercial radio formats.)

7.8   CHANGING ROLES

With their small but generally upscale audiences, noncommercial stations are concerned with survival in the new century in at least two areas—financial support and competition from other services. But they face an even more basic dilemma as well: Can they successfully continue to justify their very existence?

New Options

A common argument heard these days is that the development of alternative media, especially cable, has weakened the major rationale for public television. Programs for children, good drama, science and history, and other cultural programming—once the nearly exclusive domain of public broadcasting—now appear on several cable networks delivered via DBS as well as cable systems.

The counterargument, of course, is that these newer distribution technologies are subscription-based and many people who might otherwise want these services cannot afford them, while more than 98 percent of the U.S. population has access to at least one over-the-air public television signal. Further, audience research shows that five times as many people watch PBS prime-time programs

EXHIBIT 7.f
•••••••••••

The Mythical World of Garrison Keillor

Every Saturday evening, increasing numbers of public radio listeners across the country tune in to the soft and zany homespun humor of Garrison Keillor, creator and host of *A Prairie Home Companion*. Produced by Minnesota Public Radio and distributed by Public Radio International, the program regularly airs on more than 485 stations to a dedicated (fanatical?) listenership of about 2.8 million.

The program's appeal centers on Keillor and a cast that includes "The Guy's All-Star Shoe Band" and a sound effects expert, Tom Keith. The live variety show runs for two hours and features comedy sketches, music heard nowhere else, guests, and Keillor's monologue on the news from all those people in mythical "Lake Wobegon."

To a considerable degree, the characters and style of Lake Wobegon grow out of Keillor's own background. A Minnesota native who grew up in a small town, Keillor had aspirations—and some success—as a writer early in life. He worked as both a writer and staff announcer for the campus radio station while earning a degree in English at the University of Minnesota. He wrote occasionally for *The New Yorker* after college, and in 1974, while he was doing a story on radio's

Grand Ole Opry, the notion of what would become *A Prairie Home Companion* came to him. The show first aired in Minnesota that year and went national in 1980.

Keillor explains why he likes radio:

> The beauty of radio is its comparative simplicity. We count on this. It means I can work on a show, be inventing things, right up until airtime and not be so beset with theater or transportation problems that I can't pay attention to the real job at hand.

Television, he feels, is just "the Wal-Mart of the mind. Radio is infinitely sexier."

More than 875 live broadcasts over a span of 25 years testify to the popularity of *A Prairie Home Companion*. Commenting on the show's longevity, Keillor acknowledges that "when the show started, it was something funny to do with my friends, and then it became an achievement that I hoped would be successful, and it's a good way of life." Just like life in Lake Wobegon, where Keillor emulates the creativity of radio's golden age every Saturday night.

Source: Cheryl Walsh Bellville, 1998. Courtesy Minnesota Public Radio.

as watch The Discovery Channel, and seven times more see PBS than A&E, two of the most-often-cited channels offering PBS-like programming.

Still, it seems clear that cable, DBS, and Internet competition—and the increasingly widespread use of VCRs, DVDs, and CD-ROMs for educational material—makes public broadcasters more aware of the need to specify clearly what role they will play amid a wider variety of viewer choices. But it remains just as clear that, for the foreseeable future, newer media cannot provide noncommercial-system-type services to all who want them.

Critical Views

In addition to citing an increasing variety of media options providing educational and cultural fare, critics in recent years have broadened their attacks on the current system of noncommercial broadcasting on several related but distinct fronts:

- The funding shortage always cited by public broadcasters is largely a matter of evasion: Lack of sufficient audience appeal (and thus support) lies at the root of their problem.
- Public participation in public television is in too many cases a sham because professional managers make most of the decisions, just as they do in commercial operations.
- Public television has little public appeal. Its vaunted fine arts and high culture merely serve privileged groups that are easily able to afford other sources of such material without resorting to publicly supported broadcast channels.
- Conservatives have long targeted public broadcasting as being too liberal in news and documentary programs, presenting a one-sided view of American life—this despite regular appearances by such alternative spokespersons as William F. Buckley and John McLaughlin.

Outlook

Some of these criticisms reflect the deregulatory philosophy that dominated discussions of U.S. media policy during the conservative political ascendancy of the 1980s. Similar attacks on public broadcasting surfaced in other countries, also stimulated by a market-oriented, *laissez-faire* approach to media regulation. Even so widely esteemed a service as the BBC came under attack as elitist and lacking in the fiscal responsibility that, according to deregulatory theory, only the "discipline of the market" can impose.

In Europe, as in the United States, these attacks raise a basic question: Should all broadcasting be regarded strictly as an *economic* undertaking, with programs treated as ordinary consumer goods? Or should at least some part of any national electronic media system be regarded as a cultural undertaking, with programs treated as a significant aspect of national life? Must everything depend on that slogan of free marketers, "consumer choice"? Not everyone chooses to attend great museums, galleries, and libraries that public funds support—often in locations not far from festering slums. Yet few would advocate dismantling all such cultural treasures and diverting their government grants to public housing.

Nevertheless, market-oriented thinking and the deregulatory policies that ride its crest cannot be ignored. Leaders of national public-service broadcasting organizations readily understood and agreed with the need for in-depth self-assessments in the early 1990s. CPB worked within a constantly updated five-year plan. NPR undertook a review of its role and options.

As discussed earlier in this chapter, the funding for public broadcasting increasingly is shifting to nonfederal sources, as federal tax dollars now represent less than 15 percent of total public broadcasting revenues. Many stations report that fundraising appeals based in part on reduced federal funding are effective. Further, additional noncommercial sources of programming are available and are slowly increasing. While the future of public broadcasting might not necessarily be considered bright, these developments suggest that noncommercial services are important to the public and that the public, when faced with a diminution of those services, will respond with increased financial support in an attempt to retain them.

CHAPTER **8**

Programs and Programming Basics

Ask anyone for a personal opinion about programs, and, quite properly, every listener/viewer proves to be an expert on the subject. This chapter, however, deals not with personal preferences but with anticipating group preferences. The collective term *programming* refers to practices used in selecting, scheduling, and evaluating programs. Whether viewed from the perspective of terrestrial broadcast, direct satellite broadcast, cable, or the Internet, the overriding purpose of commercial programming is either to attract advertisers by delivering audiences of sufficient size and appropriate composition or to attract audiences who pay subscription fees—or both.

8.1 PROGRAM COSTS

Programs must be both affordable and relevant to the programming goals of particular schedules. Programmers never end their search for what they refer to as usable *product*—a term, derived from the motion-picture industry, that gives a hint of the programmer's neutral point of view as to program quality.

Audience Targeting

As the numbers of stations, cable systems, satellite-delivered program services, and networks grew, many program services gave up aiming at the mass audience that was the original target of broadcasting. The terms *narrowcasting* and *niche services* came into vogue, suggesting a specific alternative goal. Cable networks such as CNN, MTV, Nickelodeon, and the Home Shopping Network deliberately limit their appeal to specific audience segments. Radio stations adopt rigid music formats (a particular radio format may repel most listeners but nevertheless prove irresistible to a narrow audience segment).

Advertisers influenced the trend toward narrowcasting. They are not inclined to pay to reach people who have no money to spend on their products or no interest in buying them. The need to target people willing and able to pay the bills, whether as advertisers or as subscribers, has led to *audience targeting* and segmentation throughout the electronic media.

Even though noncommercial broadcasting and subscription program services sell no advertising,

they also target specific audience groups. They try to attract audience segments most likely to support them by paying subscription and membership fees. Public television favors programs that appeal to middle-to-upper-income, well-educated families. Subscription program services tend to select PG-13 and R-rated movies that attract teens and adults. Pay-per-view (PPV) programs on cable and direct broadcast satellite (DBS) focus heavily on movies, often those with adult content, and events such as boxing matches and music concerts.

The major broadcast television networks target the largest groups of viewers. Some cable/DBS networks target the same broad audience; others program for more narrowly defined groups. Each service defines audiences in terms of *demographics* (age range and gender) or *psychographics* (lifestyle and interests). Targeting women 18 to 34 years is a demographic goal; targeting sports fans is a psychographic goal.

Radio has further refined the process of targeting by using *segmentation*, defining extremely narrow subsets of the potential radio audience. Radio usually segments audiences in both demographic and psychographic terms: teenagers who want to hear only hit songs or 25- to 44-year-old adults who prefer the music of the 1980s, for example. Audience segmentation makes attracting an economically desirable audience of adequate size more difficult. Many media owners feel that it is essential for them to control multiple program channels if they are to prosper or even survive.

Parsimony Principle

Product that can attract audiences of adequate size and desirable composition costs so much that filling all channels continuously with brand-new—not to mention "good"—programs would be economically impossible.

To cope with high program costs, writers, producers, and programmers resort to a variety of strategies based on what might be called the *parsimony principle*. This rule dictates that program material must be used as sparingly as possible, repeated as often as possible, and shared as widely as possible.

Sparing use of material means, for example, stretching a dramatic plot over many episodes instead of burning it up in a single performance—the technique of the soap opera. Often material collected for a news feature is released in a series of minifeatures. An expensive live sports event is not programmed as a single item: It is stretched to the utmost, with a pregame show, halftime interviews, and a postgame wrap-up. The most dramatic example, of course, is the Super Bowl. Pregame and postgame shows often fill more airtime than the contest itself.

Repeated use is illustrated by the standardization of openings, closings, and transitions in daily newscasts, quiz shows, weekly dramatic series, and the like. Cartoons, movies, and television series replay endlessly over time. The broadcast networks typically rerun programs in the summer months rather than scheduling expensive original programming when viewership is lower. This saves money and pleases viewers who have missed episodes of their favorite programs. Airing summer reruns at the same time new episodes of the show will later appear also can help develop viewing habits early and boost fall ratings. Some cable networks repeat the same movie at several different times on the same day. Successful syndicated shows may also double-run (air twice the same day) on broadcast stations. In 1999 producer David E. Kelley stretched his Emmy Award–winning *Ally McBeal* program over two nights by also offering a half-hour version of the one-hour show, using outtakes and material from earlier shows.

Shared use is best illustrated by the networks, which enable the same programs to appear on hundreds of different stations and thousands of different cable systems. Such sharing by stations and systems spreads high costs among many users, bringing prices down to a reasonable level for each. The sharing does not stop there, because network entertainment programs reappear in syndication (the mechanics of which are explained later in this chapter); this enables further sharing worldwide among stations, cable systems, and networks. Stations in many markets also share their newscasts with local cable systems or reuse source materials to produce another version of the newscast for cable distribution or broadcast on another (usually independent) station.

Product sharing across several media became a contentious issue in the late 1990s. Off-network syndication typically involves reuse of a program years after its initial network broadcast. But the networks now want to *repurpose* (find new ways to profit from) some programs much faster through distributors other than their own affiliates. The

broadcast networks want to distribute some programming on cable/DBS within days (or even hours) of its broadcast by affiliated stations. The stations protest this type of repurposing because it threatens their position as exclusive distributors of major network programming. Repurposing is sometimes the subject of heated debate when negotiating affiliation contracts.

From Prime Time to Any Time

The best-known and most expensive entertainment programs appear during the network portion of *prime time*, between 8:00 and 11:00 P.M. EST and PST, 7:00 and 10:00 P.M. CST and MST. The Big Four networks (ABC, NBC, CBS, and Fox) spend more than $7 billion annually on programs, well over half of it for sports and prime-time shows alone. Budgets on that scale put enormous pressure on those who conceive, develop, and schedule the programs. The networks spend hundreds of millions of dollars each year just on *pilots* for new entertainment series. Only a fraction of them survive long enough to go on the air as first episodes in series. Exhibit 8.a depicts the high attrition rate in the progress from program concept to on-the-air prime-time product. First-run syndicated programs (discussed in Section 8.2) run a similar gauntlet in their bid for distribution over the air or by cable.

Deficit Financing

Outside companies have traditionally produced many network programs; then licensed networks to show them a limited number of times (usually twice). Following the elimination of the financial interest and syndicatin rules (discussed in Section 8.2), the networks began taking an ownership interest in the programs they schedule. This helps them recover their initial costs for programming by participating in the syndication of a program after its network broadcast. Although costs can vary, in the late 1990s the broadcast networks reportedly paid about $600,000 per episode for the rights to broadcast a typical half-hour prime-time situation comedy (sitcom) during its first year on the network. An hourlong drama costs about $1 million per episode.

High as such payments may seem, they rarely cover the full cost of production. Again, costs can

EXHIBIT 8.a

Prime-Time Series: Concept to Air

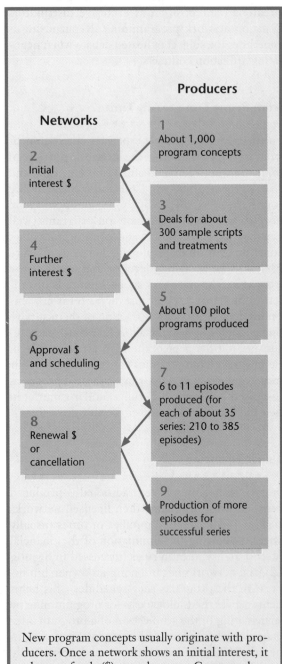

Producers

Networks

1
About 1,000 program concepts

2
Initial interest $

3
Deals for about 300 sample scripts and treatments

4
Further interest $

5
About 100 pilot programs produced

6
Approval $ and scheduling

7
6 to 11 episodes produced (for each of about 35 series: 210 to 385 episodes)

8
Renewal $ or cancellation

9
Production of more episodes for successful series

New program concepts usually originate with producers. Once a network shows an initial interest, it advances funds ($) stage by stage. Concepts drop out at each stage. In the hypothetical year depicted, nearly 29 program concepts were discarded for each one actually produced. The numbers listed here typify program development from many concepts to relatively few completed episodes.

vary, but in the late 1990s it cost approximately $1 million per episode to produce a typical sitcom and about $1.5 million to produce an hourlong drama. Producers of prime-time entertainment series count on subsequent relicensing and resale fees to bring in the profits. This maneuver, known as *deficit financing*, capitalizes on the peculiar dynamics of the program market. The producer of a highly successful network program can negotiate a much higher licensing fee once the original contract expires. The cost to NBC for airing *ER*, for example, jumped to $13 million per episode in 1998. Increased production costs, however, can offset higher licensing fees. A producer still may not realize a profit during the network run if star performers successfully negotiate for much higher salaries following a successful season. After seeing the fees paid for *ER*, Paul Reiser and Helen Hunt demanded and got $1 million per episode to continue performing in the popular sitcom *Mad About You*. Jerry Seinfeld reportedly was offered $5 million per episode for another season of *Seinfeld*, a figure that would have dramatically increased production cost had he accepted the offer.

A long, successful run on a broadcast network is very important to a producer, however, because it partially offsets production costs and, most important, enhances the resale value of a series. National exposure on a broadcast network confers recognition and a ratings track record.

After reverting to their producers, such shows become known as *off-network programs*. Many go on to earn handsome profits when licensed for exhibition by broadcast stations, DBS, cable systems, cable networks, and foreign broadcast systems. In fact, the staggering profits that can be made from the distribution of entertainment programs have caused some of the major movie studios in Hollywood to pay more attention to distributing television programs than to producing them.

8.2   SYNDICATION

The distribution method just described, known as *syndication*, offers both a supplement and an alternative to networks as a mechanism for financing centrally produced, high-cost programs. Syndication occurs worldwide wherever broadcasting or cablecasting exists.

The FCC defines a *syndicated* television program as

> any program sold, licensed, distributed, or offered to television stations in more than one market within the United States for noninterconnected [that is, nonnetwork] television broadcast exhibition, but not including live presentations. (47 CFR 76.5[p])

In practice, programmers classify as syndicated all *nonlocal* programs not currently licensed to a network, including movies, and even some live presentations.

There are two types of syndicated programming. *Off-network syndication* refers to programs that have appeared (or may still be running) on a broadcast network; *first-run syndication* refers to programs originally designed for syndication and never seen on a broadcast network. *Seinfeld, Mad About You,* and *Home Improvement* are examples of off-network programs in syndication. *Wheel of Fortune* and *Jeopardy!* are first-run syndication programs. While a successful network run is important in creating a strong market for the show in syndication, first-run programs such as *Oprah* and *Entertainment Tonight* appear in most of the nation's 200-plus markets. This reach rivals that provided by the national broadcast networks.

How Syndication Works
• • • • • • • • • • • • • • •

Rather than *selling* programs, the syndicator *licenses* the "buyer" (actually the lessee) to use a syndicated product (program or series). The license permits a limited number of *runs* or *plays* of a program, or each *episode* or *title* of a program series, within a limited time period. The buyer pays a fee to the syndicator—in cash (also over time), in free advertising time (*barter*, described later in this section), or in a combination of the two. After the license period expires, the programs return to the syndicator, who can recycle them over and over again.

Syndicated distribution resembles network distribution insofar as the same programs go to many outlets. But syndication differs as to both financial arrangements and timing. After a station leases a syndicated program, it usually may schedule the program at any time. Network affiliates, however, contract to broadcast programs, usually in exchange for compensation, and normally carry them at network-stipulated times.

Syndicators deliver their programs by various means. A station might receive one syndicated program through a satellite feed and either broadcast it at the feed time or record it for later playback. The same station might receive another program by *bicycle*—a system whereby the syndicator ships perhaps two weeks' worth of program tapes to station A, which airs the shows and then forwards them to station B, and so on. The bicycle system was predominant in the early years of television, but satellite distribution is far more common today.

Major television syndication firms often operate as units within vertically integrated companies—those that engage in several related activities. Disney, for example, has cable networks as well as studio and broadcasting interests. King World Productions ranks as the top syndicator, in terms of hit shows, with *Wheel of Fortune, Jeopardy!* and *Oprah*. All three consistently rank among the four or five most popular syndicated programs on TV. King World was an independent syndication company until it was acquired by CBS in 1999. This multimillion-dollar acquisition enhanced vertical integration within CBS, although CBS already controlled one syndication operation, Eyemark Entertainment. Later the same year CBS announced plans to merge with Viacom, continuing an industrywide trend toward increased consolidation and ever greater vertical integration.

Long-lived syndicated television series run season after season for decades, replayed scores of times. *I Love Lucy* (1951–1956), the quintessential off-network syndicated series, dates back to precolor days and has been syndicated in virtually every country in the world. At times as many as five *Lucy* episodes have been available on the same day in a single American city.

Syndication firms also provide *packages* of programs dealing with a particular theme or sharing a common characteristic. A syndicator may bundle a collection of westerns, crime dramas, or science fiction or action movies and offer them for one price. Often syndicators offer programs featuring a particular performer, such as Clint Eastwood.

Television syndicators showcase new productions at annual meetings of the National Association of Television Program Executives (NATPE) and at other national and international program trade fairs. The elaborate NATPE convention sometimes draws criticism for being both too expensive and unnecessary. Some question its value following the widespread merging of program buyers and sellers under consolidated ownership, but its survival does not appear immediately threatened.

The Nielsen rating service documents the track records of syndicated programs already on the market, issuing special reports on the size and composition of the audiences attracted by current syndicated series (see Section 10.2, page 296).

Syndication Exclusivity (Syndex)

Stations may obtain from distributors syndication exclusivity—the sole right to show a product within the buyer's own broadcast market for the term of the syndication deal. Cable program services, especially superstations, often distribute the same programs, bringing them into markets where broadcast stations have paid extra for local exclusive rights to the same programs. That duplication, of course, divides the audience and diminishes the programs' value to the broadcast station.

Responding to this program-duplication dilemma, the FCC in 1988 reintroduced the *syndication exclusivity (syndex) rules* it had rescinded eight years earlier in the first wave of deregulation. The syndex rules empower broadcast television stations to force cable systems in their coverage areas to delete duplicate syndicated programs to which the stations hold exclusive rights. Cable systems must block out an offending superstation or other program service, substituting another program or an on-screen message that explains the deletion.

In some cases a superstation's satellite carrier handles this task by inserting alternative programs before downlinking the station's signal to cable systems. When it operated as a superstation, Turner Broadcasting's WTBS resolved the issue by acquiring only those shows that had no syndex problem and thus ran a "blackout-proof" schedule. This is a less serious problem today than previously because there are fewer superstations and more original programming on the cable networks.

Syndex rules do not apply to two affiliates of the same network carried by a cable system or to very small cable systems, nor are all syndication contracts exclusive. In fact, obtaining exclusivity increases the price of syndicated programs; sometimes stations prefer forgoing exclusivity for the sake of a lower price.

Financial Interest and Syndication (Fin/Syn)

For years ABC, CBS, and NBC owned relatively few off-network syndicated programs—the result of an FCC decision and a U.S. district court antitrust consent decree. Designed to prevent network domination of program production and distribution, the *financial interest and syndication (fin/syn) rules* severely limited network freedom to participate in production and ownership of prime-time programs or in their domestic syndication.

The FCC modified the fin/syn rules slightly in 1991 and effectively repealed them in 1993. The financial interest portion of the rules ended immediately, permitting ABC, CBS, and NBC to increase their involvement in program production. The Commission delayed repeal of the syndication portion, however, until two years after dismissal of the court's consent decree. As part of its 1993 action, the FCC exempted the Fox network from the rules entirely.

Later in 1993 a federal judge in Los Angeles lifted the consent decree, removing the last barrier to full participation by ABC, CBS, and NBC in program ownership and syndication, although the Commission's two-year delay of its repeal of the syndication rule remained in effect. As expected, the networks moved rapidly to take advantage of their new opportunities. By the 1998–1999 season CBS had acquired a financial stake in about half the network's prime-time programs, with ownership ranging from 30 to 100 percent. CBS also had financial interests in programs airing on other networks, such as NBC's *Caroline in the City*. Similarly, NBC Studios, the in-house production unit of NBC, produced almost half of the original made-for-television movies aired on the network during the 1998–1999 season. Elimination of the fin/syn rules helped smooth the way for the melding of networks, production companies, and syndication firms. During the first six months of 1999 alone, large media conglomerates with interests in the networks and Hollywood studios acquired four previously independent syndication companies.

Prime-Time Access

Before 1971 the three major commercial television networks filled nearly all the best evening hours of their affiliates' schedules. This network monopoly left little opportunity for producers to sell programs aimed at the national market but not good enough (or lucky enough) to be selected for network exhibition. On affiliated stations only the fringe hours (late afternoon and immediately following prime time) remained open for syndicated material. Of course, independent stations have prime time available, but in the early 1970s few could afford to pay for recently produced, high-quality syndicated shows.

In part to encourage local programs and to enlarge the market for producers of new programs, and in part to diminish the network hold on the highest audience hours, the FCC adopted the *prime-time access rule* (PTAR), effective in 1971. It limited network entertainment programs to no more than three of the four prime-time hours—the evening hours when the television audience reaches maximum size and hence when stations can pay the most for nonnetwork programs. As defined by the FCC, *prime time* consists of the four evening hours between 7:00 and 11:00 P.M. Eastern and Pacific time (one hour earlier in the Central and Mountain time zones, with variations during daylight-saving time).

In practice, the networks had already abandoned the 7:00 to 7:30 P.M. slot to their affiliates. The PTAR therefore gave the affiliates only the additional prime-time half-hour between 7:30 and 8:00 P.M.

With the PTAR, the entire 7:00 to 8:00 P.M. hour became known as *access time*. Affiliates in the top 50 markets could fill access time with either locally produced programs or nationally syndicated *first-run* programs, but they could not schedule either regular network feeds or former network programs. This provided syndicators an enormous opportunity.

After a quarter-century, however, the PTAR had outlived its reason for being. Television, with its multiple outlets, was no longer a medium of scarcity, and the Big Three networks had lost much of their dominance. In July 1995 the FCC scrapped all aspects of the PTAR, effective in 1996. Although there was some dispute about this action, most observers agreed that affiliates would not allow their networks to retake any part of the highly valuable access hour.

The increasing control of syndication by a few media giants, however, raises old concerns. Some fear that the consolidation of distribution will make it impossible for independently owned stations to compete for product against those owned by the media conglomerates. The creative community also fears that there will be fewer opportunities to get new ideas accepted when a relatively few corporate executives exercise veto power over any product proposed for syndicated distribution.

Barter Syndication

• • • • • • • • • • • •

Because program costs are high, stations sometimes run short of ready cash with which to buy syndicated programs. Distributors and stations therefore work out deals for trading advertising time for programs. The practice, known as *barter syndication*, has become well established.

At first barterers mostly offered only once-a-week programs of little interest to most stations. Generally, these were *straight* (or *full*) *barter* deals, meaning the producer/syndicator sold all the commercial time nationally and stations received a free program but no commercial time to sell locally—and hence no opportunity for profit. Later nearly all first-run access programs, including such hit series as *Wild Kingdom*, *Hee Haw*, and *The Lawrence Welk Show*, were sold as *partial barter*, meaning the station retained some minutes for local advertising sales.

Then syndicators discovered that former network hits could command both cash payments and advertising time. In the late 1980s most barter deals for off-net sitcoms were of this *cash/barter* or *barter-plus-cash* type. By the 1990s, however, as more and more stations became cash poor, syndicators had returned to partial barter arrangements. *Designing Women* was one of the first programs offered on the newly revived *barter–no-cash* basis.

Barter continued with newer first-run access shows, notably hit game shows such as *Wheel of Fortune* and *Jeopardy!* Soon barter included most daytime talk shows, as well as access magazines such as *Entertainment Tonight*. The original *Cosby Show*, for years NBC's top-rated show, sold on a cash/barter basis and shattered all revenue records for a syndicated program. Licensed mostly to network affiliates as a lead-in to the early-evening news, it earned more money than any previous syndicated series.

Barter occasionally presents stations with a dilemma. A station that carries a barter program typically agrees that, if the program is moved to another, less desirable time period, the station will continue to air the syndicator's commercials in the original time period. This could mean that the replacement show, if it too were on a barter deal, would be filled with both syndicator's commercials, leaving little if any time for station sales. Stations may also find that some of the most popular barter shows are too expensive. That is, the value of the commercial time given to a syndicator may be far more than the station would have been willing to pay in cash for the right to carry the program and sell all the commercial time itself.

Sometimes buyers simply pay too much for a program. A show's prime-time network performance does not guarantee similar success in syndication. Highly rated during its 1980s network run and very expensive in syndication, *The Cosby Show* failed to produce the kind of ratings in syndication suggested by its network success. The original *Star Trek*, in contrast, enjoyed only a modest three-season network run but went on to become a successful off-network syndicated program, in addition to launching a series of theatrical features and TV programs using the same theme. Buyers and sellers of programming continue to price product based on its network performance, however, because this remains the best predictor of success in syndication.

Seinfeld dominated the ratings during its mid-1990s network run on NBC, so it was expected to fetch a good price in syndication. But some observers were astonished at the prices paid for *Seinfeld* reruns. New York's WNYW is reported to have paid about $100 million for a five-year run of *Seinfeld* starting in 2000. KCOP in Los Angeles paid even more, in excess of $300,000 a week. Ted Turner's TBS Superstation is said to have paid about $1 million per episode (plus relinquishing a minute of ad time per episode for the syndicator) for a four-year run of *Seinfeld* that will not begin until September 2002. These are remarkable prices for a program that ended its network run in May 1998. They tend to support analysts' estimates that *Seinfeld* generated revenues in the $6 million to $8 million per episode range in its second syndication cycle. Some observers question whether stations and cable networks can recover their expenses considering the prices paid for *Seinfeld*. Others say that the promotional and "branding" value of *Seinfeld* will justify the purchase even if revenues do not completely cover the costs.

Radio Syndication

Syndication and barter syndication also operate in radio. Syndicators use satellites to relay news, sports, and entertainment material to stations. Programmers distinguish between *syndicated formats* and *syndicated features*. A station might buy the use of a ready-made syndicated country music format, for example, supplementing it with syndicated news and entertainment features from other sources. Thus stations can create unique program-

ming mixes from commonly available syndicated elements. Radio feature material is often bartered, whereas syndicated formats are usually cash deals.

8.3   PROGRAM TYPES

As the previous sections indicate, programs can be classified according to their method of distribution as *local*, *network*, or *syndicated*. This section deals with other classification methods based on

- *content*, in terms of a broad two-way division into entertainment and information
- *scheduling*, in terms of frequency ("strip scheduling") and time of day ("prime-time soap opera")
- *format*, in terms either of individual programs ("the talk format") or of entire services ("the all-talk radio format")
- *genre*, usually in terms of content ("the sitcom genre") but sometimes in terms of target audience ("the children's genre")

Entertainment vs. Information

Most programs qualify primarily as entertainment. However, information programs receive special attention and deference because many of them enhance the electronic media's social importance. A hybrid class of programs sometimes called *reality shows* became trendy in the late 1980s and expanded in the 1990s. They earned the name *infotainment* because they take on the aura of information programs but aim primarily at entertaining.

Shows such as *Extra* and *A Current Affair* mimic traditional news programs in that they deal with real people and events, but the treatment is more entertaining than informative. Other reality-based shows such as *Cops*, *Real Stories of the Highway Patrol*, *World's Wildest Police Videos*, and *America's Most Wanted* also deal with real events and people and present the stories in a documentary style. The producers' objective, however, seems more focused on entertaining than informing. The same can be said of shows such as *World's Most Dangerous Animals*, *When Animals Attack*, and *When Good Pets Go Bad*. In the wake of the O. J. Simpson trial and coverage of the Clinton-Lewinsky affair, shows such as *Judge Judy*, *Judge Mills Lane*, and *Divorce Court* appeared in increasing numbers as the 1990s ended. Although it is possible to learn some things about

EXHIBIT 8.b
. . . .

TV Dayparts

The length of the TV news block varies from market to market and station to station, running as long as two or three hours on some major-market TV stations and as short as an hour elsewhere.

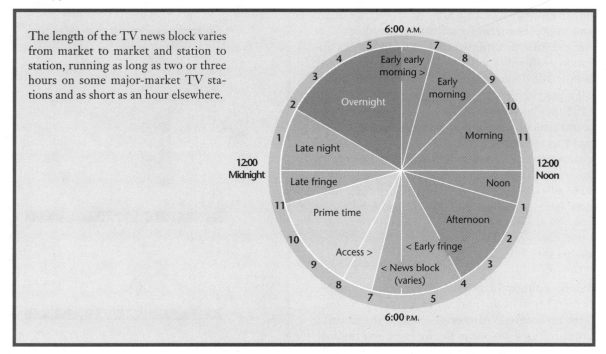

the justice system from these programs, the cases and the personalities involved seem to be selected for their ability to engage viewers emotionally rather than intellectually.

Infomercials, long-format commercials that often resemble talk shows, also mix information (or sales pitches masquerading as information) and entertainment.

Legitimate information programs have two main subtypes, news and public affairs. Some people argue that sports programs should be categorized as information rather than entertainment. However, programmers regard sports as primarily entertainment. Network sports departments differ in purpose, style, and types of personnel from news departments.

During the 1990s the traditional network news programs generally, but not always, retained an approach to content and production techniques that enabled viewers to recognize them as being informative in nature. But in some programs the line between information and entertainment became practically invisible. The resulting confusion may be causing many people to view broadcast journalism less favorably now than a decade ago, as discussed in Section 9.11 (page 285).

Dayparts
. . . .

Two programs of the same type that are scheduled at two different times of day take on different colorations. Prime-time programs have distinctive qualities, as do other programs associated with specific schedule positions. Thus a daytime soap opera differs from a prime-time soap opera, and prime-time sports shows usually differ from weekend sports shows.

For scheduling purposes, programmers break the 24-hour day into blocks they call *dayparts*. Radio programmers generally divide the day into *morning drive, midday, afternoon drive, night,* and *overnight* segments. The morning and afternoon drive periods, with commuters traveling to and from work, provide the largest audiences for most radio stations. The radio audience on weekends peaks near midday in most markets but is only about 65 percent as large as during drive periods on weekdays.

In television, broadcast and cable network programmers divide the day as shown in Exhibit 8.b. *Network prime time*, the most important segment, commands the largest audiences. *Access time*, the prime-time hour preceding network prime time, often delivers the most revenue for network affiliates.

Formats

• • • • •

A program may have a quiz show format. A cable channel may have a home shopping format. CNN has an all-news format, ESPN an all-sports format, and so on. The term *format* can refer either to the organization of a single program or to the organization of an entire service. Most radio stations have adopted distinctive formats, such as country, classical music, or all news.

Television stations and broadcast television networks tend not to adopt single formats because they need to appeal to a broad audience. As a single-channel service, a station cannot afford to narrow its audience to followers of one particular format. Cable and DBS can afford such specialization, however, because they are multiple-channel services.

Genres

• • • • •

In the context of programs, *genre* (pronounced "zhahn'-ruh") usually denotes a type of content. Familiar entertainment genres include the situation comedy, the game show, the western, and the soap opera. Sometimes target audience rather than content defines genre. Children's shows, for example, can take many different forms—not only cartoons and action/adventure drama but also news, discussion, quiz, and comedy.

Programmers identify a program by genre as a shorthand way of conveying a great deal of information about its probable length, seriousness, subject matter, visual approach, production method, and audience appeal. Hybrid programs sometimes use combined names such as *dramedy* (a drama/comedy hybrid) and *docudrama* (a documentary/drama hybrid). Exhibit 8.c illustrates how programmers can classify a program by genre, by how it is scheduled, and by the daypart in which it airs.

EXHIBIT 8.c

• • • • • • • • • • •

Genres, Schedules, and Dayparts

	The Drew Carey Show	The Tonight Show	Monday Night Football	General Hospital	Diagnosis Murder	Nightline	Wheel of Fortune
Genres							
Soap				●			
Game							●
Sitcom	●						
Drama					●		
Crime					●		
Sports			●				
News						●	
Talk		●					
Schedules							
Strip		●		●		●	●
Series	●		●		●		
Miniseries							
Special							
Dayparts							
Access							●
Afternoon				●			
Prime time	●		●		●		
Late fringe						●	
Late night		●					

Particular television shows can be classified by their genre, the way they are usually scheduled on the networks, and the time of day they are typically scheduled on the networks. In syndicated reruns, the methods of scheduling and daypart often differ from the original network run.

8.4 ENTERTAINMENT PROGRAM SOURCES

Individual broadcast stations and cable systems obtain nearly all their entertainment programs ready-made from networks (if they are affiliated) and syndicators. In-house production by stations, cable systems, and networks once consisted mostly of news and sports programs, but now networks increasingly produce entertainment shows as well. In recent years original program production by the cable networks has increased, although these networks continue to depend heavily on entertainment programming that was originally seen in theaters, or on the broadcast networks or that is widely available in syndication.

Before the mid-1990s nearly all original entertainment programming on the networks was provided by *outside* producers falling into five creative groups:

- about a half-dozen major Hollywood film studios
- a few additional studios below the "major" rank
- about a dozen major independent producers
- a host of smaller and more specialized independent producers
- some foreign syndicators distributing in the United States

Elimination of the fin/syn rules (see Section 8.2) and increased co-ownership of production studios, cable networks, and broadcast networks means there will be more in-house—or at least within-conglomerate—production and distribution of entertainment programming in the future. It would be a mistake, however, to assume that a network carries only those programs produced by a co-owned company. It is not unusual for an owner's product to air on a network that competes with the owner's co-owned network. For example, *Caroline in the City* aired on NBC, but CBS held an ownership interest in the program. This sometimes occurs in part because of contractual obligations made before a merger or acquisition, but also because of the constant demand for good programming at the lowest cost, regardless of the source. CBS Television president Leslie Moonves probably was reflecting a widely held attitude among network executives when he said, "We want to put the best product on the air regardless of who owns what" (*Broadcasting & Cable*, 7 December 1998: 47).

Major Studios

The major Hollywood studios have familiar movie-studio names, such as Paramount and Twentieth Century Fox. Best known to the public for feature films made originally for theatrical release, they also produce a less expensive type of film, made-for-TV movies, as well as program series made especially for broadcast and cable television networks.

By definition, theatrical feature films normally reach audiences first through motion-picture theaters. Thereafter distributors release them successively to video rental stores, cable and satellite pay-per-view services, and other electronic media at intervals governed by windows of availability, as shown in Exhibit 8.d. Increasing co-ownership of studios and networks is making it easier to alter windows of availability in response to market demands.

Despite the exhibition priority enjoyed by movie theaters, movie studios make more money from television, cable, videocassette, and foreign rights than from American theatrical exhibition. Feature films are a staple item in broadcast and cable network schedules and even in some individual broadcast station schedules. They have the advantage of filling large blocks of time with material that has strong audience appeal, especially for the young female subaudience that advertisers often target.

Independent Producers

Independent producers range in size from large firms that usually have several series in production simultaneously, to smaller producers with only a single series under contract at any one time. In the 1970s independent producers created some of the most innovative television series, such as *All in the Family* and *The Mary Tyler Moore Show*. By the 1980s and on into the 1990s, independent producers such as Spelling, Goldberg, Lear, Tandem, MTM, and Cannell produced an enormous number of prime-time series. As consolidation continues, independent producers are finding it increasingly advantageous to affiliate with one or more major media conglomerates.

Made-for-Cable Programs

The growth of national cable program services has created an insatiable demand for more product. The leading pay-cable network, HBO, consumes hundreds of movies a year.

To increase their program supply, to gain control over content and costs, and to create a distinctive

EXHIBIT 8.d
• • • • • • • • • •

Movie Windows of Availability

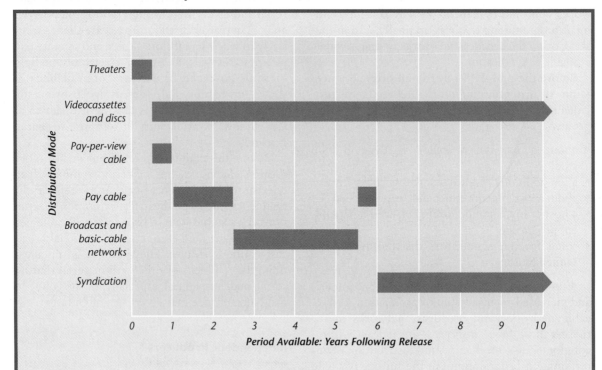

Occasionally, exceptions to this sequence occur. Sometimes distributors withhold highly popular movies from broadcast and cable for many years; sometimes a broadcast network pays extra to step in ahead of the pay-cable window. Some unsuccessful films move directly to syndication, and some movies never appear in theaters or television at all but go directly to video stores. With increasing vertical integration among producers and distributors, the ability to move a product into the most economically advantageous window has improved.

image, cable services co-finance movies with producers. This investment gives them *exclusivity*, the right to show the product first on cable. A few movies co-financed by cable, however, such as *On Golden Pond* and *Sophie's Choice*, had box-office success in theaters. HBO's parent corporation is Time Warner, a magazine publisher and cable multiple system operator (MSO). By the 1980s it had become Hollywood's largest financial backer of movies. HBO planned to produce about a dozen original movies in the year 2000.

Each year ad-supported as well as pay-cable services increase their investments in production activities. By the mid-1990s basic-cable and pay-cable networks spent about $4 billion annually for original programming. The Cable Advertising Bureau (CAB) reported that basic-cable networks alone in-

vested an estimated $5.5 billion in programming in 1999. That was about 13 percent more than in 1998, twice as much as in 1995, and almost 240 percent more than at the start of the decade. About $3 billion of that $5.5 billion was spent on original programming. Perhaps even more startling, a CAB survey of its member cable networks indicated that about 75 percent of basic cable's 1999 prime-time schedule was original programming. (CAB, 1999)

In addition to news and sports, original cable programming includes made-for-cable dramatic and comedy series and variety programs. Some products qualify as true cable-only programs; others merely add episodes to existing television series. Cable's original programs, especially on the pay channels, are often among the best available on television. HBO collected 23 prime-time Emmy

Awards in 1999, more than any other network. Cable is also sometimes the birthplace of shows that later migrate to a broadcast network. ABC's *Politically Incorrect with Bill Maher*, for example, began on cable's Comedy Central, and *The Larry Sanders Show* was first seen on HBO. Made-for-cable movies are somewhat less expensive to make than major theatrical releases, but as the decade ended, cable paid increasingly more for these movies to enhance its original programming. TNT's 1999 *Animal Farm*, for example, cost a reported $24 million.

Although some made-for-cable programs may not recover their approximately $600,000 (sitcom) to $1.1 million (hourlong drama) cost per episode, they help to counteract the image of cable as a mere parasite on broadcast television. They also make good cable's promise to enhance program diversity and help one cable service distinguish itself from another and gain subscribers seeking new program options.

8.5 NEWS SOURCES

A markedly different atmosphere usually surrounds the creation of news programs. By assuming the serious function of conveying news and information, broadcasters, at least initially, took on a more important role than that of merely entertaining. Admittedly, some broadcasters cared little about being part of a great free-speech tradition if doing so interfered with another great American tradition—making money through free enterprise. But in time news programs changed from money losers into money-makers, ensuring the survival of broadcasting's special role as bearer of information-as-entertainment.

News Agencies

Long before broadcasting and cable, newspapers established news-collecting agencies, the news field's equivalent of program syndicators. Today more than a hundred news services are available to U.S. electronic mass media. Many specialize. For example, Bloomberg Business News provides stations with reports on financial matters; Metro Networks, Inc., is known for traffic reporting in large cities.

Major news agencies such as the Associated Press (AP), Worldwide Television News (WTN), and Reuters are international in scope. They supply text, audio, video, and still pictures to subscribing networks and stations for incorporation into their newscasts and other programs.

Network News

The broadcast networks maintain tight control over most news and public-affairs production. ABC, CBS, and NBC operate news divisions separate from their entertainment divisions, each employing hundreds of people. Each also supports its own foreign news bureaus in the major world capitals, staffed by correspondents with in-depth knowledge of the regions they cover. Fox does not provide its broadcast affiliates the array of news programs provided by the older networks. Fox affiliates can, however, use material provided by Fox NewsEdge, a Fox news service, in their local newscasts.

The broadcast networks traditionally avoided news footage or documentaries from outside sources. This policy began to erode in the 1980s when, to cut costs, networks began accepting news footage from affiliates and from *pool feeds*—cooperative arrangements whereby media rotate responsibility for coverage of news events, or at least share costs.

The networks also cut back their news department operating budgets, reduced staffs, and consolidated or closed some of their news bureaus, both overseas and in the United States, relying more than ever on affiliates and on outside organizations such as Reuters and Worldwide Television News (WTN). In contrast, Cable News Network (CNN) expanded its presence abroad in the 1990s by opening bureaus in Asia, Africa, and Latin America. In 1997 CNN opened a bureau in Havana, the first U.S. network to establish a news operation in Castro's Cuba since 1969. CNN's success in challenging the major networks resulted in further cable specialization, with entire channels devoted to news genres such as finance, sports, weather, and court trials.

Local News

Local TV stations increased the time devoted to news in the 1990s, sometimes dramatically. By 1999 many stations aired three to four hours of local news daily.

National and international news is the natural province of the networks, local news the province of stations. But stations have increasingly expanded their coverage beyond local and regional news. They have many sources to call on beyond those of their local staffs and regional news services.

Both television and radio affiliates can obtain what amounts to syndicated news services from their own networks. Provided by the networks' news divisions, these services feed hard news and features over regular network relay facilities during hours when these facilities carry no scheduled network programs. Affiliates can record these feeds, selecting items for later insertion in local newscasts. They can obtain the right to record regular network news programs as sources of stories for insertion in local programs.

Affiliates and independents alike can draw on established news agencies—AP, Reuters, and others. With the advent of electronic news gathering (ENG) and satellite news gathering (SNG), local television news teams can fill longer newscasts and provide more on-the-scene coverage of both local and distant events. The availability of modern news gathering technology and satellite uplinking by local stations also allows the networks to call on their affiliates to provide network coverage of breaking news. For example, Denver TV stations provided extensive live coverage of the 1999 school shootings in Littleton, Colorado, for their respective networks long before network news teams could reach the scene.

In addition, the higher-quality images produced by modern consumer camcorders provides another news source for local stations and network news organizations—the amateur.

CNN, though originally a news network only for cable, now acts also as a syndicated news source for television stations, affiliates as well as independents. CNN exchanges news stories with stations and supplies them with *Headline News* in 5- or 30-minute blocks, or continuously overnight. Another station source, Conus Communications, coordinates SNG by more than 100 stations, facilitates exchange of news footage and stories, and, with Viacom, produces its own 24-hour news service, All News Channel.

Cable systems select from a bewildering array of sources to meet their news and sports needs: CNN, CNBC, ESPN, ESPN2, ESPNews, MSNBC, Fox News Channel, CNN/SI, CNNfn, plus regional networks supplying news and sports of mostly local interest. All News Channel supplies material to some regional cable news networks (New England Cable News, for example), which in turn supply newscasts for cable systems.

8.6 SPORTS PROGRAM SOURCES

Networks

Except for professional football games, the broadcast networks rely increasingly on production companies that specialize in televising sports. The networks simply pay the cost of national rights to the events and of producing live games, as well as their own announcers' salaries and travel costs. Mega-events such as the World Series, the Olympic Games, the Super Bowl, and regular-season NFL football remain exceptions to this trend. Cable networks and individual broadcast stations also employ sports production companies. ESPN produces most of the games it carries.

Sports Sponsorship

Advertisers gave up most program sponsorship long ago, but the practice has been revived for some lesser sports events. Sponsorship means that the advertiser obtains the rights to broadcast or cablecast the event and controls the program.

A network carrying a sponsored sports event supplies only play-by-play and color announcers. The sponsors hire nearby television stations or sports production houses to do the rest of the production work. They also often participate in lining up celebrity guests, promoting the program, and even selling tickets. They sometimes also recover some of their expenses by selling spots to other advertisers and peddling subsidiary coverage rights to the event via radio stations, cable/DBS systems, and the Internet.

Both the networks and advertisers benefit from this division of labor. The sponsors retain control of the costs and ensure maximum promotional value; the networks gain hundreds of program hours of minor sports events too risky financially for them to cover by themselves.

8.7 NETWORK SCHEDULE STRATEGIES

Whatever their program sources, stations and networks strive to structure their selections into coherent schedules. Effective scheduling requires, among other things, coordinating program types and production tempos to complement typical audience activities.

For example, during the busy early-morning period, listeners/viewers get ready for their day's activities at work, at school, in the home, or at play. This period calls for light, up-tempo treatment of news, weather, traffic reports, and short entertainment features. A more relaxing tempo and longer program units suit the less-structured evening period.

Audience Flow

Schedulers try to draw audience members away from rival channels and to prevent rival channels from enticing away members of their own audience. These efforts focus on controlling *audience flow*—the movement of viewers or listeners from one program to another. Flow occurs mostly at the junctions between programs or, on radio, after one block of songs ends and before the next begins. Audience flow includes both *flowthrough* (on the same stations or channel) and *outflow* or *inflow* (to or from competing programs).

Audience research can measure the extent and direction of audience flow. Such data give programmers guidance on how to adjust schedules and program types to retain audience members and gain new ones at the expense of competing channels.

However, remote-control channel selectors and VCRs, now in more than 90 percent of all households, have made tracking television-audience flow more difficult. A special vocabulary has evolved to describe how some audience members use these devices: they *surf* or *graze* through the available channels, *zap* the sound of unwanted commercials, *jump* between pairs of channels, *flip* around to see what's happening on nearby channels, and *zip* through the boring parts of prerecorded cassettes. The increasing popularity of more sophisticated personal video recorder products (discussed in Chapter 5) also contributes to the difficulty of tracking audience flow and controlling it through programming strategies. The growing availability of video-on-demand as cable systems convert to digital technology will further erode the ability to control audience flow with programming strategies.

Other new technologies, such as *picture-in-picture*, allow viewers to divide their attention among several competing on-screen channels simultaneously. Stations distributing data, such as supple-

mentary web pages embedded in their broadcast signals, also provide viewers with more, potentially distracting information on-screen, along with the primary program content. Not all viewers take advantage of these technologies, however, and programmers continue to employ scheduling strategies designed to influence audience flow.

The continuousness of electronic mass media differentiates them from other media, whose products reach consumers only at intervals, coming in individually packaged physical units—a videocassette, a compact disc, an edition of a newspaper or a book, or an issue of a magazine. Only broadcasting, DBS, and cable television afford the consumer an instantaneous choice among many continuously flowing experiences.

Specific Strategies

Taking into account the line-ups on competing channels, programmers adjust their own program schedules, fine-tuning them to take advantage of opponents' weak points. Following are some typical scheduling strategies used by the networks (broadcast and cable) and by individual systems and stations to exploit audience flow.

- *Counterprogramming* seeks to attract the audience to one's own station or network by offering programs different from those of the competition. For example, an independent station might schedule situation comedies against evening news programs on the network affiliates in its market.
- *Block programming* tries to maintain audience flowthrough by scheduling programs with similar appeal next to each other—for example, by filling an entire evening with family-comedy programs.
- *Stripping* tries to create viewing habits by scheduling episodes of a series at the same time every weekday.
- A strong *lead-in* attempts to attract the maximum initial audience by starting a daypart with a particularly strong program in the hope of retaining the audience for subsequent programs.
- *Leading out* with a strong program may similarly attract viewers to the program that precedes it.
- *Hot switching* fine-tunes the above lead-in strategy through the use of "seamless," commercial-free transitions from one show to another. Running crunched-together credits at warp speed while viewers are enticed to stay tuned by promotions occupying two-thirds of the screen

further enhances a seamless transition. An NBC study that found almost a third of the audience stopped watching when the closing credits appeared prompted this practice.

- A *hammock* tries to establish an audience for a new program, or to recover the audience for a show slipping in popularity, by scheduling the program in question *between* two strong programs. Flowthrough from the previous (lead-in) program should enhance the initial audience for the hammocked program, and viewers may stick with the weak (or unfamiliar) show in anticipation of the strong (lead-out) program that follows it.

- *Bridging* attempts to weaken the drawing power of a competing show by scheduling a one-hour (or longer) program that overlaps the start time of the competing show. Turner's TBS Superstation uses a version of this strategy routinely, typically ending and starting programs at five minutes past the hour and half-hour.

- *Repetition*, a pay-cable strategy, makes it convenient for viewers to catch a program, such as a movie, because repeat showings are scattered throughout the schedule. Local stations may also repeat syndicated talk or reality-based programs during the same day.

- *Stunting* seeks to keep the opposing networks off balance in the short term by tactics such as making abrupt schedule changes, opening a new series with an extra-long episode, and interrupting regular programming frequently with heavily promoted specials. Networks resort to stunting especially during the February, May, and November ratings periods. Unfortunately, stunting may confuse and frustrate not only competing networks but also viewers when a favorite program is suddenly replaced or moved in the schedule.

Two general theories sometimes get in the way of these scheduling strategies. One suggests that no matter what is offered, viewers will watch the *least objectionable program* (LOP) in the time period. The other, *appointment television*, holds that successful programs may be scheduled anywhere because serious-minded fans will follow wherever they go.

Another increasingly popular practice also conflicts with these scheduling strategies. Some programmers try to extend their control over the audience by diverting them to a related information source at the program's conclusion. Many programs now invite viewers to visit their associated web site to learn more about the topic, interact

with the program's performers or production staff, comment on the show, enter a contest, or explore the topic in greater detail through links to other web pages. This *extended viewing period* allows the program sponsors additional opportunities to present their messages. This strategy has obvious advantages for sponsors, but it is antithetical to the scheduling stategies just discussed, which attempt to hold viewers through a series of shows. While diversion helps prevent audience flow to a competing channel, it also diverts the audience from the next program on the same channel. At least this is true for those viewers who do not, or cannot, use the Internet while watching TV.

Network Prime-Time Arena
● ● ● ● ● ● ● ● ● ● ● ● ● ● ● ● ● ● ●

The main arena of broadcast network rivalry is the weekly 22 hours of network prime time—the three hours from 8:00 to 11:00 P.M. each night of the week in the Eastern and Pacific time zones (an hour earlier in the Central and Mountain time zones), plus an extra hour from 7:00 to 8:00 P.M. on Sundays (EST/PST).

Prime time gives broadcast networks access to the largest and most varied audiences of any medium. It therefore demands programs with the broadest appeal. Because networks pay so much for prime-time programs and because they need scheduling maneuverability, they normally run prime-time series once a week. NBC's *Dateline*, a prime-time news magazine running four or five nights a week, is a remarkable exception to this general rule. But even *Dateline* does not run at the same time each day it airs. If the networks scheduled all of prime time with daily half-hour series, each network would have only six weeknight programs to schedule, providing too few chances for exploiting scheduling strategies.

In other dayparts, however, broadcast networks "strip" most of their shows—scheduling the same series at the same times, Monday through Friday. Network morning talk programs, afternoon soap operas, and evening newscasts, for example, occur at the same times each weekday.

Cable Network Strategies
● ● ● ● ● ● ● ● ● ● ● ● ● ● ● ● ● ● ●

In the early 1980s most cable networks showcased their best product in dayparts other than prime time. They were satisfied with reaching demographic subgroups in prime time. By the mid-1980s, however, the largest cable program services, such as ESPN, TBS Superstation, and USA Network, had begun to compete head-on with broadcast networks for the prime-time mass audience.

Advertiser-supported cable networks commonly adopt habit-forming strategies to build loyal audiences. The broad-appeal cable services strip their programs across the board both in daytime and in prime time. USA Network and TNT, for example, strip costly off-network series such as *JAG* and *ER* in prime time, hoping to draw viewers away from broadcast stations and to build a cable-watching habit among first-time viewers.

Pay-cable networks rely heavily on *repetition*, scheduling repeat showings of their movies and variety shows in various time periods, cumulatively building audiences for each program. Each movie plays at varying start times on different days. A movie may be recycled as many as a dozen times a month. For this reason, and also because cable program guides come out once a month and cable companies need to encourage monthly subscription renewals, pay-cable networks plan their schedules in monthly cycles.

Pay cable also uses the bridging strategy, scheduling across the start times of other programs. HBO and Showtime movies, for example, usually start at 8:00 P.M. bridging the 9:00 P.M. station break. This is the period with the largest number of people watching television. Sometimes pay-cable networks try to get the jump on the broadcast networks by starting their movies earlier in the evening (at 7:00 or 7:30 P.M.). This strategy works best when broadcast schedules have been disrupted by late-running sports or political programs.

8.8 LOCAL SCHEDULE STRATEGIES

TV Stations

ABC, CBS, and NBC fill about 70 percent of their affiliates' schedules. The newer networks, especially The WB and UPN, provide fewer program. During time not filled by network programming, affiliates apply the strategies previously described to maximize their audience. For affiliates the programmer's most important decisions concern the choice of programs for the early fringe and access dayparts.

Programmers at independent stations have charge of their total schedules. Their chief stratagem, counterprogramming, capitalizes on the inflexibility of the affiliate's schedule because of its prior commitment to network programs. For example, an independent station can schedule sports events at times when affiliates carry major network shows. Networks can afford to devote prime time to only a few top-rated sports events of national interest. Independents, however, can schedule sports events of local interest, even during prime time.

Both network affiliates and independent stations utilize stripping on weekdays. Saturday and Sunday programs are scheduled only weekly. Monday-through-Friday stripping of syndicated and local programs has three advantages for stations. First, daily same-time scheduling encourages the audience to form regular viewing habits, such as the 6:00 P.M. news habit. Second, a single promotional spot can publicize an entire week's schedule for a given time slot. Finally, purchasing many episodes of a syndicated series in a single transaction earns quantity discounts from syndicators.

The practice of stripping off-network programs led to an enormous demand for television series with many episodes already "in the can." Stripping ideally requires 130 episodes for a half-year run. Off-network series that have generated such large numbers of episodes have, by definition, earned good ratings on a network over more than one season. They therefore command the highest prices among syndicated series. Popular long-running shows such as *Murphy Brown*, with more than 200 episodes, are ideal off-network products for stripping in syndication.

In recent years competition made network schedules more volatile. The networks nervously canceled shows at the first sign of weakening ratings. Runs became shorter, building up too few episodes for strip scheduling over the long period needed for best results. This in turn meant increasing scarcity of, and thus intensified competition and higher prices for, successful off-network syndicated series.

Besides syndicated entertainment programming, local stations increasingly turn to paid programming such as infomercials and home shopping to fill time when viewing levels are typically low (often late night and Sunday mornings). These programs provide a reasonable profit for the stations, even if they attract a small audience or none at all.

Cable Systems

• • • • • • • • •

Technical considerations play a major role in determining cable program line-ups. Despite having multiple channels, most systems cannot carry all of the program services available to them.

Cable networks compete among themselves for carriage on cable channels because there are more networks wanting carriage than there are channels available. This has caused some new cable networks to offer cable systems *launch fees* to gain access. These fees are based on the number of households subscribing to the cable service. It was widely reported that News Corporation, for example, paid more than $10 per subscriber to ensure that the Fox News Channel would reach a subscriber audience numbering in the tens of millions soon after its launch in 1996. Cable networks also may offer programming to cable systems at low initial license fees to gain access.

Carriage on a digital rather than analog channel is another approach employed by new cable networks seeking access. By 1999 more than 50 cable networks were distributed on cable systems offering digital channels. AT&T BIS (formerly Tele-Communications Inc.) offers a mix of advertising and subscription networks for digital cable distribution through its satellite-delivered *Headend in the Sky* (HITS) service. Time Warner provides a similar service for use by both its own cable systems and other cable operators. Some cable networks offer programming on both analog and digital channels. Discovery Networks, Inc., and HBO are two well-established analog program providers that have moved aggressively into the *digital tier* to provide new services. Program providers see early digital service as a way to ensure themselves "shelf space" as competition for digital carriage increases. They also believe that a strong subscriber response on the digital tier may convince cable system operators to migrate a network to the analog tier. Many more digital channels and digital networks are expected as cable systems roll out services. Eventually, all cable channels may be digital, dramatically increasing cable system channel capacity through digital compression technology.

Cable systems usually include signals of most television stations operating in their service areas. Subscribers find convenient, and stations strongly advocate, *on-channel* carriage—that is, for example, a station broadcasting on channel 2 appearing as well on the cable system's channel 2. Occasionally, this proves impossible, especially in cases where a strong station signal would cause *ghosting* (double images) if carried by the system on its own channel.

Most established cable networks charge systems a fee to carry their programs. Were systems to carry them all, they would have to charge subscribers an unacceptably high monthly rate. Understandably, a system's MSO often dictates that its own systems carry those networks in which the MSO has an ownership interest. Similarly, a cable network may offer its programming at reduced prices to those MSOs that agree to carry it on all of their systems. Systems also favor ad-supported networks that offer the most time for local advertising. As cable systems seek new revenue sources and technology increases channel capacity, cable operators are dedicating more channels to pay-per-view services.

Some cable systems group channels together by program genre, a practice referred to as *clustering*. They place pay-cable networks and other services that target similar audiences on adjacent channels. Cable operators also consider program content and the tastes of their subscribers.

Radio Stations

• • • • • • • • •

Radio stations use counterprogramming, stripping, and blocking strategies even more than television stations do. Most radio stations schedule their program elements, whether songs or news items, in hourly rotations, creating 60-minute cycles. As the day progresses, the hourly pattern is altered by daypart to match changing audience activities. Exhibit 8.e shows an hourly plan for a Top 40 format. Radio stations pay special attention to their programming during drive-time hours, the periods when stations reach the largest audiences and commercials command top dollar.

Long before technology brought increased diversity to TV programming, competing local radio stations battled each other for small segments of the audience. With more than 12,000 radio stations on the air, the number of possible variations in programming boggles the mind. Greatly relaxed ownership restrictions during the late 1990s have allowed individual stations to target ever smaller segments while still attracting a large total audience for a particular owner.

EXHIBIT 8.e
. . . .
Radio Station Format Clock

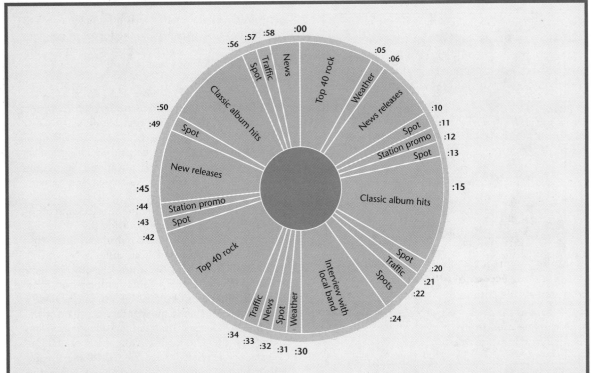

In the beginning radio programmers planned their schedules hour by hour, drawing a circle around a 45 rpm record and dividing the circle into segments to indicate where program elements would air. Many continue to use this concept, employing different clocks to adapt to audience activities in various day-parts by incorporating changes in both content and tempo of presentation. The example given here represents a typical afternoon drive-time clock for a commercial station.

8.9   PROGRAM PROMOTION

Having the best programs in the world means little if audiences don't know about them. Promotion therefore ranks as a major aspect of programming strategies.

On-Air Promotion
. . . .

Broadcast stations and the broadcast and cable networks consider on-air promotional spots the most cost-efficient way to advertise their programs. Breaks between and often within programs usually contain promos for upcoming shows. Credits at the end of programs often include audio-only *voice-over* announcements urging viewers to stay tuned to what follows. Stations and the broadcast networks also air *teasers*—brief mentions of upcoming news stories—during the hour preceding newscasts. The networks and affiliated stations wrestle to claim precious seconds in which to promote their respective programming. So fierce has this contest become that popular shows may sometimes run 21, or even 20, minutes rather than the traditional 22 to provide more time for promotional (and commercial) announcements.

Television stations and cable systems obtain from program distributors, or produce themselves,

EXHIBIT 8.f
· · · · · · · · · ·

TV Guide

Millions of viewers turn to *TV Guide* weekly to see what's on television and to read articles about the television and cable industries. *TV Guide* comes in a dis-

tant third, behind weekly television supplements and daily listings in newspapers, as the source of program information most frequently cited by viewers. But these local publications cannot match *TV Guide*'s nationwide readership.

Walter Annenberg (later a U.S. ambassador to Great Britain) combined three local weekly television program guides to create the first edition of *TV Guide* in 1953. Media baron Rupert Murdoch took control of *TV Guide* in 1988 when he paid $3 billion for its parent company, Triangle Publications; he already owned half of *TV Guide*'s Australian counterpart, *TV Week*.

Though remaining an icon in the television industry, *TV Guide* lost circulation during the 1990s and, compared to some Murdoch acquisitions, did not become an extremely profitable asset in the News Corporation portfolio. In 1998 Murdoch sold *TV Guide* to a subsidiary of Tele-Communications, Inc., (now AT&T BIS) for a reported $2 billion, but retained a significant voting and equity interest as part of the deal. In 1999 Gemstar International announced plans to acquire TV Guide, Inc., for about $9 billion. Gemstar had previously been *TV Guide*'s major competitor in developing the predominant interactive electronic program guide (EPG). Industry observers say that the combination of Gemstar's interactive technology experience and *TV Guide*'s reputation

promos for syndicated shows. Announcements that promote an entire series are referred to as *generic*, while those that highlight one episode in a series are called *specific* or *episodic*.

The pay-cable networks schedule *billboards* of upcoming programs as filler between the end of one show and the start of the next. These networks periodically descramble their signals for several days (often over a weekend) so that nonsubscribers can sample their wares. Usually, the networks offer special promotional rates to new subscribers during the free preview periods. Special discounts are often pitched to current subscribers adding another channel of the provider's multiplexed offerings.

Cable systems and DBS typically dedicate one or more channels entirely to program listings. Some form of on-screen "navigation device" is increasingly important to viewers as they try to find pro-

gramming of special interest among dozens of available channels. New technologies allow viewers to see program titles on-screen as they graze channels, to search for programs by theme or genre, to call up a program by title, to scan a seven-day program guide organized by channel and time, and even to program their VCRs and PVRs for desired programs, program types, or program characteristics.

Other Media
· · · · · · · · ·

To reach nonviewers, networks and stations advertise in newspapers and magazines, on radio, on outdoor billboards, and in other media. Radio stations promote television programs, and television stations promote radio stations—sometimes as paid advertiswith cable system operators and advertisers should give the company a commanding advantage in the interactive EPG marketplace. Development of an interactive EPG and an enhanced Internet presence are prime objectives of the new company. However, there will probably always be those who want to plan their weekly viewing by highlighting a printed schedule with a marking pen. For those consumers, *TV Guide* magazine, and the larger-format *TV Guide Ultimate Cable* for up-scale cable subscribers, will remain coffee-table fixtures.

Each week television stations throughout the country send in upcoming program schedules, and each week magazine editors call stations to update program information. Refusing to list program content based on press releases from networks and other program providers, magazine staffers screen every program in advance, read the script, or, at the very least, talk to the show's producers, writers, or talent.

Cable systems operating within a region covered by a single *TV Guide* edition often carry different channel line-ups, so the magazine usually lists cable programs not by system channel but by program service. However, *TV Guide* is moving toward system-specific listings in some editions. Still, more cable networks exist than could be listed in a magazine of its size. *TV Guide* currently solves this problem by including only the most-watched services.

Moving into the twenty-first century, TV Guide Interactive offers its services on-screen. This electronic version of the magazine's listings promises not only to permit home viewers to check program schedules but also to remind them when a favorite program is on, to tune directly to the show, to record it on the VCR, to search for programs by theme or by title, and even to lock out those shows they don't want their children to see.

Source: Reprinted with permission from News America Publications, Inc., publisher of *TV Guide* magazine. Copyright January 22-28, 2000, News America Publications, Inc.

ing and sometimes as *tradeouts*, whereby media exchange airtime with no cash involved.

The World Wide Web is an important promotional medium that developed rapidly in the 1990s. Web site designers use anything they think will generate more interest or viewership.

Still other promotion forms abound. Radio and TV stations sponsor rock concerts and ice shows and give away T-shirts and bumper stickers. Television networks send news anchors abroad to report from the locales of major events. Daily newspapers and Sunday supplements devote considerable space to broadcast and cable listings. Cable systems mail customized program guides to their subscribers; some are free, others are not. Magazines such as *Satellite Orbit* list which transponders carry which programs for satellite dish owners not subscribing to a DBS service such as DirecTV or Echostar.

TV Guide is the most popular source of printed program listings in the country and the nation's largest weekly circulation magazine. In addition to providing localized broadcast and cable program listings, *TV Guide* publishes articles about the telecommunications industry and TV stars. Today *TV Guide* is much more than the paper-based product started by Walter Annenberg in the early 1950s. In addition to the magazine, there is now the TV Guide Channel "passive" program listing that is available on screens in more than 50 million cable homes. There is also an on-screen interactive program guide that will replace the passive version as more cable systems offer digital services, and future versions are expected to provide features such as search functions and electronic mail. Exhibit 8.f provides more information about *TV Guide*.

CHAPTER 9

Programs: Network, Syndicated, Local

Programs readily fall into three broad categories—network, syndicated, and locally produced. Subcategories are based on paired differences: prime time versus other times, radio versus television, broadcast television versus cable television, first-run syndication versus off-net syndication, network affiliates versus independents, and entertainment versus information content. Each subcategory has its own characteristic audience; its own production, distribution, and delivery methods; and its own type of content.

9.1   PRIME-TIME NETWORK TV ENTERTAINMENT

The best-known and most popular programs appear on television broadcast networks in prime time. They consist mostly of light entertainment—comedy and drama. In the 1990s NBC, CBS, and ABC began filling more prime-time hours with newsmagazines such as *Dateline, 60 Minutes/60 Minutes II,* and *20/20.* Prime-time entertainment network programs come as weekly series, with a sprinkling of movies and occasional one-time specials. A series can run for an indefinite number of episodes. Those designed typically with three to eight episodes are known as *miniseries.*

Movies include both theatrical feature films and made-for-TV movies. After being shown in movie theaters, feature films are licensed for release to the networks and other outlets in sequence, as shown in Exhibit 8.d (page 238). But it is important to remember that the windows of availability illustrated in Exhibit 8.d are not inflexible. Made-for-TV movies come in both 90-minute and 2-hour lengths. Typically, though not universally, they have lower budgets than theatrical films and are adapted to the limitations of the small television screen.

Audience Share

During prime time, the broadcast networks vie for huge audiences—larger than any in the history of entertainment. The percentage of viewers watching the broadcast networks, however, has been declining for two decades. In the 1970s the top-rated network typically attracted a third or more of the prime-time TV audience. As the new century began, a top 10 prime-time broadcast network program needed to attract only about 20 percent of the viewing audience. A network averaging 16 or 17 percent of the weekly prime-time audience could be ranked number one.

Despite a decline in their combined audience share, the broadcast television networks still draw the most massive audiences of any media. Combined prime-time cable ratings exceeded those of the Big Four networks for the first time in 1998, but the average viewership per broadcast network was estimated to be more than 5 million, while the average cable network had an audience of only around 110,000 (Media International, February 1999). The rise and fall of broadcast network prime-time programs makes headline news. And after (often even during) their contractual network runs, those same programs travel throughout the world as syndicated off-network shows.

Situation Comedies

Among television prime-time series, situation comedies (sitcoms) are the staple product. Sitcoms such as *Seinfeld, Frasier,* and *Friends* topped the charts in the 1990s. Hit sitcoms attract huge audiences and tend to earn higher ratings than do hour-long dramas.

Writers of situation comedies create a group of engaging characters who find themselves in a particular situation—often a family setting. Plots spring from the way characters react to a new source of tension injected into the situation week by week. The characters have marked traits—habits, attitudes, and mannerisms—that soon become familiar to the audience. *Frasier, Everybody Loves Raymond,* and *Dharma & Greg* are sitcoms that lean heavily on the unique personas of principal characters for their success.

Early TV sitcoms, such as *Ozzie and Harriet* and *Father Knows Best,* portrayed only stereotypical modular family units—with a handsome middle-class father, loving mother/homemaker, and adorable kids—and emphasized idealized "wholesome" values. *The Honeymooners* was a little different—temperamental working-class husband, no kids—but was still rather conventional. *All in the Family* broke new ground in the 1970s by dealing with generational conflict, civil rights, racial and religious prejudice, homosexuality, and gender issues.

In the 1980s and 1990s some sitcoms moved out of the living room, and the definition of "family" changed. Hits such as *M*A*S*H*, *Cheers*, and *Murphy Brown* revolved around character interaction in the workplace, but surrogate family relationships and support networks remained basic to the plots.

Some 1990s sitcoms increasingly probed the boundaries of acceptable home viewing. *Empty Nest* featured a motherless family, and *Seinfeld* offered a variety of offbeat plots ranging from an attempt to author a television show about "absolutely nothing" to a contest to determine which of the four leading characters could survive the longest without masturbating. *Ellen* created controversy when the lead character, and the actress portraying her, revealed that she was gay. But *Home Improvement*, *Dave's World*, and other 1990s sitcoms successfully continued patterns established decades earlier with only minor changes.

A variant, sometimes called the *slobcom*, depicted less-than-ideal family situations. *Married . . . With Children* featured a frustrated housewife, a husband who failed as a provider, and outrageously undisciplined children. *Roseanne*, especially during its most popular years, featured an aggressively earthy, blue-collar family, truly different from *Ozzie and Harriet*.

In the early 1980s producers began injecting comedy elements, formerly the exclusive province of sitcoms, into action/adventure shows. Series such as *Moonlighting* included lighthearted comic scenes. Critics referred to some of these shows as *dramedies*—blends of drama and comedy. The late-1980s shows *thirtysomething* and *Beauty and the Beast* added nostalgia and fantasy to the mix. In the 1990s programs such as *Xena: Warrior Princess* combined fantasy, action, adventure, and tongue-in-cheek humor. *Twin Peaks* and *Northern Exposure* added the word *quirky* to the vocabulary of millions of viewers. *The X-Files* uniquely blended several themes to generate ratings and attract a cult following worldwide (see Exhibit 9.a).

Spinoffs and Clones

A successful program with a new angle instantly begets brazen imitations. Programmers speak of such copies as spinoffs and clones. A *spinoff* stars secondary characters from a previous hit on the same network, placing the characters in a new situation. *Buffy the Vampire Slayer*, for example, spun off *Angel*.

A *clone* closely imitates an already popular program, often on another network, changing only the stars and details of plot and setting. Examples include sitcoms featuring stand-up comics (*The Drew Carey Show* and *Everybody Loves Raymond*) that hoped to duplicate the popularity of *Seinfeld* and *Home Improvement*. (Indeed, this latter form of clone runs in cycles, dating back at least as far as *Make Room for Daddy*, with Danny Thomas, through *Cosby* and *Roseanne*.)

Crime Shows

Police, courtroom, and detective dramas such as *Dragnet* and *Perry Mason* peaked in the ratings around 1960. In the mid-1980s a new breed of more authentic crime shows captured top ratings. *Hill Street Blues* and *Cagney and Lacey* began a trend toward crime dramas dealing with tough social issues, using multilayered plots involving many characters.

NYPD Blue tested the limits of acceptable program content for broadcast networks by using limited nudity and strong language. Although the series enjoyed some success with viewers and received numerous awards and critical praise, many advertisers withheld their support. In contrast, *Diagnosis Murder* featured sanitized homicide mysteries, usually set in upper-class surroundings, with a well-bred, mature physician as the unlikely detective.

Medical Dramas

For years programs in hospital settings were staple fare on prime-time network television. Shows such as *Ben Casey* and *Dr. Kildare*, exploring the lives of both patients and practitioners, drew millions of viewers every week. Then, bowing to the cyclical nature of audience taste, they disappeared. After some successes, such as *St. Elsewhere*, the genre returned in the mid-1990s with *Chicago Hope* and *ER*, the latter often finding itself at the top of the Nielsen ratings. *City of Angels* appeared in January 2000 and was unique among medical dramas in that it featured minorities in the leading roles, and also employed many minorities on the production staff.

EXHIBIT 9.a
• • • • • • • • • • •

The Truth Is Out There

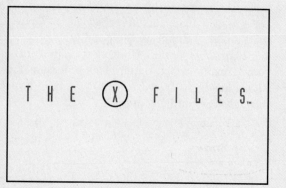

Following its debut on Fox in the fall of 1993, *The X-Files* developed into both an unqualified network ratings success and a cult program with fanatical fans worldwide. Chris Carter's mod-Gothic melodrama combines two popular themes—a belief that contemporary science cannot explain everything and a nagging concern that government officials do not always level with the public. The show owes much of its success to the main characters, FBI agents Fox "Spooky" Mulder (David Duchovny) and Dana Scully (Gillian Anderson), who together investigate bizarre top-secret cases involving the paranormal. Mulder is an intense, brooding individual who believes in the paranormal; Scully is a physician trained in forensic medicine and a skeptic. The show also benefits from topnotch production values and its music. It was initially shot in Vancouver, Canada, where the fog and mist contributed to an enveloping atmosphere of dread. As critic James Wolcott said, "even the sunlight looks a little ill" (*The New Yorker*, 6 January 1997: 76).

In the mid-1990s, *The X-Files* reached an estimated 10 million American TV households and had a large following internationally. It became the focus of intense interest among fans (who called themselves X-Philes) on the World Wide Web. The series inspired a bewildering array of promotional merchandise, and at *X-Files* conventions fans interacted with people associated with the program, sometimes even Chris Carter himself.

As early as the mid-1990s some critics were saying that the show was losing some of its appeal. But in 1997 *The X-Files* won a Golden Globe Award as best drama and was nominated for a dozen Emmy Awards, including outstanding dramatic series. The following year Duchovny and Anderson appeared on the big screen in a movie based on the TV program.

The X-Files ended the decade as one of Fox's most popular programs, but by the 1999/2000 season the series' future was uncertain. Gillian Anderson publicly said she wanted out, although her contract ran through 2001. Both Duchovny's and Chris Carter's contracts expired at the end of the 1999/2000 season, and Duchovny did not seem excited about returning for an eighth season. The show's prospects were not improved when Duchovny sued Fox for $25 million, claiming that Fox had, among other things, sold the show in syndication to the studio's cable network for less than the fair market price. Eventually he agreed to appear in about a dozen new episodes for the 2000/2001 season. The show's remarkably successful run through the 1990s and into the new century ensures that it will enjoy a long life in syndication. And both Anderson and Duchovny also have indicated that they would be interested in doing additional movies based on the TV series.

Movies
• • • • •

By the 1970s the television networks were paying increasingly astronomical prices for licenses to exhibit hit feature films. A decade later ABC set a new record by paying $15 million for the right to air *Ghostbusters*, even though it had already been available in the video rental market and had played for two years in theaters and on pay cable. By the 1990s VCRs and pay cable had devalued network showings of movies to the point that the average network licensing fee dropped to about $3 million per showing.

Nevertheless, both broadcast and cable networks still pay huge fees for the movie blockbusters. NBC's $50 million payment in 1995 for multiple runs of *Jurassic Park* set a new record. These costly films rarely earn back their full rental fees in advertising revenue but can be useful for clobbering opposing networks in the ratings.

Typically, a broadcast network's rental fee for a single showing of a major theatrical feature would cover all or most of the $4 million typically required to produce a brand-new, modest-budget, made-for-TV-movie. Often the top made-for-TV movies command higher audience shares than most televised theatrical movies. By 1990 two-thirds of network movies were "made-fors." The broadcast networks ordered almost 100 made-for-TV movies and miniseries for the 1998–1999 season, about two-thirds as many as were ordered by the cable networks. Made-fors also sometimes serve as pilots for prospective prime-time network series. For plots, producers often turn to actual, usually sensational events.

This practice reached new heights—or exploitive depths—in 1993 when ABC, CBS, and NBC each aired a made-for-TV movie based on the true story of the attempt by a teenage girl, dubbed the "Long Island Lolita," to murder her adult lover's wife. By the following year, these networks had begun to back away from this strategy, turning more to what some called "uplifting, feel-good" movies. Fox, on the other hand, stayed on the crime-and-tragedy path with, among others, its O. J. Simpson docudrama. Waiting until the jury had been selected and sequestered, Fox aired the Simpson movie during the February 1995 ratings sweeps, where, to the network's certain disappointment, it ran fourth in its time period. Fearing negative reactions, in 1999 the networks canceled or delayed broadcasting movies dealing with teen violence in the wake of school shootings in Kentucky, Arkansas, Georgia, and Colorado.

NBC, however, easily won its time period in 1997 with the first prime-time network broadcast of Steven Spielberg's acclaimed black-and-white docudrama, *Schindler's List*. Sensitive to the program's serious content, NBC aired it largely unedited and without commercials, thanks to underwriting by Ford. It received the then-new parental advisory system's most restrictive TV-MA (mature audience) rating based on its violence, full frontal nudity, and profanity. There were some complaints, but most people praised NBC's handling of the broadcast, including politicians often critical of TV programming.

Beginning in the mid-1980s, the major pay-cable networks (such as HBO and Showtime), as well as some other cable services (USA and TNT), also plunged into the financing of *made-for-cable* movies. Pay cable still relies on theatrical films as its bread-and-butter entertainment, but original movies are increasingly important components of the schedule. HBO, for example, ordered about a dozen original movies for 1998, about twice the number ordered only three years earlier.

Miniseries

Roots, a 14-hour adaptation of a bestseller about the evolving role of blacks in American life, ran for eight successive nights in 1977. It started a trend toward miniseries. At the time experts doubted the drawing power of the series because of the subject matter and questioned its ability to sustain viewership for so many hours, but *Roots* took them by surprise. The audience increased for each episode, breaking all records on the eighth night. A western miniseries, *Lonesome Dove*, scored a similar surprise success in the late 1980s. Miniseries proved able to compete well against pay-cable movies and to attract new viewers, especially upscale professionals, who may otherwise watch little entertainment television.

Lengthy miniseries such as *Roots* were less successful in the 1990s. Early in 2000, NBC's lavishly produced prime-time miniseries *The 10th Kingdom* attracted only about 12 percent of the viewing audience. The hectic lifestyles of viewers and the multiple channels competing for attention make it difficult for miniseries covering a dozen or more evenings to be successful, the VCR notwithstanding. Miniseries today typically are limited to fewer than a half-dozen episodes.

Music and Variety

Except for variety shows and a few programs such as *American Bandstand*, which in the 1950s became one of television's first hits, and *Your Hit Parade*, which aired from 1950 to 1974, television paid little attention to popular music. That changed in 1981 with the formation of the Music Television (MTV) cable network (see Exhibit 9.b). MTV quickly became a 24-hour rock-video powerhouse that targets teens and young adults ages 12 to 24. A co-owned network, VH1, programs to attract 18- to 49-year-olds, who grew up with music videos.

EXHIBIT 9.b
• • • • • • • • • • •

MTV

MTV ushered the music-television era into the electronic-media world on August 1, 1981, at 12:01 A.M. Its very first video clip featured an obscure British band known as the Buggles.

Says Tom Freston, head of MTV Networks, "From the very beginning, we made a lot of hay out of the fact that MTV was meant to alienate a lot of people. It was *meant* to drive a 55-year-old person crazy" (Williams, 1989). MTV may well have driven the older generation crazy with worries that, like Elvis's pelvis in the 1950s, MTV would seduce their children with its hypnotic sexuality, but it does delight the young. Today MTV is available in almost 70 million U.S. homes and reaches more than 300 million homes worldwide.

MTV's music videos—essentially commercials for songs—rescued the record industry, whose sales in 1981 had dropped 30 percent from their 1978 peak.

Over the ensuing decade, sales hit new highs, doubling by 1991. With its fast edits, bizarre camera angles, and exaggerated colors, the MTV "look" has also influenced nearly all areas of the entertainment world. Commercials, TV shows, even movies have adopted MTV's visual style.

In 1985 a sister channel, VH1, was born, intended for an audience somewhat older than the 12- to 24-year-olds targeted by MTV. In 1996 another channel, M2: Music Television, was added. The network has also expanded to include game shows, newscasts, stand-up comedy, documentaries, a soap opera, and an action/adventure series.

MTV's animated series *Beavis and Butt-Head*, came under attack in 1993 when a woman blamed the program for a fire, set by her five-year-old son, which killed his two-year-old sister. She said her son began playing with matches after watching the cartoon, in which the characters often talked about or played with fire. Although MTV denied any responsibility, it agreed to delete from the show all references to fire. The controversy did no damage to the commercial viability of the show's characters. Besides remaining on MTV, they starred in *Beavis and Butt-Head Do America*, a 1996 feature-length movie. MTV moved the show from its original 7:00 P.M. slot to 10:30 P.M., but later moved it back to 8:00 P.M.

Most MTV airtime remains devoted to music, and not just for Americans. Versions of the network are now available in Europe, Asia, Central and South America, Australia, Russia, Japan—the list goes on and on.

Source: *Real World*, artwork provided courtesy of MTV: Music Television. © 1999 MTV Networks. All Rights Reserved. "MTV: Music Television" and all related titles, characters, and logos are trademarks owned by MTV Networks, a division of Viacom International, Inc.

Originally intended to promote record sales, music videos became a television genre in their own right: Performers act out song lyrics, interpret them, or otherwise create imaginative visual images for songs. As promotional tools, videos originally came free of charge to stations and networks. But MTV changed the ground rules in 1984 by *paying* for exclusive rights to Michael Jackson's much-publicized *Thriller* video. MTV

now contracts for exclusive early *windows* (periods of availability) for some videos. These strategies essentially demolished music videos as a source of free program material.

MTV's success had wide ramifications. As one commentator put it,

the fast pace and kaleidoscopic style of music video has ricocheted across popular culture, changing the way people listen to music and leaving its frenetic mark on movies, television, fashion, advertising and even TV news. (Pareles, 1989)

In 1996 Viacom/MTV launched yet another channel devoted to music videos, M2: Music Television. By the end of the 1990s MTV and VH1 had moved away from a heavy reliance on music videos and were offering more music-related features, such as *Behind the Music* and *Before They Were Rock Stars*, in addition to live concerts, music documentaries, and other original programming.

Today there are about two dozen music video channels distributed nationally featuring many kinds of music. For example, Country Music Television (CMT) provides country-western music videos and music-based features, as does Great American Country (GAC). BET on Jazz: The Cable Jazz Channel programs a broad variety of innovative jazz productions, celebrity interviews, and documentaries. HTV is a 24-hour, all-Latin, all-music service in Spanish.

Variety programs once played a role both in viewers' habits and in the world of music. Hosted by stars such as Ed Sullivan (who introduced Elvis and the Beatles to America), Judy Garland, Frank Sinatra, Dean Martin, Perry Como, and Sonny and Cher, these programs once brought hours of music into the home. But their popularity faded as production costs increased and as musical tastes changed. By the 1990s music-based variety programs had disappeared as series from network television, reappearing only occasionally as specials. Today talk shows such as *The Tonight Show with Jay Leno* occasionally feature recording artists and groups and even have their own in-studio bands, but music is not the central focus of these programs.

9.2 NON-PRIME-TIME NETWORK TV ENTERTAINMENT

Although the broadcast networks put their best creative efforts into prime-time programs, daytime programs yield a higher profit margin. The huge production costs and the smaller number of commercial minutes in prime-time shows make them less efficient revenue earners.

Daytime programs air about 5 to 7 more commercial minutes an hour than do prime-time programs—as many as 16 commercial minutes each hour. Constant switching of soap opera plot lines from the doings of one character to those of another accommodates interruptions that would be intolerable in most dramas. Daytime talk shows also carry heavy commercial loads but do not always provide convenient breaks, and sometimes their hosts reveal frustration when they must interrupt guests to insert commercials.

Dayparts

Network non-prime-time breaks down into the dayparts depicted in Exhibit 8.b (page 235). A characteristic type of program tends to occupy each daypart:

- Networks fill the early-morning daypart with newscasts and talk, weekends with sports, all discussed later in this chapter.
- Soaps, games, and talk shows dominate daytime network television just as they once dominated daytime network radio.
- Late-night fare consists mostly of talk and comedy/variety programs.

Soap Operas

In the early days of radio, soap companies often sponsored daytime serial dramas whose broad histrionics earned them the nickname "soap operas." This genre is a classic case of parsimonious use of program resources. Notorious for their snail-like pace, soaps use every means imaginable to drag out each episode's story. Most minimize scenery costs, relying heavily on head shots of actors in emotion-laden, one-on-one dialogue. In addition, a typical soap features perhaps a dozen sets depicting the workplaces, living rooms, and bedrooms of the principal characters.

Contemporary soaps have responded to changing public tastes. They include story lines based on once-forbidden subjects such as drug addiction, social diseases, and family violence. Women and members of minority groups began appearing in more varied roles by the mid-1970s. In 1989 NBC began *Generations*, a trailblazing if unsuccessful soap featuring close relations between a black family and a white family.

Changes in audience composition encouraged soap opera writers to deal with controversial topics and new social roles. First *General Hospital* and then *The Young and the Restless* stimulated a faddish interest on the part of younger viewers, including males, during the 1980s. ABC launched the promotion *My Time for Me* during the summer of 1988 to lure young viewers during the school vacation.

Sexual attraction and conflict continue to be key themes in soaps, with bedroom scenes perhaps steamier than ever. In the 1990s female leads increasingly held positions of financial power, often cast as owners or top-level managers of businesses. Most women portrayed in soaps had careers outside the home and exercised greater control over their lives than female soap opera characters in earlier decades. Responding to criticism of the blatant sexuality depicted, soaps began incorporating at least tangential references to safe sex and sexual responsibility. Alternative lifestyles also are treated with greater sensitivity in story lines.

Even a modestly successful soap generates profits amounting to tens of millions of dollars annually, thanks in large measure to relatively low production cost. However, the audience for soaps dropped dramatically in the second half of the 1990s. Nielsen estimated that total viewing declined more than 25 percent and viewing by women in the key 18-to-49 demographic dropped about 30 percent. Observers note that there are more women in the workforce today, and even those who work at home are busier than ever. Competition from cable, DBS, and the Internet also is blamed. Still, at the end of the decade it was estimated that the typical one-hour soap takes in about three times what it costs to produce.

Not surprisingly, the economic importance of soaps to daytime programming continues to inspire efforts to launch new shows, but often with little success. Between 1987 and early 1997 only one new soap was launched successfully, CBS's *The Bold and the Beautiful*. During the same period two NBC soaps (*Generations* and *Santa Barbara*) and three ABC soaps (*Ryan's Hope*, *The City*, and *Loving*) disappeared. NBC dropped both *Another World* and two-year-old *Sunset Beach* in 1999. NBC immediately replaced *Another World* with a new soap, *Passions*. ABC's *Port Charles*, launched in 1997, was also lagging in the ratings as the decade ended. In spite of the failures, some soaps enjoy long lives. Before being canceled, *Another World*, for example, had been on the air for more than 30 years. *The Young and the Restless*, *All My Children*, *One Life to Live*,

and *Days of Our Lives* were all on the air before 1980. *Guiding Light* began as a radio soap in 1937 and has been on TV since 1952 (see Exhibit 9.c).

Soaps can generate enormous rapport between viewers and the characters/actors that carries on year after year. Magazines featuring performers who play the roles of enduring soap characters are popular, and performers are often mobbed when they make personal appearances. The entire nation gave a sigh of relief when Susan Lucci, after repeated nominations, finally won an Emmy in 1999 for her performance as Erica Kane in ABC's *All My Children*.

Some viewers treat soap opera characters almost as real people. Soap producers exploit this relationship by providing supplemental information about the programs on their web sites. They hope that this will increase interest and stem the loss of viewers. Producers are also looking for ways to repurpose (reuse on another medium) their products. In January 2000, Disney/ABC launched a 24-hour cable channel called SoapNet that features ABC soap operas and soap-related programming. Sony-owned Columbia TriStar Television announced plans to launch its own 24-hour soap channel tentatively called SoapCity later in the year. SoapNet launched with only about a million subscribers and some observers question whether there is enough viewer interest (and cable carriage capacity) to justify two soap channels. Some believe that the two ventures could merge to form a single channel at some point.

Soap operas have highly successful counterparts on ethnic and foreign-language outlets. Two competing U.S. Spanish-language cable/broadcast networks, Univision and Telemundo, feature imported *telenovelas*. These enormously popular soap operas reach tens of millions of viewers throughout the Hispanic world. In contrast to American soaps, which can endure for decades, telenovelas burn themselves out in a few months. Though primarily entertainment, they often carry educational messages, typically promoting socially approved conduct and family values.

Game Shows
• • • • • • • • •

Another parsimonious format, the audience participation game show, became a staple of network radio more than a half-century ago. One of the cheapest formats, game shows cost little in time, talent, effort, and money once a winning formula

EXHIBIT 9.c

Guiding Light

Guiding Light, television's longest-running soap opera, celebrated its 48th anniversary in June 2000. The program, which began on January 25, 1937, as a 15-minute radio serial, made the transition to television on June 30, 1952. Its longevity attests to the enduring popularity of the soap opera genre.

Sources: (top, left and right): CBS Photo Archive; (bottom, left): Robert Milazzo/Televest, courtesy of Procter & Gamble Productions, Inc.; (bottom, right): Arthur Cohen/Televest, courtesy of Procter & Gamble Productions, Inc.

has been devised and a winning emcee selected. Talent expenses are limited to the host's salary and, in some cases, fees for show business personalities, who usually work at minimum union scale because of the publicity value of game show appearances. Taping five or more episodes in a single day reduces production costs further. Once a successful game show formula is perfected, a producer can often duplicate that success in foreign markets by hiring a native emcee and perhaps making a few changes to accommodate the local culture.

Opportunities for enhanced commercial content also contribute to the profitability of game shows. The giveaway format justifies supplementing the normal daytime limit of 16 minutes or so of advertising spots with *plugs*—short paid-for announcements on behalf of advertisers who donate prizes and other services such as transportation and wardrobe items. The prizes come free of charge from advertisers, who write them off as business expenses. Game shows were among TV's most popular programs in the 1950s and remained on the broadcast networks well into the 1980s. But by the mid-1990s game shows had practically disappeared from broadcast network schedules.

In the summer of 1999, however, ABC introduced *Who Wants to Be a Millionaire*, hosted by syndicated talk-show celebrity Regis Philbin. Encouraged by the program's summer performance, ABC added it to the prime-time schedule during the following November ratings sweeps period. The show got spectacular ratings and, early in 2000, ABC aired the program five nights a week during prime time. In typical copycat fashion, *Millionaire*'s success prompted Fox to counter with *Greed*, and NBC brought back *Twenty One*, a 1950s game show that died after government investigations revealed it was rigged. CBS announced plans to resurrect *The $64,000 Question*, another game show tainted by the 1950s scandal, but dropped the idea after seeing the pilot episode.

Game shows continued to fill slots in the cable networks' schedules well into the late 1990s. In 1997 The Family Channel (now Fox Family Channel), Lifetime, Nickelodeon, and MTV each scheduled at least two game shows. The Family Channel ran a half-dozen. Even Odyssey and MSNBC ran game shows. By the end of the decade there were fewer cable networks running game shows, but one network specialized in them. Launched in 1994, the Game Show Network (GSN) by the end of the decade served about 20 million subscribers with game show programming 24 hours a day. GSN fills its schedule mostly with syndicated product that had its debut elsewhere, but it also features new episodes of shows such as *Wheel of Fortune* and *Jeopardy!* plus games especially for children, including *Juvenile Jury* and *The Quiz Kids Challenge*. GSN offers limited interactivity, allowing viewers to play along with certain shows. Viewers can win prizes such as vacation packages, Sony Playstations (Sony has an ownership interest in GSN), CDs, furniture, appliances, and jewelry.

Game shows such as *Wheel of Fortune* and *Jeopardy!* remain among the most successful programs in first-run syndication. *Wheel* benefits both from the suspense element of the wheel's unpredictable stopping places and from the winning talent combination of emcee Pat Sajack and his assistant, Vanna White. She won celebrity status for her unique letter-turning "talent." The producers later made her task even less demanding by replacing the puzzle board's cards with multiple high-tech computerized touch screens, but Vanna remains popular. As she herself candidly admits, her primary role is simply to look beautiful.

Most games run the half-hour length favored for syndication to enable flexible scheduling in all dayparts. CBS runs *The Price Is Right* in the late morning, a traditional slot for game shows on the broadcast networks. Network affiliates typically purchase first-run game shows to schedule in the hour before prime-time network programming begins.

Part of a game show's appeal is that it encourages viewers to play along with the contestants. The popularity of interactive games available on CD-ROM has not been lost on producers, and they are working to enhance viewer loyalty and involvement by providing ways viewers can participate in a version of the game. Game show producers, of course, are not alone in seeking audience participation. The Internet is a popular vehicle for such activities, and many web sites promoting programming provide varying degrees of interactivity.

Magazine Shows

In the 1950s programmers began extending network television into hitherto unprogrammed early-morning and late-evening hours, a radical move at the time. For those innovative shows, NBC developed the *magazine format*—a medley of short features bound together by a personable host or group of hosts. *Today*, NBC's pioneer early-morning magazine show, started in 1952. Continuing into the 1990s, the *Today* show shared the 7:00 to 9:00 A.M. spotlight with CBS's *This Morning* (in various guises) and ABC's *Good Morning America*.

Magazine shows are not centered on the host's personality, but the host is an important factor. The format's creator, Pat Weaver, demonstrated his understanding of the host's importance when he selected the urbane and laconic Dave Garroway for NBC's *Today* show at its inception. ABC discovered that little had changed 45 years later, in 1997, when *Good Morning America* began losing ground after the departure of its popular female host, Joan Lunden, followed by co-host Charles Gibson. Early in 1999 ABC brought Gibson back, along with Diane Sawyer, one of ABC's star personalities. Both Gibson and Sawyer initially described their jobs as temporary, but the show's ratings improved, and a year later both were still hosting the program.

Talk Shows

• • • • • • • •

Although related to magazine shows, the talk show more closely resembles an essay than a magazine. It emphasizes the talker's personality, which colors the interviews and other segments of the show. NBC's *Tonight* (later called *The Tonight Show* and now *The Tonight Show with Jay Leno*), a late-night companion to the morning magazine show, started in 1954 as a showcase for the comic talents of Steve Allen. After a series of other hosts, including Jack Paar, Johnny Carson took over in 1962.

For 30 years Carson reigned as the "King of Late Night." He hosted some 5,000 shows, talked with more than 22,000 guests, and ran about 100,000 commercials. Carson's reign finally came to a close in 1992. NBC estimated that a record 55 million people watched his final show. The lead-in gave David Letterman the highest-ever rating for his *Late Night* program, which followed.

Letterman, who moved to CBS in 1993 to compete directly against the program, had aspired to move up to *The Tonight Show*. But it was Jay Leno who succeeded into the host's chair. Leno had first appeared as a *Tonight Show* guest in 1977, becoming one of several guest hosts in 1986 and exclusive guest host in 1987. Although Letterman and Leno had been friends, they became competitors for late-night viewers—and the millions of dollars in network revenue they represented.

The networks have tried a number of late-night talk programs over the years. For several years ABC offered shows similar to *The Tonight Show* hosted by Dick Cavett and Arsenio Hall. In the late 1990s ABC enjoyed considerable success with *Politically Incorrect with Bill Maher* (a program originally seen on cable's Comedy Central), and NBC scheduled *Late Night with Conan O'Brien* following *The Tonight Show with Jay Leno*. O'Brien's show follows a pattern similar to Leno's, with the host interviewing guests sequentially. *Politically Incorrect* typically features celebrities seated in a semicircle as they talk about controversial issues, with Maher playing devil's advocate and injecting humor to keep the mood light and irreverent.

Cable television also takes advantage of the relatively low production cost and flexibility of talk shows. Lifetime, for example, specializes in talk programs, carrying a dozen or so on health, consumer services, and the like. In addition to *Larry King Live*, CNN regularly schedules talk shows on money management, as well as news interviews and discussions. Fox News Channel features many talk shows that focus on news and politics. Religious networks rely heavily on inspirational talk programs, and sports channels depend on sports talk segments and interview shows.

Home Shopping

• • • • • • • • • • •

Not all viewers find commercials boring, at least not those that appear on the various home shopping networks. These broadcast/cable operations market consumer items such as clothing, jewelry, home appliances, and novelty ware, claiming that bulk purchasing and low overhead cost result in drastically reduced prices compared with those in stores. A cable system carrying a shopping network receives a percentage of each sale in its service area.

Such nonstop commercialism represents a radical break with established broadcasting tradition. The pre-deregulation FCC enforced an arm's-length relationship between programs and advertising by penalizing stations for program-length commercials—programs that so interwove ad and program content that audiences could not tell one from the other. The FCC reasoned that full-time advertising violated the public interest by displacing normal program functions.

The conversion to digital television is expected to make home shopping an option for viewers during any programming. Viewers will be able to click on icons displayed during programs or commercials and actually purchase a product while watching. Alternatively, viewers may click on a "Want to know more?" icon to see a recorded "pitch" for the product, such as those used on traditional home shopping channels. The viewer may prefer to have the set-top box record the information for viewing later. In either case the viewer will have the opportunity to click on another icon to order the product.

Home shopping programs in the 1990s accustomed viewers to buying products they see on TV directly rather than purchasing them from local retail outlets. Consumer attitudes about shopping at home have changed in the last decade, and digital broadcast technology promises to make home shopping even more popular in the new century.

9.3   NETWORK TV SPORTS

For true fans, sports are ideal television and radio subjects: real-life events that occur on predictable schedules yet are filled with suspense.

Professional football, basketball, and major-league baseball attract by far the largest audience. The numbers of viewers for other sports fall off rapidly. Super Bowl football and World Series baseball rank among the all-time hit programs because their excitement and spectacle appeal to a broad audience.

The Olympic Games are also a major TV sports attraction. Atlanta's 1996 Summer Olympics, for example, generated high ratings for NBC. Unfortunately, some of that success was attributable to additional viewer interest created when a bomb exploded in the Olympic Park. Olympic Games coverage gives a network prestige and promotional opportunities, but it no longer ensures huge ratings victories. Mediocre performances by U.S. athletes, bad weather, problems with scheduling because of time zone differences, and strong counterprogramming by competing networks can cut into viewership. ABC's programming, for example, beat CBS's coverage of the closing ceremonies of the 1998 Winter Olympics in Nagano, Japan. Overall, however, CBS's Olympic coverage beat other programs during the period, even though it sometimes failed to meet the audience size guarantees CBS made to advertisers.

Network television strategists value sports programs because they appeal to middle-class males, an audience not well reached by most other programs. The ability of sports to capture such elusive consumers justifies charging higher-than-normal advertising rates for commercials on sports programs.

Evolution of Network Sports

By 1990 combined broadcast/cable rights payments for on-the-spot sports events exceeded a billion dollars annually. Yet both professional and college football ratings, long a sure thing, had begun to decline. Sports fans increasingly divided their attention among dozens of televised sports events.

The broadcast networks responded by purchasing stock in cable networks. In 1984 ABC acquired control of ESPN (originally Entertainment and Sports Program Network). The 24-hour all-sports network furnishes full-length coverage that ABC cannot provide as well as enhanced bidding power for rights to sports events, a significant advantage. ESPN2 was launched in 1993, providing even greater sports coverage, and three years later ESPNews appeared. ESPN and ESPN2 focus mostly on covering sports events, while ESPNews is a studio-based service providing scores, interviews, and features related to sports. In 1997 ESPN added ESPN Classic to its roster of cable sports networks. ESPN Classic reaches about 20 million subscribers and features the greatest sporting events and sports heroes in history, while adding a current perspective.

Following ABC's lead, NBC in 1989 moved into basic cable to create supplementary sports and news outlets. In 1990 NBC's SportsChannel America introduced high school sports to national television. In 1997 NBC assumed a 25 percent interest in Rainbow Programming Holdings, Inc., owner of several SportsChannel regional networks and also Bravo, American Movie Classics, and the Independent Film Channel. CBS was slower to become involved in cable. CBS's participation in cable, including the regional sports networks Home Team Sports and Midwest Sports Channel, picked up after its purchase by Westinghouse. Following its merger with Viacom, the network came under the same corporate umbrella as cable powerhouses MTV and Nickelodeon.

In 1996 Time Warner-Turner launched CNN/SI, a joint effort of Turner's CNN and Time Warner's *Sports Illustrated* magazine. CNN/SI competes directly with ESPNews, serving as a source of sports-related information and features rather than live coverage. Other sports cable networks were developing, some of them, such as The Golf Channel, devoted to a single sport or sports genre.

During the 1990s newcomer Fox emerged as a major player in sports programming, both on-air and on-cable. In 1993, Fox made a spectacular entry into sports broadcasting by outbidding CBS for four years of National Football League (NFL) games. The Fox bid topped that of CBS by $400 million, showing the seriousness of Fox's desire to become a power in sports programming. When a six-year deal signed in 1994 between NBC and ABC for major-league baseball fell through, thanks largely to a baseball strike, Fox stepped in. A dramatic come-from-behind win by the New York

Yankees in the 1996 World Series gave Fox its first-ever weekly win against ABC, NBC, and CBS during prime time. The following January Fox attracted another huge sports audience with Super Bowl XXXI coverage. Between these two major sporting events Fox launched the Fox Sports Net, based on regional cable sports networks covering much of the country.

By 1999 the National Cable Television Association identified more than 40 cable networks offering sports exclusively or routinely featuring sports programming. Many cable networks serve only a regional audience. Typical examples are the New England Sports Network, the Empire Sports Network, and of course, about a dozen regional Fox Sports Net services. Most sports networks focus on popular team sports, but others, such as The Outdoor Channel, feature hunting, fishing, camping, hiking, and similar outdoor activities.

The proliferation of sports channels on cable and the desire of the broadcast networks to increase sports coverage attest to the importance of the genre. Cable network TNT finished first in prime time among basic-cable networks in 1996 by running a heavy schedule of NFL and National Basketball Association (NBA) games. NBC announced early in 1997 that it would add weekly coverage of Women's National Basketball Association (WNBA) games to its schedule. By 1997 major-league baseball was out of its earlier slump and enjoying multiyear broadcast, cable, and World Wide Web rights contracts worth several billion dollars. ABCs *Monday Night Football* ended the decade by frequently beating competing programs in its time slot, sometimes being the most-watched prime-time program of the week. In September 1999 it also gained added distinction by being ABC's first live sporting event produced in high definition, although only about two dozen stations broadcast the HDTV signal.

The most significant testament to the importance of sports to the broadcast and cable networks is the price they pay for carriage rights. Stunned by its loss of NFL games to Fox earlier in the decade, CBS in 1998 pledged a reported $500 million a year for eight seasons of AFC games and two Super Bowls. Disney simultaneously agreed to pay more than $9 billion to secure professional football rights for ABC and its ESPN cable networks, a deal including three Super Bowls. Fox signed on for $550 million annually, more than $4 billion over

eight years, also including three Super Bowls. The total NFL package amounted to more than $17.5 billion. NBC, faced with staggering program cost increases for hit shows such as *ER*, passed on professional football at these prices. TNT also balked at paying rates twice what it had paid under its previous contract.

Though dwarfed by the NFL package, the NBA in 1998 signed broadcast and cable networks to four-year contracts worth more than $2.6 billion. Major-league baseball raked in almost $1.7 billion over five years. Sports events such as the Daytona 500 and Indianapolis 500 cost networks more then $10 million each annually. College and other non-professional sports are also costly for the networks. The National Collegiate Athletic Association (NCAA) pulls in millions of dollars for college games during the regular football season and millions more for the bowl games that determine the national championship. College basketball also brings in hundreds of millions annually from the broadcast and cable networks. In November 1999 CBS agreed to pay the NCAA $6.2 billion for media rights to the college men's basketball tournament through the year 2013. That amounted to about $550 million annually, more than the network was paying at the time for rights to NFL games. As discussed later in this section, many critics blame the money paid by broadcasting and cable for turning college athletic departments into cash cows for their respective institutions, calling into question the concept of the "amateur student athlete."

Even as the cost of major sports programming continued to increase in the 1990s, audiences shrank. Ratings for NFL games in 1999 were significantly lower than at the start of the decade. The departure of superstars such as Michael Jordan and bizarre behavior by some players both on and off the court resulted in dramatic drops in the ratings for professional basketball. Strikes for higher salaries by professional athletes already making millions of dollars also cooled the enthusiasm of many fans. But greater interest in sports such as figure skating, auto racing, tennis, and golf helped balance the losses in traditional professional team sports. Interest in women's basketball and soccer also boosted sports viewership.

Although some people would question its qualifications as a sport, professional wrestling appeals strongly to the 18-to-49-year-old male demographic so coveted by advertisers who buy time in sports programming. Wrestling often wins five or six of the top 10 slots in the weekly basic cable ratings. TV wrestling is dominated by two organizations, the World Wrestling Federation (WWF) and World Championship Wrestling (WCW). WWF programming is carried principally on cable's USA Network, and WCW appears mostly on TNT, but also on other Time Warner–Turner cable outlets. Wrestling is also distributed on pay-per-view channels. Wrestling was a staple of early network broadcast television but had practically disappeared by 1990. The costumes, gestures, language, and (apparent) violence of contemporary wrestling make its return to an over-the-air broadcast network unlikely. But wrestling appears to have found a secure, and highly profitable, home on cable. Boxing, another controversial sport, appears with varying frequency on cable and pay-per-view channels, but seldom on the broadcast networks, except during the Olympic Games.

In spite of huge increases in fees for major sports rights and smaller audiences, a strong economy in the second half of the 1990s allowed networks to continue passing on higher prices to advertisers. Companies that sell products purchased predominantly by males know that team sports are among the few programs that continue to attract men in significant numbers. For breweries and automobile makers, major sports are a "must-buy." But even with the strong ad revenue stream, the higher costs of professional football forced networks not only to significantly decrease or, in some cases, discontinue providing network compensation (discussed in Section 6.3, page 171) but also to demand that affiliates help offset the increased cost by paying the networks. ABC, CBS, and Fox will each receive about $50 million annually from their affiliates for at least three years starting in 1999, and ESPN raised the fee it charges cable operators.

Even if increased advertising rates charged by the networks, affiliated stations, and cable systems do not match the higher costs of sports programming, owners may still consider sports a wise investment. In addition to attracting hard-to-reach adult male viewers, sports programming is key to providing lead-ins to entertainment programming and establishing a brand-name image for a network. Although the decline of CBS and the rise of Fox during the mid-1990s was not entirely attributable to Fox's acquisition of NFL rights from CBS, many analysts list it high among the contributing factors.

Pay-Cable and PPV Sports

Increases in rights charges for sports events forced some events to pay cable, which can afford the high costs. Despite its relatively small subscribership, HBO, for example, receives stable, predictable revenue that handily covers the costs of sports rights.

Events such as championship boxing and professional wrestling have strong appeal for relatively small but intensely loyal and willing-to-pay audiences. Such events can be profitably scheduled on national pay-per-view (PPV) television. A 1997 heavyweight rematch between Evander Holyfield and Mike Tyson grossed more than $90 million from almost 2 million PPV subscribers. A successful PPV boxing event requires a match between two sharply contrasting and therefore easily promoted personalities, preferably in the heavyweight division.

In addition to high-cost sports events, PPV is a viable delivery vehicle for sports programming that lacks the wide appeal necessary for carriage on a broadcast or cable network. PPV programs are seen mostly by means of addressable cable technology in homes, bars, and hotels. The average cable home has access to about a half-dozen PPV channels, but the number will increase as more system operators offer digital services. DBS subscribers also have access to PPV programs.

Scheduling and Buying Sports

The seasonal nature of sports and the limited control that stations and networks have over their timing create scheduling complications. ABC made a daring innovation in 1970 when it started scheduling *Monday Night Football* in network prime time. This scheduling risked devoting a long stretch of extremely valuable time to a single program with only selective audience appeal. And once the football season ended, replacement programs had to be found.

With the ABC Monday night exception, the major broadcast networks usually keep sports out of prime time. Regular-season sports generally get lower ratings than do prime-time entertainment programs but capture the largest audiences on weekends.

Over the years the broadcast and cable networks sometimes have paid for sports programming rights by sharing some of their advertising revenue with the sponsoring sports organization. In 1993 the broadcast networks entered into unprecedented arrangements for coverage of major sporting events. First NBC signed a four-year deal for NBA basketball at $750 million *plus* an agreement to share advertising revenues with the league once the network reached a certain sales level (about $1 billion). Next ABC and NBC announced a six-year contract for major-league baseball with no rights fees at all, but with an immediate sharing of advertising revenue from televised games. This deal dissolved after the 1995 season, when disappointing revenue triggered an escape clause in the contract.

Cable followed the revenue-sharing trend later in 1993 when TBS renewed its exclusive NBA basketball cable rights for four years (through 1997–1998) for $350 million, a 27 percent increase over the previous contract, plus agreement to split advertising revenues 50-50 after sales exceed that $350 million.

Sometimes a network may pay nothing up front for coverage of a sports event nor have to share advertising revenues exceeding a certain amount with the sponsoring sports organization. It was reported that ABC and ESPN were given free rights to the Women's World Cup soccer tournament in 1999 because the organization hoped the coverage would lead to the creation of a professional women's soccer league. In other instances the networks may be paid by the sponsoring organization for carriage of the sports event, with the organization selling advertising and sponsorship fees, or they may pay a rights fee but share in the profits generated by the event with the event's sponsor. The range of possible financial arrangements between the networks and sponsoring organizations is almost limitless.

The growth of cable has obviously had a major impact on the scheduling and financial structuring of sports programming. Unlike the broadcast networks, which must appeal to the largest possible audiences, the cable networks can cater to the special interests of smaller but highly desirable audiences. They have dramatically improved sports coverage not only of professional "big-event" sports but also of nonprofessional and "minor" sports. Viewers now see a variety of sports once largely ignored by the broadcast networks. Team sports such as volleyball, rugby, and soccer appear frequently. Individual-event sports such as track, gymnastics, tennis, golf, swimming, diving, and skiing receive unprecedented coverage. More automobile, motorcycle, and boat races are available. Even events such as dog and cat shows, that many would question as being sports, receive coverage.

Cable also merits credit for reducing the gender gap in sports programming. Although sports programming remains firmly focused on male competitors and targeted to male audiences, cable's increasing coverage of women's sports has demonstrated that these sports attract loyal fans of both genders.

Issues in Sports Broadcasting

Television has increased the popularity of sports and opened new opportunities for professional sports careers. The downside is that the enormous fees exacted for television rights have brought up troubling issues:

- When coverage of sports events moves to cable, and especially when it moves to pay-per-view, it becomes unavailable to millions of viewers.
- Astronomical player salaries, induced by television stardom and revenues, invite press and fan criticism.
- The millions of dollars in television rights fees that go to colleges have heightened the temptation to commercialize recruiting, to tolerate low graduation rates among players, and to manipulate college and NCAA rules.
- Television influences event scheduling and even how some games are played. Stations and networks sometimes try to overcome the unpredictability of sports events by staging them expressly for television or by presenting videotaped coverage of a presumably live event—practices that are innocent enough when the events are clearly labeled, less innocent when the manipulation is concealed.
- Over the protests of fans, East Coast games are played under lights to maximize Pacific-time audiences. Also controversial is the delay of college

basketball start times to 9:30 P.M. so that networks can carry double-headers on weekends. Such late starts mean that players and student fans arrive home as late as 2:00 or 3:00 A.M.

- Referees call arbitrary time-outs about every 10 minutes to accommodate advertising spots within football and basketball games. Such artificial breaks interrupt a team's momentum and undermine coaching strategies.
- Teams and sports associations often insist on controlling the hiring of play-by-play and color announcers. Critics regard such control as an illegal surrender of broadcaster responsibility for programs.
- Critics deplore beer sponsorship of televised sports, which inappropriately links the consumption of alcoholic beverages with athletic performance and encourages underage drinking.

9.4 CHILDREN'S PROGRAMS

Children's programming encompasses virtually all genres. Though cartoons dominate, most adult program types have their counterparts in children's programs.

In the early 1990s cable provided about 60 percent of total television hours devoted to children; Fox and independent stations about 17 percent; and ABC, CBS, and NBC only 3 percent. (Section 7.6, page 218, includes a discussion of children's programs on PBS, which supplied the remaining 20 percent.)

By the mid-1990s broadcasting's share of the children's programming audience had declined, falling to less than half the total number of children viewing. In 1997 broadcasters sometimes attracted only about a quarter of the audience watching children's programs. Broadcast television's share of this audience continued to decline as new cable networks catering to kids appeared in 1998 and 1999.

Broadcasting

Congress boosted children's programming in 1990 when it passed legislation requiring that broadcasters air an unspecified amount of programming designed specifically to meet children's needs. In 1996

the FCC clarified the obligation by mandating a minimum of three hours of educational or informational children's programs weekly between 7:00 A.M. and 10:00 P.M. Shows meeting the FCC's requirements are flagged on-screen with an E/I icon. Stations not meeting the requirement must explain their failure at license renewal time, as discussed in Section 13.9 (page 399).

For decades ABC, CBS, and NBC scheduled several hours of children's programming every week for their affiliates. This programming consisted mostly of Saturday morning animated cartoons. They were joined by the Fox Children's Network, which by 1994 had become the highest-rated supplier of children's programs on weekdays as well as Saturdays.

In 1992 NBC began a phaseout of its Saturday morning animated-cartoon line-up, moving instead to live-action programs that the network hoped would attract a somewhat older audience—so-called teens and 'tweens—as well as expanding its *Today* program to seven days a week.

Four years later CBS, suffering massive erosion of its Saturday morning audience, also considered quitting the kid business and focusing on teen viewers. Ultimately, CBS added four new shows to the Saturday morning schedule, but all were live-action programs rather than animated fare. Right in the middle of Saturday morning the network scheduled a two-hour block of *CBS News Saturday Morning*, followed by *The Sports Illustrated for Kids Show*. ABC, in spite of its ownership by family-oriented Disney, was down to fewer than a dozen hours of children's programming weekly by 1997. The WB, in contrast, doubled the number of hours devoted to kids, from 9 to 19, and Fox added seven more children's shows to its already kids-friendly schedule. UPN took a more conservative approach, scaling back to one hour its original plan to launch a new two-hour weekday program targeted at teens.

As the decade ended, the commercial broadcast networks attracted only a small fraction of the audience for children's programming. Of the original Big Three networks that once dominated all TV viewing, only ABC remained a competitive player in children's programming, with NBC out of the running and CBS pulling extremely low ratings. Fox and The WB remained important sources of over-the-air children's programming through the 1990s. The WB increased its share of the total kids audience from about 3 percent in 1998 to more than 8 percent in 1999 thanks largely to *Pokémon*, a Japanese-based cartoon series (*Broadcasting & Cable*, February 7, 2000). UPN and PAX TV are less significant broadcast network competitors in the battle for young viewers.

Pressure remains on the networks to help affiliates meet the FCC-imposed three-hour minimum of educational and informational programming for children. But reaching the kids is increasingly difficult because of competition from CD-ROM games, computer services such as America Online, and, most seriously, the cable networks.

Children's programs earn a relatively small percentage of broadcast television's total advertising revenues, but most run at times that might otherwise go unsold. Toy, candy, and cereal manufacturers have been the traditional sponsors of commercial children's programming. In the late 1990s the range of sponsors widened as commercials for clothing, personal-care products, fast food, and computer software became increasingly important.

Cable
• • • •

Traditionally a strong competitor for young viewers, Nickelodeon became the principal channel for kids in the mid-1990s. By 1997 Nickelodeon accounted for almost 60 percent of TV viewing by kids ages 2 to 11. Nickelodeon targets younger children in the daytime. After 8:00 P.M. (Eastern time) the network runs Nick at Nite, featuring a heavy schedule of classic off-network sitcoms such as *Happy Days*, *The Brady Bunch*, and *The Mary Tyler Moore Show*. Besides sitcoms, cartoons, and game shows, Nickelodeon also originates high-quality programs that avoid violence as entertainment and feature a broad range of role models. In 1996 Nickelodeon began another 24-hour channel, Nick at Nite's TV Land, featuring shows such as *Sonny and Cher*, *Hogan's Heroes*, and *Hill Street Blues*.

In 1992 Ted Turner launched his new 24-hour-a-day Cartoon Network. Many of its programs come from the Hanna-Barbera library, which includes old television series such as *The Jetsons* and *The Flintstones*. Turner also owns about 800 MGM half-hours, including *Tom and Jerry*. The merger of Turner with Time Warner in 1996 improved Turner's access to the well-stocked Warner Brothers cartoon library, further ensuring the Cartoon Network of programming well into the future.

In 1997 News Corporation CEO Rupert Murdoch acquired a 50 percent interest in International Family Entertainment, Inc., owner of The Family Channel. The following year the network relaunched as the Fox Family Channel, with a renewed emphasis on family-friendly programming and more programs especially for children.

Several cable channels offer children's programs that, while entertaining, have an educational or informational objective. The Learning Channel, an advertising-supported Discovery Communications, Inc., channel, programs commercial-free, nonviolent shows for preschoolers on weekdays. In 1996 Discovery Communications dramatically increased the number of children's programs on basic cable by launching two new advertising-supported networks: Animal Planet, featuring animal-related programs designed specifically for children, and the Discovery Kids Channel, a 24-hour programming service targeted at young people ages 7 to 14. Animal Planet is an ad-supported basic-cable network, while the Discovery Kids Channel was launched as a digital cable channel, one of several digital channels dealing with special-interest topics such as health, science, and aviation initiated by Discovery Communications. In early 1999 MTV Networks launched Noggin, a commercial-free digital cable network that targets children ages 2 to 11. Late in 1999 the Fox Family Channel launched two gender-specific, ad-supported digital cable networks. The girlzChannel programs for girls ages 2 to 14 during the day, and for their parents during the evening. The boyzChannel follows the same strategy, but the programming is designed specifically to interest boys.

Children's programming also is available on the pay-cable channels. HBO offers some children's programming, and HBO Family caters specifically to the special interests of children during the day, shifting the programming to target older demographic groups with family-friendly fare in the evening.

Long a premium-cable channel, the Disney Channel began migrating to basic cable during the mid-1990s but remained a commercial-free service. The Disney Channel programs for children in the daytime and for the family in the evenings. After years of rumors about a new basic-cable network, in 1998 Disney finally launched Toon Disney, which showcases the company's vast library of animated programs stretching back to the 1930s.

Using materials in the Disney library dramatically reduced the financial risk of launching a new cable network catering to kids at a time when the number of competitors for those viewers was increasing. In the 1990s Disney also expanded its operation to offer Disney Afternoon, a two-hour block of syndicated programs for broadcast stations. The lack of available time slots during peak children's viewing periods, caused Disney to shorten the package of rotating cartoons to 90 minutes in 1997.

Industry consolidation and increased vertical integration of production and distribution during the 1990s meant that children's program syndicators faced increasing difficulty selling their products to individual stations as the decade ended. Ironically, the FCC mandate requiring stations to program three hours of children's programming weekly may have done syndicators more harm than good. Shows satisfying the FCC's criteria were often not those most popular with kids, prompting programmers to place them in unfavorable time slots and thus almost ensuring their commercial failure.

Children's Television Issues

Children have such easy access to television, they consume so much of it, and it exerts such a powerful hold on their attention that society has a special stake in the quality of programs made especially for them.

Most foreign countries regulate children's programs in considerable detail; some forbid advertising to children altogether. The FCC, however, has imposed somewhat less regulation (see Section 13.9, page 399). Advocates of greater control raise issues such as these:

- the inappropriateness of much television viewed by children in terms of their needs and vulnerabilities
- the negative impact of violent and aggressive program content
- the absence of a wide range of suitable role models on television
- exploitation of children by advertisers, especially by those that encourage eating candy and sugarcoated cereals
- the shortage of age-specific programs, especially for very young children

Congress reflected concern about these issues in the Telecommunications Act of 1996 by requiring that manufacturers install a V-chip (*V* for violence) in new television receivers that parents can use to block programs containing objectionable content. The V-chip senses codes transmitted by the broadcast station or cable network that reflect ratings assigned to each program. The law gave manufacturers two years to begin installing the chip and gave program distributors a year (until February 1997) to devise their own ratings system or have one developed by an FCC advisory committee. The broadcast and cable industry reluctantly chose to develop its own system rather than have one imposed on it, as discussed in Exhibit 9.d.

Many TV critics and the Clinton administration hailed the V-chip legislation, and its associated program rating codes, as an important step in providing parents the tools needed to exercise greater control over their children's viewing. Skeptics questioned whether the legislation would achieve its goals. Content creators and distributors viewed the new requirements as infringing on their First Amendment rights.

As previously noted, in 1996 the FCC adopted rules mandating that broadcast stations air a minimum of three hours of educational children's programming weekly. This new regulation and the V-chip legislation were efforts to address a growing concern that much TV content has a harmful effect on children. (See Section 11.7, page 337, for additional discussion of the effects of televised violence.)

9.5 NETWORK TV NEWS AND PUBLIC-AFFAIRS PROGRAMS

The strict network separation of news and public-affairs programs from entertainment began to erode as competition for audience attention increased beginning in the 1970s and continuing through the 1990s. The networks reacted by squeezing more value from their heavy investment in their news division—more frequent news and newsmagazine programs and more syndication of network news product to affiliate and foreign organizations. At the same time, and deplored by

EXHIBIT 9.d

• • • • • • • • • • •

The TV Parental Guidelines

Late in 1996, under political pressure to do so, the television industry adopted this six-category age-based rating system for programs. The system allowed programs to be rated both as series and as individual episodes. The ratings applied only to entertainment programs, not news, sports, or home shopping channels. Infotainment shows received ratings unless they involve hard news stories or interviews with recognized news figures such as the first lady. Program guides began publishing the ratings, and they appeared superimposed on-air at the beginning of programs for 15 seconds.

Critics immediately objected to the system. Most, including outspoken Congressman Edward Markey (D-Mass.), said that the ratings lacked the specificity needed by parents, arguing instead for a system that ranked each program according to the levels of violence, sex, and inappropriate language the program contained. Objections increased when most prime-time shows received a TV-PG rating. Critics said that TV-PG was an especially vague category that neither set age limits nor provided guidance regarding the level or specific type of objectionable material in the program. Television industry leaders, saying it was a work in progress, asked critics to give the system a fair chance before adopting a government-designed alternative. But the outcry was too broad and too shrill to ignore.

After only seven months, the cable and broadcast industry, except for NBC, agreed to add content warnings for violence (V), sexual content (S), strong language (L), suggestive dialogue (D), and fantasy violence (FV) to the existing rating categories. The cable network Black Entertainment Television (BET) refused to rate its programs at all. Additionally, it was agreed that the ratings symbols would be made larger on the screen for easier reading, but the time on-screen would remain the same as before, 15 seconds. The new ratings system went into effect October 1, 1997.

Unedited movies on pay-cable channels are not rated under this system. Instead, they carry the original movie-theater rating assigned by the Motion Picture Association of America (MPAA).

The following programs are designed solely for children:

TV-Y   Appropriate for all children. The themes and elements in this program are specifically designed for a very young audience, including children ages 2 to 6. This program is not expected to frighten younger children.

TV-Y7   Designed for older children, age 7 and up. This program is appropriate for children who have acquired the developmental skills needed to distinguish between make-believe and reality. Programs in which fantasy violence may be more intense or more combative than in other programs in this category will be designated **TV-Y7-FV.**

The following programs are designed for all members of the audience:

TV-G   General Audience: This program is appropriate for all ages. It contains little or no violence, no strong language, and little or no sexual language or situations. Most parents would find this program suitable for viewing by children of all ages.

TV-PG   Parental Guidance Suggested: This program contains material that parents may find unsuitable for younger children. The theme itself may call for parental guidance and/or the program contains one or more of the following: moderate violence (V), some sexual situations (S), infrequent coarse language (L), or some suggestive dialogue (D).

TV-14   Parents Strongly Cautioned: This program contains material that many parents would find unsuitable for children under the age of 14. The program contains one or more of the following: intense violence (V), intense sexual situations (S), strong coarse language (L), or intensely suggestive dialogue (D).

TV-MA   Mature Audiences Only: This program is specifically designed to be viewed by adults and therefore may be unsuitable for children under 17. This program contains one or more of the following: graphic violence (V), explicit sexual activity (S), or crude indecent language (L).

critics, show-biz culture invaded news, making some news offerings increasingly superficial.

Broadcast News
• • • • • • • • • •

Back in 1963 CBS and NBC expanded their prime-time news programs from 15 minutes to a half-hour (actually only about 22 minutes, after time for commercials and openings and closings is subtracted). The text of an entire half-hour network newscast fills less than a single page of a full-size newspaper. Nevertheless, moving to the half-hour format played a role in elevating network television to the status of the country's most widely accepted news source.

A CBS movement to expand the evening news to a full hour was defeated in 1975 by affiliates who wanted to retain both their own highly profitable local newscasts and the 7:00 to 8:00 P.M. access hour for revenue-producing syndicated fare. CBS, unwilling to invade network prime time with news at the expense of entertainment programs, gave up the plan.

In 1979 ABC started an experimental late-night network news program, *Nightline*, featuring Ted Koppel. The experiment became a permanent half-hour at 11:30 P.M. Each evening Koppel concentrates on one or two current news stories.

In 1992 Fox announced plans to establish a national news operation. The following year, however, the network backed away from its national news ambitions, although it did expand its local news presence on its owned-and-operated stations.

In 1995 Fox concluded a deal with the international news service Reuters to provide news feeds for its stations. In the mid-1990s stations with strong local news operations became Fox affiliates, and in 1997 Fox announced that it was making local news programming a condition for continued network affiliation. Fox helped affiliates develop local news operations by creating Fox NewsEdge, a service unit that provides news content to affiliated stations. ABC NewsOne, CBS Newspath, and NBC News Channel provide a similar service to their affiliates. The networks and many of their affiliates also subscribe to news services such as CNN Newsource, Reuters TV (RTV), and Associated Press Television News (APTN).

Cable News Networks
• • • • • • • • • • • • • •

Ted Turner launched Cable News Network (CNN) as a fourth major television news service in 1980. With a 24-hour schedule to fill, CNN can supply in-depth reporting as well as continuous coverage of breaking news stories.

Early doubts about CNN as a serious news service began to disappear when in 1986 it was the only network to carry live coverage of the explosion of the space shuttle *Challenger*. Those doubts dissolved completely in January 1991 as government officials and even ABC, CBS, and NBC turned to CNN for its coverage of the war in the Persian Gulf. In 1992, after having received abysmal ratings in 1984 and 1988, the major networks conceded to CNN, PBS, and C-SPAN gavel-to-gavel coverage of the Democratic and Republican conventions, limiting their own coverage to a few hours each day.

Between news roundups, CNN schedules interviews (*Larry King Live* is the best known), features on managing money, stock market analyses, sports news, and public-affairs discussions. Today more than 77 million cable and DBS television subscribers have access to CNN.

Turner's companion news service, CNN Headline News, provides news headlines and frequent updates in continuous half-hour cycles. It resembles all-news radio with pictures and reaches more than 72 million subscribers.

Other news channels joined CNN on cable in the mid-1990s, including CNN International (CNNI) in 1995. Available in Europe since 1985, CNNI is a 24-hour global news and information channel serving more than 150 million subscribers worldwide. In 1996 News Corporation launched the Fox News Channel, and NBC collaborated with Microsoft to start MSNBC, a cable network with an exceptionally strong Internet presence. Heated public clashes between CNN's Ted Turner and News Corporation's Rupert Murdoch regarding cable system carriage made the launch of Fox News Channel especially memorable.

CBS launched Eye on People in 1997 with only about 2 million subscribers. CBS did not bill Eye on People as a news network but rather as an entertainment and information network focusing on people and personalities. In spite of early promises to allow Eye on People to operate at a loss for up to five years while adding subscribers, CBS sold the network in 1999 to Discovery Communications, Inc.

Informational Networks

Many cable networks provide informational programming. NBC's Consumer News and Business Channel (CNBC) moved to become a major cable player when it acquired its prime competitor, Financial News Network (FNN), in 1991. A strong economy and a raging bull stock market during much of the 1990s made CNBC popular with investors wanting the latest financial information. At the end of the decade CNBC reached almost 70 million U.S. subscribers (only about 5 million fewer than cable icon CNN) and about 160 million subscribers worldwide.

Drawing upon its considerable news resources, CNN began a competing financial news network, CNNfn, in 1995. Five years later CNNfn trailed well behind CNBC in number of subscribers. CNNfn suffered a major blow in 1999 when Lou Dobbs, veteran host of the network's popular *Moneyline News Hour*, left to pursue other interests.

Also launched in 1995, Bloomberg Information Television provides financial and other news featuring a multiscreen format with a newscaster in the upper corner and text filling the remainder of the screen. Court TV features live coverage of courtroom trials from around the country, with analysis and commentary by experienced legal journalists. Court TV's ratings skyrocketed during the O. J. Simpson criminal trial.

The Weather Channel was among the first (1982), and remains one of the "purest," information channels. Many others offering healthy doses of information along with entertainment have either joined it or announced launch dates. They include The Learning Channel; the Food Network; The Travel Channel; Home & Garden Television; the Hobby Craft Network; The Love Network (a planned education channel regarding relationships, self-esteem, and self-improvement); and the Recovery Network (information for people dealing with alcohol and other chemical dependencies).

Public-Affairs Programs

News and public-affairs programs tend to overlap, but the Federal Communications Commission makes a distinction, defining "public affairs" as local, state, regional, national or international issues or problems, including, but not limited to, talks, commentaries, discussions, speeches, editorials, political programs, documentaries, minidocumentaries, panels, roundtables and vignettes, and extended coverage (whether live or recorded) of public events or proceedings, such as local council meetings, congressional hearings and the like.

The Commission stressed public-affairs programs because of the traditional view that broadcasters in a democracy have a special obligation to serve the needs of citizen-voters.

Most commercial broadcast television networks and most large stations maintain at least one weekly public-affairs discussion series, sometimes also a news documentary series, and often minidocumentaries within newscasts (especially during sweeps periods).

One of the most striking developments in network public-affairs programs has been the continuing success of *60 Minutes*, the weekly CBS magazine-format documentary series (Exhibit 9.e). It ranks as one of television's most-watched programs. That success violated all conventional wisdom, which said that documentaries repelled the mass audience. One reason previous documentaries had low ratings, aside from not being as well done as *60 Minutes*, is that the networks usually scheduled them in unfavorable time slots and denied them the long-term stability they needed to build an audience.

After years of wandering through the CBS schedule, *60 Minutes* finally arrived at a good hour—a byproduct of the prime-time access rule, which left the Sunday 7:00 to 8:00 P.M. time slot open for nonentertainment network programs. Another factor may have been the CBS counterprogramming strategy of scheduling *60 Minutes* against children's programs and movies.

The success of *60 Minutes* generated corporate pressure to expand the program beyond its single hour on Sunday night. For years reporters and producers associated with *60 Minutes* resisted expanding the program to other nights, fearing that the new program would not live up to its reputation for quality. Finally, in 1999, CBS launched another hour of the program, calling it *60 Minutes II*. In addition to original reporting, *60 Minutes II* includes fresh reports on classic *60 Minutes* stories and updates on stories presented on the Sunday program that are of particular interest or urgency.

EXHIBIT 9.e
• • • • • • • • • • •

60 Minutes

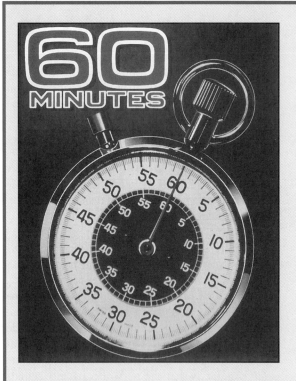

A stellar team of correspondents, originally consisting of Mike Wallace, Harry Reasoner, and Morley Safer, contributes to the *60 Minutes* success story. As a *New York Times* commentator put it,

> their gray or graying hair, their pouched and careworn countenances, the stigmata of countless jet flights, imminent deadlines, and perhaps an occasional relaxing martini, provide a welcome contrast to the Ken and Barbie dolls of television news whose journalistic skills are apt to be exhausted after they have parroted a snippet of wire service copy and asked someone whose home has just been wrecked by an earthquake, How do you feel? (Buckley, 1978)

Later, Dan Rather joined the team; later still, Ed Bradley and Diane Sawyer improved its ethnic/gender balance. Rather left the show in the mid-1980s to anchor *CBS Evening News*, the most prestigious news position a network can offer. In 1989 Sawyer moved to ABC, which gave her a multimillion-dollar contract, and in 1991 Harry Reasoner retired. Today Leslie Stahl and Steve Kroft, together with commentator Andy Rooney, round out the team.

The magazine format allows the program to treat a great variety of subjects in segments of varying lengths. With executive producer Don Hewitt and a staff of some 70 producers, editors, and reporters, the *60 Minutes* team develops about 120 segments annually, each of which, on the average, requires 6 to 10 weeks to produce. The *ambush interview*, a Mike Wallace specialty, adds drama to investigative reports. Presenting his victims on-camera with damning evidence of wrongdoing, Wallace grills them unmercifully. The victims' evasions, lies, and brazen attempts to bluff their way out of their predicaments—often captured through extreme-close-up lenses—fascinate some viewers, while others question these tactics. Tabloid TV programs have imitated the Wallace-type confrontation to the point of turning it into a cliché.

60 Minutes is generally acknowledged to be the most successful news program—in terms of both ratings and revenues—in television history. It may well be the most successful of *any* program, news or otherwise. It is the only broadcast to have finished in the top 10 Nielsen ratings for more than 22 consecutive seasons, and the only program to be rated number one in each of three different decades. The program finished the 1998/1999 season ranked seventh in the Nielsen ratings.

The success of *60 Minutes* caused network executives to pressure the producers to expand the program to other nights. But as clones sprouted on competing networks, *60 Minutes* continued to maintain its unique position as network television's most prestigious newsmagazine. Those who had built the program's journalistic reputation, especially Don Hewitt, staunchly resisted expanding the program, expressing concerns about harming the brand name by offering a show that did not live up to the original's high standards. By the late 1990s, however, NBC's *Dateline* and ABC's *20/20* were nipping at the heels of *60 Minutes* and threatening its position as TV's highest-rated newsmagazine. The commercial success of *Dateline* and *20/20*, both airing multiple evenings each week, put even more pressure on *60 Minutes* executives to expand the program. In January 1999 CBS launched *60 Minutes II*, the long-awaited spinoff of the network's showcase news program. Hewitt and others with strong professional and emotional ties to *60 Minutes* gave the new program their public blessing, but some believed that they still had reservations. During its first year on the air *60 Minutes II* frequently rated as one of prime-time television's top 30 programs.

Source: CBS Photo Archive.

In 1988 CBS started *48 Hours*, another public-affairs series, which turns cameras on a single topic, such as a hospital or an election. ABC's *20/20*, a magazine-format show similar to *60 Minutes*, is a respected and successful prime-time program. In the late 1990s the show gained increased popularity, due in part to the successful chemistry of its hosts, Barbara Walters and Hugh Downs, both of whom began their careers at NBC. Downs retired in 1999, but the show's success prompted the network to continue programming it as often as four nights a week.

NBC had difficulty developing a prime-time newsmagazine, but not for want of trying. It attempted 18 times to make the format work, beginning with *First Tuesday* in 1971, before 1992's *Dateline*, with former *Today* anchor Jane Pauley and former *20/20* correspondent Stone Phillips, attracted enough viewers to survive. *Dateline* eventually went on to become one of NBC's most successful programs. NBC gradually expanded the program until, by 1999, the show was sometimes on five nights a week.

Fox did not enter the arena until 1993, when it launched its first prime-time newsmagazine, *Front Page*. In May 1994 Fox canceled the show and aborted others—variously named *Assignment* and *Full Disclosure*—before they even went on the air.

The networks favor prime-time newsmagazines because, compared with most entertainment programs, they are relatively inexpensive to produce. They often use and reuse resources already in place in network news divisions. However, many of them compete for the same stories and for the same audience. Still, the 1990s were very good to prime-time newsmagazines. Often the various episodes of *Dateline*, *20/20*, and *60 Minutes/60 Minutes II* were among the top-rated programs on the broadcast networks. Even when they do not rank among the top 10 or even top 30 programs, their low production costs allow the networks to offer prime-time advertising opportunities at attractive rates while still guaranteeing the network a comfortable profit. And because they are produced in-house, the networks do not face increased demands for higher licensing fees when a show becomes successful.

ABC, CBS, Fox, and NBC schedule public-affairs question-and-answer sessions with newsworthy figures on Sunday mornings. NBC's *Meet the Press* started in 1947 and is the oldest continuously scheduled program on network television. *Meet the Press* made history in 1997 when it became the first live network television program broadcast in digital high definition. CBS launched *Face the Nation* in 1954, and ABC began *Issues and Answers* in 1960, later replacing it with *This Week with David Brinkley*. Brinkley retired in 1996, and the title was shortened to *This Week*. In the mid-1990s Fox joined the older networks with *Fox News Sunday*.

Beginning in 1979 *Sunday Morning* provided a vehicle for CBS's much-loved Charles Kuralt. Although the show has never drawn high ratings, the time slot suited Kuralt's leisurely, reflective, low-key style. Many considered the program to be one of the very best on television. After 37 years with CBS, 15 of them as *Sunday Morning* anchor, Kuralt retired in 1994. Charles Osgood, who followed him on the anchor stool, said, "Although I am succeeding Charles Kuralt, I know that I cannot replace him. Nobody could" (*Miami Herald*, 2 April 1994: A2).

Public Affairs on Cable

In 1979 cable developed a unique public-affairs vehicle specifically to cultivate a positive image for its industry. C-SPAN (Cable Satellite Public Affairs Network), a nonprofit corporation, originates a full 24-hour public-affairs service. Together with its companion, C-SPAN II, it offers live coverage of congressional floor sessions, hearings, political conventions, and other informative programs.

Although C-SPAN and C-SPAN II contribute to cable's positive image by offering perhaps the purest form of public-affairs programming available anywhere, the same cannot always be said of other cable programs that sprang up in the wake of the O. J. Simpson trials, the Clinton-Lewinsky affair, and President Clinton's subsequent impeachment trial. Interview-discussion programs such as *Hardball with Chris Matthews* and *Rivera Live!* filled many prime-time hours on CNBC late in the decade. Fox News Channel scheduled *The O'Reilly Factor* and *Hannity & Colmes* on weekday evenings and during various dayparts on weekends. On Sundays during prime time Fox added *The Beltway Boys* and *Drudge*, hosted by Internet muckraker Matt Drudge. These programs often focused on the same issues as their broadcast network counterparts, but the hosts exhibited more "attitude" in their approaches to both the topics and guests appearing on the shows. Interviewees frequently became embroiled in heated arguments with each other and with the hosts. The hosts claimed they were simply discarding the hypocritical mask of objectivity worn by more established broadcast reporters to reveal the grittier, more emotional side of issues glossed over by traditional public-affairs programs. Critics accused the shows of trivializing serious issues, of selecting guests based more on their celebrity (or notoriety) than their expertise, and of exploiting conflict for its own sake, further blurring the line between journalism and entertainment.

9.6 SYNDICATED TV PROGRAMS

The mechanics of syndication and how stations and cable television use syndicated programs are discussed in Section 8.2 (page 230). Here we focus on the syndicated programs themselves, divided into three categories: off-network programs, first-run programs, and theatrical films (movies). Some additional syndicated material comes from religious and noncommercial sources.

Off-Network Syndication

After having completed their contractual network runs (usually two showings), network entertainment programs revert to their owners, who then make them available for licensing to broadcast stations or cable networks as off-network syndicated programs.

Increasingly, early episodes of network series go into syndication while new episodes are still being produced for the network. Off-network programs are highly prized, because their previous exposure on networks confers recognizability and a ratings track record. An off-network series newly released in syndication typically earns a rating about 10 points lower than it earned in its initial network appearance. On cable networks, ratings for such programs usually fall still lower.

Most stations schedule half-hour syndicated series because they attract a desirable demographic group (typically women 18 to 49 years) and permit scheduling flexibility. Hourlong programs, on the other hand, do not attract many new viewers during their second 30 minutes.

Sitcoms attract both children and adult women. Sixty-minute action/adventure series have had some success in late afternoon because of their male demographics. Action/adventure shows are also used on weekends to lure and retain hard-to-reach male viewers before, between, and following sports programs.

Cable networks have found that some hourlong dramatic series appeal to cable viewers in access-time, prime-time, and late-fringe-time periods. FX, for example, successfully stripped *NYPD Blue*, and USA stripped *Walker, Texas Ranger*—both in prime time.

First-Run Syndication

Shows that never appeared on a network—*first-run* programs—are brand new when first sold to stations. They are produced especially for the syndication market. Casual viewers can hardly distinguish the best first-run syndicated programs from network programs. For example, the daily syndicated magazine show *Entertainment Tonight* (described in Exhibit 9.f) is so timely and slickly produced that it could easily be taken for a network program.

By the mid-1990s producers were offering even high-priced dramatic programs in first-run syndication, such as *Hercules* and *Star Trek: Deep Space 9*. Sometimes programs discarded by the networks later turn up in first-run syndication (*Baywatch*, for example, which eventually became the most-watched program worldwide).

Producers of first-run syndicated programs most often employ low-budget genres such as quizzes and games. The reigning king of first-run programs, *Wheel of Fortune*, typically gets higher ratings than any competition, including off-network shows.

Syndicators and stations also favor first-run talk shows. Most find homes in morning or afternoon time slots. *Oprah* ranks at the top of the syndicated talk program ratings ladder. In the late 1990s *The Jerry Springer Show* generated both ratings and controversy by featuring sensational topics and bizarre guests who invariably came to blows.

A much-discussed first-run syndicated program genre emerged in the late 1980s. Known variously as *reality shows*, *trash TV*, and *tabloid TV*, these shows have in common a pseudo-journalistic approach to real-life topics, usually items currently or recently in the news. Such programs cater to morbid inter-

EXHIBIT 9.f

Entertainment Tonight

Entertainment Tonight (ET) revolutionized the syndication business by proving that expensive, original, nonnetwork programs—other than game shows—could be profitable for stations as well as for producers and syndicators. The economics of such an effort

are all the more unusual in that each episode is an original—with no reruns to help spread out the cost.

Introduced in the late 1970s, *ET* took off in 1981 when Group W began delivering the topical show by satellite. Mary Hart joined *ET* in 1982, and John Tesh arrived in 1986. Tesh later left the show to develop his career as a musician and composer. Bob Goen now shares anchor duties with Hart.

Entertainment Tonight provides a classic example of the constant innovation essential to keep a series from going stale. Initially, the program capitalized on the audience's appetite for gossip, personality exploitation, and show-biz fluff. Later the producer countered the lightweight image by introducing brief think pieces, some hard news, and more in-depth stories. By the 1990s, to compete with a glut of tabloid shows, the program had begun to turn more and more to sexually oriented items.

Among the best elements of the show are the contributions of Leonard Maltin, who became a regular contributor in 1982 and obviously knows and loves "the business." His consistently perceptive pieces are as rare on popular commercial television as is an episode of *Entertainment Tonight* that uses neither the word *exclusive* nor the phrase *behind-the-scenes*.

est in the sensational and the bizarre and exploit scandals, sex, and violence. Their blending of entertainment and journalism were symptomatic of a broad trend toward treating news from an entertainment perspective. *A Current Affair* started the genre in 1988 and remained in production through 1996. Its success inspired clones such as *Inside Edition*, *National Enquirer TV*, and *Extra*. Other reality-based programs include *Rescue 911*, *Cops*, and *Survivor*, CBS's controversial prime-time game show.

In the late 1990s reality programming moved into the courtroom, or at least a TV studio designed to look like one. By 1999 there were about a half-dozen programs available in first-run syndication based on real cases being tried by real judges. Most shows assumed the name of the presiding judge, an indication of the importance of the judge's person-

ality to the show's success. Unlike the O. J. Simpson murder trials, which gave many Americans their first view into an actual courtroom, these shows focused on less weighty issues. An episode of *Judge Mills Lane* involving the distribution of jointly owned property following the breakup of a lesbian couple is representative of the cases heard.

Movies

In addition to off-network and first-run programs, television syndicators offer movies, singly or in packages. Distributors often sell "all or none" packages—a questionable practice resembling block booking, a practice legally banned in the motion-picture industry. The buyer has to accept second-rate movies to get the more desirable films in the group of films sold as a package. Broadcast network affiliates usually strip movies in late-night and Saturday afternoon slots. Independent stations and cable networks are more likely to schedule them in prime time and on Sundays.

9.7   RADIO NETWORK AND SYNDICATED PROGRAMS

Competition from television devastated the original radio networks as full-service program suppliers, but they survived by reducing their role to that of supplements to radio music formats adopted by their affiliates. In the 1970s new networks began to offer their own music formats, made possible by inexpensive satellite relays.

Statistical Research, Inc., which twice each year conducts its radio network listener survey called Radio's All-Dimension Audience Research (RADAR —see Chapter 10), defines a network as a program service that has continuity of programming; written, contractual agreements with its affiliates; the capability of an instant feed to all affiliates; and a clearance system so it can determine which affiliates carry which programs. This does not keep any number of station alliances from calling themselves "networks." RADAR reports refer only to those satisfying the definition devised by Statistical Research. Tracking network radio is difficult because "radio networks" form and dissolve frequently to better target a desired audience.

Some group owners have "network affiliates" nationwide, thanks to consolidation. A program carried by the more than 800 radio stations owned by Clear Channel Communications, for example, would enjoy national coverage similar to that of the broadcast TV networks.

Network News

Radio networks associated with television networks get the benefit of daily newscasts voiced by well-known television news personalities. Some television journalists anchor radio news bulletins such as morning-drive reports; others prerecord news stories for later inclusion in scheduled radio newscasts.

ABC Radio Networks has five full-service networks, each targeted to a particular demographic group and each designed to complement the affiliated station's particular program format. Fox News Radio is the newest TV-related network providing radio news services. Fox News Radio provides one-minute news updates during the morning drive time, coverage of crises and breaking news stories, and features by Fox News Channel personalities such as Brit Hume and Paula Zahn. The CNNRadio Network provides news, sports, business, and feature reports to about 1,900 affiliated radio stations. CNNRadio Noticias is a full-service Spanish-language network with affiliates throughout Latin America and the United States. About 750 stations use Associated Press Network News, and more than 50 take AP's All News Radio, a 24-hour news radio network.

Network radio news is noted for its news features hosted by TV anchors and reporters such as Peter Jennings, Dan Rather, and Charles Osgood. However, the most popular news reporter/commentator is Paul Harvey, who is heard exclusively on radio (see Exhibit 10i, page 305). Famous for his distinctive phrasing and use of pauses, Harvey began his radio career in 1933 at KVOO-AM in Tulsa, Oklahoma, while still in school. Harvey's *News and Comment* program is carried by more than 1,200 radio stations and heard by more than 19 million people. In 1976 Harvey added *The Rest of the Story*, a feature based on little-known or forgotten facts about famous people and events. In 1998 Harvey won the National Association of Broadcasters' prestigious Marconi Award as Network Personality of the Year. He had previously received Marconi Awards in 1989, 1991, and 1996.

Talk Shows and Sports

The Mutual Broadcasting System pioneered all-night talk on network radio with *The Larry King Show*, predecessor of the CNN television program *Larry King Live*. Later Art Bell began a late-night radio talk program heard on more than 400 stations. Bell's show specialized in the paranormal and often featured guests who claimed to have sighted UFOs or met extraterrestrials. Bell originated the program from his home in Pahrump, Nevada, rather than a traditional broadcast studio.

In the 1990s talk radio turned strongly conservative. Best known of the conservative talkers is Rush Limbaugh. Oliver North and G. Gordon Liddy also attract devoted supporters. Both men owe their celebrity status to political scandals, North for the sale of arms to Iran during the Reagan administration, and Liddy for the Watergate break-in during the Nixon era. Liddy also appears in *18 Wheels of Justice*, an action-adventure show on cable's TNN. In 1999 another controversial figure, former Speaker of the House Newt Gingrich, joined the conservative chorus by offering a daily 90-second feature program through Premiere Radio Networks. Gingrich's program was not popular, however, and was canceled in February 2000.

Some people think that Limbaugh and other conservative hosts helped Republicans take control of the House of Representatives in the 1994 election. Their political influence seemed diminished by 1996, although Limbaugh continued to draw a devoted audience. In addition, new revelations about possible government mismanagement leading to the Branch Davidian disaster in Waco, Texas, combined with the Clinton-Lewinsky affair and Clinton's impeachment, breathed new life into conservative talk radio during the last years of the decade.

There are few blatantly liberal radio talk programs, but some programs, such as one hosted by Jim Bohannon, often deal with current events, politics, and popular culture minus the conservative tone characteristic of much talk radio.

Many radio talk shows cater to interests other than politics and current affairs. Dr. Laura Schlessinger, for example, offers culturally conservative advice to callers having relationship problems with spouses, children, in-laws, and bosses, while Bruce Williams specializes in helping callers with financial concerns. There are also shows dealing with computers and the Internet, health and fitness, automobile maintenance, gardening, home repairs, and a variety of other topics. Talk America Radio Networks offers two services that provide more than 400 affiliates with talk programming 24 hours a day, 7 days a week.

Several radio talk show hosts have also tried working in television. In 1994 King dropped his syndicated radio show, electing instead to simulcast on Westwood One the audio portion of his CNN television program. Westwood One now offers radio stations a daily two-minute digest of the previous evening's *Larry King Live* program. Both Limbaugh and Paul Harvey also migrated to television, but their styles are more effective on radio.

The sports component of radio networks is sufficiently important to affiliates for CBS to pay $50 million for radio baseball rights for the early 1990s, nearly doubling the previous contract payment. The importance of radio sports became even more evident with the advent in 1992 of ESPN Radio, one of the services offered by ABC Radio Networks. ESPN Radio started as a weekend program provider but expanded to daily programming in 1998. The following year ESPN Radio became a 24-hour-a-day, 7-day-a-week service with about 375 affiliated stations, more than 70 of them programming ESPN Radio full time. ESPN Radio offers network coverage of National Basketball Association games and major-league baseball.

CBS-managed Westwood One and its associated radio networks offer a wide variety of sports programming. In the late 1990s Westwood One/CBS Radio Sports offered NFL games, including the NFL playoffs, the AFC and NFL Championship Games, the Super Bowl, and the Pro Bowl. In addition, Westwood One/CBS Radio Sports offered college basketball with regular-season Game-of-the-Week broadcasts, plus several conference championships, the NCAA Championship Tournament, and the Men's and Women's Final Four. Westwood One also provided network radio coverage of the NHL All-Star Game, NHL Conference Finals, and Stanley Cup Championships, as well as golf's Masters Tournament, U.S. Open, U.S. Senior Open, and U.S. Women's Open.

Music Networks
• • • • • • • • • •

Westwood One is a major provider of network music. Stations can choose from among formats such as Adult Rock & Roll, Bright Adult Contemporary, Soft Adult Contemporary, Adult Standards, Hot Country, Mainstream Country, and two oldies formats. ABC's Galaxy Radio Network is a 24-hour-a-day programming service offering music formats ranging from rock to golden oldies to country. Some providers offer one specialized music format. Reach Radio Network, for example, provides only southern gospel music full time via satellite.

Syndicated music programs for radio are provided by the networks and independent syndicators. Westwood One, for example, has *Out of Order*, a two-hour countdown show hosted by KROQ-FM (Los Angeles) disc jockey Jed the Fish. ABC Radio Networks syndicates *American Gold with Dick Bartley*, a four-hour show targeted to aging baby boomers and featuring music from the 1960s and 1970s. *Out of Order, American Gold*, and similar countdown programs grew out of *American Top 40*, a program hosted by legendary radio personality Casey Kasem that first aired in 1970. Kasem's countdown technique, incorporating brief stories about artists and their music, eventually evolved into multiple programs with different formats to accommodate a fragmented Top 40 audience. Kasem moved the show to Westwood One in 1989 after running on ABC for nearly 20 years, but he had to change the name because ABC retained the rights to *American Top 40*. Nine years later, amid lawsuits filed by Westwood One to prevent his departure, Kasem jumped to a network operated by AMFM (now Clear Channel Communications), where he once again hosted the syndicated program under its original name.

Commercial radio networking and syndication are likely to appear indistinguishable to listeners, but each has its own type of content, delivery means, advertising procedures, and payment practices. The major networks supply news, sports, and specials, or all-talk or all-music, relaying programs to affiliates by satellite, accompanied by national advertising. Syndicators concentrate on popular music formats and features. Radio networks often pay compensation to major-market affiliates, but radio syndicators usually charge for their programs. Often syndicated programming is offered through a barter arrangement, with the station agreeing to broadcast commercials inserted into the program by the syndicator in exchange for use of the program.

The availability of music programming from the networks and through syndication allows individual stations (or, increasingly, a group of co-owned stations) to adjust their formats rapidly as listener or advertiser preferences shift in local markets. Because of industry consolidation, there is increasing pressure on stations to use the music programs offered by co-owned networks and syndication organizations.

9.8   LOCALLY PRODUCED TV PROGRAMS

Both broadcasting and cable disappointed many who saw them as playing a strong role in giving opportunities for producing local programs. The economics of program production, however, inevitably favored program centralization—networking and syndication.

Local News Production

The one program category to escape the centralizing tendency is local/regional news. Audience interest in local television newscasts escalated during the 1970s, converting news from a loss leader into a profit center. It is not unusual for a network affiliate to make up to half its annual profit from advertising during local newscasts. It helps if the affiliate is the top-rated news station in the market, but even a modestly successful news operation attracts sponsors who would not otherwise advertise.

In the 1970s major-market stations developed their own investigative reporting and documentary units. Satellite technology enlarged the reach of local stations even further, minimizing time and distance constraints. By the mid-1980s large-market stations routinely dispatched local news teams to distant places to get local angles on national news events. They send, or *backhaul*, live stories back to the home base via satellite.

Multimillion-dollar budgets for local television news departments became commonplace in large markets, enabling stations to invest in high-tech equipment such as that shown in Exhibit 9.g. As the new century dawned, local news executives braced themselves for the expensive conversion to digital news gathering technologies. The first news vans equipped to transmit digital signals back to the station carried price tags starting at around $100,000, not including the costs of digital cameras and portable editing equipment.

Local news departments sometimes capitalize on the availability of home camcorders to turn amateur photographers, at least temporarily, into *stringers* (self-employed professionals who sell individual stories to radio and TV stations). Stations have used amateur footage of everything from

EXHIBIT 9.g

Electronic News Gathering Equipment

This lightweight ENG (electronic news gathering) van is typical of those used by local stations to cover live news events. The device atop the van is a microwave antenna for relaying sound and pictures back to the studio.

Many metropolitan stations use helicopters to cover news events when ground-based units cannot do so or when an aerial perspective is more desirable. A dramatic example was the helicopter coverage of police "chasing" O. J. Simpson's white Bronco on the Los Angeles freeway in 1994.

Sources: (top) Photo courtesy WTVJ-TV, Miami, FL; (bottom) photo © Tom Martin, Aspen Photography/The Stock Market.

earthquakes and tornadoes to train wrecks, and, perhaps best known, scenes of the Rodney King beating by police officers whose subsequent acquittal on assault charges set off the 1992 riots in Los Angeles. News professionals use such donated video cautiously, knowing the donor may have less than honorable motives. Most amateur video seen on-air involves natural disasters or provides information independently verified by other sources.

Responding to criticism that local news followed an "if-it-bleeds-it-leads" philosophy, some stations during the mid-1990s adopted a policy of not using sensational or emotionally charged video in their early-evening newscasts. Some viewers praised the broadcasters for their sensitivity, while others complained that the news was artificially sanitized. One station promoting its "family-*centered*" newscast received angry calls objecting to the "family-*censored*" news. Faced with apathetic viewer response and even criticism, stations soon abandoned promotion of the policy on-air.

Although many Americans continued to criticize televised violence, including live news coverage of events such as the 1999 school shootings at Columbine High School in Colorado, news directors at the decade's end appeared unaffected. A 1999 survey of TV news directors found that 95 percent had made no effort to reduce their coverage of crime or violence because of the Columbine tragedy or similar events (*Broadcasting & Cable*, 27 September 1999: 38).

News Origination by Affiliates

Most network affiliates originate an early-evening newscast and a late-fringe newscast. Affiliates' local evening news shows either lead into network news or both precede and follow it to form a "sandwich." In the Eastern and Pacific time zones, network evening news usually starts at 6:30, preceded by local newscasts ranging from a half-hour to two hours in length.

Late newscasts typically appear at 11:00 P.M. (Eastern time). For years they lasted a half-hour. By the 1990s, however, they had expanded to 35 minutes—a network concession to affiliates' desire for more local advertising time and to the network's own efforts to improve clearances of their programs that followed.

Fox stations often schedule their late newscast an hour earlier than affiliates of other networks. Modern lifestyles make staying up past 11:00 P.M. for the news unappealing, especially when many people can get national news, sports, and local weather anytime from cable or DBS. Fox's early news is a more successful programming strategy on the East and West coasts than in the nation's midsection, where most network stations air late news at 10:00 P.M. local time.

Many affiliates also schedule a half-hour or hour of noon news. Most affiliates originate early-morning newscasts or magazine/talk shows, preceding the network morning programs. The network shows provide slots into which affiliates can insert local news segments. Some stations broadcast brief *updates* or *news capsules* during breaks in prime-time programs, many of which actually provide little news, serving mostly as promotions for their late newscasts. Some also offer "news-all-night," a mix of network overnight news programs, CNN Headline News and other alternative news services, and repeats of the affiliate's own late newscasts.

To increase their reach, some stations produce special newscasts for, or rerun their regular programs on, independent stations or cable systems. They also offer the audio portion of their newscasts to local radio stations. Some TV stations produce weather reports for local insertion into cable networks such as The Weather Channel. Others provide local radio stations frequent weather updates voiced by their highly promoted on-air meteorologists, expensive talent many radio stations cannot afford.

News Origination by Independents

With seven TV broadcast networks (if PAX TV is included), there are now fewer "pure" independent stations ("indies"). Relatively few of them, or stations affiliated with one of the "weblets," produce early-evening newscasts. They usually schedule their most popular syndicated entertainment shows against local news programs on network affiliates.

Many produce late local newscasts, usually at 10:00 P.M. (Eastern and Pacific times) to counterprogram network entertainment on competing stations. Independents also carry national news programs provided by services such as CNN,

scheduling them immediately before or after their late local newscasts. A handful of indies have devoted enormous resources to local news. In the 1990s, for example, Miami's WSVN-TV at times had the highest-rated 10:00 P.M. newscast in the nation.

Station-Produced Non-News TV Programs

Aside from news, broadcast stations produce relatively few programs. Affiliates generally schedule local talk and public-affairs shows on weekday or Sunday mornings. During election years or at times when important local issues arise, they may increase local production, but they normally avoid scheduling local shows in valuable time periods such as early fringe, access, or prime time. Affiliates that produce regularly scheduled local non-news/public-affairs shows usually choose a magazine/talk program for women, scheduled on weekday mornings. Nonnetwork magazine shows, though actually syndicated, may give the appearance of local productions. In fact, syndicated formats sometimes acquire local color by providing for live or recorded inserts by local personalities.

Local Cable Production

About a third of the approximately 10,700 cable systems in the United States originate local programs other than automated services such as time, weather, and channel guides. Those that do produce such programs mostly offer either commercial *local-origination* (LO) channels, controlled and programmed by the cable operators themselves, or noncommercial *public-access* channels, programmed by private citizens and nonprofit institutions such as schools and municipal governments.

Although some cable systems occasionally originate coverage of local sporting events, cable systems rarely undertake the expense required to produce regularly scheduled half-hour newscasts. Many do, however, produce local news inserts in CNN Headline News.

An increasing number of large-market cable systems devote one channel full-time to regional news services. Pioneering *News 12 Long Island*, launched in 1986, provides 24-hour news service to more than 70 cities on New York's Long Island.

News 12 the Bronx, *News 12 Connecticut*, *News 12 New Jersey*, and *News 12 Westchester* serve more than 2.4 million subscribers. *New England Cable News* serves about 2.5 million subscribers in a six-state area, and *News Channel 8* reaches more than a million homes in the Washington, DC, market. *New York News 1* provides news about New York City to some 1.6 million subscribers in the area. *ChicagoLand Television News* serves about 1.7 million subscribers in an eight-county region of Illinois and Indiana. *NorthWest Cable News* (Seattle) has more than 2 million subscribers; *BayTV* (San Francisco) has more than a million; *Orange County* (California) *NewsChannel* and *Pittsburgh Cable News Channel* each have more than half a million. These services and about a dozen others launched during the 1990s indicate the increasing interest in regional cable news.

Local-origination channels are less common than access channels. The latter focus mostly on informational and cultural programs. Much of cable-access programming consists of public affairs, typically school board meetings, city and county council sessions, hearings on community issues, and discussion or documentary programs on political, environmental, and educational matters. But access channels have increasingly begun to offer somewhat more lively fare (see Exhibit 9.h.). Public-access channels have sometimes become the focus of heated debate when that "lively fare" deals with controversial issues or handles subjects in ways some feel do not meet community standards of decency.

EXHIBIT 9.h
. . . .
The Wacky World of Public-Access Television

"You, too, can be a star!" Such is the promise of some public-access channels on local cable systems around the country.

It all started in the 1970s when, as one inducement to city fathers to grant them an operating franchise, cable operators committed to opening their facilities to virtually all comers. Many such channels carry meetings and activities of local organizations, public forums on environmental issues, and local election debates.

In the 1990s, when *Wayne's World* had progressed from a segment on NBC's *Saturday Night Live* to a hit motion picture, the access concept skyrocketed. One such program, *The Late Mr. Pete Show*, began as an access show in Los Angeles and moved to USA Network. Today, cable subscribers can see most anything on their local system's access channel—so long as it is advertiser-free and not obscene. Among the offerings:

- environmentalists modeling shoes recycled from old tires
- would-be talk-show hosts with an endless parade of nervous guests and belly dancers
- Catholics, Jews, Mormons, atheists, pagans, Peruvians, Sikhs, Democrats, Republicans, right-to-lifers, and Indian gurus
- hordes of *Wayne's World* wanna-bes, hoping for their big break

Whether anybody watches is an entirely different matter.

9.9 LOCAL RADIO PROGRAMS

Most locally programmed radio stations use modified block scheduling, changing format slightly to suit each daypart. For example, many stations emphasize news and weather/traffic information during drive times, shifting to music at other times.

Music Formats
. . . .

Most parsimonious of all broadcast formats, the disc jockey show relies on recorded music, reducing local production costs to the lowest possible level. Increasingly, even the DJ's portion of the broadcast is recorded. Managers have learned that it is possible for a DJ to voice track an hour of on-air programming in perhaps 15 or 20 minutes. A computer then melds the DJ's recorded material and the recorded music to produce the show. Group owners are also using one popular DJ to do shows heard on several co-owned stations.

Until about 25 years ago, at least some DJs could select pretty much whatever music they wanted. Today stations increasingly set up strict *playlists* of songs from which the host must select or even play in the order listed.

Formats based broadly on rock music predominate, as Exhibit 9.i shows. Radio draws more fine distinctions among types of rock than among other musical genres, producing what some refer to as *format fragmentation*. Rock formats include *adult contemporary* (AC, a broad array of popular music and golden oldies); *contemporary hit radio* (CHR) or *Top 40* (playlists sometimes restricted to about 20 to 40 of the most recent hits); *classic rock* (familiar songs from popular albums of the 1960s, 1970s, and 1980s); *oldies* (hits from the 1950s through the 1970s); and *urban contemporary* or simply *urban* (a mix of rap, rhythm and blues, and jazz favoring black artists). Among these subformats, many of which themselves are subdivided, adult contemporary has been the most successful. The *alternative*, or *progressive*, format featuring avant-garde music not played by conventional commercial outlets became popular on many college stations in the early 1990s. Exhibit 9.i shows that there is some commercial interest in the format, but most stations featuring the alternative/progressive format are noncommercial. In the 1990s formats including *rap*, or *gangsta rap*, received criticism disproportionate to their popularity because of controversial messages they sometimes conveyed regarding violence, drugs, sex, and the value of human life.

In the aggregate, more stations play *country* or *country-western music* than any other single type except rock, if all rock formats are added together. Though Exhibit 9.i shows country as a single format, it has followed rock's trend of subdividing, splitting into urban, traditional, crossover, oldies, and other specialties. After years as a mostly AM format, country music today is more often heard on FM. Adult contemporary is also most often heard on FM stations. *Religious/gospel* programming is heard about as often on AM as on FM.

Variety/diverse (or *eclectic*), as a format, refers to the *full-service* station that schedules a mix of news, talk, and music. Similarly, some foreign-language, black, and other ethnically oriented stations mix music and informational programming as a commercial format, targeting specialized groups.

The *beautiful music* format, which uses mostly unobtrusive vocal or instrumental music, has now largely been replaced by *easy-* (or *EZ-*) *listening*. Syndicated easy-listening, commercial-free music services such as Muzak go to subscribers by satellite or FM subcarriers. More than 100 CD-quality music formats are provided to cable and DBS subscribers by Digital Music Express (DMX) and Music Choice.

Classical music appears mostly on noncommercial FM stations, but a few commercial FM and AM stations have adopted the format in the largest markets.

Information Formats

• • • • • • • • • • • • • •

The radio formats listed in Exhibit 9.i include six predominantly information types: *talk*, *news*, *news/talk*, *educational*, *sports*, and *agricultural/farm*. In 1996 about 3,000 stations offered some variation of these formats, but by 1999 almost 3,600 U.S. radio stations had a predominantly information format. The number of stations offering an all-sports format increased by more than 50 percent. Most information formats are on AM stations, but only about a quarter of radio's audiences listen to AM. Most under-45 listeners tune in to FM, but the top-rated morning and/or afternoon drive-time broadcasts in many of the nation's largest markets are still on AM stations.

Talk radio occurs primarily on commercial AM stations. It combines call-in and interview programs with feature material and local news. Talk content varies between the extremes of sexual innuendo and serious political or social commentary. In major markets, shock radio deliberately aims at outraging conservative listeners by violating common taboos and desecrating sacred cows. Shock radio's contempt for adult authority and social tradition tends to attract listeners younger than the usual talk radio audience.

Howard Stern, probably the best-known practitioner of the shock radio format, has had repeated run-ins with the FCC (see Section 13.5, page 382). In 1998 Stern expanded into broadcast television by offering *The Howard Stern Radio Show* in syndication. He got off to a good start by getting clearance on 12 of the 14 CBS owned-and-operated stations in the major markets (CBS also controls Westwood One, the company that distributes Stern's radio program). But some other stations that initially signed up for the TV version canceled even before airing the first episode, fearing negative audience reaction.

EXHIBIT 9.i

Radio Station Formats by Popularity and Station Type

Format	No. of Stations	Percentage of Stations	AM	FM	Commercial	Noncommercial
	Popularity in Terms of:		Station Types			
Country-western	2,590	14.5	934	1,656	2,552	38
*Religious/gospel	2,153	12.0	1,123	1,030	1,451	702
Adult contemporary	1,939	10.8	417	1,522	1,834	105
Oldies/classic rock	1,708	9.5	579	1,129	1,656	52
News/talk	1,104	6.2	931	173	975	129
News	836	4.7	387	449	451	385
Talk	712	4.0	590	122	556	56
Rock/album-oriented rock	696	3.9	38	658	522	174
Sports	644	3.6	577	67	641	3
Contemporary hit radio/Top 40	610	3.4	64	546	515	95
Variety/diverse	534	3.0	102	432	126	408
Spanish	528	3.0	336	192	491	37
Classical	514	2.9	17	497	54	460
Progressive/alternative	459	2.6	16	443	133	326
Jazz	445	2.5	23	422	112	333
Middle-of-the-road	335	1.9	259	76	313	22
Urban contemporary	340	1.9	92	248	285	55
Big band/nostalgia	307	1.7	250	57	289	18
Educational	273	1.5	24	249	17	256
Beautiful music/easy listening	171	1.0	75	96	149	22
Black	140	0.7	100	40	119	21
Agricultural/farm	91	0.5	68	23	91	0
†All others	760	4.2	346	414	565	195

*Religious/gospel includes sacred, Christian, and inspirational music.

†Included in the "all others" category are such disparate formats as blues and bluegrass, disco and drama, folk and foreign language, New Age, and polka and reggae.

Source: Based on data in *Broadcasting & Cable Yearbook* (1999): D 638–639.

Occasionally, radio shock jocks pay a personal price for their outrageous behavior. In 1999 Washington, D.C., station WARW fired Doug "Greaseman" Tracht for a racist comment he made linked to the death of a Texas black man dragged along a country road by white supremacists. Even Stern's supporters expressed concern when he mused as to why the murderers in the 1999 Columbine high school shooting tragedy did not have sex with at least the prettier girls before killing them.

Most talk programs focus on controversial issues, using guest-expert interviews and call-in questions. The two-way telephone call-in show attracts an older and generally conservative group of listeners who have both the time and the militant convictions that incline them to engage in discussions with talk-show hosts. Program directors have to be alert lest a small but highly vocal group of repeat callers kill listener and advertiser interest in telephone talk shows. As noted earlier, not all talk shows focus on controversial issues. Some help callers solve problems with their health, careers, marriages, money, even their automobiles and pets.

During the 1980s talk radio emerged as an important public forum in the black community. *Black talk* stations provide a window through which candidates for public office, community organizations, and reporters can obtain a unique perspective on black public opinion. Politicians found that black talk radio gave them political input that they could not get through mainstream media.

The *all-news* format costs a lot to produce yet earns low ratings compared with successful music formats. All-news stations count on holding listeners' attention for only about 20 minutes at a time, long enough for listeners to arm themselves with the latest headlines, the time of day, weather tips, and advice about driving conditions. To succeed, this revolving-door programming needs a large audience reservoir that only major markets can supply.

9.10 INTERNET PROGRAMMING

Programming on the Internet is in its infancy. Program providers are moving rapidly toward offering content, but the Internet remains a poor medium for transmission of programming compared to broadcasting and cable. It is likely to remain so until most people can access the Internet with something other than analog modems connected to conventional phone lines. Even if the user's connection were instantly and dramatically improved, the bandwidth demands of audio and video would require significant equipment upgrades by Internet service providers, and the entire Internet infrastructure would need strengthening.

However, great progress was made during the late 1990s to both upgrade the infrastructure and widen the pipeline into the home. Huge quantities of fiber-optic cable were installed by the major telephone and cable companies, and the FCC made new spectrum available for wireless delivery of digital data. Broadband access to the network improved as high-speed digital cable and digital subscriber line (DSL) technology began rolling out in many communities. Internet software for transmitting audio and video improved, as did the hardware available to run the software. Even as the power of computer-based technology available to consumers increased, prices dropped and use of the Internet exploded. In spite of the technical limitations, mass media managers recognize the opportunity to expand their audiences by both creating new content and repurposing programming produced as part of their traditional operations for distribution via the Internet.

Audio and Video Streaming

Several competing companies provide software that allows content providers to stream audio and video over the Internet for reception by persons with the necessary hardware, software, and Internet connection. RealNetworks reported having more than 50 million registered users of the firm's Real-Player software early in 1999. Microsoft had only 25 million users of its Windows Media Player, but that number increased dramatically when the company began distributing the software as part of its Windows operating system.

Audio and video streaming requires that the user's computer buffer the data flow to avoid loss of the signal because of network delays. Once the user clicks on the icon to begin streaming, there is a delay of several seconds before the audio and/or video begins playing as the buffer fills with data. If network congestion becomes so great that the buffer runs dry, the audio or video stops until the buffer fills again. Internet audio quality is sometimes described as equivalent to that of AM radio. Video typically is restricted to a small area on the screen, lacking in resolution, and very jerky compared to full-motion broadcast and cable TV programming. Video and audio quality improve as users acquire access to connections offering greater bandwidth.

Internet content often duplicates what the provider already offers through other distribution systems. Many radio stations, for example, stream their on-air signals. Radio broadcasters are finding that there are more computers than radio receivers in the workplace. Streaming their on-air signals expands their audiences by reaching people outside the home or automobile, the traditional locations for radio listening. Audio streaming works in the office because workers can listen to a radio station in the background through their computers while working on other tasks. Some TV stations and networks are also streaming programming in real time, at least during certain times of the broadcast day and most often during newscasts. Video requires more of the audience's attention, so most workers can devote only a few minutes to watching a program, making news and other types of information programming a popular choice for real-time video streaming. The poor quality of Internet video also works against longer-format programming. Content, such as news, that involves less motion and requires less resolution makes the most of the technical limitations.

Although an increasing amount of video is streamed live as it is distributed over the air or on cable, most is content that was previously distributed through other media and available via the Internet on demand. Short-format programming predominates, but less expensive storage capacity is prompting producers to offer longer programs for those with the time and interest to watch them.

Supplemental Content

Most program producers and distributors use the Internet to repurpose and supplement content provided through other channels and to promote their primary delivery systems. Content producers often have more material than they use. They constantly make editorial decisions about what to delete, trying to fit the content into a restrictive time frame. The Internet lets producers put more of their content out there and allows people to select what they want based on their interests and time availability. A rock concert on MTV probably will not include an interview with the producer of the event discussing site selection, booking of performers, and other pre-production planning details. Putting the interview on MTV's web site, however, does not tie up valuable time on the cable channel and allows those with an interest to get "inside" information.

There are limits to how many people can simultaneously access information on a web site, but in a sense the Internet provides the programmer many new "channels" for distribution of supplemental content.

Original Content

Although much Internet content duplicates what is available through the conventional mass media, some content is available only on the Internet. Some media organizations are putting original content on the Internet to test its popularity before migrating it to broadcast or cable. The Cartoon Network, for example, offers animated shorts specifically for Internet distribution. A cartoon feature doing well on the Internet is likely to be developed for distribution via the cable network. Dozens of other media companies are finding the Internet an inexpensive medium for testing content before risking its distribution through conventional channels. Many companies also believe that the Internet is a new medium with unique characteristics and unique audiences that merits original programming reflecting those characteristics. Warner Bros. is among the major media companies developing original programming designed expressly for Internet transmission rather than supplementing or marketing existing programming. ABC is another major media player experimenting with original Internet-only programming. In 1999 the network initiated a live, Internet-only video news program hosted by veteran ABC News correspondent Sam Donaldson.

The Internet also is a popular venue for content considered inappropriate, or marginally appropriate, for conventional mass media distribution. Broadcast and cable networks devoted much attention to independent prosecutor Kenneth Starr's allegations that led to President Clinton's impeachment, but most reports avoided the graphic language included in Starr's documentation. Putting the document (or links to it) on their web sites allowed millions of viewers wanting greater detail to get it. A program produced by the American Health Network showing a live birth attracted more than a million viewers to their PCs. The leading aggregator of streaming audio and video on the Internet, broadcast.com, had an estimated audience of about 1.5 million when it distributed a 15-minute fashion show featuring lingerie sold by Victoria's Secret. The audience would have been larger had the technology allowed more people simultaneous access. In 1999 Yahoo!, a leading Internet media company and popular portal provider, acquired broadcast.com.

In addition to established media companies and commercial organizations, an unknown number of small groups and individuals are streaming audio and video on the Internet. The low cost of streaming technology and the fact that no license is required make the Internet a popular medium for those wanting to "broadcast." Although some content streamed by groups and individuals consists of original material, much of it duplicates that distributed by the traditional mass media. Many non-broadcast Internet-only "stations" are aggregated by imagineradio.com.

Developing Rapport and Talking Back

Besides providing users with the ability to select the content they want when they want it, the Internet encourages them to interact with it. On-air contests, slogans, and promotions seek to create a sense of "belonging" among audience members, to make them feel "we're your" station, network, or channel. Web site content often extends the opportunities for interaction by offering contest clues or earlier access to promotional materials.

Many radio station web sites provide pictures of their on-air staff, along with biographical information. TV stations may include pictures and a detailed explanation of how their "exclusive Doppler radar" makes the station's weathercast

superior to other stations' in the market. National Association of Broadcasters research indicates, however, that the materials building audience allegiance and promoting the station most effectively are those that provide information that people find instantly and personally useful. Weather and traffic updates, information about concerts, opportunities to win free tickets to events, coupons for special discounts on advertised products, and similar items are more likely to bring people back to the station's web pages than photos of the staff.

E-commerce became acceptable to a large number of people during the late 1990s. This encouraged networks and stations to increase the commercialization of their web sites. In addition to simply displaying advertising messages for commercial sponsors on their web pages, many stations and networks began offering products for sale directly on-line or added links to their sponsors' web sites. The addition of e-commerce features to a web site makes it more useful and further helps promote audience identification with the station or network.

Some radio stations test new musical selections by asking web site visitors to listen to short excerpts and rate them. Or they may ask listeners to say what they like most or least about the station. Deciding how much weight to give these responses in making managerial decisions is a difficult matter, because the respondents are self-selected and may not represent the typical audience member.

Future Programming

Whether the Internet will remain largely a supplementary content vehicle or become a primary program delivery channel like broadcasting and cable will depend on how quickly, if ever, the technological and economic restraints are eliminated. Meanwhile, programmers jump in response to each new advancement in technology and shift in the marketplace. As the vice president of new media for Lifetime Television, Brian Donlon, says regarding Internet programming, "It's a brave new world and there are no rules. You just keep putting material up there and see if it works" (*Broadcasting & Cable*, 10 February 1997: 24).

9.11   PROGRAM CRITIQUE

This and the preceding chapter have examined broadcast and cable programs largely from the industry viewpoint—as vehicles to carry commercial messages or to entice paying subscribers. Here we touch on other perspectives—those of critics and consumers.

Diversity

A persistent complaint about commercial television, especially prime-time network programming, is its lack of diversity—the sameness of its program types, themes, plots, production styles, and sources. Broadcast networks risk so much on each program series that they take the safe route of copying successful shows again and again. Spinoffs and clones reduce prime-time entertainment essentially to sitcoms, crime dramas, and movies. Because mass media strive to reach the largest absolute audience possible, the broadcast networks must provide programs with the greatest appeal to the largest number of people, and they must do so most of the time.

The networks' homogenizing influence also affects production styles. Programs from one production company look much like those from another. Yet programmers desperately seek novelty. This seeming paradox comes from wanting to be different but not wanting to take chances.

Popular cable networks depend on the same mass-appeal programs as broadcast networks. Although the cable networks offer many original shows, their programs are mostly variations on successful genres. Specialized cable networks do provide viewers much more of what mainstream broadcast and cable networks offer only in snippets. As cable systems and DBS continue to increase their channel capacity, and as competing delivery technologies develop, there will be more networks serving niche audiences, with greater program diversity. Economics, rather than technology, will become a more critical limiting factor.

Journalism

As the public's primary news source, television has special obligations not always met in a competitive marketplace. Some observers express concern over reduced journalistic diversity resulting from the trend toward media concentration in the hands of a few huge corporations with no direct accountability to the public.

Others contemplate the impact of television trial coverage on the rights of defendants and victims. The criminal and civil trials of O. J. Simpson received unprecedented coverage. Some say that this coverage made the public more aware of judicial procedures. Others say that the minute coverage of the trials with their starkly different verdicts did more to confuse, frustrate, and divide people than inform or reassure them.

The announcement of the verdict in the Simpson civil trial during President Clinton's 1997 State of the Union address posed further problems for the networks. Managers had to decide whether to stick with Clinton talking about issues affecting the entire nation or switch to Los Angeles for news that, strictly speaking, had a direct impact on only a few individuals. In essence, they had to decide between giving the audience what journalists consider most important and what viewers consider most interesting.

Traditional news programs have come under attack for stressing violence and tragedy at the expense of more far-reaching but less sensational stories. Even the highly respected *60 Minutes* was accused of tabloid journalism when it broadcast video of Dr. Jack Kevorkian assisting a patient committing suicide. Many believe that almost all the media, but especially the broadcast and cable news networks, exploited the Clinton-Lewinsky affair and its aftermath. Some researchers believe that sensationalized coverage of Clinton-Lewinsky accounts in part for the relatively low opinion many Americans have of journalism today.

The need for visually appealing stories has tempted producers to create *staged news events*. Television also is susceptible to manipulation by persons or organizations who stage dramatically visual *pseudo-events* designed to attract news coverage. TV journalists may sometimes use, but fail to identify, footage from *video news releases* (VNRs) produced by outside sources such as government agencies and businesses. Sometimes broadcasters deceive viewers by not identifying apparently live coverage as being, in fact, recorded. (See Section 11.2, page 321, for a more thorough discussion of these issues.)

Digital enhancing offers high-tech opportunities for deception. This technology permits a reporter or producer to manipulate images in countless ways. The temptation to do so is not limited to the electronic media, as the print media also have access to the technology.

Critics deplore *checkbook journalism*—the practice of paying for interviews. Reputable news organizations seldom pay for interviews. Sometimes, however, their purity is questionable. In 1999 ABC, for example, broadcast a highly promoted Barbara Walters interview with Monica Lewinsky in prime time and during a national ratings period. Although the network paid Lewinsky nothing for her appearance, the program aired the evening before her book detailing her relationship with President Clinton hit the market. It is difficult to believe that Lewinsky received no financial benefit from appearing on the broadcast, even if no cash changed hands. Networks also use their prestige and the celebrity status of their top reporters to entice people to grant interviews. It is less likely that a source will refuse an interview request by Leslie Stahl, Barbara Walters, or Tom Brokaw than by an unknown reporter.

Many people dislike the way journalists go about their business, particularly when technique creates more controversy than content. In 1997 a North Carolina jury awarded Food Lion supermarkets $5.5 million in punitive damages for a story ABC aired about alleged wrongdoing involving the handling of food products. Jurors did not fault the report so much as the fact that the reporters gained access to Food Lion's internal operations by falsifying information on the supermarket's employment application forms. The award was later reduced to $315,000 and then to only $2, but the verdict reflected public concern about invasive reporting techniques.

Later in the decade many observers criticized broadcast journalists for the way they covered a number of shooting incidents. They thought that some of this coverage tended to glorify violence and paid too much attention to the perpetrators, perhaps encouraging other disturbed people to attract national attention by acting out their violent fantasies.

Programs and the Public Interest
• •

Fundamentally, the viewer/listener perspective asks whether commercial motives should be the primary factor in program choice and quality. Debate has swirled around the issue of commerce versus culture ever since broadcasting began. How should society balance these sometimes conflicting goals?

Is it enough for media to treat programs simply as "product"—articles of trade? Some people argue that programs can also be broadly cultural and contribute to the intellectual, artistic, and moral quality of national life. Seen in that perspective, programs should do more than encourage audiences to watch commercials.

"Wasteland" vs. "Toaster"
• • • • • • • • • • • • • • • • • • •

The best-known critique of the industry's program performance in public-interest terms came from an FCC chairperson appointed by the Kennedy administration, Newton Minow (1961–1963). In an address to the National Association of Broadcasters in 1961, Minow challenged station owners and managers to watch their own programs for a full broadcast day. They would, he assured them, find a "vast wasteland" of violence, repetitive formulas, irritating commercials, and sheer boredom (Minow, 1964: 52). The "vast wasteland" phrase caught on and became a permanent part of broadcasting lore.

Some 20 years later a Republican-appointed FCC chairperson, Mark Fowler, pointedly refrained from talking to the industry about program quality. In his view, the FCC had no business interfering with the workings of the marketplace. He too coined a memorable descriptive phrase when, in addresses to various broadcast industry groups, he described television as "a toaster with pictures" and the agency he headed as "a New Deal dinosaur." These dismissive phrases reflected a then-dominant theory in Washington: The economic laws of supply and demand suffice to ensure that commercial television will supply suitable programs; if programs degenerate into a vast wasteland, blame not the industry but the audience.

In 1991 Newton Minow (long since an attorney in private practice) revisited television, commenting on his famous words, "vast wasteland":

Today that 1961 speech is remembered for two words—but not the two I intended to be remembered. The words we tried to advance were "public interest." To me, the public interest meant, and still means, that we should constantly ask: What can television do for our country?—for the common good?—for the American People? . . .

If television is to change, the men and women in television will have to make it a leading institution in American life rather than merely a reactive mirror of the lowest common denominator in the marketplace. Based on the last thirty years, the record gives the television marketplace an A+ for technology, but only a C for using that technology to serve human and humane goals. (Minow, 1991: 9, 12)

Despite the often exaggerated charges of some critics, most would concede that television sometimes rises to peaks of excellence—even though between the peaks lie broad valleys (vast wastelands?) of routine programs. How green the valleys are depends on the viewer's personal tastes.

Taste
• • • •

A continuously available mass medium such as television cannot satisfy every taste all the time. *Most* of the time, but not *all* of the time, it must try to please *most* of the people, but not *all* of the people—an extraordinary demand. In meeting this demand, television has exposed the low common denominator of mass popular taste as contrasted with the more cultivated standards of high taste. As Daniel Boorstin, an authority on American cultural history, put it,

> much of what we hear complained of as the "vulgarity," the emptiness, the sensationalism, of television is not a peculiar product of television at all. It is simply the translation of the subliterature onto the television screen. . . . Never before were the vulgar tastes so conspicuous and so accessible to the prophets of our high culture. Subculture—which is of course the dominant culture of a democratic society—is now probably no worse, and certainly no better, than it has ever been. But it is emphatically more visible. (Boorstin, 1978: 19)

From the standpoint of the middle ground between the extremes of programs-as-merchandise and programs-as-culture, it is unrealistic to expect programs always to rise above the lowest common denominator. As the most democratic of media, broadcasting necessarily caters to popular tastes. That mission, however, need not preclude serving minority tastes as well. One hallmark of a democracy is that though the majority prevails, minorities

still have rights. Public broadcasting exists in part to compensate for the omissions of commercial broadcasting in this regard. But even public broadcasting has had to increase its popular appeal to attract program underwriters and to broaden its subscriber base. It has so far lacked the financial support it needs to offer an adequate alternative program service.

Promoters of the marketplace philosophy argued that deregulating broadcasting and cable would automatically bring diversity and enhance quality. So far, however, many of the new options have turned out to be merely repetitions of the old options. After all, every delivery system responds to the same marketplace imperatives that drive commercial broadcasting.

Popular Taste vs. Bad Taste
• • • • • • • • • • • • • • • • • •

Although commercial broadcasters have always catered to *popular taste*, they once refrained from catering to the appetite for downright *bad taste* in programs. In its now-abandoned Television Code, the National Association of Broadcasters (NAB) emphasized the role of television as a family medium, warning that "great care must be exercised to be sure that treatment and presentation are made in good faith and not for the purpose of sensationalism or to shock or exploit the audience or appeal to prurient interests or morbid curiosity" (NAB, 1978: 2).

That kind of sensitivity, along with the Television Code, has fallen victim to changing times. In fact, the Code's statement of what not to do accurately describes exactly what succeeded on the air beginning in the late 1980s. "Raunch' on a Roll," proclaimed a headline in a trade journal over a story about the rise of slobcom—sitcoms that "stretch the bounds of what's acceptable" (*Broadcasting*, 21 November 1988: 27). The Fox network's *Married . . . With Children* occasioned the article. Its plots dealt with topics such as premenstrual syndrome, treated with outrageously vulgar humor. Despite some complaints, such programs drew high ratings, which in turn ensured widespread imitation.

By the mid-1990s there were isolated examples of changes in some prime-time programming. CBS, in particular, tried "going home" to its traditionally older, more conservative viewers with "wholesome" shows such as *Touched by an Angel*. This show was among broadcast television's top-rated programs in the late 1990s, but it was not extremely successful with advertisers because it did not attract the proper demographic—younger and more socially sophisticated viewers, who buy the most products. CBS was not lacking programming of an entirely different type, however. The network's *Walker, Texas Ranger* consistently ranked among the most violent programs on television, averaging dozens of violent acts per episode. Notwithstanding the program rating system and V-chip technology, violence remains an integral part of many successful programs.

Although containing little or no violent content, sitcoms continue to test the boundaries with their use of language and sexual innuendo. A watchdog organization critical of TV programming reported that during the week when content-rating icons (see Exhibit 9.d) first appeared on-air, one sitcom character reminisced about having had teenage sex in a driver's-ed car; another boasted of having had sex four times in one day, including once on the hood of an automobile; another talked about having had sex with his father's fiancée; yet another joked about a man who masturbated by rubbing against a tree; and still another fantasized about eating pralines off the naked body of a single woman. The TV-G-rated and TV-PG-rated sitcoms also included words such as "ass," "suck," and "bastard," and one referred to a part of the male anatomy as "Mr. Sturdy." In 1999 CBS, the stodgiest broadcast network, allowed a character in *Chicago Hope* to use a four-letter word for excrement during prime time.

Cable, which operates on a standard less strict than that of broadcast, "double-shoots" some programs—one version with topless scenes, for the pay-cable network and foreign syndication; another without, for ad-supported cable and domestic station use. Pay-cable "adult" services such as Playboy and Spice offer full nudity and carefully edited, but nonetheless graphic, movies.

Violence, mayhem, and sociopathic behavior, long the subjects of critical debate, continued unabated throughout the 1990s. One can speculate on influences that may have had a hand in bringing about this shift in program standards. Among the many possible influences might be these:

- the FCC's laissez-faire policy during the 1980s, which enouraged broadcasters to test the limits of public tolerance
- abandonment of the NAB codes, implying to some that anything goes
- heightened competition, encouraged in part by FCC policies, requiring ever more strenuous efforts to capture audience attention
- the impact of cable television, which until the 1992 Cable Act had been unconstrained either by legislated public-interest standards or by a tradition of self-restraint
- corporate mergers that replaced experienced broadcast and cable executives with cost-conscious corporate managers saddled with huge debts
- social changes in the direction of more open and permissive behavior, marked especially by violence, sex, and rebellion against conventional standards
- the content ratings system, which, ironically, may "free" producers unafraid of the TV-MA rating to push the boundaries under the protection afforded by the label

Program producers have a difficult task. They face constant pressure to produce a product generating maximum profit regardless of its artistic quality or social impact. On the other hand, the electronic media, especially the licensed media, have a legal and moral responsibility to do more than offer programs that are simply popular and make money. Indeed, they have a responsibility to provide programs that sometimes may be very unpopular.

At least theoretically, ultimate power over programming rests with the audience. Programming must compete for survival as does any other product in a consumer-driven economy, and there are political channels of control available when the marketplace fails to produce culturally acceptable programming. This does not mean that the media do not have a responsibility to strive independently toward a higher standard than the lowest common denominator without being goaded to do so by economic or legislative threats. Nor does it mean that the media are not often leaders, rather than followers, in establishing cultural standards of taste. It does suggest that a passive public will get no better programming than it collectively accepts without complaint.

The new millennium brings with it wondrous new technologies that will dramatically improve the technical quality of programming. The new technology will also usher in a new age of audience control over what they receive, how they will receive it, and when. But there is no guarantee that program content will be the better for these technological advancements.

In 1958 Edward R. Murrow addressed television's potential in a speech to the Radio-Television News Directors Association. At one point in that famous speech Murrow said, "This instrument can teach, it can illuminate; yes, and it can even inspire. But it can do so only to the extent that humans are determined to use it to those ends. Otherwise it is merely wires and lights in a box." Murrow's words are no less true today.

CHAPTER 10

Ratings

Commercial broadcast and cable services compete to win the popularity contest that ratings measure. If people don't watch or listen—as measured by ratings—then advertisers lose interest and programs die, to be replaced by others, or stations fail and leave the air. Electronic media managers need unbiased and consistent audience information. For this, broadcasters, cable services—whether delivered via traditional systems, DBS, or wireless cable—and advertisers employ independent companies to conduct most day-to-day audience research, using scientific methods for probing into human behavior and attitudes.

10.1   RATINGS BUSINESS

Practically everyone knows what television ratings are in a general sense. In the long broadcast television network battle for ratings dominance, the rise and fall of prime-time programs always make news. But unlike most other industries, broadcasting and cable deliver no physical products. Program "publishing" goes on continuously, with audiences flowing at will from one program to another. Hence the need for highly specialized research.

Arbitron and Nielsen
• • • • • • • • • • • • • •

Two ratings firms, Arbitron and Nielsen, dominate the ratings business as sources of most audience measurements used by electronic media and their advertisers. Other competitors—Statistical Research, Inc (SRI), for example—focus more on limited or specialized types of research or have lasted only a few years. By 1994 Arbitron had ended the television (and later cable) ratings service it had begun in 1949, leaving Nielsen as the sole provider of audience research in those areas. In 1999 Nielsen announced that it would merge operations with VNu NV, an international information and publishing company headquartered in the Netherlands. Arbitron continues to dominate radio ratings.

The two services' revenues come mainly from subscriptions by cable and broadcast networks, stations, advertising agencies, sales rep firms, program suppliers, and syndicators. Station subscription rates vary according to station revenue. Major

ad agencies subscribe to both network and local-market ratings reports, with some agencies spending close to a million dollars a year.

Local-Market Ratings
• • • • • • • • • • • • •

Nielsen gathers and publishes television data for virtually all television markets. Local reports reflect the relative position of each station among its competitors and estimate local audience size for network, syndicated, and locally produced programs.

For all but the largest markets, it costs too much to collect data for local television ratings continuously. Instead, researchers gather data in short spurts known as *rating periods.*

The *Nielsen Station Index* (NSI), the major Nielsen local television market measure, covers 210 markets—162 by means of written diaries, the other 48 by a combination of diary and passive household meters (see Section 10.2). The company measures all markets four times a year in four-week-long *sweep weeks* (or simply *sweeps*); it measures larger markets up to three additional times per year. A sample page from a local-market report appears in Exhibit 10.a.

Arbitron is now the primary firm providing radio market ratings—Nielsen stopped measuring radio use in 1964, and Birch Radio left the business at the end of 1991. Arbitron covers 268 radio markets, about 95 year-round and the rest in the spring and fall. Exhibit 10.b offers a sample page from a diary used in producing radio rating reports.

Network Ratings
• • • • • • • • • • •

Broadcast television networks demand faster and more frequent reporting than do local stations. But in two respects network ratings are easier to obtain: (1) not every market need be surveyed, as a sample of network markets yields a reliable picture of national network audiences; and (2) far fewer broadcast television networks than stations compete at any one time. Nielsen issues the only network television ratings.

The *Nielsen Television Index* (NTI) is a report of television network ratings in what is called a "pocket piece" (see Exhibit 10.c). Nielsen's regular network reports use a national sample of about 5,000 peoplemetered homes. Nielsen also reports on its national Hispanic sample (National Hispanic

EXHIBIT 10.a
· · · · · · · · · · · ·

Local-Market TV Ratings Report

Nielsen uses diaries and household meters to measure 210 local-market television audiences. Shown here is part of a sample page of late-night ratings and shares for network and syndicated programs in the New York market. This report includes data from four weeks and shows a variety of demographic breakdowns so that station management and advertisers can obtain ratings for, say, women 25 to 49 years of age.

Sources: © Nielsen Media Research. *Viewers in Profile*, Nielsen Media Report, November 1999. Used by permission.

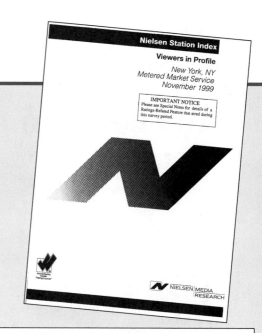

Nielsen Station Index
Viewers in Profile
New York, NY
Metered Market Service
November 1999

IMPORTANT NOTICE
Please see Special Notes for details of a Ratings-Related Feature that aired during this survey period.

NIELSEN MEDIA RESEARCH

NEW YORK, NY

WK1 11/04-11/10 WK2 11/11-11/17 WK3 11/18-11/24 WK4 11/25-12/01

METRO HH	MONDAY-FRIDAY 11:15PM - 12:00MD	DMA HOUSEHOLD								DMA RATINGS																					
		RATINGS WEEKS				MULTI-WEEK AVG	SHARE TREND			PERSONS										WOMEN						MEN					TNS	CHILD
R T G / S H R	STATION PROGRAM	1	2	3	4	R T G / S H R	MAY '99	FEB '99	NOV '98	2+	18+	12-24	12-34	18-34	18-49	21-49	25-54	35+	35-64	50+	18+	12-24	18-34	18-49	25-49	25-54	WKG 18+	18+	18-34	18-49	21-49	25-49	25-54	12-17	2-11	6-11

1/1	R.S.E. THRESHOLDS 25+% (1 S.E.) 4 WK AVG 50+%	LT	LT	LT	LT	LT				LT	LT	1	1	1	1	1	1	LT	LT	LT	LT	2	1	1	1	1	LT	2	1	1	1	1	1	3	2	3
	11:15PM																				LT	LT	LT	LT	LT	LT	LT	LT	LT	LT	LT	LT	LT	1	LT	1
8/14	WABC AVG. ALL WKS	8	8	8	7	8	14	14	17	17	4	5	2	2	3	4	4	5	6	6	7	5	1	2	4	4	4	6	3	4	5	5	5	1	1	
8/15	#11PM EYEWT NWS	8	8	8		8	14				4	5	2	3	3	4	4	6	6	6	7	5	1	3	4	4	4	6	4	5	5	5	5	1	1	
5/9	WCBS AVG. ALL WKS	4	5	5	5	5	8	8	8	8	2	3		1	1	2	2	2	4	4	5	4		1	2	2	3	3	2	1	1	2	2	1	1	
5/9	NEWS 2 AT 11PM	4	5	5		5	8				2	3		1	1	2	2	2	3	4	5	4		1	2	2	2	3	2	1	1	2	2			
5/9	#NEWS 2 AT 11PM				5	5	9				3	3				2	2	3	4	5	6	4		1	2	2	3	3	3	2	2	2	3			
<<	WLIW BBC NEWS	<<	<<	<<	<<	<<		X																												
<<	WLNY AVG. ALL WKS	<<	<<	<<	<<	<<																														
<<	TV55 NW LVE-11	<<	<<	<<		<<																														
<<	#TV55 NW LVE-11				<<	<<																														
10/17	WNBC AVG. ALL WKS	10	10	9	9	9	17	17	16	17	5	6	1	2	2	4	4	5	8	8	10	7	1	3	5	5	6	6	6	1	3	4	4	5		
10/18	NEWSCH 4 AT 11	10	10		9	10	18				5	6	1	2	2	4	4	6	8	8	10	7	1	3	5	5	6	7	6	2	4	4	4	5		
9/16	#NEWSCH 4 AT 11			9		9	16				5	6	1		2	4	5	8	7	9	7	1	3	4	5	5	6	6	5	1	3	3	3	4		
1/2	WNET AVG. ALL WKS	1	1	1	1	1	2	1		1	2	1	1							1	1	1														1
2/3	CHARLIE ROSE	1	1		1	1	2				1																										
	#CHARLIE ROSE				1	1	3				1	1	1			1	1	1	1	1	1				1	1	1	1							1	
<<	WNJN+ AVG. ALL WKS	<<	<<		<<	<<																														
<<	NJN NEWS	<<	<<		<<	<<																														
<<	#NJN NEWS		<<			<<																														
<<	WNJU NOTICRO 47-NCT	<<	<<			<<		X																												
3/6	WNYW AVG. ALL WKS	3	3	2	3	3	5	5		5	4	1	1	3	3	2	2	2	1	1		1	4	2	1	1	2	2	2	4	2	2	2	2	4		
3/6	SIMPSONS	3	3			3	5				4	1	1	4	4	2	2	2	1	1		1	5	2	1	1	2	2	2	4	2	2	2	2	4		
7/13	WPIX AVG. ALL WKS	7	6	7	6	6	12	17	18	17	4	6	6	7	5	5	5	3	3	2	5	7	9	9	7	6	5	6	3	5	4	4	4	3	4	4	
7/13	FRIENDS	7	6			6	12				3	4	6	7	6	6	7	6	4	3	4	5	9	8	7	5	5	5	3	5	4	4	4			1	1
7/13	#FRIENDS			7	6	6	11				4	4	7	6	7	6	6	4	3	4	2	5	6	9	8	7	5	5	4	6	4	5	4	4			
1/2	WPXN VARIOUS	1	1	1	1	1	2	2		2	2	1		1	1	1	1	1	1	1	1			1	1	1	1									
3/6	WWOR BLIND DATE	3	3	4	1	2	5	5		4	2	1	1	1	2	2	2	2	1	1	1	2	2	2	2	1	1	1	2	2	2	1	1	1	1	1	
2/3	WXTV AVG. ALL WKS	2	2	2		2	3	3		3	1	1	1	2	2	2	2	1	1	1	1	1	1	2	2	2	1	1	1	1	1	1	1			
2/3	#NOTICIAS 41-11				2	2	3				1	1		1	2	2	2	1	1	1	1		1	2	2	2	1	1	1	1	1	1	1			
<<	ZNY1 AVG. ALL WKS	<<	<<	<<	<<	<<																														
<<	INSDE-HALL R	<<	<<	<<		<<																														
<<	#INSDE-HALL R				<<	<<																														
56	HUT/PUT/TOTALS*	54	55	55	56	55		57	56	55	29	34	20	24	26	31	32	34	37	37	39	36	21	29	33	35	36	35	32	24	29	30	31	32	15	4	5
	11:30PM																																			

EXHIBIT 10.b

Local-Market Radio Ratings

You count in the radio ratings!

No matter how much or how little you listen, you're important!

You're one of the few people picked in your area to have the chance to tell radio stations what you listen to.

This is *your* ratings diary. Please make sure you fill it out yourself.

Here's what we mean by "listening"!:

"Listening" is any time you can hear a radio — whether you choose the station or not.

When you hear a radio between Thursday, (Month, Day), and Wednesday, (Month, Day), write it down — whether you're at home, in a car, at work or someplace else.

When you hear a radio, write down:

TIME

Write the time you start listening and the time you stop. If you start at one time of day and stop in another, draw a line from the time you start to the time you stop.

STATION

Write the call letters or station name. If you don't know either, write down the program name or dial setting.

Check AM or FM. AM and FM stations can have the same call letters. Make sure you check the right box.

THURSDAY

	Time		Station	Place				
			Call letters or station name. Don't know? Use program name or dial setting.	Check band		Check location		
	Start	Stop		AM	FM	At Home	In a Car	At Work	Other Place
Early Morning (from 5 AM)	5:45	7:15	KGTU		✓	✓			
	7:15	7:40	108.5 on the dial		✓		✓		
	9:30		WGXP	✓				✓	
Midday									
		3:00							
Late Afternoon	4:20	4:25	Jo Convery Show	✓					✓
Night (to 5 AM Friday)	11:30	12:15	KADV		✓	✓			

If you didn't hear a radio today, please check here. ☐

PLACE

Check where you listen.
- at home
- in a car
- at work
- other place

Write down *all* the radio you hear. Carry your diary with you starting **Thursday, (Month, Day).**

No listening? If you haven't heard a radio all day, check the box at the bottom of the page.

Questions? Call us toll-free at 1-800-638-7091. in Maryland, call collect 410-312-8000.

© 2000 The Arbitron Company

This instruction page from a radio diary illustrates the Arbitron system. Each individual in a household fills out a separate diary.

Source: Copyright © 2000. The Arbitron Company. Used by permission.

EXHIBIT 10.c
· · · · · · · · · · ·

Nielsen TV Network Ratings Report

A-10
NATIONAL *NielsenTV* AUDIENCE ESTIMATES — EVE.FRI. OCT.22, 1999

TIME	7:00	7:15	7:30	7:45	8:00	8:15	8:30	8:45	9:00	9:15	9:30	9:45	10:00	10:15	10:30	10:45		
HUT					46.7	48.1	49.1	50.5	51.5	52.7	53.8	55.3	55.5	55.6	56.0	56.8	55.5	55.4	54.5	53.3

ABC TV

BOY MEETS WORLD / ODD MAN OUT / SABRINA-TEENAGE WITCH / HUGHLEYS, THE / ←20/20-FRI→

HHLD AUDIENCE% & (000)	5.2 5,210	4.9 4,970	6.1 6,160	4.9 4,960	8.8 8,910	9.1*
TA%, AVG. AUD. 1/2 HR %	6.4 5.8	7.2	6.0	13.9 8.6*	17*
SHARE AUDIENCE %	10 9	11	9	16 16*	
AVG. AUD. BY 1/4 HR %	5.2 5.1 4.9 5.0	6.0 6.2	5.0 4.8	8.2 8.9 9.2 9.0	

CBS TV

KIDS SAY DARNDEST THINGS / LOVE & MONEY / ←NOW AND AGAIN→ / ←NASH BRIDGES→

HHLD AUDIENCE% & (000)	6.5 6,580	4.5 4,560	7.0 7.030	9.0 9.080	9.2*
TA%, AVG. AUD. 1/2 HR %	8.0 5.7	9.3 6.7* 7.2*	11.6 8.8* 13*	17*
SHARE AUDIENCE %	13 8	12 12*	17 16*	
AVG. AUD. BY 1/4 HR %	6.5 6.6 4.6 4.5	6.7 6.7 7.0 7.4	8.6 9.0 9.1 9.3	

NBC TV

←PROVIDENCE→ / ←DATELINE FRI→ / ←COLD FEET→

HHLD AUDIENCE% & (000)	10.7 10,800	8.7 8,800	4.0 4,010	
TA%, AVG. AUD. 1/2 HR %	13.6 10.2* 11.3*	13.0 8.7* 8.8*	6.8 4.2* 3.8*	
SHARE AUDIENCE %	20 19* 21*	16 16* 16*	7 8* 7*	
AVG. AUD. BY 1/4 HR %	9.8 10.6 11.0 11.5	8.7 8.6 8.7 8.9	4.5 3.9 3.7 3.9	

FOX TV

←RYAN CAULFIELD: YEAR ONE→ / ←HARSH REALM→

HHLD AUDIENCE% & (000)	2.1 2.090	2.5 2.490	
TA%, AVG. AUD. 1/2 HR %	3.3 2.0* 2.1*	3.9 2.6* 2.4*	
SHARE AUDIENCE %	4 4* 4	4 5* 4*	
AVG. AUD. BY 1/4 HR %	2.1 2.0 2.0 2.2	2.6 2.6 2.3 2.5	

WB TV

JAMIE FOXX SHOW1 - WB (R) / JAMIE FOXX SHOW, THE - WB / STEVE HARVEY SHOW, THE-WB / FOR YOUR LOVE - WB

HHLD AUDIENCE% & (000)	1.8 1.810	2.2 2.170	2.4 2.370	2.1 2.160	
TA%, AVG. AUD. 1/2 HR %	2.3	2.6	2.6	2.5	
SHARE AUDIENCE %	3	4	4	4	
AVG. AUD. BY 1/4 HR %	1.8 1.8	2.1 2.2	2.3 2.4	2.1 2.2	

UPN TV

←BLOCKBUSTER CINEMA→ / SAFE HOUSE

HHLD AUDIENCE% & (000)	2.0 2.030			
TA%, AVG. AUD. 1/2 HR %	3.8 2.1* 2.2* 1.9* 1.9*			
SHARE AUDIENCE %	4 4* 4* 3* 3*			
AVG. AUD. BY 1/4 HR %	2.0 2.1 2.2 2.2 1.9 1.8 1.9 2.0			

PAX TV

←LITTLE MEN→ (R) / ←TOUCHED BY AN ANGEL-FRI→ (R) / ←DIAGNOSIS MURDER-FRI→ (R)

HHLD AUDIENCE% & (000)	0.5 510	1.0 1.020	1.0 1.000	
TA%, AVG. AUD. 1/2 HR %	0.4* 0.6*	1.4 1.0* 1.1*	1.4 1.0* 1.0*	
SHARE AUDIENCE %	1 1* 1*	2 2* 2*	2 2* 2*	
AVG. AUD. BY 1/4 HR %	0.4 0.4 0.5 0.6	0.9 1.0 1.0 1.1	1.0 1.0 1.0 1.0	

U.S. TV Households: 100,800,000

For explanation of symbols, See page B

This ratings report for a weekday evening compares the prime-time appeal of the national broadcasting networks. Shown for each broadcast network program is the average audience in TV households (of almost 10 million total in 2000),

the share of audience, and the average audience rating by quarter-hour.

Source: Courtesy Nielsen Media Research. Used by permission.

Television Index), which includes about 1,000 peoplemetered homes.

Radio's All-Dimension Audience Research (RADAR), financed by networks that contract with Statistical Research, Inc., uses telephone recall interviews to provide the only ratings service for network radio. RADAR issues reports four times a year, each covering a sample week, based on surveys of 12,000 respondents. Unlike most telephone surveys, RADAR calls the same people each day, seven days in a row.

Syndicated-Program Ratings

By using data already obtained for market ratings reports, Nielsen provides regular analyses on the relative performance of nationally syndicated television programs.

The Nielsen *Cassandra* service provides detailed audience demographic information for both first-run and off-network syndicated programs (as well as for network and some local programs). Cassandra allows researchers to compare household demographic information for syndicated programs across markets and to collect and measure similar data on both lead-in and lead-out programs.

Special Studies

Many supplementary reports, drawing on data gathered in preparing regular ratings reports, are also available. Clients can order special reports tailored to their needs. For example, to help determine when to schedule its news promotion announcements, a TV station might order a special "audience flow" study to discover which of its own programs attract viewers who normally watch a competing newscast. Another might commission a study to find out how much a program appeals to specific audience subgroups.

10.2   COLLECTING DATA

Whatever electronic medium they analyze, researchers use several methods for collecting data on which to base ratings: diaries, meters, peoplemeters, and two different kinds of telephone interviews.

Special studies often use other methods and combinations. Each has advantages and disadvantages.

Diaries and Passive Meters

Arbitron and Nielsen researchers use a written diary method for daily gathering of most local-market data. To obtain radio data, Arbitron sends a separate diary to each person over 12 years of age in every sample household. It asks diary keepers to write down for one week their listening times and the stations they tune to, keeping track of away-from-home as well as in-home listening.

Nielsen uses diaries for its Nielsen Station Index television ratings, asking one person in each sample home to report all household viewing. VCR use is measured separately (see Section 10.8).

Diaries suffer from a serious flaw: People often enter inaccurate information, purposely (to show they have good taste in programs) or not (many sample families fill in a diary in "catch-up" style at the end of the week rather than as listening or viewing takes place). Some also tire of making entries, suffering from *diary fatigue*.

The *passive household meter* (Nielsen's Audimeter) was first used to gather data in 1950. Attached to each television set in a sample household, the so-called black box automatically records when the set is on and to which channel it is tuned. These meters are capable of identifying tuning on cable converters, VCRs, and DBS systems. Newer meters currently in development will be able to measure the multiple channels made possible through digital technology. Researchers refer to the household meter as *passive* because no effort is required on the part of the viewer to record its information, and to distinguish it from the newer peoplemeter. Because it provides no information as to who, if anyone, is actually watching, Nielsen supplements meter data with written diary data from some of the same households to obtain demographic data on actual viewers. Although passive household meters were phased out of network ratings use in 1987 with the activation of peoplemeters, they remain essential for individual market ratings.

Peoplemeters

Nielsen uses *peoplemeters* for its national television ratings service, the Nielsen Television Index. Like the Audimeter, the peoplemeter keeps a record of receiver use. It also offers research that formerly only diaries provided by simultaneously collecting demographic data. It does this by requiring each viewer, whenever she or he watches television, to "check in" and "check out" by pushing a special handset button (see Exhibit 10.d). Data on both receiver use and viewer identity go by telephone line to a central computer containing basic household demographic data, stored earlier when the peoplemeter was installed.

Nielsen began a three-year test of a peoplemeter sample in 1983 and started its peoplemeter-based national network ratings service in 1987. Arbitron, meanwhile, had developed *ScanAmerica*, a three-pronged single-source system providing ratings, audience demographic data, and—through the use of a *wand* that read universal bar codes—a measure of product purchasing. In 1992, lacking sufficient advertiser interest, Arbitron dropped use of both the wand and the name ScanAmerica and soon after ended all of its broadcast TV and cable ratings activities.

The introduction of peoplemeters created enormous controversy. Network ratings in all dayparts dropped dramatically, while ratings for many independent stations and cable channels rose. In 1989 the networks funded an independent study that concluded that peoplemeters appeared to slight such typical network heavy viewers as children and older people who might be intimidated by (or simply too impatient with) the computerlike button-pushing that peoplemeters required.

Passive Peoplemeters

Intense network and advertiser pressure to improve their methods pushed researchers to announce plans to replace peoplemeters with *passive* peoplemeters. Using computerized image recognition, these passive devices presumably could "recognize" regular (family) viewers. Further, the devices eventually could tell when viewers were actually watching the screen rather than doing something else.

A number of variations are under consideration and being tested—including one that counts someone as watching television only if that person is actually facing the receiver. There have been kinks—several demonstrations confused pets with people—but when perfected, the passive peoplemeter should be able to record electronically regular

family members and others, such as guests, tabulating who and how many are watching and when. Although such systems might alleviate the operational problems of active peoplemeters, they raise other, more troublesome concerns—chiefly viewer privacy; TV viewers may experience severe discomfort knowing that their television is watching back. In the late 1990s Arbitron began testing a new method of data collection that could lessen these concerns. Arbitron's system, detailed in Exhibit 10.e, uses a small meter to sense and recognize the electronic media signals being heard or watched by the wearer.

Coincidental Telephone Interviews

Some researchers consider coincidental telephone data gathering the most accurate means of obtaining audience information. The term *coincidental* means that researchers ask respondents what they are listening to or watching during (that is, coincidental with) the time of the call. Putting questions that way eliminates memory concerns and reduces the possibility of intentional misinformation. Researchers ask whether respondents have a set turned on at that moment and, if so, what program, station, or channel the set is tuned to, plus a few demographic questions such as the number, gender, and age of those watching or listening.

Because coincidental methods provide only *momentary* data from each respondent ("What are you listening to or watching *now?*"), they require many calls, spaced out to cover each daypart, to build up a complete profile of listening or viewing. Properly conducted, coincidental calls require large batteries of trained interviewers, making the method expensive. Nor can coincidental calls cover the entire broadcast day; information on audience activity after 10:30 P.M. and before 8:00 A.M. must be gathered during more socially acceptable calling hours, when researchers ask respondents to recall programs they listened to or viewed during these nighttime and early-morning hours.

Telephone Recall

Tricks of memory make *telephone recall* less reliable than coincidental calls, but it costs less because more data can be gathered per call. RADAR, the only source of radio network ratings, uses telephone recall, collecting samples daily over a period

EXHIBIT 10.d
• • • • • • • • • • • • •

Peoplemeter Ratings

The Nielsen national audience reporting system depends heavily on automation, from the home meter through the gathering, analyzing, and reporting processes. The peoplemeter used for national ratings consists of two parts: the hand-held unit on which each viewing member of the family and guests can "check in," using individual keys, and the base unit on top of the receiver that stores home viewing information.

The Nielsen system begins with the individual sample household (1), whose peoplemeter base unit results are (2) "read" over a telephone line (leased by

Nielsen) at 3:00 A.M. local time each morning by a computer that sends the viewing results of the sample household into (3) the Nielsen central ratings computers. The collected raw ratings are (4) sent to the screening and audit process, where all ratings materials are assembled and (5) combined with the detailed program information constantly gathered by Nielsen. They are then sent for (6) computer generation of actual overnight ratings. By midday the overnights are ready for (7) client (agencies, stations, networks) re-

trieval by computer terminal and (8) weekly printed ratings reports. Nielsen diary and passive household meter markets go through a similar process to generate market-by-market reports.

Source: Photo courtesy Nielsen Media Research. Used by permission.

EXHIBIT 10.e

Meters for a Digital Age

There are three simple parts:

front view   *back view (with clip)*

1 A Personal Meter is given to each member of your household six years old or older. You simply wear or carry your Personal Meter with you wherever you go. It picks up a special silent signal broadcast with the TV and radio transmissions.

The Personal Meter can be clipped to your belt or handbag, or carried in a variety of convenient ways.

"Thank you"

2 A Personal Meter Recharging Unit is placed in the bedroom of each household member. This unit recharges the Personal Meter and collects *your* viewing and listening information.

When you go to bed, you'll place the Personal Meter into its recharging unit so all your TV viewing and radio listening will be counted every day. This unit also lets us know how many hours you carried your Personal Meter with you that day. This lets you earn points which equal cash gifts!

3 The Collection Unit sends your household's information to Continental Research's main collection computer through a special freephone line once each day. The whole process takes only a few seconds late at night, so you won't even know it's happening. Your household's information is combined with data from many other households for a total picture of what's being watched and listened to in your area.

That's all you need to do!

Keep the Personal Meter with you:

At home

In your car

At work

Even on a picnic

Arbitron is developing and testing the Personal Portable Meter in Manchester, England. Arbitron's system requires respondents to carry at all times a small meter about the size of a pager. The meter detects all radio and television signals being heard or seen by the wearer and stores the listening and viewing data. At night the meter is placed in a data-collection box in the user's home. While in the box, the battery is recharged and the data are sent via telephone lines to a computerized collection center. This technology, should it prove viable, would make out-of-home measurement more dependable and significantly reduce the burden on the survey participant.

Source: Copyright © 2000. The Arbitron Company. Used by permission.

of seven days, thus minimizing memory errors while obtaining a week's coverage.

Personal Interviews

The use of in-person, door-to-door surveys has declined in recent years largely because of dangers implicit in knocking on doors in strange neighborhoods. Typically, interviewers question people on the street or in shopping centers, or, for car radio listening, at stoplights. Data gathered in these ways cannot be projected to the general population, however, because the samples are not at all representative.

10.3   SAMPLING

Around-the-clock monitoring of all the private listening and viewing behaviors of millions of people is obviously impossible. The task becomes possible only by *sampling*—studying some people to represent the behavior of all.

Using Samples to Simplify

Sampling simplifies three aspects of ratings research: behavior, time, and number of people.

- *Behavioral sampling.* Researchers agreed years ago on a series of minimum measurable behavioral responses—turning on a receiver, selecting a station or programs, and later turning off the set.
- *Time sampling.* The second simplification used in ratings takes advantage of repetitive daily and weekly cycles of most broadcast and cable programming. A sample taken every few weeks or months from this continuous program stream suffices for most purposes. Daily measurements occur only for network and major-city audiences.
- *Number of people sampled.* The most controversial ratings simplification arises from the use of only a few hundred or thousand people to represent program choices of thousands or millions of others. On a local-market level, Nielsen uses a few hundred households per market. It uses about 5,000 homes to represent almost 10 million for its national network ratings. For a very large radio market like Los Angeles (more than 9 million people ages 12 and older in the metro area), Arbitron will send out more than 14,000 diaries. By contrast, in a market the size of Lubbock, Texas (market 175; 12 and older population of 192,000), Arbitron distributes nearly 2,000 diaries. Only about half the diaries that Nielsen and Arbitron send out are returned. The RADAR national radio network surveys 12,000 people.

Random Samples

Contrary to what many people believe, small samples *can* give reasonably accurate estimates. The laws of chance, or *probability*, predict that a *randomly selected* small sample from a large population will be representative (within a predictable degree of accuracy) of the entire population. Random selection means that, ideally, *every* member of the entire population to be surveyed has an *equal* chance of being selected. Major defining characteristics of the sample (such as age and gender) will appear in about the same proportion as their distribution throughout the entire population.

However, choosing at random is not as easy as it sounds. Drawing a sample randomly from a large human population requires some means of identifying each member by name, number, location, or some other unique distinguishing label. In practice, this usually means using either lists of people's names or maps of housing unit locations. Such lists are called *sample frames.*

Selecting a Ratings Sample

Nielsen draws its national sample of metered television households from U.S. Census maps by a method known as *multistage area probability sampling. Multistage* refers to the step-by-step narrowing down of selection areas, starting with counties and ending with individual housing units, as shown in Exhibit 10.f. For local-market ratings Nielsen uses special updated telephone directories. *Random digit dialing*, a computer method of generating telephone numbers without reference to directories, can solve the problem of reaching unlisted and newly installed telephones, but it increases the number of wasted (unused or business-address) calls, thus increasing survey cost.

Ideally, each time a company takes a survey, it should draw a brand-new sample. On the other hand, if the company uses expensive sampling and data-gathering methods, it cannot afford to discard each sample after only one use. Nielsen tries to retain each peoplemeter household in its national sample for no more than two years but allows local-market diary families to stay in samples for up to five years.

Sample Size

Having established a sample universe, the researcher must next decide how large a sample to choose—the larger the sample, the greater its relia-

EXHIBIT 10.f

Multistage Sampling Method

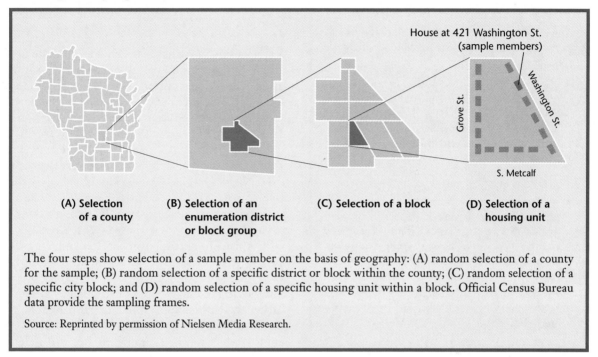

House at 421 Washington St.
(sample members)

Grove St.

Washington St.

S. Metcalf

(A) Selection of a county (B) Selection of an enumeration district or block group (C) Selection of a block (D) Selection of a housing unit

The four steps show selection of a sample member on the basis of geography: (A) random selection of a county for the sample; (B) random selection of a specific district or block within the county; (C) random selection of a specific city block; and (D) random selection of a specific housing unit within a block. Official Census Bureau data provide the sampling frames.

Source: Reprinted by permission of Nielsen Media Research.

bility. But reliability increases approximately in proportion to the square of sample size (the sample size multiplied by itself). For example, to double the reliability of a sample using 1,000 homes, you would have to increase the sample size to 4,000 homes. Thus a *point of diminishing returns* soon arrives, after which an increase in sample size yields such small gains in reliability as not to be worth the added cost.

Sources of Error

At its best, sampling yields only estimates, *never* absolute certainties. Thus the real question becomes how much uncertainty can be accepted in a given sampling situation.

Even when researchers carefully select their samples and sample sizes, two kinds of error can invalidate research findings. The built-in uncertainty of all measurements based on samples arises first from *sampling error*. No matter how carefully a sample is drawn, that sample will never exactly represent the entire population (or universe) being measured. With the use of statistical techniques, the *probable* amount of statistical uncertainty in ratings (that is, the amount of sampling error to be expected) can be calculated in advance.

The second possibly invalidating problem—*nonsampling error*—arises from mistakes, both intentional and inadvertent. Such errors produce *bias* in the results. Bias can come from lying by respondents, as well as from honest mistakes. Researchers are probably consciously or unconsciously prejudiced. The wording of questionnaires may be misleading. And mistakes can occur in recording data and calculating results.

Sometimes reports state sample size as the number of people (or households) the researcher contacted, when the key element should be how many actually *participated*. A 45 percent response rate indicates that only 45 out of 100 homes or individuals contacted actually participated. In practice, a response rate of 100 percent never occurs.

EXHIBIT 10.g

Nonresponse Errors

> Typical nonresponse errors for each of the principal methods of ratings data collection include the following:
>
> - *diaries:* refusal to accept diaries; failure to complete accepted diaries; unreadable and self-contradictory diary entries; drop-off in entries as the week progresses ("diary fatigue'); and failure to mail in completed diaries
> - *passive (household) meters:* refusal to allow installation; breakdown of receivers, meters, and associated equipment; telephone-line failures
> - *peoplemeters:* same drawbacks as passive meters, plus failure of some viewers (especially the very old and very young) to use the buttons to "check in" and "check out," having succumbed to "response fatigue"
> - *telephone calls:* busy signals; no answer; disconnected telephones; refusal to talk; inability to communicate with respondents who speak foreign languages

Depending on the research method and purpose, a response rate of less than about 85 percent for other than day-to-day ratings research is cause for concern.

Notwithstanding a series of incentives and entreaties from the ratings services, nonresponse places a serious limitation on ratings accuracy. Some age groups—teenagers, for example—have a particularly low response rate for radio diaries. Moreover, those who do return diaries may not represent those who do not. Exhibit 10.g lists typical *nonresponse errors.*

Overall, diary and meter methods yield a usable response rate of about 40 percent ("usable" being variously defined but meaning "returned on time," plus other measures of how completely the diary is filled in); the telephone method comes close to 75 percent. The number of such usable responses is termed the *in-tab* sample, the sample actually used in tabulating results and producing a given ratings report.

10.4   DETERMINING RATINGS AND SHARES

Defining Markets

A crucial step in ratings research is an accurate definition of the local market to be measured. Advertising depends on a universally recognized, national system of clearly defined, nonoverlapping markets.

Arbitron's *area of dominant influence* (ADI), first developed in 1965, was the most widely accepted system for defining TV markets until the company abandoned the TV ratings business three decades later. Nielsen has its own version, called *designated market area* (DMA). A DMA consists of one or more counties in which stations located in a central town or city are the most viewed (see Exhibit 10.h). DMAs usually extend over smaller areas in the East, where cities are closer together, than in the West. Nielsen assigns each of the more than 3,000 counties in the United States to a single DMA, updating the assignments annually. DMAs range in size from No. 1 (New York City, with about seven million television households) to No. 210 (Glendive, Montana, with fewer than 4,000).

Arbitron's ADI is still used to define local radio markets. As with Nielsen's DMA, Arbitron assigns every county in the United States to one specific ADI. In addition to measuring audience use in the ADI, Arbitron reports on two other related areas. One, the *metro* area, usually consists of one or more counties around a central city core. The metro area covers an area smaller than the ADI. The other, *total survey area* (TSA), the largest local region on which Arbitron reports, includes 98 percent of a market's listening audience, thus covering counties outside the ADI and overlapping with adjacent ADIs. Thus, although counties are assigned to only one ADI, they may be shown in more than one TSA, depending on listening in that county. Nielsen takes a similar approach with its TV ratings.

Households

Another preliminary step in ratings research is to define what will count as "one" when measuring audience size.

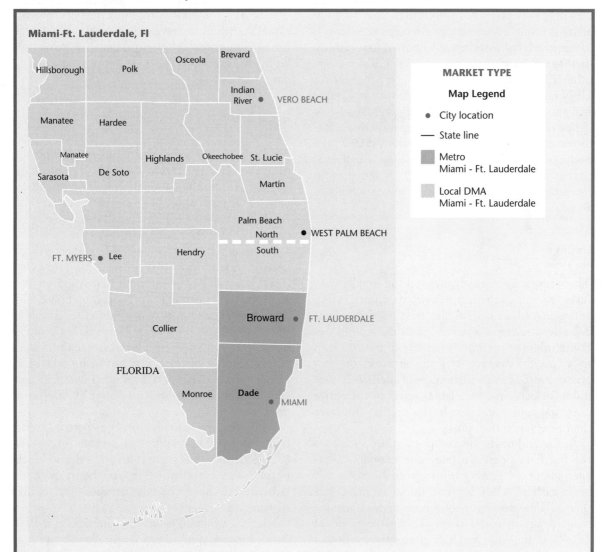

EXHIBIT 10.h
. . . .
TV Market Definition Concepts

In the Miami–Ft. Lauderdale TV market, the *designated market area* of stations located in or near Miami extends to three counties—Miami's own Dade, and Broward (also shown here as the *Metro* area), and Monroe, which includes the Florida Keys. In practice, viewers often receive programs from stations in two or more markets. In those cases, Nielsen determines which stations are viewed most frequently in the market being defined and *designate*s which counties are to be assigned to which DMAs. For example, in this illustration, when viewers in Broward County receive signals from stations in both Miami and West Palm Beach, Nielsen has determined that most of the time Broward County viewers tune to Miami stations.

No county is assigned to more than one DMA. Based on viewing patterns, however, Nielsen for some reporting purposes splits some counties into north and south (as with Palm Beach County on this map) or east and west sections. Nielsen also measures audiences in the *total survey area*, which includes all counties wherever viewership is reported. For superstations, the *total area* is virtually the entire United States.

Source: Map courtesy Nielsen Media Research. Used by permission.

Television viewing has traditionally been a family activity, making the *household* a logical unit of measure, even though a majority of households now have two or more television sets and much viewing takes place as a solo rather than a family activity. A single diary or meter records the viewing of each television receiver. Thus, if one family member is watching a program on CBS and another family member is watching a program on UPN on another set in the house, both programs will get credit for delivering that household.

For radio, each listener has a separate diary. Radio researchers prefer to count persons rather than households because (1) radio listening usually occurs as an individual activity, and (2) much radio listening takes place outside the home, especially in autos and workplaces.

Ratings
• • • • •

The term *rating* is widely misused and usually confuses two quite separate industry measures: rating and share (discussed in the next section). Specifically, a *rating* is an estimate of the number of households (or persons in the case of radio) tuned to a specific channel (station, network, or program), expressed as a percentage of *available* households (or persons). "Available" refers to the entire potential audience, even those who may not have their receivers turned on.

At the end of the decade prime-time shows on the Big Four broadcast television networks averaged about a 10 rating (down from 17 just a few years earlier), while daytime shows averaged less than 5. The most successful nonsports entertainment program of all time, the final episode of *M*A*S*H* in 1983, had a Nielsen rating of 60.2. This means that 60.2 percent of all U.S. television households watched at least part of that episode. The Super Bowl of 1982 achieved a rating of 49.1, a record for that annual football championship. By contrast, the most popular cable television programs—typically professional wrestling events—seldom garner ratings higher than 5.

Radio stations—and public television stations—often earn ratings of less than 1 and rarely more than 2 or 3. Such low ratings make no meaningful distinctions among stations; radio therefore relies more often on *cumes* (measures of cumulative audiences) and shares.

Network and syndicated radio programs typically do not garner national ratings anywhere near as large as network television programs because most radio programs do not have full nationwide coverage. A few radio programs that do have national distribution may achieve a rating of 2 or 3, but such figures are very unusual. Exhibit 10.i highlights several of the most popular network radio programs.

Shares
• • • • •

A *share* is an estimate of the number of households (or persons for radio) tuned to a given channel, expressed as a percentage of all those households (or persons) *actually using* their receivers at that time. Recall that a rating is based on all those *owning* receivers, not necessarily *using* them.

A station's share of the television audience is calculated on the basis of *households using television* (HUT). A HUT of 55 indicates that at a given time an estimated 55 percent of all television households are actually tuned in to *some* channel receivable in that market; the remaining 45 percent are not at home, busy with something else (screening videotapes perhaps), or otherwise not using television. HUTs vary with daypart, averaging about 25 percent for daytime hours and about 60 for prime time.

Radio research usually measures persons rather than households, yielding a *persons using radio* (PUR) rating. PUR reports usually refer to blocks of time—either individual quarter-hours or cumulative quarter-hours for a day or week—rather than to programs.

Shares are figured based on either HUT or PUR data. *A station always has a larger share than rating for a given time period.* For example, top prime-time broadcast television network programs usually average ratings of about 15, but their corresponding shares are closer to 25. Exhibit 10.j explains how to calculate both ratings and shares.

Television programmers use shares in making programming decisions; salespersons usually use ratings as advertising sales tools. The reason is that shares give programmers a better estimate of their competitive positions within a medium, whereas ratings more readily allow comparison of advertiser exposure on radio or television with that in other media. Also, because HUT levels vary

EXHIBIT 10.i
• • • • • • • • • • •
Radio's Top Talkers

Talk radio rules—at least for those station managers interested in achieving winning radio ratings. Arbitron reports that national shares for talk radio have risen steadily from 15.4 in 1993 to more than 17 today. The competition among the nationally syndicated talkers is strong, with local stations seeking the edge in their markets by affiliating with one or more of the popular talk shows. Leading the pack is the venerable Paul Harvey, whose news and commentary shows boast a listenership of more than 19 million, for an *average quarter-hour* (AQH) rating of 2.3 among listeners 18 and older. Rush Limbaugh, the tough-talking scourge of liberals, environmentalists, and "femi-nazis" everywhere, ranks second in overall listenership with an AQH rating of 1.9. Popular and controversial shock jock Howard Stern ranks next with a 1.7, followed by pop psychologist and rising star Dr. Laura Schlessinger at 1.3. Quickly gaining national popularity is *The Bob and Tom Show*, featuring Bob Kevoian and Tom Griswold, who have won an unprecedented four prestigious Marconi Awards, including the 1999 award for national radio personalities of the year.

Sources: (top, left): Copyright © 2000 ABC, Inc.; (top, center): Lennox McLendon/AP-Wide World Photos; (top, right): Fotos International/Archive Photos; (bottom, left): Susan Sterner/AP-Wide World Photos; (bottom, right): Courtesy Bob and Tom Radio Network.

EXHIBIT 10.j

Ratings Concepts

(A) The pie chart shows television set use information gathered from a sample of 400 households, representing a hypothetical market of 100,000 TV households.

Note that program ratings are percentages based on the entire sample (including the "no-response" households). Thus Program A, with 100 households, represents a quarter (25 percent) of the total sample of 400. The formula is $100 \div 400 = .25$; the decimal is dropped when expressing the number as a rating.

Projected to the entire population, this rating of 25 would mean an estimated audience of 25,000 households. The formula is .25 (the rating with the decimal restored) $\times 100,000 = 25,000$.

(B) The smaller pie chart, representing 55 percent of pie A, includes only the households using television, in this case $80 + 100 + 40$ households, or a total of 220 (expressed as a households using television, or HUT, rating of 55, as shown in A). Shares are computed by treating the total number of households using television, in this case 220, as 100 percent. Thus Program A's 100 households divided by 220 equals about .4545, expressed in rounded numbers as a share of 45.

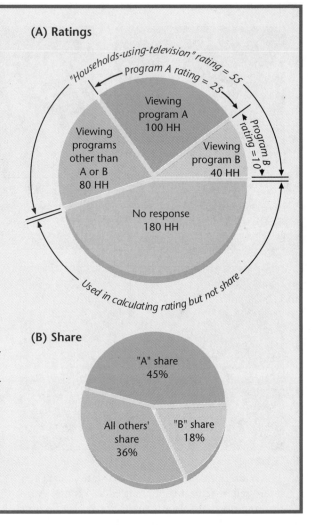

(A) Ratings

"Households-using-television" rating = 55
Program A rating = 25
program B rating = 10
Viewing program A 100 HH
Viewing programs other than A or B 80 HH
Viewing program B 40 HH
No response 180 HH
Used in calculating rating but not share

(B) Share

"A" share 45%
All others' share 36%
"B" share 18%

---

widely during different times of the year (higher in November, lower in May), programmers use share rather than rating to compare a program's competitive performance from one time of year to another.

Because radio ratings are typically very low, radio stations more often than not judge their competitive standing in a market based on audience shares. With few exceptions the top-rated stations in major radio markets, based on listening of people 12 years old and older between 6:00 A.M. and midnight, average between a 6 and 7 share.

Cumes

A radio program reaches a relatively small number of people in any given quarter-hour. *Cumulatively,* however, over a period of many hours, or during the same period over a number of days, it reaches a large number of *different* listeners. A *cume* rating gives an estimate of the (cumulative) number of *unduplicated* persons a station reaches over a period of time. "Unduplicated" means that during the two or four weeks that typically make up a ratings period, a person who listened several times to a particular station on different days would be counted as only one person in constructing a cume figure. A person listening only once during that period would also be counted as one person, because a cume shows how many *different* people tuned to the station during a given period of time. The terms *reach* and *circulation* usually refer to cume audience measurements. Cumes are useful in commercial radio and television because ads are often repeated. Consequently, the larger the cume for a station, the greater the chance that more listeners have been exposed to the advertiser's message.

Reporting Demographics
● ● ● ● ● ● ● ● ● ● ● ● ● ● ● ● ●

Ratings reports detail audience composition by gender and age. These *demographic breakouts*, or simply *demographics*, divide overall ratings into such subgroups as those for men, women, and teens. Adult audience age-group categories typically consist of decade units (such as men 35–44) for radio and larger units for television (for example, women 18–34 or 25–49), although "persons 12+" serves as the basic category for determining, for example, which radio station in a given market is number one.

Advertising agencies usually buy demographics rather than generalized audiences. Most advertisers would rather have an audience of moderate size with the right demographics for their products than a huge audience containing many members not likely to be interested.

10.5 USE AND ABUSE OF RATINGS
● ●

Ratings and shares perform a vital function for operators and advertisers—which makes them subject to misuse. Blind consideration of only the numbers could lead to one station getting a disproportionate amount of advertising dollars when the spread between that station and its nearest competitor is less than the sampling error reported. Most users of ratings data understand and accept the fact that the data are estimates. But because everyone uses the same data, users tend to shrug off the data's shortcomings and treat them as being more precise than they really are.

Facing Complaints
● ● ● ● ● ● ● ● ● ● ●

In response to complaints and investigations about ratings in the 1960s, the Broadcast Rating Council was established in 1964. Eighteen years later, in 1982, it became the Electronic Media Rating Council (EMRC), serving as an independent auditing agency representing ratings users. The EMRC accredits ratings services that meet its standards and submit to annual auditing paid for by the service provider, not by the EMRC. The procedures are sufficiently involved and costly that many smaller companies don't bother to apply.

Reliability of Ratings
● ● ● ● ● ● ● ● ● ● ● ● ● ●

Reliability in research refers to the degree to which methods yield consistent results over time. For example, the drops in broadcast television network ratings in initial peoplemeter surveys raised questions about ratings reliability, because those results clearly were *not* consistent with earlier findings. Consistent reports of lower network ratings since the peoplemeter's introduction suggest that the method is generally reliable, though not necessarily valid.

Validity of Ratings
● ● ● ● ● ● ● ● ● ● ●

Validity in research refers to the degree to which findings actually measure what they purport to measure. Ratings purport to measure the *entire* broadcast audience, but in practice they can account for only the broad middle-range majority of that audience. Thus ratings tend to underrepresent the very rich, the very poor, the very young, and ethnic minorities, and they are less valid for that limitation.

Television's use of households rather than individuals as its ratings measurement unit also affects validity. About a quarter of today's households consist of lone individuals whose lifestyle (and therefore broadcast use) patterns often differ from those of multiperson households.

Further, measuring only households excludes many other venues for watching television. Not included in the ratings data is viewing from bars, hotels and motels, hospitals, and college dormitories. Until recently, comparatively little was known about how much viewing was done away home. A 1994 Nielsen study found that about 4 percent of adult weekly viewing takes place away from home, a figure that, given the recent proliferation and popularity of sports bars, is surely rising. A 1996 study of daytime network viewing found that each week 3.7 million women ages 18 and over (about 4 percent of that demographic group) watched some daytime television from a location outside the household (Nielsen Media Research, 1998: 45). The ever-increasing amount of programming being delivered via the Internet could also significantly affect away-from-home viewing rates, especially as more office computers come equipped with TV receiver cards.

Hyping and Tampering

A widespread industry practice known as *hyping* (or, sometimes, *hypoing*) can also bias ratings. Hyping refers to deliberate attempts by stations or networks to influence ratings by scheduling special programs and promotional efforts during ratings sweeps. For example, both radio and television stations sometimes hype with listener contests.

Both the FTC and the FCC have investigated hyping, but with little effect—networks and nearly all stations do it to some degree. The arrival of 52-weeks-a-year meter ratings for television has reduced the incidence of hyping somewhat, yet many stations continue hyping during sweep weeks. In any case, Arbitron and Nielsen note in ratings reports any exceptionally blatant activities so advertisers can take them into account in market analyses.

Less commonly, ratings can be vulnerable to *tampering:* Someone who can influence the viewing habits of even a few households in a small sample can have a substantial impact on resulting ratings. Ratings companies therefore keep sample household identities a closely held secret. Still, a few cases of outright manipulation of viewers have become public, causing ratings services to junk reports for some programs, and even entire station or market reports for a given ratings period.

Qualitative Ratings

As long as ratings have dominated programming strategies, critics have complained that ratings encourage mediocrity by emphasizing sheer size to the exclusion of *qualitative* program aspects. Time and again, programs that are seemingly of above-average quality receive enthusiastic reviews and audiences but fail to meet the rigid minimum-share requirements for commercial survival. Critics question whether programs that are merely accepted by large audiences should automatically win out over programs that attract smaller but intensely interested audiences. Quantitative ratings dictate this kind of judgment—the system favors "least objectionable" majority programs over possible alternatives.

The current multichannel environment has changed this situation somewhat, though not entirely. Prime-time network programs today survive with ratings and shares that would have resulted in sure cancellation even five years ago. And because of the proliferation of cable networks eager to find product, programs that are forced off the broadcast networks may find a place on cable. Numbers are still the name of the game for the networks, but shows do not have to be as popular as in years past to survive commercially.

About the only qualitative ratings in which U.S. commercial broadcasters have shown an interest are those conducted by Market Evaluations, Inc. Using questionnaires completed by a random national sample of 1,000 families, this firm regularly estimates image (familiarity and likability—not necessarily the same thing) ratings of major performers and the popularity of specific programs among various demographic groups.

 ## 10.6 BROADCAST AUDIENCES

Over decades of intensive ratings research, a vast amount of knowledge about electronic media listening and viewing habits has been accumulated. This is surely the most analyzed media activity in history.

Set Penetration

The most basic statistic about broadcast audiences is *set penetration* or *saturation*—the percentage of all homes that have broadcast receivers. In the United States, radio and television penetration has long since peaked at more than 98 percent. Indeed, most homes have several radios and more than one television set. In short, for practical purposes the entire U.S. population constitutes the potential broadcast audience.

Set Use

HUT measurements tell us the average percentage of television households actually using sets at different times of the day. Television viewing climbs throughout the day from a low of about 12 percent of households at 7:00 A.M. to a high of about 70 percent in the top prime-time hour of 9:00 to 10:00 P.M. (Eastern time). Audience levels for television change somewhat with the seasons: Viewing peaks in January–February and bottoms out in June, reflecting some influence of weather and leisure activity on audience availability.

On the other hand, radio listening has a flatter profile than television. Radio listening typically reaches its highest peak during morning drive-time hours, recedes through the midday shifts, and climbs again during the afternoon drive-time period. Listening after 6:00 P.M. continually declines through the evening, as more listeners begin watching television news and then prime-time programming. Unlike television viewing, radio listening has very little seasonal variance.

TV as Habit

Long-term trends aside, people tend to turn on their television sets day after day in the same overall numbers, with no apparent regard for the particular programs that may be scheduled. Expressed in terms made famous by Marshall McLuhan, the *medium* matters more than the *message*. Paul Klein, former CBS programming chief, proposed a similar theory, that of the *least objectionable program* (LOP). He theorized that people stay with the same station until they are driven to another station by an objectionable program. But even if they find *all* programs objectionable, they will stay tuned to the least objectionable one rather than turn off the set.

Impact of Remote Controls

A corollary of LOP was *tuning inertia*. Whether because of viewer loyalty to a station or network, or simple laziness before widespread use of remote controls, viewers tended to leave sets tuned to the same station. Tuning inertia still strongly affects radio audiences; in large markets with as many as 40 stations from which to choose, listeners tend to confine their tuning to only two or three favorites.

By the late 1980s, however, remote-control devices (RCDs) for television—and the increased number of cable channels—began to modify old patterns. Whereas fewer than 30 percent of households had remote controls in 1985, by the mid-1990s nearly 90 percent did—and could thus very easily change channels. One result has been *restless viewers*, or *grazers*—those who use a remote control to switch rapidly among several channels. These patterns of use may be already changing. Research indicates that many RCD users are becoming much more sophisticated and deliberate. One study found that the most common use of the RCD was punching in a specific channel number, a direct substitute for on-set tuning (Eastman and Newton, 1995).

Time Spent

The total amount of time that people devote to television serves as a broad measure of its audience impact. This statistic arouses the most widespread concern among critics: Any activity that takes up more time than sleeping, working, or going to school—as watching television does for many viewers—surely has profound social implications.

In the 1982–83 television season, average daily viewing per household was 6 hours and 55 minutes (or, more precisely, at least the TV set was turned on for that length of time). Fifteen years later that average had climbed to 7 hours and 16 minutes. This total combined viewing by all household members. Exhibit 10.k details other average viewing levels by various demographic groups. According to Arbitron, radio use by all people ages 12 and over averaged 22 hours per week in 1998, down slightly from previous years (see Exhibit 10.l on p. 312).

As technology changes and new generations of consumers begin using media, new habits develop. Children may now turn to computers rather than television for entertainment. And many students today choose to listen to their favorite music via portable CD and MP3 players rather than the radio.

EXHIBIT 10.k
TV Viewing Trends

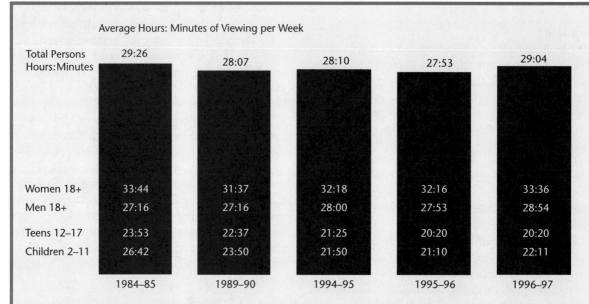

Average Hours: Minutes of Viewing per Week

Total Persons Hours:Minutes	29:26	28:07	28:10	27:53	29:04
Women 18+	33:44	31:37	32:18	32:16	33:36
Men 18+	27:16	27:16	28:00	27:53	28:54
Teens 12–17	23:53	22:37	21:25	20:20	20:20
Children 2–11	26:42	23:50	21:50	21:10	22:11
	1984–85	1989–90	1994–95	1995–96	1996–97

During the 1995–96 broadcast year the amount of time persons spent viewing television declined slightly from that of the previous year. The largest shift was among teens, followed by children. In 1996–1997 viewership rose again for all categories except teens.

Source: From Nielsen Media Research's *1998 Report on Television*. Reprinted by permission of Nielsen Media Research.

10.7 CABLE AUDIENCES

Researching cable audiences somewhat parallels the patterns familiar in broadcast research, in terms of both methods and findings.

Cable Research

Cable television determines its potential audience on the basis of basic-cable subscribers, although one might argue that the number of *homes passed* by cable feeder lines—which includes those who choose not to subscribe—is a better indicator of cable's real potential. Though *subscriber* numbers tell nothing about actual cable *use*, they do show that by the year 2000 cable was present in about two-thirds of U.S. television households.

Because of cable's many channels (with more being added all the time), its total audience subdivides into many small groups, posing severe measurement difficulties. Further, some cable-tuning converter boxes make it difficult for Nielsen to attach passive household meters. Cable systems that provide addressability (the ability to control and record access to each subscribing household from the headend) can resolve this problem, but only as the number of systems with that capability continues to expand.

Cable Ratings

Nielsen introduced audience reports for a few national cable services in 1979, gradually increasing the amount of information provided on basic and pay networks. Based on Nielsen's national people-meter sample, the *Nielsen Homevideo Index* (NHI) issues quarterly *Cable National Audience Demographic Reports* and *Nielsen Cable Audience Reports*, which provide cable household viewing data in all dayparts for the largest national cable and broadcast networks.

However, *local* cable audiences are far more difficult to measure than national audiences. Political

Average Daily Usage per TV Household—Hours: Minutes

Year	February	July	Annual	Year	February	July	Annual
1981–82	7:22	6:09	6:48	1989–90	7:16	6:24	6:55
1982–83	7:33	6:23	6:55	1990–91	7:36	6:26	6:56
1983–84	7:38	6:26	7:08	1991–92	7:32	6:39	7:04
1984–85	7:49	6:34	7:07	1992–93	7:41	6:47	7:12
1985–86	7:48	6:37	7:10	1993–94	7:51	6:53	7:16
1986–87	7:35	6:32	7:05	1994–95	7:42	6:45	7:15
1987–88	7:38	6:31	6:59	1995–96	7:38	6:59	7:17
1988–89	7:32	6:27	7:02	1996–97	7:28	6:49	7:12

Prior to 1993–94, 12 months ending August each year; after 1993–94 dates follow broadcast seasons (mid-September to mid-September). Data prior to 1987 based on NTI Audimeter sample; 1987 to present based on Nielsen peoplemeter TV audience.

boundaries define cable franchise areas, which often comprise only a single county or even just part of one, whereas broadcast signals, by ignoring such artificial lines, create larger markets that are more cost-effective to measure. Nielsen includes local cable viewing only in markets where cable audiences reach minimum television reporting levels, usually a share percentage of 3 or more.

In 1992 Nielsen previewed a new local cable ratings book designed to show market-level viewing of major national cable services. In 1993 it announced a new computer-readable service that would more equitably compare local, cable, broadcast, and syndicated program viewing patterns. Meanwhile, several cable systems turned to telephone coincidentals to measure their audiences.

Uses of Cable

By the mid-1980s research had begun to show clear evidence of national cable network influence on au-dience viewing patterns. Among households subscribing to cable at the end of the decade, basic-cable networks collectively had a viewing share of 50 percent, compared with 19 percent in 1985. Pay-cable hovered at an 8 percent share, down from 10 percent in 1985. Broadcast networks' share of viewing in those same households had declined from 56 percent to 44 percent across all dayparts (NCTA, Spring 1999). Use of television increases in households with more than one set and as more than one set is connected to cable television.

10.8   VCR AUDIENCES

Video cassette recorder (VCR) use patterns call for research approaches quite different from those for broadcasting or cable. Yet as VCRs become increasingly common (they are now in more than 80 percent of TV homes, with more than 20 percent

EXHIBIT 10.1

Radio Listening Patterns

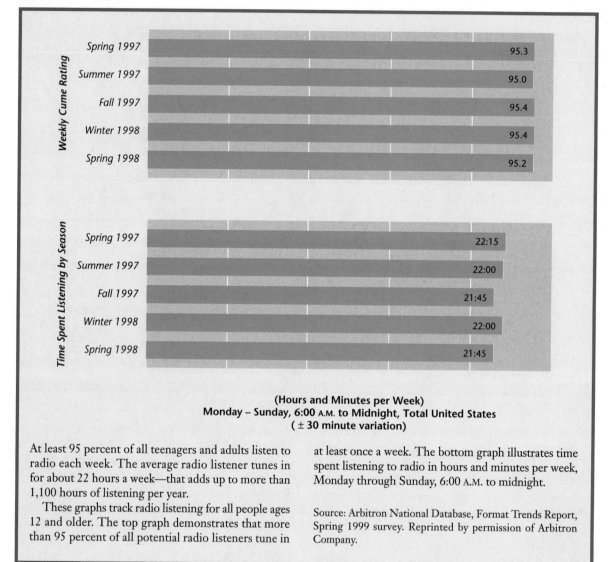

Weekly Cume Rating

Season	Value
Spring 1997	95.3
Summer 1997	95.0
Fall 1997	95.4
Winter 1998	95.4
Spring 1998	95.2

Time Spent Listening by Season

Season	Value
Spring 1997	22:15
Summer 1997	22:00
Fall 1997	21:45
Winter 1998	22:00
Spring 1998	21:45

(Hours and Minutes per Week)
Monday – Sunday, 6:00 A.M. to Midnight, Total United States
(± 30 minute variation)

At least 95 percent of all teenagers and adults listen to radio each week. The average radio listener tunes in for about 22 hours a week—that adds up to more than 1,100 hours of listening per year.

These graphs track radio listening for all people ages 12 and older. The top graph demonstrates that more than 95 percent of all potential radio listeners tune in at least once a week. The bottom graph illustrates time spent listening to radio in hours and minutes per week, Monday through Sunday, 6:00 A.M. to midnight.

Source: Arbitron National Database, Format Trends Report, Spring 1999 survey. Reprinted by permission of Arbitron Company.

of TV households owning two or more), understanding their many uses becomes vital.

VCR Research

Time shifting enables VCR owners to control when they watch broadcast or cable programs. VCRs also encourage skipping commercials by *zipping* (the avoidance of commercials during VCR playback) or *zapping* (the deletion of commercials during record-

ing). A recorded program (and possibly its advertisements) may be seen several times and by different viewers, or the ads may be seen only once if at all. To trace all these varied patterns, Nielsen publishes the quarterly *VCR Usage Study* based on special VCR diaries filled out monthly by sample households. The time broadcast material is recorded is considered the time of viewing for purposes of determining ratings—playbacks are not counted. Thus, even if no one ever watches the tape (a common phenomenon), the material is still counted as having been viewed.

EXHIBIT 10.m
• • • • • • • • • • • •

Patterns of VCR Use

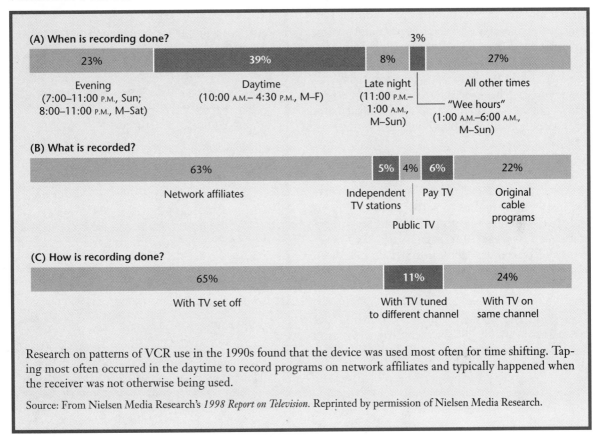

(A) When is recording done?

23%	39%	8%	3%	27%

Evening
(7:00–11:00 P.M., Sun;
8:00–11:00 P.M., M–Sat)

Daytime
(10:00 A.M.– 4:30 P.M., M–F)

Late night
(11:00 P.M.–
1:00 A.M.,
M–Sun)

All other times

"Wee hours"
(1:00 A.M.–6:00 A.M.,
M–Sun)

(B) What is recorded?

63%	5%	4%	6%	22%

Network affiliates

Independent
TV stations

Pay TV

Original
cable
programs

Public TV

(C) How is recording done?

65%	11%	24%

With TV set off

With TV tuned
to different channel

With TV on
same channel

Research on patterns of VCR use in the 1990s found that the device was used most often for time shifting. Taping most often occurred in the daytime to record programs on network affiliates and typically happened when the receiver was not otherwise being used.

Source: From Nielsen Media Research's *1998 Report on Television*. Reprinted by permission of Nielsen Media Research.

Uses of VCRs
• • • • • • • •

VCRs are almost twice as likely to be used to play tapes than to record them. Although they are still used mostly to supplement real-time broadcast and cable program viewing, VCRs do give viewers more control over their time in at least four ways:

• to time shift broadcast or cable programs for later viewing (sometimes more quickly, as in sports events, by fast-forwarding through dull spots and commercials) but not for retention after that viewing
• to develop a permanent home video library
• to view purchased or rented prerecorded tapes
• to watch tapes made with home camcorders

Most playback occurs during television prime-time hours. Exhibit 10.m offers more data on VCR usage.

10.9 MEASURING INTERNET USE

As the Internet, particularly the World Wide Web, becomes a more important advertising medium (see Section 6.8, page 187), more interest is placed on discovering how people use the web. A number of companies now attempt to measure Internet use. According to one of these companies, Nielsen// NetRatings, by the beginning of 2000 approximately 76.5 million people actively searched the Internet. The company estimated that 123 million people ages 2 and older in the United States had access to the Internet but did not necessarily go online. Web site providers and advertisers on those sites want to know how many people access their sites and for how long. Nielsen//NetRatings used a sample of 20,000 computer-equipped households to calculate its Internet ratings.

According to the February 2000 report from Nielsen//NetRatings, active Internet users averaged almost 9 hours and 20 minutes per month. This translates into roughly 18 sessions per month at approximately 31 minutes per session. The company also reported that during a typical session the average user visited 10 unique sites and spent almost 1 minute on each page viewed. Exhibit 10.n provides a snapshot view of the most popular Internet sites during one month in 2000. (Nielsen//NetRatings, 2000)

In October 1999, Arbitron launched a new service, dubbed InfoStream, designed to measure streamed audio services on the Internet. The service measures all tuning to participating channels and then compiles monthly cume audience estimates and average time spent with each service. As of December 1999, InfoStream measured listening to 290 Internet audio channels. At that time, the top-rated audio services on the Internet, two adult alternative channels out of Austin, Texas, and Monterey, California, garnered between 56,000 and 58,000 different listeners during the month (Arbitron Internet Information Services, 2000).

10.10 OTHER APPLIED RESEARCH

Nonratings research tries to find out what people like and dislike, what interests or bores them, what they recognize and remember, and what they overlook and forget. To study such subjective reactions, investigators usually use *attitudinal* research methods, which reveal not so much people's actions (such as set use) as their reactions—their *reasons* for action, as revealed in their attitudes toward programming.

Focus Groups

Commercial attitudinal research often makes no attempt to construct probability samples, because it usually does not try to make estimates generaliz-

EXHIBIT 10.n

Measuring the Internet

Top ten web properties for the month of February 2000, home.

Property	Reach %	Time per person
1. AOL Websites	60.29	0:28:20
2. Yahoo!	55.97	1:00:44
3. MSN	39.60	0:54:05
4. Excite@Home	32.59	0:27:44
5. Lycos Network	30.49	0:20:42
6. GO Network	22.98	0:21:11
7. Microsoft	20.52	0:10:25
8. Time Warner	17.27	0:13:00
9. NBC Internet	16.99	0:13:00
10. Amazon	14.61	0:13:53

Nielsen//NetRatings used a sample of 20,000 computer-equipped households to measure the 76.5 million people who accessed the Internet. Internet content providers and advertisers are interested in both the number of people visiting their sites and the length of each visit. In the above chart, "reach" refers to the percentage of Internet users who accessed a site at least once during the month.

The reported Internet usage estimates are based on a sample of households using the following platforms: Windows 95/98/NT, and MacOS 8 or higher. The Nielsen//NetRatings Internet universe is defined as all members (2 years of age or older) of U.S. households that currently have access to the Internet.

Source: Copyright 2000 NetRatings, Inc. Reprinted by permission of Nielsen Media Research.

able to whole populations. Instead, investigators choose respondents informally. They assemble small panels, called *focus groups*, and gain insights about people's motivations through informal discussion interview sessions.

Program concept research, for example, tries out ideas for potential new shows. A focus group's reactions to a one-page program description can help programmers decide whether to develop an idea further, to change details, or to drop it entirely. Advertisers often test concepts for commercials before making final commitments to full production. These tests may use simple graphic storyboards, or they may employ *photomatics*, videotaped versions of original storyboards with camera effects and audio added to make them look and sound something like full-scale commercials.

To test new or changed programs, producers often show pilot versions to focus groups. People watch, give their reactions, and sometimes discuss reasons for their attitudes with a session director. Frequently, producers, writers, and others study reactions by watching these discussions through one-way mirrors or by screening videotapes of the session. Increasingly, web site designers are testing new sites in similar ways.

Program Analysis

Minute-by-minute reactions to a program can be studied using a *program analyzer*—a device first developed in the 1940s and now thoroughly computerized—which enables test-group members to express favorable, neutral, or unfavorable reactions by pushing buttons at regular intervals on cue. The machine automatically sums up the entire test group's reactions, furnishing a graphic profile. A follow-up discussion can then probe for reasons why audience interest changed at given moments in a script, as revealed by peaks and valleys in the graph.

Theater vs. In-Home Testing

The movie industry has long used theater previews to gauge audience response. Several firms specialize in staging similar theater previews (sometimes called *auditorium testing*) of television programs and commercials. Investigators sometimes test advertisements under a pretext of testing programs, with commercials seeming to appear only incidentally. Viewer reactions come out in questionnaires or discussions.

Physiological Testing

Most of the methods described thus far depend on self-analysis by panel members. In an attempt to eliminate subjectivity and to monitor responses more subtly, researchers have identified a number of involuntary physical reactions that give clues to audience thinking. Reactions measured for this purpose include changes in brain waves, eye movements, eye pupil dilation, breathing rates, pulse rates, voice quality, perspiration, and sitting position (the "squirm test").

For example, a number of researchers have capitalized on the two-sided nature of the human brain. Each side has its own specialized functions and reacts to different stimuli. Reasoning ability seems to be located in the brain's left side and emotions in the right. It follows that product ads with emotional appeal should, if correctly designed, stimulate primarily the right side of the brain more than would commercials that appeal mainly to logic. Commercials shown to viewers wired for brain-wave recording can be tested to determine whether the messages draw the desired brain-wave responses.

Audience Response

Telephone calls and letters from listeners or viewers about programs, performers, and commercials provide additional audience information. Indeed, early broadcasters relied entirely on voluntary listener mail for audience information. But people who write or call a station are not a good representative sample of the entire audience. Research has shown that letter writers differ significantly from the general population in terms of race, education, income, type of job, age, and marital status—all differences important to advertisers and programmers. Further, letter-writing campaigns for or against a given point of view, product, or service can give misleading impressions about general audience reaction. For these reasons, stations take phone calls less seriously than systematically gathered data. Likewise, unsolicited e-mail comments concerning a company's web page may not be considered as carefully as responses to a systematic survey.

CHAPTER

Effects

The time we spend with electronic media—typically more each day than with any other leisure activity—underlines our need to understand their varied impacts. This chapter reports some of what research has discovered about the many ways electronic media affect our lives and raises questions for which we don't yet have complete answers. Researchers now strongly feel that the background audiences bring to their use of electronic media is more important to understanding media "effects" than any study of media content.

11.1 DEVELOPING EFFECTS RESEARCH

People have long assumed that spending so much time with electronic media simply had to have some impact, good or bad. Such assumptions take on practical significance when policy decisions are based on supposed or real media impact. For that reason, and because of scientific curiosity, researchers have spent considerable effort to determine media effects.

Early Findings

Early media research focused on the impact of *mass propaganda* efforts used during World War I (1914–1918), before broadcasting arrived. After the war, the extent of deception by propagandists on all sides showed how thousands had been manipulated. The postwar advent of radio broadcasting—along with the emergence of the politically threatening new communist regime in Russia—made fear of increased propaganda manipulation more alarming. This led to more research on the social and psychological dynamics of propaganda.

The 1920s concept of media impact saw messages as so many bullets of information (or misinformation) aimed at passive groups. Because researchers assumed that messages penetrated and caused specific reactions, the concept became known as the *bullet* or *hypodermic-injection* theory.

By the late 1930s researchers realized the bullet theory was oversimplified in treating message receivers as mere passive targets. New and more so-

phisticated studies discovered that audiences react to messages as individuals. Media effects, therefore, depend on many variables unique to individual audience members.

Researchers labeled such factors (many not directly observable) *intervening variables* because they come between messages and effects. They vary a message's impact because of each person's previously acquired attitudes, traits, experiences, social situation, and education. Intervening variables explain why an identical message often has a different impact on different people.

One important intervening variable studied in the 1940s was the personal influence of *opinion leaders*, as contrasted with media's *impersonal* influence. One theory suggested that media influence often passes through leaders to followers rather than to all individuals directly. For most people, opinion leaders play a greater role in various decisions than does the direct influence of radio, newspapers, magazines, or books. Other studies confirmed and refined this *two-step flow* hypothesis of media influence (Katz and Lazarsfield, 1955), which had great impact on research for two decades.

Selective Effects

Yet even opinion leaders are not influenced in direct proportion to the amount of persuasive media content they receive. People pay attention to messages that fit their established opinions and ignore those that don't. Researchers discovered that because of this *selective exposure*, media tend to *reinforce* existing views rather than convert people to new ones. Existing mindsets color how people perceive media they select. Those who select the same message may interpret it in different ways. Though the stimulus is constant, the response varies. Called *selective perception*, this variable shows that people interpret messages based on their opinions and attitudes rather than receiving them passively (as the old bullet theory had supposed).

Selective perception also accounts for a *boomerang effect*. Experiments have shown that those with strong beliefs tend to misinterpret messages with which they disagree—they distort evidence or retain only those elements that reinforce their existing attitudes, rather than allowing the

EXHIBIT 11.a
· · · · · · · · · · · ·

A Simple Communication Model

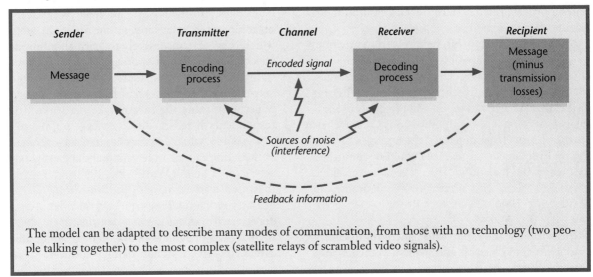

The model can be adapted to describe many modes of communication, from those with no technology (two people talking together) to the most complex (satellite relays of scrambled video signals).

message to change even slightly their present thinking. Thus propaganda can boomerang, producing exactly the opposite of the intended effect (Cooper and Jahoda, 1947). As a classic example, many television viewers misread the antibigotry message of the Archie Bunker character in *All in the Family* in the 1970s. They saw an endorsement of the very prejudices that producer Norman Lear intended to satirize.

This refusal to forsake an existing mindset suggests that people have a kind of internal gyroscope that tends to maintain a consistent set of attitudes, opinions, and perceptions. Psychologists developed several versions of a concept known generally as *congruence theory* to account for this tendency, which is an important intervening variable in determining communication effects. Such theories hold that a person's internal state of mind is usually in balance or congruence. A message that contradicts established opinions causes dissonance, lack of congruity, or imbalance. An effort to restore balance (conscious or unconscious) follows. It might take the form of rejecting the message (saying the source is unreliable, for example), distorting the message to make it fit the existing mindset, or, least likely, adjusting the balance by accepting the new idea—conversion to a new point of view.

These concepts all came out of the social sciences. A very different way of looking at communication came from engineering with the publication of *The Mathematical Theory of Communication* (Shannon and Weaver, 1949). Here information is seen as a transmission system, using such familiar engineering terms as channel capacity, noise, encoding, and decoding (see Exhibit 11.a). *Information theory* contributed valuable insights to media research, including the concept of *feedback* (Wiener, 1950), whereby communicators modify their messages in response to information that comes back from audiences. But because they lack the immediate give-and-take of face-to-face conversation, media operate at a disadvantage. It takes time for them to receive and analyze feedback and to modify their product accordingly. This may change for some electronic media as interactive capabilities speed up the feedback process.

More recent *accommodation theory* argues that media are but one part of a person's information system as evidenced by the importance of what background factors any individual brings to his or her use of media. Personal experiences and interpersonal contacts heavily color the way all of us learn from and make use of media (Anderson & Meyer, 1988: 44–45).

Research Topics

• • • • • • • • • • •

Because communication involves a chain of events, it is best explained as a *process*. A communicator initiates a process that produces an end result—if definable effects occur.

Study of the process can be directed to any one, or any combination, of five different elements: (1) *originators* of messages, (2) *contents* of messages, (3) *channels* through which messages travel, (4) *audiences* that receive messages, and (5) *effects* of messages. A pioneer communications researcher, Harold Lasswell, summarized these stages by saying that the objects of research could be identified by posing the question, "*Who* says *what* in which *channel* to *whom* with what *effect*?" (adapted from Smith et al., 1946: 121).

- Researchers study the "who" of communication to learn about sources and shapers of media content—those who act as *gatekeepers*. FCC regulation and station or network "clearance" offices are types of gatekeeping, as is the selection, placement, and editing of all content presented. Gatekeeping studies examine how controls operate, where gates in information flow occur, and what effects they have on content when it reaches its destination.
- "What" studies usually consist of content analyses of electronic media programming and thus concern us only insofar as they indicate potential impact—as with measures of the amount and types of violence present in programs of different kinds aimed at different audiences.
- As to the "which channel" question, research shows that media channels differ in their psychological impact because audiences form expectations for each medium and interpret what each delivers accordingly. Since 1959 television broadcasters have sponsored biennial Roper Organization (now Roper Starch Worldwide) image studies of the medium. The surveys regularly ask respondents which of several media they would believe in case of conflicting news reports. Since 1961 they have consistently chosen television over other media by a wide margin. In the 1998 study more than half the respondents considered television the most credible news source, followed by newspapers (23 percent), radio (7 percent), magazines (4 percent), and on-line news sources (1 per-

cent). The survey also revealed that 69 percent of adults got most of their news from television, 37 percent from newspapers, 14 percent from radio, 7 percent from other people, and 5 percent from magazines. On the advertising front, 62 percent of adults cited television as the medium from which they were most likely to learn about products or brands. Twenty percent of the survey respondents cited newspapers, 12 percent noted magazine ads, and 4 percent listed radio. Among households using the Internet, 15 percent of the respondents listed on-line services as their top news source, while 14 percent cited such services as their first choice in learning about new products. (Roper Starch Worldwide, 1998)

- Ratings are an example of the "to whom" factor of Lasswell's model inasmuch as they detail media use with *media exposure* or *time-spent* data. Breakdowns of audiences into demographic subcategories give further details about the "to whom" of broadcasting. But in researching the "to whom" question, scholars use more detailed personal and social-group indicators to study the composition of audiences than does ratings research. In particular, the child audience has been extensively analyzed, using variables such as race, intelligence, social class, home environment, and personality type. These analyses relate audience characteristics to effects. Researchers ask questions such as, What types of children will be most likely to believe what they see on television, and what types will imitate what they see?
- Finally, Lasswell's "with what effect" culminates his question, because communicators, content, channels, and audiences all help determine the ultimate outcomes of communicating. Joseph Klapper summarized effects theories as of 1960 in his influential *The Effects of Mass Communication*. Summarizing more than a thousand research reports, Klapper concluded that ordinarily communication "does not serve as a necessary and sufficient cause of audience effects, but rather functions among and through a nexus of mediating factors and influences." Media largely reinforce existing perceptions and beliefs. They may persuade people to buy a product, but not to change a political allegiance or religion. The broadcasting industry welcomed this conservative conclusion, known as the *law of minimal effects*, because it gave apparent scientific sanction

to the industry's argument that programs or advertisements could not be blamed for causing antisocial behavior.

In the 1970s opinion began to shift away from minimal effects concepts, largely because of intensive research on violence (discussed in Section 11.7). Researchers now even avoid talking about *effects* as such. The very word implies an oversimplification of what is now understood to be an extremely complex process. Without denying that specific media content might under specific conditions have specific effects on some specific people, researchers prefer to speak in terms of *associating* certain inputs with certain outputs. They avoid reliance on a simple, straight-line, cause-effect relationship. Exhibit 11.b summarizes the stages of research development, from the simplistic cause-effect concept of early studies to our present interest in looking more deeply into the many variables involved in impact.

The People Problem

In designing studies on media impact, researchers face the frustrating problem that impact is usually made up of *subjective responses* hard to determine by direct observation and measuring instruments.

Most impact research today relies in whole or in part on questioning people about their subjective experiences—how they think and feel about something—rather than on observing their reactions. Self-reporting, of course, is not altogether reliable. People are sometimes unwilling—or unable—to tell the truth about their inner experiences, or they may be forgetful or unaware of their own subconscious motivations.

Methods of Research

Investigators usually select one of five major types of research:

- *Sample surveys.* The research strategy most familiar to the public is the sample survey used in opinion polls and audience ratings reports (see Section 10.3, page 300). Such surveys can estimate characteristics of entire populations through the use of very small random samples. Sample surveys tell us nothing, however, about the *causes* of tuning—a major weakness of the survey strategy.

- *Content analysis.* Classification of programs into various categories constitutes one form of content analysis. On a more sophisticated level, content analysis categorizes, enumerates, and interprets message content. Media researchers have used content analysis to study advertising copy, censors' actions, television specials, violent acts in programs, and the portrayal of minorities in television dramas.

- *Laboratory experiments* allow researchers to control some experimental factors precisely while excluding others. Both subjectively reported and objectively observed data can be derived from such experiments. Lab experiments do, however, put people in artificially simplified situations unlike the complex ways in which most of us actually use media.

- *Field studies* record behavior in "the real world" without unduly intruding on or otherwise influencing participants. In studying the impact of violent programming, for example, field-study researchers can watch children in their normal home or school environment rather than in a lab setting. However, these experiments require a great deal of time and effort to arrange, and some of their methods can be just as questionable as those used in laboratories.

- *Ethnographic studies.* Here the researcher becomes part of the (usually small) group being studied, and this *participant observation* over extended time periods allows a realistic "window" on media consumption patterns to develop, along with some sense of subtle media effects. Ethnographic studies are expensive because of their intensive and time-consuming nature and are not readily generalizable to larger groups in society because of their narrow study-group base.

11.2 NEWS

Because most Americans depend primarily on television for news, TV journalism clearly plays an important role. Most of us see the world beyond our own lives much the way cable and television news present it to us.

EXHIBIT 11.b

Development of Media Effects Research

Time Period	Prevailing Viewpoint	Empirical Basis
1. 1920s–1930s	Mass media have strong effects	Observation of apparent success of propaganda campaigns Experiments show immediate attitude change after exposure to messages Evidence of selective perception—persons ignore messages contrary to existing predispositions
2. 1940s–1950s	Mass media largely reinforce existing predispositions, and thus outcomes are likely to be the same in their absence	Evidence of personal influence—persons are more influenced by others than by the mass media Evidence of little influence on voting No relationship observed between exposure to mass media violence and delinquent behavior among the young
3. 1950s–1960s	Mass media have effects independent of other influences, which would not occur in the absence of the particular mass media stimuli under scrutiny	Evidence that selective perception is only partially operative Evidence that media set the context and identify the persons, events, and issues toward which existing predispositions affect attitudes and behavior Evidence that television violence increases aggressiveness in the young
4. 1970s–1980s	Process behind effects so far studied may be more general, suggesting new areas for research	Research finds that under some circumstances TV may influence attitudes and behavior *other than* aggressiveness Many agenda-setting studies
5. 1980s–1990s	Individual differences in audience background and needs are found central in defining media impact	Qualitative studies (including ethnographies) of media as one of many social resources

Source: First four segments adapted from George Comstock et al., *Television and Human Behavior.* Copyright 1978. Used by permission of RAND Corporation.

Gatekeeping

Media can report only a tiny fraction of everything that happens in a day. On its way to becoming neatly packaged pieces on the TV screen, raw reporting of events passes through many editorial *gatekeepers*. News directors, news producers, photographers, and reporters open and close gates deliberately, deciding which events to cover in which places and how stories should be written, edited, and positioned in the news presentation. Some gatekeeping occurs inadvertently, depending on accessibility of news events or availability of transportation or relay facilities.

Some gatekeeping has an institutional or an ideological bias—or both. Institutional biases can develop from news organization priorities ("If it bleeds, it leads"); ideological biases grow out of individual political, social, economic, or religious beliefs ("We don't reveal the names of rape victims").

Television demands pictures. This tends to bias the medium toward covering events that can be visualized. Impacts of this visual bias include a preference for airing stories that have dramatic pictures (fires, accidents, disasters) and a forced effort to illustrate nonvisual stories with often irrelevant shots (as when file pictures of bidding on the floor of the stock exchange illustrate a story on financial trends).

Agenda Setting

Gatekeeping focuses our attention on selected events, persons, and issues that temporarily dominate the news. The list changes frequently as old items drop out and new ones claim attention.

No more vivid example of this process exists than the 1994–1995 coverage of the arrest and prosecution of football and media star O. J. Simpson for the murder of his ex-wife and her male friend—hours of live helicopter-based shots of Simpson's Ford Bronco on Los Angeles freeways and of the ensuing courtroom drama, all of which preempted regular programming, from basketball playoffs to prime-time programs to newscasts to soap operas. More recent examples of this process include media coverage of the deaths of Princess Diana and John F. Kennedy, Jr.

Researchers term this overall process of selection and ranking *agenda setting*, one of the primary ways in which media shape our perceptions of the world.

In short, although media may not tell us *what* to think, they do tell us what to think *about*.

A related impact is *status conferral*. An event's very appearance on the air gives it importance. Well-known anchors and correspondents lend glamour and significance to the events and people they cover personally. If the story were not important, would Dan Rather, Tom Brokaw, Peter Jennings, or Bernard Shaw be covering it? Conversely, can an event really matter if television chooses *not* to be there?

Covering Controversy

Similar agenda setting is evident in coverage of the AIDS epidemic. Constant television coverage of the disease's spread after 1980, health agency efforts to counter that spread, case studies of individuals who are infected and become spokespersons to warn others (especially celebrities such as basketball star Magic Johnson)—all serve to focus more attention on AIDS and the search for its cure.

Concern about the environment has been encouraged by television coverage of bad news (burning tropical forests, disappearing animal species, air and water pollution) and good (planting of new forests, fining polluting companies, enacting tougher laws).

Coverage of the abortion controversy features highly vocal groups and speakers on both sides of this emotional, political, religious, and personal issue. Here again, television news largely *reflects* divisions evident in society at large. The camera and microphone merely record events—though in so doing, they keep a controversy at the top of the current agenda.

No recent controversy has consumed the news media and captivated the public more than the affair between President Bill Clinton and White House intern Monica Lewinsky. From the first hint of scandal, to the president's televised denial and later admission, through the web publication of special prosecutor Kenneth Starr's titillating report to Congress, the news media covered the story with such relish that many media critics complained of overkill. For example, Tom Dorsey, media critic for the *Louisville (Ky.) Courier-Journal*, reported,

> Whatever you think of the Bill Clinton–Monica Lewinsky scandal, it was massively overreported by television, which has kept the story front and center [since it first broke in January 1998]. Cable was the

worst offender. In a greedy grab for ratings [several all-news channels] drown the audience in coverage day in and day out. But we were only reporting the facts, spokesmen for these networks say. Talk about misleading the public. There were precious few facts in all the hours of coverage. Most of what the cable news networks aired over the past . . . months was nothing more than "experts" talking off the tops of their heads. (Dorsey, 1998: 2c)

Whether or not the media overplayed the story, they had an eager audience. When Clinton delivered his explanation of the affair in August 1998, the speech drew an estimated audience of 45.9 million households and 67.6 million viewers on 10 networks. In September of that year, 9 networks interrupted their normal daytime schedules (at a loss of millions of dollars in ad revenue) to present four hours of Clinton's videotaped grand jury testimony to an estimated audience of more than 22 million people.

Following Starr's report to the House of Representatives, public interest in the television proceedings began to wane. When the House voted to impeach President Clinton on December 19, 1998, CBS decided to stay with its broadcast of a football game, at times interrupting the game with updates from Dan Rather. Later, CBS officials declared that more people had learned about the impeachment from Rather on CBS than on any of the other networks. By the time the trial of the president began in the Senate, most people who watched did so out of curiosity, knowing before the trial even began that a conviction was not likely.

In the late 1990s school shootings, church burnings, and the commission of so-called hate crimes increased media coverage of violence in our society. Ironically, this issue forced the media to cover serious questions and complaints about its own role in the spread of violent behavior. Some observers believe that the problem results at least in part from our exposure to so much violent content on television, in films, in video games, and on the Internet. This topic is addressed in Section 11.7.

Media News Staging

Television's need for images creates the temptation to enhance the pictorial content of news stories artificially. Even when news crews make no move to provoke reactions, the presence of cameras in tense situations can escalate or sensationalize ongoing action.

As a practical matter, some artificiality is accepted as normal in news coverage. In televised interviews, for example, a camera usually focuses on the interviewee. "Reverse angle" shots of the interviewer are usually taken afterward and spliced into the interview to facilitate editing and to provide a visual give-and-take. This tactic allows a single camera to cover interviews. Reporters also sometimes ask a source to do something (such as ask a police officer to turn on a patrol car's flashing lights) simply to provide more dramatic video.

Deliberate staging by reporters, however, is a far more serious violation of normal news practice. All too often we hear of a station or network carrying as news something set up by a news crew showing a drug bust, illegal betting, or the like, without any indication of the staging process. Early in 1993 NBC News got into trouble over what turned out to be a staged and partly faked story concerning General Motors truck safety. The head of the News Division was forced out, and several others were fired. But the damage to NBC's credibility was still considerable.

More recent and troublesome is the ability to change photographs, film, or videotape digitally to show things that never happened. This kind of staging may merely modify a sky to make it more dramatic or may totally alter a real event to suggest something quite different. Early in 2000, CBS News was heavily criticized for its New Year's Eve Times Square coverage because the network digitally replaced an NBC billboard that was actually in Times Square with a CBS logo. Stung by the criticism, CBS News personnel later apologized and admitted that such digital deceit had no business in the newsroom.

Pseudo-events

Outright staging of events by the *subjects* of news occurs when press agents or public-relations counselors seek to plant stories in media or to create happenings to attract media coverage. Daniel Boorstin (1964) coined the term *pseudo-event* to describe these contrived happenings. He analyzed the many forms they take, such as press conferences, trial balloons, photo opportunities, news leaks of confidential information, and background briefings "not for attribution." Free video news releases tempt stations short on photographic material.

These brief items supplied by business and government PR offices are often visually interesting, with the real message subtly buried.

Court Trials

The former long-time ban on cameras and microphones in virtually all courtrooms arose from assumptions that broadcast coverage would affect the behavior of participants (witnesses, lawyers, and defendants) and that these effects would be detrimental to the judicial process. Later, experience showed that once the novelty of being photographed wore off, subjects of coverage betrayed little reaction.

Three factors account for this minimal impact. Equipment became smaller and less obtrusive (able to operate with existing light levels, for example). News crews became more professional and sensitive to the need to avoid disruption. Finally, society became more tolerant, even expectant, of electronic media access to official activities. (For more on cameras in the courtroom, see Section 13.4, page 378).

11.3   WORLD EVENTS

During World War II (1939–1945), broadcast radio's first war, network news and entertainment played a highly supportive role in building both civilian and military morale.

Vietnam

Television's first war coverage came with the Vietnam conflict (the Korean War of 1950–1953 occurred during the formative years of television news, when live coverage from such a remote distance was impossible). Television coverage for the 1965–1975 decade made Vietnam a "living room war," in the words of *New Yorker* critic Michael Arlen (1969).

In total, this longest war in U.S. history played in living rooms for 15 years. CBS sent its first combat news team to Vietnam in 1961, and news photography of the final evacuation of Saigon, showing desperate pro-American Vietnamese being beaten back as helicopters lifted off the landing pad atop the U.S. embassy, came in 1975.

At first, Vietnam coverage tended to be "sanitized," stressing U.S. efficiency and military might, and playing down the gore and suffering of actual combat. Although little direct military censorship was imposed, as had been the case in previous wars, in this undeclared war generals and presidents had public relations uppermost in their minds. Thus control of access to events could serve the same end as more blatant censorship.

A more violent phase of news coverage came as a result of the early 1968 Tet offensive, which brought fighting to the very doors of the Saigon hotels where correspondents stayed. Broadcast news from a host of network and local-station reporters turned increasingly to combat realities, and Vietnam became a real war in American living rooms, not the sanitized version of military public relations.

Following the Tet offensive by the North Vietnamese, several events combined to turn U.S. opinion against the war. Television's leading journalist, CBS's Walter Cronkite, gave a sharply negative assessment of the war after a visit to Vietnam. Loss of Cronkite's support for the war was said to have solidified President Lyndon Johnson's decision not to run for reelection. At about the same time, vivid photographic images from the battlefields became icons of American disillusionment.

More Wars

Television's effect on public perception of the Vietnam conflict raised troublesome questions about future war reporting. Could a nation at war afford to allow television freely to bring home the horror of combat night after night?

A 1983 terrorist bombing that killed some 240 Americans at a U.S. Marine barracks in Beirut seemed a case in point. Television coverage of the bombing's aftermath undermined public support for the "peacekeeping" role of the Marines in Lebanon, making the venture politically untenable. American troops pulled out shortly thereafter.

On the other hand, would the public support a war that was made temporarily invisible by rigid military censorship? To the dismay of journalists, a U.S. invasion of the tiny Caribbean island of

Grenada in 1984 suggested that it might. Alleging security concerns, the military barred all media access to the initial assault. For the first 48 hours the world knew what happened only from military press releases. Subsequent press disclosures of military bumbling in the invasion and apparent official misrepresentation of circumstances leading to the attack came too late to dispel entirely the aura of success and righteousness surrounding the Grenada "rescue" of American civilians allegedly endangered by the volatile political situation on the island.

The public largely supported the action and applauded the administration's decision to limit media access—and impact. Strong media criticism of this abrupt departure from prior practice, however, led to an agreement that in future actions the military would set up a news pool to cover events from the outset.

In the 1980s, Reagan administration efforts to bring down the Sandinista regime in Nicaragua and related guerrilla fighting in several Central American countries served as testing grounds for television journalism's post-Grenada maturity. Controversy surrounded coverage of these conflicts. New Right politicians claimed that television weakened efforts to gain public support for "freedom fighters"; others, remembering Vietnam, said that aggressive television coverage might serve to keep American soldiers out of another prolonged, undeclared conflict.

When President Bush sent troops to Panama in 1989 to help remove President Manuel Noriega from power, the media were critical of the press pools set up by the military. According to many correspondents covering that action, media personnel once again were not allowed to accompany the combat troops and were admitted to battle areas only after most of the fighting had ceased.

Persian Gulf Crisis
• • • • • • • • • • • •

Lessons learned in all these events heavily influenced television's coverage of and impact on the 1990–1991 Gulf War, when the United States led a successful UN effort to throw Iraqi troops out of occupied Kuwait (see Exhibit 11.c). The Pentagon invoked strong censorship and pooling of reporters, making it almost impossible for the media to report from troop sites during the months-long

buildup unless a military public-affairs officer was present (SGAC, 1991).

An intense air war began in mid-January 1991 and was seen live on American East Coast television screens in prime time as CNN cameras in Baghdad focused on falling bombs and antiaircraft fire. Reporters Peter Arnett and Bernard Shaw continued live reports from Baghdad by means of a high-tech satellite telephone.

When reporters questioned Pentagon officials the next day as to the impact of initial bombing runs, officers made clear that much of what they knew had come from the same CNN reports seen by everyone else. That pattern was evident in the first few days of the air war, when (as shown by research) 44 percent of those turning to television watched CNN coverage, while 41 percent turned to one of the three traditional broadcast networks (Birch Scarborough, 1991: 9). Later analyses often attributed CNN's "arrival" as an accepted news source to its stellar role in reporting all aspects of the Gulf War.

Even more than had been the case with Vietnam, Operation Desert Storm—thanks to time-zone differences and satellite relays, as well as around-the-clock reporting—was a "living room war" in American prime time. But it was a far more *controlled* television war than Vietnam had been, the result of tight Pentagon limits on media access to troops and events—and of censorship, often directed more at the possible political impact of negative coverage than at any military security need.

Only months after the war was over did disquieting television reports about Iraqi civilian losses, allied soldiers killed by "friendly fire," and attempted cover-ups of other military mistakes become widely reported—but by then few were paying attention. Pentagon and administration officials called the Gulf War press pools a model for future conflict coverage, while reporters argued the opposite— that censorship had prevented their usual multifaceted job of reporting.

Most public-opinion surveys found that viewers were generally satisfied with what they had seen and believed military controls to have been justified. The controversy arose again when television images of starving women and children pressured the United States to send troops to Somalia to help stabilize that country in 1992. Military personnel attempting to disembark on the beaches of Somalia under cover of darkness were met by the bright

EXHIBIT 11.c
· · · · · · · · · · · ·

A Tale of Two Television Wars

Television coverage of war has changed dramatically since Vietnam became the first "television war." In that conflict, correspondents traveled freely with the troops and covered the war from their vantage point. Scenes showing the horror of war showed up on the TV screens of U.S. viewers night after night. In picture A the chief of South Vietnamese police executes a Viet Cong suspect during the 1968 Tet offensive. Picture B shows a Marine tank piled high with American dead. Concerned that this unlimited coverage turned public opinion against the conflict, military leaders later restricted the ways in which the news media could cover armed conflicts. Reporters no longer moved with the troops, but were held behind and allowed in the area only after much of the fighting had been completed.

The military highly orchestrated the ways in which television viewers experienced the Persian Gulf conflict in 1990–1991. The news media were limited to traveling in groups escorted by military personnel, and the military itself controlled much of the news available to the media. Here (picture C) General Colin Powell explains the campaign during a news conference. Picture D provides an infrared night view from a Stealth F-117 bomber of a "smart" bomb homing in on its target. Such pictures were regular features of Gulf War coverage. Very little of the death and destruction caused by the conflict was ever seen by TV audiences. Critics of the coverage suggested that it sanitized the war; military leaders and other supporters of the coverage believed that the censorship was necessary to protect the lives of combat troops and reporters and to maintain public support for the operations. Regardless of who was correct, the Gulf War set the standard for media treatment of later conflicts, as evidenced by television's similar coverage of the NATO action against Yugoslavia in late 1998.

Sources: (A) AP–Wide World Photos; (B) UPI/Corbis Images; (C) UPI/Corbis Images; (D) Reuters/Corbis Images.

lights of the television cameras, recording their every action. Public opinion turned against U.S. participation after several of the troops were killed and their bodies were shown being abused by their killers. It seemed likely that the contest of wills over the "proper" role of news media in wartime would continue into the *next* war.

Terrorism
• • • • • • •

News media problems took a vicious turn when terrorist organizations began committing crimes to gain news coverage. Starting in the 1980s small and desperate political or religious groups seeking world attention increasingly employed violence against usually innocent third parties. Such *publicity crimes* paradoxically transform pseudo-events into real events.

Bombings of airport terminals and subway stations, hijackings, and kidnappings, the most common kinds of terrorist stories, pose difficult ethical dilemmas for all news media. The very act of reporting a publicity crime transforms the media into accomplices, and the eagerness with which the public awaits news about it makes all of us inadvertent accomplices as well.

Observers complain that television's massive coverage encourages future terrorists by providing them with an international forum. Criticism focuses especially on coverage of news conferences staged and controlled by terrorists, whereby media appear captive to terrorist manipulation. Cable and broadcast news people often counter that they are merely accommodating the public's insatiable demand for more coverage. And surely these criticisms do not apply to terrorist acts in the mid-1990s, such as the Oklahoma City or Atlanta's Olympic Park bombings, where no terrorist group came forward to claim "credit" for the actions.

Revolution in Eastern Europe
• • • • • • • • • • • • • • • • • • • •

Television covered the dramatic changes in Eastern Europe and the USSR in and after 1989. Indeed, television's impact seemed to speed the collapse of the Soviet system.

The highlight—really an icon for a revolution—was the televised scenes, many at night under harsh lights, of Berliners tearing down the hated wall that

had divided their city for nearly three decades. The hard-line East German regime collapsed in days. In the following weeks, similar upheavals were reported in Poland, Hungary, Czechoslovakia, Bulgaria, and Romania.

In mid-1991 television highlighted the tension of what turned out to be a short-lived but frightening coup attempt in Moscow, when hard-liners tried to revive the former rigid communist regime that had prevailed before Mikhail Gorbachev's more open government. For two days television showed tanks moving in the streets of Moscow and highlighted the brave front put up by Russian leader Boris Yeltsin, leading those who challenged the coup plotters. The weakened Gorbachev soon stepped down, as reaction to the coup hastened the end of the nearly 75-year-old USSR. The world watched in amazement as television showed the red hammer-and-sickle flag being lowered over the Kremlin on New Year's Eve, 1991, to be replaced with the Russian flag—a scene previously imaginable only in spy novels.

But just as lingering and even more haunting in the years that followed was television's coverage of rising ethnic unrest in former communist-dominated countries. As had often happened in centuries past, Balkan political divisions broke into bitter civil war in what was once Yugoslavia. Viewers were treated in mid-1992 to oddly juxtaposed pictures of a war-torn Sarajevo and its devastated population (the city had been the site of the 1986 Winter Olympic Games) and prosperous Barcelona, which was hosting the 1992 games.

Night after night, television news showed the destruction of once-handsome cities and their hapless populations. Television pictures of concentration camps and reports of massive forced population shifts (ethnic or religious "cleansing" as it was called) increased pressure on other European governments to intervene to stop what struck outsiders as senseless tribal warfare.

Several years later, carnage began again when ethnic Albanians clashed with Serbs in the Serbian province of Kosovo, resulting in attacks on Yugoslavia by NATO forces. Television cameras caught the massive Albanian exodus from Kosovo and recorded the displaced Albanians' anguish over leaving their homes. Satellite photographs provided evidence of mass graves and wartime atrocities. Much like the coverage of the Gulf War, television viewers learned about the NATO action against Yu-

EXHIBIT 11.d
• • • • • • • • • •

Television Covers a Troubled Peace Process

Although most of the peace process went on privately, Palestine Liberation Organization president Yassir Arafat shook hands with Israeli prime minister Yitzhak Rabin after signing a peace accord in Washington, D.C., in front of a delighted President Bill Clinton and a global television audience. Television coverage then followed the peace process as Palestin-ian police took local control of areas once patrolled by the Israeli soldiers. Within a year television also recorded the assassination of Rabin by an Israeli student upset at his peace initiatives.

Source: Reuters/Corbis Images.

goslavia through staged briefings and videos of "smart" bombs surgically striking their targets.

Atrocities committed during the war in Bosnia and later in Kosovo led to the establishment of a war crimes tribunal in The Hague, Netherlands. These proceedings, covered by television, were the first war crimes trials since the famous Nuremberg trials following World War II.

Middle East Peace Process
• • • • • • • • • • • • • • • • • •

Following many years of conflict in the Middle East, television covered the beginning of a new era in 1993 as Israel signed a peace accord with the Palestine Liberation Organization (see Exhibit 11.d), followed shortly by a peace treaty in 1994 between Israel and Jordan. Thereafter, newscasts covered the peace process as Palestinian-populated areas in Gaza and the West Bank town of Jericho prepared for Palestinian self-rule, which began in 1994. Television cameras also captured the 1995 assassination of Israeli Prime Minister Yitzhak Rabin following a peace rally in Tel Aviv.

Media attention then turned to the Israeli elections between Rabin's successor, Shimon Peres, a major architect and supporter of the peace process, and opposition leader Benjamin Netanyahu, a frequent critic of many aspects of the peace accords. After the election of Netanyahu in 1996, and his government's decision to continue building new settlements in the West Bank, violence between the Israelis and Palestinians increased. Once again, as before the peace accords, nightly newscasts frequently carried scenes of demonstrations, terrorist attacks, and families on both sides of the conflict grieving the deaths of the victims.

In May 1999 Ehud Barak defeated Netanyahu in national elections and vowed to reinvigorate the peace process. Although the violence continues, television cameras caught events that did not seem possible only a year earlier: the disassembling of Israeli settlements on the West Bank.

11.4 POLITICS

Early on, politicians recognized that radio could have a significant impact on elections. So in writing laws controlling broadcasting in 1927 and again in 1934, they saw to it that they would have access to radio (and later to television) when running for election. But beyond this, the electronic media have acted as facilitators and reporters of democratic processes.

Election Campaigns

Radio exerted an impact on political campaigning almost from the start. Radio speeches by Calvin Coolidge, whose low-key delivery suited the microphone, may have been a factor in his 1924 reelection. Radio became especially important to Democratic candidates because it gave them a chance to appeal directly to voters, bypassing Republican-controlled newspapers. Franklin Roosevelt used radio masterfully in his four presidential campaigns, starting in 1932.

Television brought hard-sell advertising techniques to the presidential campaign of 1952, when a specialist in such commercials, Rosser Reeves, designed television spots for Dwight Eisenhower. By 1968, when Richard Nixon won the presidential election, television had become the most important factor in national political campaigning.

After the 1960s parties redesigned their conventions to become more effective television presentations. Convention managers began to time events to the second, with set minutes allowed for "spontaneous" demonstrations. By 1988 the networks

had begun to reconsider just how much news value remained, given the conventions' declining appeal to viewers. The 1992 conventions were the first the major networks did not cover fully (C-SPAN and, to a lesser extent, CNN did), limiting themselves to an hour or two each evening. Subsequent conventions have been covered similarly.

Research and experience make clear that candidates using more television do not always prevail. One reason may be that television visualizes campaigns in two very different ways—with paid partisan advertising that candidates control, but also with objective news and public-affairs coverage. Bona fide news programs about candidates have a credibility that 30-second advertising spots and candidate-controlled appearances can never obtain. Politicians understand that news coverage provides more credibility than political ads. Thus candidates actively try to get as much favorable news coverage as possible during their campaigns.

Even so, politicians know that having a large amount of money for buying spots on broadcast and cable television is essential though not always the determining factor (consider Steve Forbes's unsuccessful runs for the Republican nomination, for example) for a successful campaign. Advertising analysts predicted that candidates for public office would spend between $500 and $600 million on TV ads alone during the 2000 election campaign. Lack of funds often leads otherwise promising candidates to end their bids for election. As the cost of campaign advertising rises and the public's trust in political ads declines, parties, politicians, and public-interest groups increasingly use the Internet to move their messages. Campaign contributions via the Internet are increasing as well. Some nonprofit, public-interest web sites, such as the Freedom Channel, offer free studio time for politicians to present 90-minute, no-frills presentations to voters.

Televised Candidate Debates

Starting in 1960, televised debates between presidential candidates began to steal the spotlight from conventions and other candidate appearances. The first debates, four 1960 confrontations between candidates John F. Kennedy and Richard M. Nixon, were made possible by a special act of Congress exempting the televised event from the "equal opportunities" law (see Section 13.7, p. 392). Often called "The Great Debates," these carefully choreographed face-offs probably decided that close race. Exhaustive research suggests that Kennedy seemed more precise and visually crisp to television viewers, while Nixon sounded more persuasive to radio listeners (Rubin, 1967).

The two 1988 televised "debates" between President George Bush and Democratic nominee Michael Dukakis (a panel of journalists questioned both candidates, who rarely addressed each other directly) seemed to have little overall impact on that election's outcome. Bush came across as "lean and mean," whereas Dukakis seemed wooden and cold.

In 1992 Bush faced Democrat Bill Clinton, who was seasoned after months of hard primary campaigns studded with televised debates among the Democratic candidates. A third candidate—independent Ross Perot—took part in the debates, adding often witty commentary on and sharp criticism of both major parties.

In 1996 the independent commission overseeing the presidential debates refused to allow third-party candidate Perot to participate, leaving Democratic candidate Bill Clinton and Republican candidate Bob Dole as the only debaters. The commission, made up of Republicans and Democrats, ostensibly excluded Perot because he did not garner enough support in the 1992 election to warrant inclusion. For a discussion of the legality of excluding candidates from television debates, see Exhibit 13.e (page 394).

After it appeared that Vice President Al Gore and Texas governor George W. Bush had wrapped up their parties' nomination for president in 2000, Gore challenged Bush to refrain from purchasing political ads and instead to engage in a series of candidate debates. This proposal represents a clear departure from the way candidates have used electronic media for the past 50 years. At the time this book went to press, it was unclear whether Bush would accept Gore's challenge, though most political analysts consider it highly unlikely. Indeed, pundits suggested that Gore did not expect Bush to accept and that the proposal was merely a political strategy for the candidate with less money to spend.

Talk Show Syndrome

During a February 1992 appearance on CNN's *Larry King Live*, Texas billionaire Ross Perot began his grassroots independent candidacy by announcing his willingness to be drafted if volunteers placed his name on all 50 state ballots. Later, in

EXHIBIT 11.e

Talking Politics

The 1992 presidential election campaign saw widespread use of a new campaign device—the topical talk show. In a pioneering example, independent candidate Ross Perot appeared with CNN's Larry King to announce his original candidacy, to explain his withdrawal six months later, and again to discuss his reentry in the campaign.

The use of talk shows to convey one's message to the public took an interesting twist during the 2000 election as competing candidates appeared together. Here, three candidates for the Republican nomination for president, Senator John McCain, Alan Keyes, and Governor George W. Bush, appear with Larry King.

Source: Eric Draper/AP-Wide World Photos.

July, Perot returned to King's show to explain why he had abruptly pulled out of the race before formally declaring his candidacy. Then, in a yo-yo campaign, he came back yet again to explain his reentry just a month before Election Day.

Every candidate appeared on local and national call-in talk shows during the 1992 primaries, giving viewers an unprecedented chance to pose questions. Bush, Clinton, Perot, Quayle, and Gore all appeared on network morning programs as well as other talk shows as the campaign reached its height in October. Democratic candidate Bill Clinton even played the saxophone on the late-night *Arsenio Hall Show*, and George Bush tried to appeal to young voters by appearing on MTV.

The use of talk shows by candidates continued in the 1996 election. As Exhibit 11.e illustrates, the use of talk shows by presidential candidates continued during the 2000 campaign, though not to the extent seen in earlier years. In fact, the 1992 election may very well have been the high-water mark for the political use of talk shows.

Election Night Predictions

Early television reports of voting trends in eastern states have been blamed for lower voter turnout in the West, where polls remain open. Members of Congress have been concerned that early predictions of winners and losers—or candidate concessions—in national races adversely impact western state and local elections.

The networks agreed in 1985 not to air predictions of any state's vote until after polls in that state had closed. They largely adhered to that promise in reporting the results of the national elections beginning in 1986.

In 1996 most network news organizations announced that President Clinton had been reelected hours before many polls had closed on the West Coast. This did not technically violate the pledge they had made earlier, because their projections were based only on results in states whose polls had already closed. The debate continues over whether such announcements affect races in states where the polls remain open.

11.5 GOVERNMENT

Aside from coverage of seemingly endless election campaigns, television has dramatically affected the governing of America.

Presidential Television

A newly elected president enjoys a huge advantage over political opposition in obtaining media coverage. The chief executive has endless opportunities to manufacture pseudo-events to support his policies or to divert attention from his failures. No matter how blatant this exploitation may seem, editors rarely dare ignore presidential events. Virtually everything the head of state says or does has news value.

For the greatest impact, the White House can call on broadcast and cable news networks to provide simultaneous national coverage of a presidential address. National addresses can give the president a gigantic television audience—although independent television and cable channels increasingly siphon off potential viewers with entertainment programs.

No law requires broadcast or cable networks to defer to presidential requests for time. Indeed, not until President Lyndon Johnson's administration (1963–1969) did these become customary. Perhaps because of the Nixon administration's Watergate troubles, networks began to evaluate such requests more critically after 1974. Presidents since Nixon have all been turned down at least once when they asked for time.

Such rebuffs might prove inconsequential in the future as more people access the Internet for presidential coverage. In November 1999 President Clinton became the first president to engage in a live Internet chat. Although the chat was limited to 50,000 participants, some commentators saw the president's action as a groundbreaking event highlighting the increasing importance of the Internet in politics. Indeed, the Internet may become much more than a political communication channel. In March 2000, Democrats in Arizona used the Internet to cast votes in a primary election. This represented the first binding Internet election. After the election, Democratic officials noted that more people voted via the Internet than those voting at the polls. Proponents of the experiment opined that the new procedures provided expanded opportunities to participate. Opponents of online voting argue that using the Internet discriminates against those who do not have ready access to computers.

Congressional Television

The Senate began allowing television coverage of its committee hearings (subject to committee chair approval) as early as the 1950s. The first televised House committee deliberations came in 1974 when the Judiciary Committee debated the Nixon impeachment resolution.

For decades, however, neither congressional body allowed broadcast access to its legislative sessions. Finally, the House agreed to live television and radio coverage in 1979. Even then, it refused to let outsiders run the show, establishing its own closed-circuit television system run by House employees. Broadcasters could carry the signal, live or recorded, at will. Unenthusiastic about coverage they do not themselves control, few broadcasters used even excerpts of House debates. Cable subscribers could see gavel-to-gavel coverage on C-SPAN, the cable industry's noncommercial public-affairs network.

Not until 1986 did the Senate approve television coverage of all floor sessions. C-SPAN added a second channel to its service to cover the Senate when in session (see Exhibit 11.f). Congress had at last given itself a degree of video parity with the White House.

Both Congress and the White House have developed sophisticated facilities for exploiting television. The House and Senate maintain fully equipped studios for the personal use of members, and the White House can make news feeds and interviews available on short notice from its own studio facilities, as well as handle call-in interviews from local stations.

Court Television

While presidents and members of Congress often use television for political purposes, televising court proceedings primarily serves the function of informing the public about the administration of justice. Nonetheless, the widespread fascination with the O. J. Simpson murder trial and the success of the cable network Court TV demonstrate that courtroom proceedings can make compelling TV viewing. Some local court systems are bypassing the traditional media altogether and televising trials to worldwide audiences via the Internet.

Each state decides the extent to which trials and other court proceedings can be covered by television. Most states now allow some television coverage, although the extent of coverage varies widely. The United States Judicial Conference, led by the chief justice of the United States, has the power to allow television coverage of federal judicial proceedings. The conference has permitted limited coverage of some civil cases on an experimental basis, but opposition by the chief justice and other members of the federal judiciary has kept television cameras out of federal courtrooms. Thus the third branch of our federal government is closed to the electronic media.

11.6   ENTERTAINMENT

Any major news story that captures headlines may become grist for comment on late-night talk shows or the subject of a special drama or miniseries; it also may influence future episodes of established series. When President Bush suggested in his acceptance speech at the 1992 GOP convention that the country needed families more like television's *The Waltons* (1972–1981) and less like *The Simpsons*, the Fox program included the comment in an animated episode that appeared just 48 hours later. Similarly, when Vice President Dan Quayle criticized the television series *Murphy Brown* for having the lead character, an unmarried woman, become pregnant and decide to be a single parent, the writers of the show responded to Quayle's comments via the characters in the show.

Docudramas, controversial blends of fact and fiction, such as the 1996 production on the ill-fated voyage of the *Titanic*, change facts around to suit dramatic needs, thus adding more distortions to already-simplified versions of reality presented to broadcast and cable audiences.

Stereotypes

Fiction influences audience perceptions by reinforcing stereotypes, versions of reality that are deliberately oversimplified to fit preconceived images, such as the stock characters of popular drama: the Italian gangster, the inscrutable Asian, the mad scientist, the bespectacled librarian, the befuddled father, or the absent-minded professor.

Authors resort to stereotypes when writing for television to save time, both their own and that of the medium:

> Television dramas have little time to develop situations or characters, necessitating the use of widely accepted notions of good and evil. Since the emphasis is on resolving the conflict or the problem at hand, there is little time to project the complexities of a character's thoughts or feelings or for dialogues which explore human relationships. To move the action along rapidly, the characters must be portrayed in ways which quickly identify them. Thus the character's physical appearance, environment, and behavior conform to widely accepted notions of the types of people they represent. (U.S. Commission on Civil Rights, 1977: 27)

EXHIBIT 11.f

Congressional Television

In 1951 televised hearings on organized crime catapulted Senator Estes Kefauver (D-Tenn.) onto the Democratic ticket in the following year's presidential campaign. The televised Army-McCarthy hearings in 1954 led to the Senate's censure of Wisconsin senator Joseph McCarthy and ultimately to the end of McCarthyism, the far-right communist witch-hunt that he had led. The 1973 Watergate hearings chaired by folksy but astute Senator Sam Erwin (D-N.C.) made him a star, while the 1974 Nixon impeachment proceedings in the House propelled many members of Congress to fame—and hastened regular television coverage of congressional sessions.

The C-SPAN networks came into being originally to cover the House of Representatives (beginning in 1979) and the Senate (C-SPAN 2, beginning in 1986). Cameras, controlled by Congress, rarely show that floor debates often take place in nearly deserted chambers.

In 1991 the nation watched transfixed as the Senate reopened nomination hearings on Judge Clarence Thomas, a prospective member of the U.S. Supreme Court. Thomas was accused of sexual harassment by Anita Hill, an articulate law professor who had once been on his staff. Although the Senate, by a close vote, approved Thomas for the post, the hearings raised consciousness about sexual harassment. Television had indeed reset the agenda of millions of Americans.

Although there were many congressional hearings that were televised following the Thomas hearings, none caught the interest of the public until the topic again centered on sex, this time relating to the affair between President Clinton and White House intern Monica Lewinsky. Millions watched at home and in their offices as Special Prosecutor Kenneth Starr presented his report to the House. Fewer watched the actual impeachment debates, and fewer still watched the trial in the Senate, as the result seemed to be a foregone conclusion. The trial coverage, though historic in its own right, also offered the American public something never before seen on television: the chief justice of the United States as the presiding judge. Federal court rules prevent televising Supreme Court proceedings; thus, unless the Court changes its rules or there is another impeachment, this may be the only time the public has an opportunity to see via television the chief justice at work.

Sources: (top) AP-Wide World Photos; (bottom) Terry Ashe/*Time Magazine*.

Stereotypical images on television clearly help perpetuate those same images in the minds of viewers. As the U.S. Commission on Civil Rights put it, "To the extent that viewers' beliefs, attitudes, and behavior are affected by what they see on television, relations between the races and the sexes may be affected by television's limited and often stereotyped portrayals."

World of Fiction
• • • • • • • • • •

As researchers examine roles in television series, they invariably find fictional characters to be markedly different from real people in the real world. Compared with life, for example, the world of fiction has far more men than women, most of them young adults, with few very young or elderly people. Many have no visible means of support, but those who do work have interesting, exciting, action-filled jobs.

Fiction therefore contains an unrealistically high proportion of detectives, criminals, doctors, lawyers, scientists, business executives, and adventurers compared with the real world, where unglamorous, dull, and repetitive jobs predominate. Most people in the real world solve their personal problems undramatically, even anticlimactically or incompletely, using socially approved methods. Fictional characters tend to solve their problems with decisive, highly visible acts, often entailing violence. Of course, none of this should surprise us. Fact may be stranger than fiction, but fact does not occur in neatly packaged half-hour episodes with regular commercial breaks.

Children's Socialization
• • • • • • • • • • • • • • •

The make-believe world of the electronic media serves as a model of reality for countless people—especially children at the very time when they are eagerly reaching out to learn about the world. Those too young to read, those who never learned the habit of reading, and those who have little access to printed sources of information and entertainment all depend heavily on the electronic media to tell them about the world outside their immediate surroundings.

Viewers and listeners identify with heroes, participating vicariously in their adventures. Research indicates that young children tend to believe what they see on television, making no distinction between fact, fiction, and advertising.

The electronic media have become major agents of socialization—the all-important process that turns a squalling infant into a functioning member of society. Socialization, though a lifelong process, occurs intensively during the first few years of life, the time when children begin to learn the language, detailed rules of behavior, and value system of their culture.

In the past, socialization was always the jealously guarded prerogative of family and religion, formalized by education and extended by peer-group experiences. Intrusion of a new, external agent of socialization—television—has been a profound change. Of course, all electronic media function as part of national culture, too, but they come from beyond the immediate family circle and its community-linked supports. The media import ideas, language, images, and practices that may be alien to the local culture.

The question of how the intrusion of broadcasting and cable has affected socialization has been and continues to be widely researched and debated. The electronic media can, of course, have both good (prosocial) and bad (antisocial) effects. Producers designed such programs as *Sesame Street* and *Mr. Rogers' Neighborhood* (see Exhibit 7.e, p. 221) with prosocial effects in mind. Follow-up research indicates that such programs do in fact succeed in achieving prosocial results, though some (especially *Sesame Street*) reinforce a sense that the world comes in short segments with 30-second breaks.

Much more effort, however, has gone into research to prove the antisocial consequences of television, especially the effects of violence. These effects are studied in Section 11.7.

Influence of Time Spent
• • • • • • • • • • • • • • • •

"There is no more clearly documented way in which television has altered American life than in the expenditure of time" (Comstock, 1978: 29). At the very least, critics argue, time spent watching or listening could have been spent in some other way—perhaps on some more useful, constructive activity. Some critics take it for granted that anything active would be more beneficial than passive absorption. Somehow it seems "wrong" for people to waste their time staring like zombies at the television.

Most researchers readily accept that by the time a young person graduates from high school, he or she has watched an average of 22,000 hours of television—more time than that spent on any other activity except sleeping. Some futurists predict that when the 500-plus channel superhighway is fully open for travel, people will exponentially increase the time they spend in front of a video monitor. Others suggest that most will divide available leisure time among the many new options.

A 1999 study by the Henry J. Kaiser Family Foundation found that children spent an average of 5 hours 29 minutes per day with the media. Television claimed the most time, with an average of 2 hours 46 minutes per day, while music came in second. On average, children spent only 21 minutes per day outside of school on the computer, although this figure may be deceptively low because many children do not have access to a computer at home. (Henry J. Kaiser Family Foundation, 1999) Another study, funded by the National Institute on Media and the Family, found that teenage boys spend an average of ten hours per week playing video games (Walsh, 1999).

In any case very little research has demonstrated that listening and watching necessarily displace more useful and active forms of recreation. In the absence of radio and television, people would do other things with their time, of course, but these would not necessarily be more beneficial. One exception to this thinking may apply to very young children. Arguing that such children need direct interactions—whether with parents or siblings, making a puzzle, or playing in the dirt—for healthy brain growth, the American Academy of Pediatrics in 1999 urged parents to ban television for all children under the age of two. (American Academy of Pediatrics, 1999)

Importance of Television

For some groups television is especially important. A team of nine psychologists, who spent five years assessing the medium, concluded in 1992:

> Children, the elderly, ethnic minorities and women are the heaviest users of television (the elderly being the highest) because television is a default option used when other activities are not available. These populations are limited by physical restrictions and/or lack of resources. The more time people spend at home or near a television set, the more likely they will watch television. (APA, 1992: 2)

For the elderly, on which more research attention has been focused in recent years as the overall population

ages, news and information programs were most popular among those who used television as a "social network to replace the informal network previously supplied by the workplace" (APA, 1992: 3). Clearly, for some groups television is virtually indispensable—nothing else can perform the same function.

Such findings raise questions about "heavy viewers" or "television addicts." Studies suggest that anywhere from 2 to 12 percent of the population may watch up to twice the average weekly amount of television, are unhappy about that viewing, and yet seem unable to do much about it. Further, they usually watch when they are sad, angry, or lonely (or just plain bored), are subject to "attentional inertia" (simply watching what's on rather than being selective), and become more passive as they watch more television.

"Glow-and-Flow" Principle

The play theory helps explain the widely acknowledged notion that programs seem of secondary importance as long as something fills the screen. Early in television history, one observer suggested that "it is the television set and the watching experience that entertains. Viewers seem to be entertained by the glow and the flow." (Meyersohn, 1957: 347)

A landmark study of television audience attitudes found that most people surveyed said that they felt more satisfied with television as a *medium* than with specific *programs*. The researcher noted that "a large number of respondents were ready to say television is both relaxing and a waste of time" (Steiner, 1963: 411). Similar studies of attitudes conducted a decade or two later showed that this mixture of reactions persisted (Bower, 1973, 1985).

The glow-and-flow principle becomes especially important when electronic media are the only companions people have. Upon asking respondents to describe satisfactions they derived from watching television, the researcher sometimes received moving testimonials such as this:

> I'm an old man and all alone, and the TV brings people and talk into my life. Maybe without TV, I would be ready to die; but this TV gives me life. It gives me what to look forward to—that tomorrow, if I live, I'll watch this and that program. (Steiner, 1963: 26)

At extreme levels of deprivation—in hospitals and similar institutions—television has a recognized therapeutic function as the most valuable non-chemical sedative available.

In short, the electronic media answer a compelling need of the mass audience simply to kill time painlessly, to fill an otherwise unendurable void. The media give people a highly effective way of *performing leisure*. But social critics worry that while we relax with our guard down watching television, violence and other antisocial activities portrayed in many programs may produce antisocial effects.

11.7   VIOLENCE

No aspect of electronic media impact raises more worry than the amount of violence portrayed on the air and the presumed effect of that cumulative violence. But concern that portrayals of violence and crime might have antisocial effects preceded television by many years. The first systematic research on media violence dates to film studies in the 1930s (Jowett, 1976: 220). Accusations about the potential effects of violence shifted to radio and comic books in the 1950s and later to television, rock videos, computer games, and the Internet. Along the way, emphasis progressed from merely measuring violent content and deploring its presumed effects toward buttressing conclusions about effects with explanatory theories.

Television viewers see a steady stream of violence, both in entertainment programs and on the news. A 1999 study conducted by the Center for Media and Public Affairs found that television viewers see a violent act about every four minutes. Often, according to the study, media violence carries no negative consequences. The most violent programs on television can carry as many as 80 violent acts per episode (Lichter et al., 1999). With respect to television news, critics have long maintained that stations follow an "if it bleeds it leads" policy of news selection.

Direct Imitation

We occasionally see news reports of real-life violence apparently modeled on similar actions in films or television programs. One particularly re-

pellent example of apparent imitation led to an unprecedented lawsuit. In 1974 NBC broadcast a made-for-television film called *Born Innocent*, in which inmates of a detention home for young delinquents "raped" a young girl with a mop handle. Four days after the telecast, older children subjected a nine-year-old California girl to a similar ordeal. Parents of the child sued NBC, asking $11 million in damages for negligence in showing the rape scene, which they alleged had directly incited the attack on their daughter.

Network attorneys persuaded the trial judge to define the issue as a First Amendment one, rather than as one of negligence. The case then collapsed: It was impossible to show that NBC had *deliberately* incited the children to attack their victim and thus lose its First Amendment protection. (Cal., 1981: 888)

Other examples include reenactment of a scene from the movie *The Program*, in which people lie in the middle of the road as automobiles speed by. Such mimicry led to the deaths of several college students. Similarly, a young boy who had watched the characters in *Beavis and Butt-Head* play with fire and say it was cool subsequently set his house on fire, killing his baby sister. It seems natural to assume that such imitation proves that televised violence sometimes causes violent behavior. Exhaustive research demonstrates, however, that viewing a violent act serves as a *contributing factor* to any subsequent imitation of that act. In other words, the act of viewing takes place within a larger context—one's background, education, predispositions, and the like. Taken together, all these influences may, in some circumstances, lead to violent behavior.

General Effects

Public concern about media violence arises primarily from fear of its general impact rather than from the risk of possible direct imitation. This point of view emerged in another much-publicized court case in 1977. The state of Florida charged a 16-year-old boy, Ronnie Zamora, with murdering an elderly neighbor during an attempted robbery. The boy's attorney tried to build his defense on the argument that Zamora could not be held responsible for his violent behavior because he had become a television addict, "intoxicated" by the thousands of murders he had seen enacted on the screen. The

EXHIBIT 11.g

Government Studies of TV Violence

The Surgeon General's Scientific Advisory Committee on Television and Social Behavior sponsored a group of studies in 1969–1971. Congress allotted a million dollars for the project, which in 1972 resulted in five volumes of findings and a final report. When questioned by a Senate committee, the surgeon general said flatly:

> Broadcasters should be put on notice. The overwhelming consensus and the unanimous Scientific Advisory Committee's report indicates that televised violence, indeed, does have an adverse effect on certain members of our society. . . . It is clear to me that the causal relationship between televised violence and antisocial behavior is sufficient to warrant appropriate and immediate remedial action. (Senate CC, 1972: 26)

A comprehensive review of previous research, commissioned by the committee, indicated that television's linkage to aggressive behavior had been intensively analyzed. That every research method available had been employed in these studies made evidence of television's impact even more convincing. Researchers summarized:

> The evidence is that television may increase aggression by teaching viewers previously unfamiliar hostile acts, by generally encouraging in various ways the use of aggression, and by triggering aggressive behavior both imitative and different in kind from what has been viewed. Effects are never certain, because real-life aggression is strongly influenced by situational factors, and this strong role for situational factors means that the absence of an immediate effect does not rule out a delayed impact when the behavior in question may be more propitious. (Comstock et al., 1972: 13)

As a result of this work and of research done since—including a two-volume revisit a decade later by some of the same researchers involved in the 1972 report—Congress and private groups pressured the FCC to limit televised violence. Chiefly for fear of violating the First Amendment and because of the trend toward deregulation, the FCC declined to act.

Still, pressure continued. At the invitation of the Department of Health and Human Services, many researchers involved in earlier studies gathered in Washington late in 1992 to review two decades' worth of research and to assess the impact of changing technology. One study suggested that larger screens and improved picture definition made televised events more real and thus potentially more dangerous. Others hinted that excessive television watching could impair brain development in young viewers because they would not face a sufficiently wide variety of experiences (*Washington Post*, 12 October 1992: D5).

In 1996 Congress passed legislation addressing the violence issue. The bill required the industry to develop a television rating code that would inform parents about the content of programs. Additionally, new television sets must be manufactured with a *V-chip* (*V* for violence) that will allow parents to program their televisions to exclude shows with a certain rating. For more on this aspect of television regulation, see Exhibit 9.d (p. 267) and Section 13.9 (p. 399).

Following a series of high school shootings and highly publicized hate crimes around the country in the late 1990s, government officials once again called for studies about the links between violence in the media and societal violence. As part of this effort, the Federal Trade Commission announced in 1999 that it will study the extent to which media companies market violent programs, video games, and music videos to youth. The commission wants to find out whether these marketers use violent imagery to lure children. The FTC hopes to complete its inquiry and issue a report by the end of 2000. (Federal Trade Commission, 1999)

trial judge rejected this argument, and the jury convicted the boy of murder. Though ill-considered, his lawyer's attempt to blame television for the crime drew its inspiration from the findings of research on the generalized adverse effects of televised violence that had accumulated during the 1970s (see Exhibit 11.g).

Violence vs. Reality

More than three decades ago, during the 1967–1968 television season, George Gerbner and his associates at the University of Pennsylvania began conducting annual analyses of television violence. From these data they constructed a violence "profile" that counted every violent act in a sample week of prime-time and weekend-morning network entertainment programs. In this way they tracked changes in the level of violence from year to year according to network and program type. The Gerbner data indicated, for example, that animated cartoons had a higher percentage of violent acts than any other program category (the coding system counted comic as well as serious acts of violence). Still being collected, these data contribute to what has become a valuable long-term indicator of television content.

Some recent television violence research goes beyond counting violent acts and considers the *context* of the situation in which the violence occurs. Researchers doing this work believe that the context of a violent act may somehow modify the effect of the act on the audience.

For instance, Gerbner theorized that violence in programs creates anxiety in viewers because they tend to see the real world in terms of their television experience. Viewers identify with those victims of violence in fiction who resemble themselves. Gerbner found that elderly, poor, and black people have high "risk ratios," or expectations of becoming victims. This *anxiety effect*, he said, may be a more important by-product of television violence than the imitation effect.

Desensitization
. . . .

The Gerbner risk-ratio hypothesis or *cultivation theory* seems more persuasive than older hypotheses predicting that people exposed to fictional violence would become *desensitized* to it. According to this view, when television dramatic violence becomes routine, people grow indifferent to real-life violence. Many instances of urban violence and nonresponse by passers-by lend credence to this hypothesis.

A related notion, the *sanitization hypothesis*, holds that self-imposed program codes cause television to depict violence unrealistically because graphic literalness would encounter viewer backlash. As a possible result, televised consequences of fictional violence seem so neat and clean that viewers remain indifferent. They never see or hear the revolting, bloody, painful aftermath of real-life violence.

Violence and Urban Unrest
. . . .

Some of the most extreme televised violence appears not in entertainment but in news—especially during coverage of live events. When south-central Los Angeles erupted in urban violence in April 1992 in the wake of a controversial trial verdict, two examples of the violence problem emerged.

First, critics argued that the jurors in the case—who acquitted police officers of a vicious assault on a black man—had become desensitized to violent behavior despite the existence of a graphic videotape of the beating taken by a bystander. Second, television news editors again faced the balancing act of how much riot coverage to show without contributing to the violence happening in front of the cameras. Television had faced this problem in the late 1960s during repeated episodes of urban unrest nationwide brought on by racial and economic inequality, on the one hand, and by protests against the Vietnam War, on the other.

Television's technology improved between the 1960s riots and those of the 1990s. Cameras today are less intrusive and thus less likely to incite trouble by their presence. But the basic problem remains— how much to show while the rioting is actually taking place. Ironically, technological improvements have made this problem worse in that stations and networks can now put these dramatic pictures on the air almost instantly with little time to consider their potential audience impact.

In Defense of Violence
. . . .

Given the steadily increasing range and depth of research showing antisocial effects of television violence, the media business has found it increasingly difficult to defend its widespread use. Yet for television to serve as a serious medium for adult viewers, it cannot ban violence. After all, violence occurs in all forms of literature, even fairy tales for children. Bodies litter the stage when the curtain falls on some of Shakespeare's tragedies.

Writers would face a difficult challenge if they had to meet television's relentless appetite for drama without resorting to violent clashes between opposing forces. A playwright illustrated the problem by explaining that authors have only four basic conflicts around which to build any plot (Baldwin & Lewis, 1972):

- *man against nature:* "This is usually too expensive for television."
- *man against God:* "Too intellectual for television."
- *man against himself:* "Too psychological, and doesn't leave enough room for action."
- *man against man:* "This is what you usually end up with."

Only the last form of conflict can easily be shown on television; it also happens to be the one most likely to involve personal violence.

Under growing pressure from the public, interest groups, and Congress to address and curb the amount of violence aired, the broadcast and cable industries took limited action in 1992–1993. The Television Program Improvement Act of 1990 had granted cable and broadcast networks and programmers a temporary antitrust exemption to encourage the development of common standards to limit the total amount and some specific types of violence on the air. The broadcast networks agreed to a generalized common set of standards late in 1992, and a few months later major cable programmers agreed to try to do the same, joining the networks in developing a national research conference to air the issues.

In 1994, under a growing threat of congressional action if they did not take more positive steps, the four broadcast television networks agreed to designate UCLA's Center for Communications Policy as an independent monitor to evaluate programs for violent content and summarize findings in an annual public report. Broadcasters also agreed to supply a violence warning, including a Parental Advisory symbol (PA), to newspapers and *TV Guide*.

Several television stations, partly in reaction to viewer criticism and partly in an effort to improve ratings, began in the mid-1990s to downplay violence in at least some of their local newscasts. They offered what they variously called "family," "family-centered," or "family-safe" programs. These programs reduced reliance on graphic depictions of crime and violence and often included a specific "good news" segment. Some stations, however, limited this policy to early-evening newscasts; at 11:00 P.M. it was violence as usual.

By the decade's end, most broadcasters had abandoned this effort entirely. In fact, a 1999 study of television news directors found that only 5 percent had taken steps to reduce the level of violence in their newscasts. (*Broadcasting & Cable*, 27 September, 1999: 38)

In 1996, despite strong opposition from broadcasters, Congress passed legislation requiring the television industry to adopt a program ratings code. Television manufacturers were required to install V-chips in all new TV sets (see Section 13.9, p. 399). Taken together, these actions were intended to give parents more information about the programs their children were watching and allow them to block out programs they found objectionable. Ironically, television content may have become even more violent as a result. A 1999 study by the Parents Television Council found that with the ratings system in place, sex, violence, and offensive language in television programming had increased 31 percent from 1996 to 1998. (Parents Television Council, 1999)

Whereas some television programmers may be using the ratings system to push the envelope of good taste, other media are reacting differently to government and public calls for greater responsibility. Owners of movie theaters agreed to begin enforcing the Motion Picture Association of America (MPAA) ratings system by requiring photo IDs for teenagers seeking admission to R-rated films. The video game industry has pledged to take a closer look at its ratings system, and several Internet service providers (ISPs) now offer a "filtered" service for families hoping to keep objectionable material off their computer screens.

11.8 ADVERTISING

Other than violence, no aspect of electronic media impact has been more measured than advertising. Advertisers demand and receive vast amounts of applied research on the reach and effects of their messages. Here we focus on the broader, long-range social impact of advertising.

Advertising as Subsidy?

At first glance, advertising appears to play a useful social role by reducing the direct cost of media to the public. In traditional commercial broadcasting, for example, advertising appears to cover the entire cost of this service, making it "free" to the public. In fact, consumers eventually pay the full cost, because companies include advertising expenses in the final prices of their goods and services. Advertiser-supported basic cable combines two viewer charges, adding direct subscriber fees to indirect advertising costs. And for all electronic media, the audience must purchase and maintain receivers, which can be thought of as a kind of consumer subsidy of advertisers.

Advertising's General Impact

Some critics suggest that advertising actually *disserves* consumers by creating a desire for unnecessary purchases—what economist John Kenneth Galbraith termed the *synthesizing of wants.* Electronic media advertising often stimulates widespread demand for goods and services for which consumers may have little real need. Advertising can build overnight markets for virtually useless products or "new and improved" versions of old products.

Critics often assume from these successes that advertising can overcome most consumer resistance. Advertising practitioners find themselves wishing that were true. Failure of a high proportion of new products every year belies the assumption that advertising is all-powerful.

Not only do many products fail to catch on, but leading products also often give way to competitors. Marketers recognize the transfer of *brand loyalty* as an ever-present threat. For this reason, much cable, television, and radio advertising aims not at moving merchandise, but at simply keeping brand names visible and viable in the marketplace.

Advertising to Children

The possible impact of commercials in children's programs raises special issues. Children start watching television early in their lives and often find commercials just as fascinating as programs. Action for Children's Television (active until 1992) and other consumer organizations have long believed that commercials take unfair advantage of young children—especially preschoolers, who are not yet able to differentiate between advertising and programs. Psychologists and child-health professionals have questioned the use of psychological techniques by advertisers of children's products. They seek to shield children from such "commercial manipulation and exploitation."

During the 1970s academic researchers, the industry, and government agencies conducted extensive research on the impact of children's television—both programs and advertising. Industry lobbying fended off effective action by either the Federal Trade Commission or the Federal Communications Commission. The FCC adopted rather ineffectual guidelines but refused to impose hard-and-fast rules. Deregulation in the 1980s further watered down the FCC's guidelines.

Finally, under pressure from Action for Children's Television and others, Congress in 1990 passed a law limiting advertising in children's programs and calling for more programs that address the educational and informational needs of children (see Section 13.9, p. 399).

CHAPTER **12**

Regulation and Licensing

CHAPTER CONTENTS

The previous chapter surveyed the pervasive social effects of the electronic mass media—some real, some probably only imagined. As long as people believe that effects actually occur, they want to have some control over them. They want to maximize beneficial effects (from wholesome entertainment and useful information, for example) while minimizing perceived harmful effects (from pornography and misleading propaganda, for example).

Direct social control comes primarily from government licensing, which determines who is allowed to operate stations, cable systems, and other program outlets. Licensing not only selects operators but also enables the monitoring and regulating of their performance. Regulation starts with the Constitution, on which all laws and administrative controls are based. These controls, in turn, are embedded in the Communications Act, which defines the role of the Federal Communications Commission.

Regulation of wireline and cable services, however, differs significantly from that of radio-based services. Local exchange telephone companies and cable systems are usually limited to a specific geographical service area within a state. Because such services are *intrastate*, they are subject to state regulation. States have their own *public utility commissions* (PUCs) that approve changes in in-state telephone rates and services. Long-distance telephone carriers that cross state lines, thus providing *interstate* service, have traditionally needed federal approval for rate and service changes.

Cable (discussed in detail in Sections 12.8 and 12.10) is a *hybrid* service. Because a cable system must use city streets and rights of way, it is subject to local and state jurisdiction. A cable system also distributes signals that originate outside the state, so it is subject to federal jurisdiction as well. Once controlled primarily at the state and local levels, cable is now subject to extensive federal regulation. Direct broadcast satellite (DBS) and other services that provide multichannel video using microwave frequencies [multichannel multiservice distribution service (MMDS) and local multipoint distribution service (LMDS), for example] are regulated exclusively at the federal level.

12.1 FEDERAL JURISDICTION

Specific constitutional justification for Congress's taking control of the electronic media comes from Article 1, Section 8(3), which gives Congress power "to regulate commerce with foreign nations, and among the several states." This *commerce clause* has played a vital role in American history, preventing states from erecting internal trade barriers to national unity. The provision forms a link in a chain of responsibility from Constitution to citizens, as shown in Exhibit 12.a.

Interstate vs. Intrastate Commerce

The commerce clause gives Congress jurisdiction over *interstate* and *foreign* commerce but not over commerce within individual states. The law regards radio communications, including all kinds of broadcasting, as *interstate by definition*. Radio (including television) transmissions, even when low in power and designed to cover a limited area within a state, do not simply stop at state boundaries.

Delegated Congressional Authority

It is impossible for Congress to dictate regulatory details in dozens of specialized fields. Therefore, it has delegated to a series of *independent regulatory agencies* the authority to oversee such often complex areas as power, transportation, labor, finance, and communication. In the realm of communication, Congress delegated authority to regulate most aspects of the electronic media to the Federal Communications Commission (FCC).

Although the president appoints FCC commissioners, with the advice and consent of the Senate, the Commission remains what some call a "creature of Congress." Congress defined the FCC's role in the Communications Act of 1934, and only Congress can change that role by amending or replacing the Act. Though commissioners have considerable leeway in determining regulatory direction, House and Senate subcommittees on communications constantly monitor the FCC, which must come back to Congress annually for budget

EXHIBIT 12.a
• • • • • • • • • • •

Chain of Legal Authority

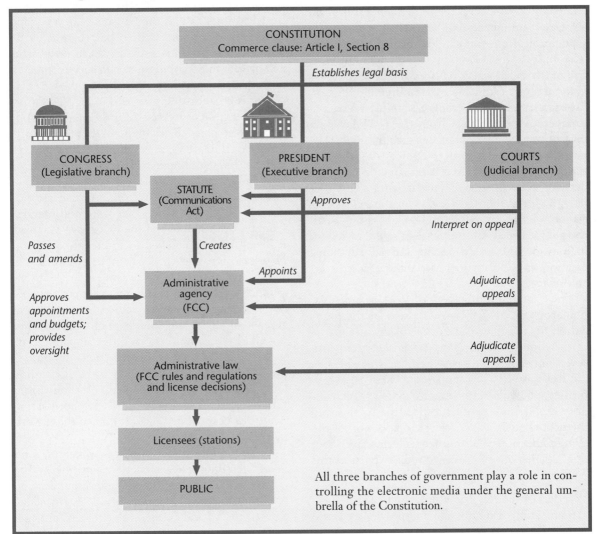

All three branches of government play a role in controlling the electronic media under the general umbrella of the Constitution.

appropriations. Since 1983 Congress has reauthorized the very existence of the Commission every two years.

Congress gave the FCC the power to adopt, modify, and repeal regulations concerning *interstate* electronic media, with the overarching goal of serving the public interest, convenience, and necessity. These regulations carry the force of federal law, deriving their power from Congress through the Communications Act. An understanding of the Act is vital to understanding the FCC's day-to-day role.

• • • • • • • • • • • • • • • • •
12.2 COMMUNICATIONS ACT

As described in Section 2.6 (p. 33), the Radio Act of 1927 ended a period of chaotic radio broadcast development. Congress for the first time crafted a statute and created an agency concerned explicitly with broadcasting. The 1927 act gave the Federal Radio Commission the responsibility of defining what the public interest, convenience, and necessity would mean in practice.

Passage

• • • • • •

The Radio Act of 1927 imposed order on broadcasting but left control of some aspects of radio and all interstate and foreign *wire* communication scattered among several federal agencies. The Communications Act of 1934 brought interstate wire and wireless communication under the control of the FCC, which replaced the Federal Radio Commission. This change had minimal effect on broadcasting because Congress simply reenacted broadcasting provisions from the 1927 law as a part of the 1934 Act. Thus electronic media law dates back to the early development of radio. Although the Act has often been amended, its underlying concepts remain unchanged. Its very first paragraph specifies the reasons for the FCC's creation: "to make available, so far as possible, to all the people of the United States a rapid, efficient, nationwide and worldwide wire and radio communication service with adequate facilities at reasonable charges."

The Communications Act of 1934, as amended numerous times by Congress, provides the foundation for the regulation of the electronic media in the United States. The Act is separated into seven units, called *titles*, which focus broadly on different aspects of regulation. These titles are further subdivided into separate *sections*, which address regulations more specifically. (In this and the next chapter, "Sec." refers to a section of the Communications Act or other indicated law. "Section" spelled out continues to refer to parts of this book.)

- Title I of the Act establishes the Federal Communications Commission and provides general guidelines on its organization and operation.
- Title II sets out the regulatory framework for communications common carriers such as the telephone companies.
- Title III contains provisions relating to over-the-air communications, including radio, television, and various forms of microwave transmission.
- Title IV focuses on additional procedural regulations covering Commission inquiries and trial-like proceedings.
- Title V outlines penalties for violation of FCC rules.
- Title VI establishes the regulatory framework for cable television.

- Title VII contains miscellaneous provisions, such as the unauthorized disclosure of private communications and telephone service for the disabled.

Specific sections in Title III of the Act formalize broadcasting's role as the "dissemination of radio communications intended to be received by the public directly." This excludes from *broadcasting* any private radio communication service aimed at individuals or specific groups of individuals.

Nor is broadcasting supposed to be used for private messages not intended for the general public. People who greet their families over the air during broadcast interviews, for example, technically violate the law—although the practice occurs regularly and without consequences.

The Act forbids *disclosure* of nonbroadcast messages to people for whom they were not intended. Congress, the courts, and the FCC have also made it illegal for people even to intercept some nonbroadcast signals, such as pay cable, unless they are subscribers.

Finally, Sec. 3 of the Act defines *radio communication* as "transmission by radio of writing, signs, signals, pictures, and sounds of all kinds, including all instrumentalities, facilities, apparatus, and services . . . incidental to such transmission." By giving the term *radio* such a broad definition, Congress made it possible for the Act to incorporate television when it became a licensed service nearly 15 years after the adoption of these words in 1927.

As communication technology changes, Congress often finds it necessary to change the law regulating that technology. In 1984, for example, Congress amended the Communications Act by adding a new Title VI and defined cable television as neither a common carrier nor a broadcasting service. Congress left the FCC with little responsibility for cable, far less than it had for broadcasting and interstate wire services, and limited the power of states to regulate cable programming and subscriber rates. But Congress amended the Act again in 1992, reinstating many cable regulatory powers to the FCC and local authorities.

Probably the most significant changes to the Communications Act came with the passage of the Telecommunications Act of 1996. This legislation changed many aspects of how the electronic media industries in the United States are regulated. The law allows companies once permitted to operate certain specified businesses to move into other areas not previously open to them. For example, telephone companies are now allowed to provide cable television services in their local service areas. Local telephone services are permitted to enter the long-distance arena, and long-distance companies may now offer local services. The 1996 act also minimized many of the ownership restrictions that limited the number of radio and television stations any one entity could own. Later amendments to the act addressed the problems of new technologies such as DBS and the Internet. The specifics of these changes are addressed throughout this and the next chapter.

Provision for FCC

The president appoints the five FCC commissioners to five-year terms, subject to Senate confirmation, and designates which of them shall serve as chair. Congress sought to minimize political bias by allowing no more than three commissioners from the same party.

Throughout much of the FCC's history, Senate confirmation of presidential appointees to the Commission was virtually automatic. This changed in the 1980s, when the Senate began using the confirmation process to chastise the Commission for actions it disapproved of. In some cases, the Senate even refused to schedule confirmation hearings for appointees, effectively blocking their appointments.

Sec. 4 of the Communications Act assigns the Commission broad power to "perform any and all acts, make such rules and regulations, and issue such orders . . . as may be necessary in the execution of its functions." Originally, in only a few instances did Congress tie the Commission's hands with hard-and-fast requirements, such as specific station license terms and restrictions concerning foreign ownership of broadcast stations.

Then, in the 1990s, as Congress became more involved in the oversight of the electronic communications industries, legislation amending the Communications Act became increasingly specific. Provisions of both the 1992 and 1996 amendments to the Act direct the Commission to regulate in very specific ways, with little room for interpretation or discretion.

PICON Standard

Congress created a highly flexible yet legally recognized standard—*public interest, convenience, or necessity* (PICON)—to limit FCC discretion. PICON occurs regularly in the Act's broadcasting sections. For example, Sec. 303 begins: "Except as otherwise provided in this Act, the Commission from time to time, as *public convenience, interest, or necessity** requires, shall . . ." and goes on to list 19 powers, ranging from classification of radio stations to making regulations necessary to carry out the Act's provisions. The PICON phrase similarly occurs in sections dealing with the granting, renewal, and transfer of broadcast licenses.

In sharp contrast, important 1984 cable amendments to the Communications Act nowhere used the phrase "public interest." Because of their strongly deregulatory nature, the 1984 amendments relied far more on marketplace than on PICON decision making. Cable amendments in 1992 did make reference, in a new Sec. 628(a), to promoting "the public interest, convenience, and necessity by increasing competition and diversity in the multichannel video programming market," thus signaling a return to a more regulatory climate.

Similarly, the 1996 amendments to the Act contain several references to the public-interest standard, particularly in relation to television broadcasters' use of a second video channel for the delivery of digital broadcast services such as high-definition television.

12.3 FCC BASICS

While theoretically an independent regulatory agency, the FCC, as a creation of Congress, acts on behalf of the legislative branch. When it makes regulations, the Commission acts in a quasi-legislative capacity. It functions as an executive agency when it puts the will of Congress and its own regulations into effect. And when the FCC interprets the Act, conducts hearings, or decides disputes, it takes on a judicial role.

---

* Authors' emphasis.

Budget and Organization

• • • • • • • • • • • • • • • • •

In the 1990s Congress appropriated between $200 million and $250 million annually for the FCC. This budget makes the Commission one of the smaller federal agencies, employing about 2,000 persons.

Throughout most of its existence, the FCC's budget came out of general tax revenues. In 1989 Congress began requiring the FCC to collect part of its budget directly from the industries it regulates. In 1999 Congress expected the FCC to collect $172.5 million, more than half of its budget allocation for that year, from regulated industries. This money comes from annual regulatory fees and application fees. Fees defray the costs of the Commission's enforcement, policy- and rule-making, user information, and international activities.

Annual fees vary widely, depending on the service. In 1999, for example, Class C AM stations paid as much as $1,750 or as little as $225, depending on the number of homes in their service areas. VHF television stations in the top 10 markets paid $41,225. DBS operators were assessed a fee of $130,550 per satellite, and cable systems paid 48 cents per subscriber. Clearly, these fees are passed on to consumers as higher rates for subscription services and higher costs of consumer goods stemming from increased advertising costs.

Exhibit 12.b depicts the FCC's organizational structure. The Mass Media Bureau, one of the units of most interest here, has three divisions:

- *Audio Services.* Processes applications for construction permits, licenses, and license renewals for radio stations.
- *Video Services.* Processes the same for television stations and MMDS operations.
- *Policy and Rules.* Handles FCC proceedings that produce new rules and conducts studies needed for policy-making decisions.

Because of greatly increased cable regulatory responsibilities, the FCC in 1994 established a Cable Services Bureau (an earlier cable bureau had operated from 1972 to 1981). It has four divisions:

- *Financial Analysis and Compliance.* Responsible for all cable television rate-related cases before the bureau.
- *Consumer Protection and Competition.* Handles consumer and competition issues, such as controversies over leased access requirements and

rules governing the availability of program networks to competing multichannel providers.
- *Engineering and Technical Services.* Provides support on technology policy and oversees cable signal leakage regulations and other general operational requirements for cable systems.
- *Policy and Rules.* Handles all rule-making activities.

The Common Carrier Bureau remains the unit most closely involved in matters associated with telephony. The bureau has six operating divisions:

- *Accounting Policy.* Develops policies and regulations for the Universal Service programs. These programs, mandated by the Telecommunications Act of 1996, seek to make telecommunication services available and affordable to all Americans, schools, libraries, and rural health care providers.
- *Accounting Safeguards.* Develops and administers accounting and reporting programs and uniform systems of accounts for common carriers, and conducts field audits and investigations of carriers' financial and operating practices.
- *Network Services.* Oversees telephone numbering issues, network reliability, and telephone hearing aid requirements.
- *Industry Analysis.* Collects and distributes data on the telecommunications industries.
- *Competitive Pricing.* Develops pricing policies for interstate services and administers telecommunications tariffs (agreements concerning service offerings and rates).
- *Policy and Program Planning.* Develops rules and regulatory programs to facilitate competition.

In 1994 the Commission added two additional bureaus: the *International Bureau*, to handle all FCC international telecommunications and all satellite programs and policies (both international and domestic), and the *Wireless Telecommunications Bureau*, to handle all FCC domestic wireless telecommunications programs and policies (including cellular and personal communications services, LMDS, and amateur radio services, but excluding satellite communications). The Wireless Bureau is also responsible for implementing competitive bidding for spectrum auctions. Also under the 1994 reorganization, what had been the Office of Small Business Activities was renamed the *Office of Communications Business Opportunities*, and its function was expanded to include promotion of participation in communication by minorities, small businesses, and women.

EXHIBIT 12.b
. . . .

FCC Organization

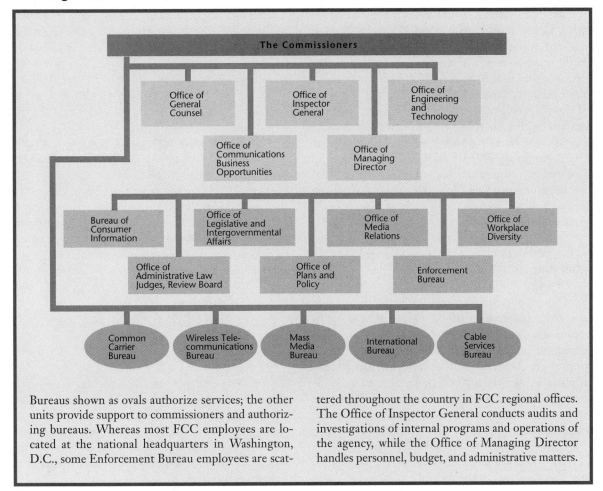

Bureaus shown as ovals authorize services; the other units provide support to commissioners and authorizing bureaus. Whereas most FCC employees are located at the national headquarters in Washington, D.C., some Enforcement Bureau employees are scattered throughout the country in FCC regional offices. The Office of Inspector General conducts audits and investigations of internal programs and operations of the agency, while the Office of Managing Director handles personnel, budget, and administrative matters.

The Commission reorganized further in 1999 by creating two more bureaus, the *Bureau of Consumer Information* and the *Enforcement Bureau*. The Enforcement Bureau combines the functions of several other bureaus and offices and now has primary responsibility for enforcing the Commission's rules.

Commissioners
. . . .

Most of the work of the FCC is done by the staff. Only the most important decisions are made by the commissioners themselves, although they establish overall policy that must be followed by Commission employees.

The five commissioners (seven until 1982) serve five-year terms and may be reappointed. Commissioners must be citizens, may not have a financial interest in any type of communications business, and must devote themselves full-time to the job. In practice, many commissioners never finish their terms of office because other opportunities entice them into higher-paying positions outside the government. There are exceptions, however. James Quello, a Democrat from Michigan, was first appointed to the Commission in 1974 by President Nixon and served until 1997.

The president appoints one commissioner to serve as chair. Although the chair of the Commission has only one vote when the Commission makes decisions, he or she exercises considerable power within the agency by setting the agenda and making important staff appointments.

The first woman commissioner, Frieda Hennock, served from 1948 to 1955. The next woman was not appointed until 1971, but since 1979 at least one commissioner at any one time has been a woman. The first African American commissioner was Benjamin Hooks (1972–1977), later the head of the National Association for the Advancement of Colored People (NAACP). Hispanics as well as blacks have been appointed since, including the appointment in 1997 of William Kennard as the first African-American chair of the FCC. In 1994 Rachelle Chong, a San Francisco–based attorney, became the first Asian American to serve as a commissioner.

Staff Role
· · · · · · ·

References to "the Commission" usually include not only the five commissioners but also staff members. Staff members handle inquiries and complaints, which seldom come to the commissioners' attention, as well as thousands of letters, applications, and forms from FCC-regulated industries. *Processing rules* spell out which decisions staff members may settle and which must go to the commissioners.

Except for top administrators (bureau and office chiefs), whom the chair appoints, Commission staff members are part of the federal civil service. Many serve for decades, developing in-depth expertise on which the commissioners depend. Because of this expertise, the professional staff often exerts considerable influence not only on the day-to-day operations of the Commission but often on long-term policy as well.

Rule-Making Process
· · · · · · · · · · · · · ·

The rule-making function generates a large body of FCC administrative law called *rules and regulations* (there is no difference between the two— the phrase is traditional). Whether the Commission acts on a petition to consider a certain action or acts on its own, it often begins with a *notice of inquiry* (NOI) for a subject that needs preliminary comment and research, a *notice of proposed rule making* (NPRM) when offering specific new rules for public comment, or a combination of both (see Exhibit 12.c). These notices invite comment from interested parties, mostly attorneys representing affected individuals, stations, companies, or industries. On rare occasions, proposed rule changes of special significance or of a controversial nature may be scheduled for oral argument before the five commissioners.

After digesting outside comments, the staff prepares a proposed decision for the commissioners, usually as a recommended *report and order* with background discussion explaining and defending the action. Once a proposed report and order is adopted, that action becomes subject to petitions to the Commission for reconsideration and/or appeal to a U.S. court of appeals. Increasingly in recent years, one party or another appeals nearly every "final" FCC decision, sometimes delaying implementation by months or years.

Adjudication
· · · · · · · · ·

A second type of FCC decision making is adjudication, which settles specific disputes—whether between outside parties (for example, a broadcaster and a political candidate arguing over equal opportunities [see Section 13.7, p. 392]) or between the FCC and an outside party (such as a broadcaster who protests a fine).

The staff may settle simple disputes quickly, but others become the subjects of hearings. Electronic media owners avoid hearings if at all possible. Such hearings require expensive legal representation and can take a long time.

FCC Critique
· · · · · · · · ·

Over the years the FCC has been among the most frequently analyzed and scathingly criticized of all federal regulatory agencies. Past official investigations and private studies of the Commission and its methods have reached negative conclusions with monotonous regularity. A few criticisms have been recurrent.

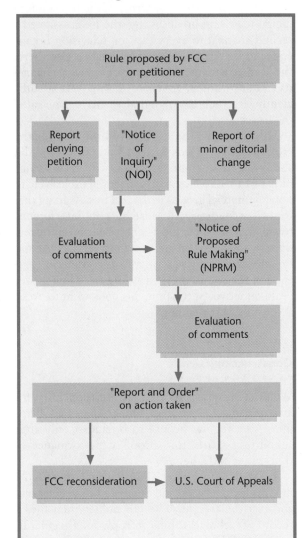

EXHIBIT 12.c
∙ ∙ ∙ ∙ ∙ ∙ ∙ ∙ ∙ ∙ ∙

FCC Rule-Making Process

Rule proposed by FCC
or petitioner

Report denying petition

"Notice of Inquiry" (NOI)

Report of minor editorial change

Evaluation of comments

"Notice of Proposed Rule Making" (NPRM)

Evaluation of comments

"Report and Order" on action taken

FCC reconsideration → U.S. Court of Appeals

A petition for a new rule or a change of existing rules can come from the public (rare), the regulated industries (common), another part of the executive branch, or a unit within the FCC itself. Parties who are dissatisfied with denials or new rules often appeal for reconsideration by the FCC and, if they are still not satisfied, to the courts. Each step of the process must be documented in the *Federal Register* so that all interested parties can keep informed of rule-making actions.

- The political process for choosing commissioners often fails to select qualified people. FCC appointments do not rank high in the Washington political pecking order. The president often uses appointments to regulatory commissions to pay off minor political debts.
- As a consequence, commissioners usually lack expertise and sometimes the dedication assumed by the Communications Act. Most appointees have been lawyers. People with experience in engineering, the media, or common carriers—or in relevant academic specialties—rarely receive appointment because they lack the political constituencies required for serious consideration.
- Commissioners' hopes for future employment with regulated industries may underlie their usually narrow reading of "the public interest." Not many stay in office long enough to attain great expertise; the more ambitious and better-qualified appointees soon move on to higher-paying positions, usually in private legal practice, specializing (of course) in communications law.

Although some of these criticisms remain valid today, others do not. Most recent Commission appointees have been highly qualified and dedicated public servants from state public-service commissions, legislative staffs, communications law firms, and the ranks of the FCC's staff. With the ascendance of telecommunications as a hot topic in Washington, being named to the Commission is now seen as a prestigious and important appointment.

More recent criticism of the FCC focuses not so much on the competence or expertise of the commissioners but on the FCC's timing and policy directions. Some members of Congress, for example, fault the Commission for not acting as swiftly as they would like in deregulating telecommunication services. Others question some Commission policy initiatives, such as proposals to implement a low-power radio service or require broadcasters to provide free airtime to political candidates.

However, it is somewhat misleading to speak of *the* FCC. As its membership (along with the administration in Washington) changes, the Commission takes on varying complexions—sometimes reform minded, sometimes more interested in the welfare of the businesses it regulates than in the public interest.

12.4 BROADCAST LICENSING

The *authorizing of service*—licensing of broadcasting stations—is the FCC's single most important function. (Local authorities that franchise cable systems play a parallel role. See Section 12.8.) Most other regulation grows out of this licensing process. The FCC acts as gatekeeper, using its licensing power to offer station operation to some and not others. No one can legally "own" any part of the electromagnetic spectrum. Conscious of both real and potential claims to the contrary, the Communications Act repeatedly codifies the ban on channel ownership.

Finding a Channel

A would-be broadcast licensee applies for specific facilities (channel, power, coverage pattern, antenna location, time of operation, and so on). Most desirable commercial channels had been assigned by the 1970s, so a new licensee nearly always had to buy an existing station. Later, several new opportunities developed.

In 1982 the Commission approved low-power television (LPTV), whose channel use could fit in among existing outlets, inasmuch as LPTVs are limited in power so as not to interfere with full-power television stations. LPTV stations were considered secondary services, which meant that if interference developed between a low-powered station and a full-powered station, the low-powered station would have to resolve the interference or stop broadcasting altogether. After the FCC allocated a second channel to all full-powered stations in 1997 for DTV services, many LPTV operators objected that they would be forced off the air by the new full-powered stations. Congress reacted by passing legislation that would protect some LPTV stations by giving them permanent status (see Section 4.9, p. 109).

In the mid-1980s the FCC created nearly 700 new FM allotments, made possible by improved technology limiting interference with stations already on the air. In the early 1990s the FCC extended the AM band to 1705 kHz, adding 10 new medium-wave channels to which existing AM outlets could migrate, thus cutting down on existing interference. But in a strong shift away from earlier policy, the Commission restricted access to some of these "new" channels to old players—existing broadcasters. In 1997, by virtue of the transition to digital television broadcasting, the FCC assigned a second channel to every existing television broadcaster in the United States. As detailed elsewhere, the licensee is supposed to return the analog channel to the government in 2006 or whenever a sufficiently large number of digital sets have penetrated the market.

Most recently, the FCC paved the way for the creation of hundreds of new low-powered radio stations. These stations, which will have service radii of 1 to 3.5 miles, will be licensed exclusively to noncommercial groups with local ties. Arguing that these new stations will cause harmful interference to existing stations, most incumbent broadcasters oppose the creation of this new service. At the broadcasters' request, Congress is considering legislation that would overturn the FCC's low-power radio initiative. At the time this book went to press, those bills were pending.

Construction Permit

Would-be licensees for new facilities apply first for *construction permits* (CPs). The holder of a CP applies for a regular license to broadcast only after submitting satisfactory proof of performance of transmitter and antenna.

A CP gives its holder a limited time (usually 24 months for television and 18 for radio) to construct and test the station. The CP holder then files a record of technical testing to apply for a regular license. With FCC permission, a permittee may (and most do) begin on-air program testing pending final approval of the license.

Licensee Qualifications

Sec. 310(b) of the Communications Act forbids foreign control of a broadcast license. Some government and business leaders in the mid-1990s began questioning that restriction. These leaders suggest that U.S. licenses could be opened up to foreign ownership in return for greater participation by U.S. companies in foreign markets. Sec. 308(b) limits licensees to those U.S. citizens, or corporations owned primarily by U.S. citizens, who qualify in these areas:

- *Character.* Applicants should have personal and business histories free of evidence suggesting defects in character that would cast doubt on their ability to operate in the public interest. Criminal records and violations of antitrust laws may constitute such evidence. Any misrepresentation to the FCC would be an almost-fatal defect.
- *Financial resources.* Applicants must certify that they have "sufficient" financial resources. The FCC has issued varied definitions of "sufficient," but currently it means the ability to construct and operate facilities for 90 days without reliance on station revenue.
- *Technical ability.* Most applicants hire engineering consultants to prepare technical aspects of their applications. They show how a proposed station will get maximum physical coverage without causing objectionable interference to existing stations.

Mutually Exclusive Applications

Until the mid-1990s, when two or more qualified applicants sought the same channel, the act required the FCC to make a choice among the applicants only after conducting comparative hearings. These hearings took on the appearance of full-blown trials, with attorneys for all parties presenting evidence, calling witnesses, and engaging in cross-examination—an often drawn-out and costly process for all concerned.

In 1993 a court of appeals declared the FCC's comparative licensing criteria illegal (F, 1993) and the Commission was either unwilling or unable to formulate new criteria. Consequently, almost two thousand mutually exclusive applications for radio and television stations were put on hold until the Commission figured out another method of awarding contested licenses. To help clear the logjam, the Commission waived its rules (which generally do not allow one applicant or an outside party to buy off other applicants) and allowed anyone interested in the contested channels to negotiate deals with the applicants and buy them out. When the dead-

line for dealmaking expired, settlements had been reached among applicants for approximately 50 television stations and almost 800 radio stations.

Congress then stepped in and ordered the Commission to deal with the remaining 968 applications and any new comparative broadcast licensing situations through a competitive bidding, or auction, process. The FCC held the auction in September 1999 and began granting construction permits to the winning applicants, many of whom had been awaiting a decision for more than five years. The auction raised $57 million, all of which, when collected, will go directly to the federal treasury.

License by Lottery

So many thousands of competitive applications piled up for LPTV stations in the early 1980s that, responding to the FCC's plea, Congress amended the Act in 1982 to authorize selection of new (not renewal) LPTV licensees by means of a *lottery* rather than time-consuming comparative hearings (47 USC 309[i]). Using lotteries, the FCC could check an LPTV applicant's qualifications to be a licensee after the applicant had already been chosen by the lottery, going back to choose another applicant if the first winner was deficient in some way.

Responding to the demands of Congress to increase the diversity of LPTV ownership among minorities and others, the Commission assigns statistical preferences to minority applicants and those applicants who own no or few other media outlets. This gives these preferred applicants a greater chance of winning licenses through the lottery process.

License by Auction

In addition to the auctions noted previously, Congress, as part of its effort to balance the federal budget, will require the auction of much of the existing television spectrum when TV broadcasters give up their analog channels after the transition to digital. Some lawmakers estimate the value of this spectrum, primarily in the existing VHF band, to be as high as $30 billion to $35 billion. The Commission also auctions spectrum for many other services, including satellite communications and cellular telephone.

Services Requiring No License

The FCC requires virtually all spectrum-using electronic mass media and common carriers to obtain licenses to operate. The few exceptions include the following:

- radio station auxiliary services transmitted on subcarriers and neither related to broadcasting nor decoded by regular radio receivers
- data and closed-captioned services broadcast by television stations
- carrier-current radio stations such as those used on some college campuses
- very low power AM radio stations designed to serve small, discrete areas

The FCC also regulates the manufacture of many nonbroadcast devices that use part of the electromagnetic spectrum (for example, remote garage-door openers, microwave ovens, and cordless telephones) but does not require that consumers be licensed to use them.

12.5   OPERATIONS

Ideally, the FCC should monitor station operations constantly to ensure operation in the public interest. But it simply is not feasible for the Commission to keep watch over all the programs of thousands of stations. In the course of normal operations, broadcasters experience little official supervision.

Employment Practices

In 1969, the FCC began monitoring broadcast employment practices, and over the years added equal employment requirements. Under the regulatory regime adopted by the Commission, broadcast licensees, cable system operators, and other multichannel video programming distributors (such as DBS and MMDS operators) were required to establish and implement an *equal employment opportunity* plan, guaranteeing that qualified females and minorities would be recruited, hired, trained, and promoted. The Commission established hiring benchmarks for females and minorities, and monitored compliance with its requirements by mandating the submission of annual employment reports.

The Commission enforced its equal employment requirements strictly, as was evidenced by the numerous fines levied against broadcast stations and cable systems that violated the guidelines.

In April 1998 a U.S. court of appeals struck down the FCC's broadcast equal employment opportunity rules (technically, the rules still applied to cable systems and other multichannel video providers), declaring that the Commission had not proved that the regulations served the public interest (F, 1998). Later that year the Commission proposed a new set of equal employment rules to replace the ones that were thrown out. In the meantime, at FCC chair William Kennard's urging, most broadcasters pledged to continue their efforts to hire qualified minorities and females.

The Commission adopted new equal employment rules in January 2000. The new rules focus on broadcasters' recruitment practices by requiring that licensees widely disseminate information about job openings so that qualified minorities and women are aware of the best jobs around the country. Broadcasters with more than five full-time employees must detail their outreach efforts and the results of those efforts by placing an annual report in their public files. Every two years during the license term these stations must also certify to the Commission that the employment rules are being met. Radio stations with more than ten full-time employees, and all TV stations, must submit a copy of their recruitment report to the FCC midway through the license term and at renewal. The Commission also amended the equal employment requirements for cable operators and other multichannel video providers to be consistent with the new broadcast regulations.

In March 2000 the National Association of Broadcasters and all fifty state broadcast associations appealed the FCC's new regulations to the U.S. Court of Appeals for the District of Columbia Circuit. According to the broadcast associations, the new rules are substantially more burdensome than the former rules due to the increased recruiting, record keeping, and reporting requirements. A resolution to this litigation may take several years.

Public File

The FCC requires stations to keep license-renewal applications and other relevant documents in a file readily available for public inspection. All stations, commercial and noncommercial, must maintain such a file, retaining some documents for seven years in the case of a radio station and five in the case of television. Cable systems have to maintain a public file containing only employment-related documents. The more inclusive broadcast file includes the following:

- the latest construction permit or license application, including any for major changes
- the latest license renewal application, as well as ownership reports and annual employment reports
- the pamphlet titled "The Public and Broadcasting—A Procedural Manual," issued in 1974 and updated in 1998
- a record of the disposition of any political-broadcast time requests for the preceding two years
- a quarterly listing of programs the licensee believes provided the most significant treatment of local community problems of public importance
- records required by the Children's Television Act of 1990 concerning programming for and advertising to children under the age of 16
- any time-brokerage agreements in force
- the annual EEO recruitment report
- copies of (television) station decisions concerning must-carry and retransmission consent governing its relationship with any local cable systems
- letters received from members of the public (to be kept for three years) and any agreement with citizens groups

Deregulatory decisions such as the abandonment of formal ascertainment of community needs, deletion of programming guidelines, and reduction of renewal applications to postcard form reduced the public file's size. Although members of the public rarely ask to see it, the file receives a thorough examination when FCC inspectors visit a station, and poor record keeping is becoming a more common source of rule violations.

Keeping Up with Washington

Electronic media managers must keep current with changes in FCC regulations. Trade organizations and publications and an army of communications attorneys in Washington help licensees with this task. Well over 1,000 attorneys representing electronic media clients are members of the Federal Communications Bar Association (FCBA). Personal contacts with FCC staff members enable Washington lawyers to get things done faster than can distant licensees unfamiliar with the federal bureaucracy. The National Association of Broadcasters, National Cable Television Association, and other trade associations offer legal clinics and regular publications on FCC rules.

From all of these sources, stations and systems receive continuous advice on what is new and what not to forget. For example, licensees often seem unaware of differing federal or state requirements on things such as employee drug use and over-the-air use of live or recorded telephone conversations. The timely filing of regular ownership, employment, and other reports required by the Commission is a mundane but important aspect of using legal counsel.

Monitoring Performance

The FCC rarely monitors electronic media. Inspectors from the Enforcement Bureau check technical aspects of station and cable system operations, but only occasionally and in random fashion. Questions about programming and commercial practices, if they arise at all, usually come to FCC attention through complaints from the public, consumer groups, or competing media. Few complainants seem to understand the FCC's legal limitations—most complaints are discarded because they ask the Commission to violate the First Amendment by censoring material that the writer personally dislikes. The leading topics vary only slightly from year to year, often influenced by organized letter-writing campaigns as well as by program trends.

12.6 RENEWAL

Under the Communications Act, licenses may be awarded for only "limited periods of time," currently a maximum of eight years, and must regularly come up for renewal. Although the FCC renews more than 98 percent of all licenses without asking any searching questions, possible nonrenewal seems always to be lurking in the background. Challenges

EXHIBIT 12.d
• • • • • • • • • • • •

Renewal Paperwork Before and After Deregulation

(A) In 1971, before deregulation, it took more than 16 pounds of paperwork to file for renewal of four Nebraska radio stations. (B) In the 1990s this small stack of papers along with a simple five-page form did the same job.

Source: Reprinted from *Broadcasting & Cable*. Copyrighted Cahners Business Information.

to renewal applications can come from the FCC but may also be lodged by dissatisfied citizens in the licensee's community. Even if the incumbent wins a contested renewal (and they usually do), defending it can be both expensive and time-consuming.

Application Routes
• • • • • • • • • • • •

Sec. 309 of the Communications Act stipulates that "licenses shall be renewed if the public interest, convenience, and necessity would be served" by renewal. Before deregulation of renewal procedures, applications often required piles of supporting documents showing how the licensee had served or would serve the public interest (see Exhibit 12.d). In 1981 the FCC began using a simple postcard-size renewal form, later expanded to a still relatively simple five-page form, which most applicants supplement with additional multipage exhibits.

Renewal applications take one of three paths: uncontested, FCC staff contested, and the petition to deny.

• Some 98 percent of all renewal applications fall in the *uncontested* category. The FCC staff uses

its delegated authority to renew the station almost automatically. In fact, consumer advocates complain that the Mass Media Bureau merely rubber-stamps uncontested applications, no matter how mediocre a station's past performance may have been.

• Occasionally, the FCC staff will initiate a proceeding to deny a license renewal if it finds serious or frequent violations of U.S. law or FCC regulations. Allegations of criminal law breaches, antitrust activity, or concealing information from the Commission could lead to such action.

• The now-rare *petition-to-deny* renewal comes from a citizen group or other party that opposes an incumbent licensee. Such groups claim that incumbents have failed to meet public-interest standards. Although the petition process is important in giving citizens the right to challenge licensees who may not be serving the public interest, in practice the petition-to-deny process rarely results in the denial of renewal.

Until the Telecommunications Act of 1996, another type of renewal scenario known as the *comparative renewal* was possible. This situation arose when a would-be licensee tried to displace the incumbent licensee, claiming that it could do a better job serving the public interest. Comparative renewals put the Commission in the odd position of having to judge the worthiness of two applicants based on entirely different qualifications: The incumbent was judged on its past record, while the challenger was judged solely on the paper promises contained in its application. In practice the incumbent broadcaster nearly always won its renewal, but often after a long and expensive hearing and appeal process. Congress changed the renewal law in 1996 by requiring that the Commission, when considering a broadcast license renewal, could not consider a competing application unless it had first determined that the original license should not be renewed.

License Transfers

Other than uncontested renewal, the most frequent licensing process is license transfer, which occurs when an owner sells a station to another party. As noted in Section 6.12 (p. 200), station sales have increased dramatically since the passage of the Telecommunications Act of 1996. A large majority of station owners have acquired their stations by buying them rather than by securing new stations from the FCC.

All station ownership transfers require Commission approval. After the parties to the transfer agree on the terms of the sale, they submit their transfer application to the FCC. The Commission issues a public notice of the application, and interested parties have 30 days to submit comments. In most instances, no one objects and the Commission staff routinely approves the license transfer. Individuals or groups may, however, file petitions to deny the sale. If the FCC finds that a petition raises serious issues, it will order a hearing to determine whether the proposed transfer would serve the public interest. Such findings rarely occur; in most cases, the Commission dismisses petitions to deny.

Because large mergers (such as the CBS/Viacom and AOL/Time Warner proposals) involve the transfer of licenses, the FCC must approve them. In reviewing these mergers the Commission often takes into account the antitrust implications of the transactions as well as their public-interest ramifications. Such comprehensive reviews may take many months to complete. Some members of Congress have criticized the FCC's actions in such cases, arguing that the agency is duplicating the ef-

forts of the Justice Department, which has primary jurisdiction over antitrust matters. Several bills were introduced in 1999 that would require the FCC to finish its review of media mergers within a stated period of time—typically six months—and would strip the Commission of its authority to consider the antitrust implications of media mergers and acquisitions. As of June 2000 none of these bills had passed.

12.7 ENFORCEMENT

The FCC can impose a variety of penalties on licensees for infractions of its rules. They range from simple letters of admonition to fines, short-term license renewal, or—the ultimate penalty—no renewal at all. When a licensee proves incorrigible, the FCC can either refuse to renew the license or revoke it outright. The FCC prefers the nonrenewal option because it puts the burden of proof on the licensee, whereas revocation places the burden of proof on the Commission. None of these sanctions can be invoked without legal due process, often beginning with a formal hearing.

Due Process

A fundamental safeguard of individual liberties under the Constitution, the Fifth Amendment *due process clause*, guarantees that government may not deprive a person of "life, liberty, or property without due process of law." Among many other applications, this means that the FCC may not use its powers arbitrarily. Fairness, the goal of due process, requires that applicants and petitioners have ample opportunity to argue their cases under nondiscriminatory conditions and that parties adversely affected by decisions may appeal for review by authorities other than those that made the initial decision. Many due process rights are detailed in the Administrative Procedure Act of 1946, which specifies how agencies such as the FCC must conduct their proceedings to ensure due process for all participants. For example:

- The Commission must advise unsuccessful license applicants of its reasons for rejecting their applications. The applicants may reply, and if the Commission still decides against them, it must then set the matter for a hearing.

- On the other hand, if the Commission grants a license application *without* a hearing, for the ensuing 30 days (after a public notice announcing this action) the grant remains open to protest from "any party in interest." If the FCC finds that some party raises pertinent issues, the Commission must then postpone the effective date of the grant and hold a hearing.
- If the Commission wishes to fine a licensee, it must invite the licensee to "show cause" why such action should not be taken.

FCC Hearings
• • • • • • • • • •

Senior staff attorneys, called *administrative law judges* (ALJs), preside over initial FCC hearings. They conduct the proceedings somewhat like courtroom trials, with sworn witnesses, testimony, evidence, counsel for each side, and so on. Appeals of initial decisions of ALJs are reviewed, first by the FCC's Review Board and then by the commissioners themselves. Procedural rules head off frivolous interventions and intentional delays by carefully defining circumstances that justify hearings and qualifications of parties entitled to *standing*—that is, the right to participate.

Court Appeals
• • • • • • • • • •

Even after all safeguards in FCC hearings and re-hearings have been exhausted, the Communications Act gives parties adversely affected by FCC actions still further recourse. The Act provides that all appeals concerning station licenses must go before the U.S. Court of Appeals for the District of Columbia Circuit in Washington, DC. Appeals from FCC decisions in cases that do not involve licensing may be initiated in any of the 12 other U.S. courts of appeals. Each serves a specific region of the country and is known as a *circuit court of appeal.*

The court may confirm or overturn Commission actions in part or in whole. It may also *remand* a case, sending it back to the FCC for further consideration in keeping with the court's interpretation of the Communications Act and other laws (see Exhibit 12.e).

From any of the federal circuit courts, final appeals may be sought before the Supreme Court of the United States. A request for consideration by the Supreme Court, called a *writ of certiorari*, is most often turned down ("cert. denied"). If that happens, the appeal process is over. Refusal to hear a case does not necessarily mean that the Supreme Court agrees with lower court findings, but the earlier decision holds nonetheless, though without standing as a compelling nationwide legal precedent.

Loss of License
• • • • • • • • • •

Since the FCC's creation in 1934, fewer than 200 stations have involuntarily lost their licenses. *Nonrenewal* accounts for two-thirds of these losses, outright *revocation*—not waiting for the current license to expire—about one-third. Though an ever-present background threat, loss of license only rarely becomes a reality among the thousands of stations on the air.

Stations that have lost their licenses have largely been obscure radio outlets whose licensees have accumulated long lists of willful misdemeanors. Transgressions accumulate because the FCC usually treats them leniently as long as the guilty licensee candidly admits error and contritely promises to reform. Eventually, another violation causes the FCC to lose patience, and some specific misdeed results in nonrenewal or revocation.

Lesser Penalties
• • • • • • • • • •

Not all offenses, of course, warrant the capital punishment of license loss. For lesser offenses, the FCC inflicts milder sanctions:

- *Short-term renewal* (usually a year or two instead of the full renewal term) puts a licensee on a kind of probation pending correction of deficiencies evident during its preceding license period.
- *Conditional renewal* is granted pending correction of some specific fault.
- *Forfeitures*, or *fines*, ranging up to a maximum of $250,000 for repeated violations may be assessed for broadcaster infractions of FCC rules. Fines against common carriers can be as high as $1 million. Most forfeitures come from technical violations, few from program violations, and most are for much less than the maximum amount. Because cable systems have no license that the FCC can threaten, fines are the agency's chief means of enforcing cable rules.

EXHIBIT 12.e

Citizen Involvement: The Landmark WLBT Case

Until the late 1960s, a broadcast station's audience members had no right to take part in regulatory proceedings concerning the station. That situation changed as a result of a court case that began in 1955 when a group of citizens made the first of a series of complaints to the FCC about the conduct of WLBT, a VHF television station in Jackson, Mississippi. The group accused the station of blatant discrimination against blacks, who formed 45 percent of its audience. The FCC dismissed the citizens' complaints, saying that they had no legal right to participate in a licensing decision. When WLBT's license again came up for renewal in 1964, local groups obtained legal assistance from the Office of Communications of the United Church of Christ (UCC) in New York.

The UCC petitioned the FCC on behalf of the local groups for permission to intervene in the WLBT renewal proceeding, but the FCC again rejected the petition, saying that citizens had no *legal standing* to intervene. At that time the Commission recognized only signal interference or economic injury (to another broadcaster) as reasons to give parties the right to participate in renewal hearings. Thus only other broadcasters had standing to challenge existing licensees.

The UCC went to the court of appeals, claiming that the FCC had no right to bar representatives of the viewing public from intervening in renewals or to award a renewal without a hearing in the face of substantial public opposition. The court agreed, directing the FCC to hold hearings on WLBT's renewal and to give standing to representatives of the public (F, 1966). The FCC held a hearing and grudgingly permitted the UCC to participate as ordered. However, it once again renewed WLBT's license.

The UCC returned to court, and in 1969 an appeals court reconsidered the case—14 years after the first complaints had been recorded. In the last opinion written by Warren Burger before he became chief justice of the United States, the court rebuked the FCC for "scandalous delay." It ordered the FCC to cancel WLBT's license and to appoint an interim operator pending selection of a new licensee (F, 1969). But 10 more years passed before the FCC finally selected a new permanent licensee. Altogether, the case dragged on for more than a quarter of a century.

As the FCC had feared, the WLBT case triggered many petitions to deny the renewal of other licenses. This "reign of terror," as a trade magazine put it, resulted in few actual hearings and still fewer denials. Of the 342 challenges filed in 1971–1973, only 16 resulted in denials of license renewal. An exacting standard of evidence established by the FCC and approved by the court ensured this high rate of petition failure. Only after an opponent presented overwhelming evidence would the FCC schedule a license-renewal hearing (F, 1972).

- Relatively minor infractions may result in a *letter* being placed in the station's file for consideration when it applies for license renewal.

12.8 CABLE

Cable "licensing," or *franchising*, follows a pattern totally different from broadcast licensing (see Exhibit 12.f). Local (and, in a few instances, state) rather than federal authorities issue franchises, because cable systems use streets and other public property that is subject to municipal jurisdiction, rather than federally controlled airways. Federal cable laws contain provisions covering subscriber fees, ownership, equal employment opportunities,

programming, customer service, and technical standards. These laws, passed in 1984, 1992, and 1996, amended the 1934 Communications Act.

Cable Acts: General Provisions

The Cable Communications Policy Act of 1984 created a new Title VI of the Communications Act that established a loose federal *de*regulatory framework for cable television. Key parts of that act were reversed in the strongly *re*-regulatory Cable Television Consumer Protection and Competition Act of 1992, which amended Title VI. This 1992 legislation grew out of public and congressional anger over sharp increases in cable subscription rates—in some cases three times the rate of inflation—in the five years before it became law. The 1996 act modified a few

EXHIBIT 12.f

• • • • • • • • • • •

Broadcast Licensing and Cable Franchising: A Comparison

Broadcast Licensing	Cable Franchising
Licensed by FCC	Franchised by local or state authority
License term: 8 years	Term set in franchise: 10–15 years standard
"Comparative" renewals not permitted	Comparative renewals allowed but unusual
Obligations set by Congress and FCC	Obligations set by Congress, FCC, and franchising authority
Station transfers approved by FCC	System transfers approved by franchising authority
Licensing process uniform for all licensees	Franchising process varies from community to community

provisions and authorized a new provider for the delivery of cable services—the telephone companies. The combined result of the three acts includes the following key provisions:

- The 1984 act defined cable television as a one-way video programming service; neither it nor the 1992 and 1996 revisions regulate two-way (interactive) services.
- The 1992 act transforms the relationship between broadcasters and cable systems. Under this act, each commercial television station has the right every three years (beginning in October 1993) either to charge cable systems within its coverage area a negotiated fee for use of its signal (termed *retransmission consent*) or simply to require that cable systems carry its signal (the *must-carry rule*). If a station and cable system cannot reach agreement on terms, that station will not be carried on that system for at least three years. Stations that fail to choose between must-carry and retransmission consent will, by default, come under the must-carry option. Local noncommercial stations may claim must-carry rights but do not have the option of seeking retransmission consent. MMDS, LMDS, and satellite master antenna television (SMATV) systems are also subject to retransmission consent rules.

Cable systems with 12 or fewer channels must devote at least three to local broadcast signals. Systems with more than 12 channels may be required to devote up to a third of their capacity to such signals. None of this applies to the relatively few systems with fewer than 300 subscribers.

The must-carry requirement, once an FCC regulation, had been twice overturned on First Amendment grounds by a federal appeals court in the late 1980s. In 1997 the Supreme Court upheld the requirements, finding that the protection of over-the-air broadcasting was an important government interest.

- A television station may also demand carriage on the same channel number on cable that it uses over the air.
- Reversing a key aspect of the 1984 law, the 1992 act required local authorities or the FCC to regulate subscriber rates for basic tiers of cable service. Congress retreated a bit in 1996 by ending all rate regulation of upper-tier basic services effective 31 March 1999. No rate regulation of pay-cable channels such as HBO and Showtime is allowed. Cable systems are exempt from rate regulation if they serve fewer than 30 percent of the households in their area or are subject to "effective competition" from some other multi-channel competitor (MMDS, SMATV, DBS, or telephone companies) that is available to at least 50 percent—and actually subscribed to by at least 15 percent—of an area's households.
- Cable program providers (for example, ESPN or HBO) that are owned even in part by cable system operators must make their services available to

multichannel cable competitors (such as SMATV, MMDS, and DBS) at the same prices offered to cable systems. This provision, which expires in 2002, is designed to give competing delivery systems a chance to get an audience foothold.

Cable Acts: Franchising
• • • • • • • • • • • • • • • •

Whereas the 1984 act severely limited regulatory powers of state and local authorities, the 1992 law reinstated many of those same powers and added new FCC oversight of many aspects of cable franchising.

- Local franchising authorities may require public, educational, and government (PEG) access channels over which the cable operator has little editorial control.
- The 1992 act added protection for children against indecent programming on commercially leased and public-access channels. Much of this regulation was subsequently declared unconstitutional by the U.S. Supreme Court.
- The 1984 act called for cable systems with more than 36 channels to set aside 10 to 15 percent of those channels for leased access to outside parties. The cable owner sets rates for leasing access channels but has no editorial control over their content—other than enforcement of obscenity and indecency limitations (as discussed later). The 1992 act requires the FCC to set maximum rates for use of such channels.
- Reinforcing existing practices and FCC guidelines, Congress allowed local franchise authorities to charge annual franchise fees of up to 5 percent of cable system gross revenues.
- When granting franchise renewals, local authorities may require operators to upgrade cable facilities and channel capacity.
- The FCC may require that specific customer service levels and technical standards be included in new and renewal franchise agreements. Initial detailed FCC rules were issued early in 1993.
- The 1992 act requires the FCC to evaluate every franchise-granting authority in the country (some 20,000 cities, counties, and other political units) and to certify that each has the authority and capacity to issue and enforce such franchises—and that the franchise provisions agree with all requirements in the amended Communications Act. If any franchise authority is not so certified, then the FCC itself is to act as the authority for cable in that market until a local franchise authority is certified.

Franchise Process
• • • • • • • • • • • •

When a local franchising authority (city, town, or county) decides it wants cable service, it first develops an *ordinance*, or legal codification, describing the conditions under which a cable system will be allowed to operate. Drawn up in many cases with the advice of outside experts, ordinances typically stipulate the following:

- the term of the franchise (usually 10 to 15 years, but sometimes longer)
- the quality of service to be provided
- technical standards, such as the minimum number of channels to be provided, time limits on construction, and interconnection with other systems
- the franchise fee
- PEG channel requirements, if any

Bidders base their offers on the design and timetable outlined in the franchise authority's *request for proposals* (RFP). Although franchisers usually grant a franchise to only one bidder, multiple awards (termed *overbuilds* if they cover the same region) are possible. In an attempt to stimulate local competition, the 1992 act states that local authorities may not unreasonably refuse the grant of a competing franchise. Still, the vast majority of cable systems operate as local monopolies.

Renewals
• • • • • • •

A local authority need not find that renewal will serve the public interest or meet any other standard, but rather may simply renew a franchise without ceremony. If, however, the local authority wants to deny renewal, the Communications Act requires that it hold a hearing, in effect raising public-interest issues by deciding whether the incumbent operator has (1) complied with the law; (2) provided a quality of service that is "reasonable in light of the community needs"; (3) maintained the financial, legal, and technical ability to operate; and (4) prepared a renewal proposal that is "reasonable to meet the future cable-related community needs and interests" (Sec. 626). The Act does not require consideration of proposals from competing would-be franchisees during the course of renewal grants or renewal hearings, but the franchising authority may consider other proposals if there are other applicants for the franchise.

12.9   OTHER ELECTRONIC MEDIA

While broadcasting and cable television certainly are the largest electronic mass media in terms of the number of people served, other electronic media are also regulated by the FCC. Based on the characteristics of those media, FCC regulation may vary.

Satellite Master Antenna Television Systems

Satellite master antenna television (SMATV) systems are functionally similar to cable television systems except that they serve multiunit dwelling complexes without crossing public streets or rights of way. As such, they are sometimes termed "private" cable systems, and they are not subject to local regulation. Thus, SMATV systems do not need franchises to operate, nor are they subject to local access channel and equipment requirements. The FCC does, however, require SMATV systems to follow broadcast television must-carry rules.

"Wireless" Cable

Wireless cable—sometimes referred to as multichannel television—is a subscription service much like cable television except that it uses over-the-air microwave frequencies to deliver its signals rather than coaxial cable and fiber. Currently, wireless cable utilizes multichannel multipoint distribution service (MMDS) to distribute up to 30 channels of programming to subscribers. The FCC licenses MMDS systems via a lottery in cases where more than one applicant seeks frequencies in the same markets. Equal employment opportunity regulations and broadcast signal carriage rules apply to MMDS systems. Because these systems use over-the-air frequencies, local governments have no jurisdiction over MMDS services.

In 1997 the FCC authorized a new service for the delivery of multichannel television known as local multipoint distribution service (LMDS). This service, which also uses microwave frequencies but reuses them in a way similar to cellular telephone service, is capable of offering subscribers a variety of broadband services, including video program distribution, video teleconferencing, wireless local loop telephony, and high-speed Internet access. The Commission believes that LMDS has the potential to become a major competitor of local-exchange telephone and cable television services. In a series of auctions held in 1998 and 1999, bidders offered more than $600 million for the available LMDS frequencies. The Commission has not yet decided what public-interest requirements will apply to LMDS licensees.

Direct Broadcast Satellite

The FCC licensed the first direct broadcast satellite (DBS) systems in the early 1980s much the same as they granted broadcast licenses. More recently, the Commission has granted new licenses through the spectrum auction process. To date, DBS systems operate on a subscription basis and deliver much the same programming as cable systems, but generally without local television signals.

As noted in Section 6.10 (p. 191), Echostar's Dish Network began offering local broadcast signals in some major markets in 1998, but does not have the channel capacity to offer all local stations in all markets. Broadcasters generally argue that if a DBS operator provides any television station in a local market, it should offer all local stations in that market. Congress enacted legislation in 1999 that permits DBS operators to carry local broadcast stations in selected markets. By January 1, 2002, DBS providers must carry all local TV signals in the markets they choose to serve. The law also permits DBS operators to provide selected television broadcast signals to households located in so-called white areas—locations where over-the-air broadcast signals are unavailable. Thus subscribers living in those areas can receive broadcast network affiliates as part of their DBS service, usually for a small extra fee. In addition, FCC equal employment opportunity (EEO) rules apply to DBS licensees, and the 1992 amendments to the Communications Act require DBS service providers to set aside up to 7 percent of their channel capacity for public-interest programming.

A new satellite service authorized in 1997 promises to deliver multichannel, CD-quality audio service directly to homes, automobiles, and portable radios nationwide. Although service requirements had not been set by the Commission when this book went to press, licensing of the two available satellite radio slots was conducted in April 1997 via a spectrum auction. Service was scheduled to begin sometime in the year 2000.

Telephone Companies

From the mid-1960s to 1996 the FCC did not allow telephone companies to provide cable television services in the same area in which they provided telephone service. The Commission enacted this crossownership restriction so that the powerful telephone companies would not dominate the nascent cable television business. Congress followed the FCC's lead and codified the crossownership ban in the 1984 amendments to the Communications Act.

Following several successful court challenges of the cable/telco crossownership restrictions in the mid-1990s, Congress removed the ban in the Telecommunications Act of 1996 as part of its effort to stimulate local competition in the multichannel video market. According to the 1996 legislation, telephone companies may now offer video services in their local telephone service areas under four different regulatory regimes: as over-the-air providers such as MMDS operators, subject to Title III regulation; as common carriers subject to Title II regulation; as cable television operators subject to Title VI regulation; or as a newly created hybrid service called an *open video system* (OVS).

Operation as an OVS allows the telephone company to deliver programming of its own choosing while giving other programmers access to the company's distribution system. Essentially, the telephone company acts as a cable system operator and common carrier at the same time. The Act requires that if there is more demand for system capacity than exists, then the telephone company may control no more than one-third of the system's capacity; the other two-thirds of the system

capacity must be apportioned among other users. Although OVS operators must satisfy some of the requirements imposed on cable operators—signal carriage and EEO, for example—they are exempt from some aspects of cable regulation. OVS systems are certified by the FCC and are not franchised locally. Also, most rate regulation does not apply to OVS operations. Local authorities are permitted to exact a share of the OVS system revenues, much like the franchise fee levied against cable operators.

 ## 12.10 DEREGULATION

Many motives underlie efforts to deregulate—in communication and several other economic sectors. Least controversial is the recognized need to discard outdated rules, to simplify unnecessarily complex rules, to ensure that those rules that remain can actually achieve their objectives, to lighten administrative agency loads, and to encourage rapid development of new technologies. Deregulation based on these motives began in the 1970s with wide political support.

A more controversial drive to deregulate stems from ideological motives growing out of a specific vision of a limited governmental role in national life. This view opposes government intervention, advocating instead reliance on the marketplace as a nongovernmental source of control over private economic behavior. This approach first emerged as a force on the national agenda in the late 1970s under President Jimmy Carter, escalated when Ronald Reagan came to the White House in 1981, and continued during the Bush and Clinton administrations. With respect to the regulation of media content, however, many of President Clinton's goals, and those of former FCC chairman Reed Hundt, were distinctly *re*-regulatory. These efforts to increase government oversight of programming content are addressed in the next chapter.

Theoretical Basis

Deregulatory theory holds that marketplace economic forces can stimulate production of better, more varied, and cheaper consumer goods and services without official guidance from above.

Where government regulation may be necessary, it should be tested by a cost-benefit formula to make sure that the costs of such regulation do not outweigh its gains.

The FCC did not go so far as to advocate abandoning all regulation. It divided rules into behavioral and structural categories. *Behavioral* regulation, which deregulators seek to discard or at least minimize, controls what licensees may or may not do in conducting their businesses; *structural* regulation controls the overall shape of the marketplace and terms on which would-be licensees can enter. Rules requiring a licensee to limit advertising in children's programs are one example of behavioral regulation; rules preventing a licensee from owning more than x number of stations are an example of structural regulation.

Deregulation, however, tilts toward the structural regulation approach, which enhances competition by making marketplace entry easier, preserves competition by preventing monopolies, and stays away from program-based decisions. Critics argue that deregulation serves mainly to enrich a few companies and individuals, often at the expense of broader public concerns.

Even theorists favoring deregulation admit that *market failure* can occur. Sometimes competition fails to produce expected favorable results. Indeed, uncontrolled competition may produce corporate giants that suppress competition (major airline control of most departure gates at "hub" airports is one example). And some public "goods," as economists call desirable services, may fall outside the realm of marketplace economics and therefore fail to materialize. If, for example, public television is thought desirable but costs too much to produce with limited private support, marketplace economics may have to be supplemented by giving such broadcasters government aid.

Deregulation in Practice
● ● ● ● ● ● ● ● ● ● ● ● ● ● ● ● ●

In the mid-1970s early deregulation did away with many incidental bureaucratic rules. During the 1980s a more zealously deregulatory FCC began dismantling or softening more basic rules, first in commercial radio (1981), next in educational broadcasting (1984), and then in commercial television (1985). In the 1980s the FCC also began loosening its structural regulations by allowing sta-

tion owners to acquire more stations. At the same time Congress also decreased regulatory burdens by lengthening the terms of broadcast licenses and permitting licensing by lottery in some situations.

In the 1990s the deregulatory trend continued in most instances as Congress further lengthened broadcast license terms to a maximum of eight years and eliminated all national ownership restrictions for broadcast stations (with a national audience cap for television owners). (See Section 13.6 p. 387, for further discussion of the ownership rules.) Congress also permitted licensing via auction in the mid-1990s. In 1995 Congress established a goal of raising more than $14 billion by auctioning parts of the electromagnetic spectrum. By the middle of 1999 the FCC had raised more than $23 billion in spectrum auctions. Exhibit 12.g summarizes many of these deregulatory changes.

With respect to cable, the FCC had dropped most of its cable regulation by 1980, as shown in Exhibit 12.h on p. 368. Congress followed suit with the 1984 cable act, limiting both local franchise and FCC authority. The cable television regulatory pendulum swung back fairly quickly as Congress, reacting to subscriber rate hikes and allegations of poor service, passed a highly re-regulatory bill in 1992. By 1996 the mood of Congress had changed again, and much of the rate regulation established in 1992 was rescinded effective in 1999.

Although many industry leaders praise deregulation in theory, they sometimes argue against it in practice because regulation often protects them from increased competition. For example, a 1980 FCC staff report on the regulation of television networks indicated that network regulation had utterly failed to bring about its stated goals of increased program diversity. Instead, the staff indicated that the regulations had stifled competition, and they recommended that the FCC undo much of its existing network regulation. When the FCC attempted to carry out that recommendation, industry lobbyists, fearful of losing their protected positions, brought the efforts to a standstill. Significant network deregulation was not accomplished for almost a decade, and then much of it was attained through litigation. Likewise, leaders in the cable television business argue for deregulation when it is in their best interests—regarding subscriber rates or signal carriage rules, for example. But the industry strongly opposed the deregulation of the cable-telco crossownership rules, not wanting the increased competition from the telephone companies.

Technical Standards

In the past the FCC assumed responsibility for setting technical standards for a new technology when it was made commercially available. Universally mandated standards protect the public—at least to a point—from investing in new products that might later prove incompatible with existing products or of less than optimum quality. The deregulatory approach, however, tends to reject official standard setting, leaving choices to the marketplace.

The issue was defined when the Commission declined to select a specific standard while approving AM stereo in 1982, a precedent for a decade of refusals to impose standards. Similar "nondecisions" followed regarding DBS, teletext, DTV, and other standards, indicating that economists had superseded engineers in Commission policy making, even in largely technical matters.

The "marketplace" in each case actually consisted not of consumers buying receivers but of station owners, operators, and manufacturers choosing equipment. Broadcasters would have to decide for themselves which transmission standard to use, hoping that set makers would eventually gear up to supply receivers that consumers would buy.

In AM stereo's case, that did not happen. A decade later few stations were broadcasting in stereo and few consumers had stereo-equipped AM receivers. Both the industry and the Commission recognized that deregulation had failed to achieve its goals. Finally, acting on a congressional mandate, the FCC in 1993 selected Motorola's C-Quam system, although few observers expected the decade-late decision to make much difference for beleaguered AM licensees.

Industry debate over the FCC's approach to technical standards rose to a fever pitch over high-definition television (HDTV). At issue were the same questions: Should government help industry select specific standards, and would selection of a standard stifle technical development? But HDTV's financial stakes were far higher—the eventual replacement of all the country's broadcast and consumer television equipment if HDTV succeeded, and concerns about further imbalance of international trade if HDTV devices were largely imported. Perhaps having learned a valuable lesson from its AM stereo debacle, the Commission in the 1990s began the arduous but vital process of setting standards for this new technology, with billions of dollars riding on the outcome. Finally, after years of testing and compromise, the FCC adopted an industry-developed transmission standard for advanced television services in 1996 (see Section 5.7, p. 146). But even in the case of advanced television services, including HDTV, the FCC left image shape, image resolution, and the method of scanning up to the marketplace. And the United States still lacked a technical standard for terrestrial digital audio broadcasting as the new century began.

Deregulation Critique

Deregulation has had many positive effects. It greatly speeded FCC actions, eliminated outdated rules, encouraged the development of new technology, and gave audiences more program choice.

But deregulation has had negative results as well. Although too much regulation can prove harmful, some areas still need closer governmental oversight. Congress, often cool to deregulation, has shifted fundamentally toward a more traditional oversight role for government, and in several areas it has forced the FCC to take a more active regulatory stand. The following policy decisions illustrate the unintended consequences of deregulation.

- Abandonment by the FCC of many record-keeping requirements as a money-saving tactic in the early 1980s left blanks in knowledge about the electronic media. For example, discontinuation of annual broadcasting and cable financial reports limited the FCC's own understanding of the changing financial health of broadcasters and cable systems amid substantial transition.

- Deregulation—no matter what the source—tends to put business interests ahead of those of the public. With its 1984 cable act, for example, Congress curbed the right of cities to control cable subscription fees, allowing systems to raise rates freely, given their monopoly status in most markets. Subsequent cable price increases and poor service pushed Congress to reinstate FCC cable oversight eight years later, though the Commission argued against the renewed powers, fearful of the multimillion-dollar cost and staff time they would require.

EXHIBIT 12.g

. . . .

Before and After Deregulation

The FCC modified or eliminated many broadcasting and cable rules during its deregulatory drive in the 1980s. Moving against the deregulatory tide, Congress mandated limits on advertising in children's programming and required television licensees to serve the special needs of children in its programming (1990). The FCC strengthened that requirement in 1996 with a rule requiring television stations to air three hours of children's programming per week. Congress substantially re-regulated cable television in the 1992 cable act but removed some of those regulations in the 1996 Telecommunications Act. The FCC further reversed its deregulatory course on technical standards when it set new standards for the next generation of television in the United States. The Telecommunications Act of 1996 is both deregulatory and re-regulatory: Many structural regulations were eliminated, thus allowing more competition in telecommunications markets, but several aspects of programming content, most notably violence and sexually explicit material, are now subject to greater regulation.

Before 1980	Changes Since 1980
Licensing	
License runs three years for any broadcast station	License lasts five years for TV, seven for radio (Congress, 1981); all broadcast licenses now last eight years (Congress, 1996)
Comparative hearings are required for choosing among competing applications	Applicants for some services (LPTV, MMDS, etc.) chosen by lottery (Congress, 1982; FCC, 1983); other services (e.g., PCS, LMDS, radio and TV broadcast) chosen by spectrum auction (Congress and FCC, 1995, 1998)
Ownership	
Limit of seven AM, seven FM, and seven TV outlets	Limit raised to 12-12-12 (FCC, 1985); radio limits raised again to 18-18 and then to 20-20 (FCC, 1992, 1994); all numerical limits on ownership removed in 1996; TV coverage limitation 35 percent of U.S. TV households (Congress, 1996)
Duopoly rule: Only one station of each kind per market	May own more than one radio station in larger markets (FCC, 1992); sliding scale now in effect based on market size; may own up to eight radio stations in largest markets (Congress, 1996)
	May own two TV stations in a market (FCC, 1999)
Trafficking rule: Must hold a station at least three years	May sell license (with FCC okay) at any time (FCC, 1982)
TV networks may not own cable systems; telephone companies may not own cable systems in their local service areas	Network-cable and cable-telco crossownership provisions eliminated (Congress, 1996)
Programming	
FCC application processing guidelines (not rules) call for minimum amounts of nonentertainment programming	No qualitative program guidelines (FCC for radio, 1982; for TV, 1985)
Specific rules requiring determination of local program interests, needs (ascertainment)	Rules dropped; must file a quarterly "programs/problems" list in public file (FCC for radio, 1982; for TV, 1985)

Before 1980	Changes Since 1980
1974 policy (no requirement) to serve the educational needs of children	Requirement to program some educational and cultural programs for children (Congress, 1990; FCC, 1991); rule strengthened to mandate three hours of educational children's programming per week (FCC, 1996)
Controversial issues must be aired and opposing sides treated fairly (fairness doctrine)	Fairness doctrine dropped (FCC, 1987)
Network affiliates in the top 50 markets may carry no more than three hours of network or off-network programming during prime time, with exceptions for some program categories	Prime-time access rule dropped (FCC, 1996)
ABC, CBS, NBC cannot syndicate TV programming and cannot take a financial interest in prime-time television programming	Syndication and financial interest rules eased (FCC, 1991) and later repealed (FCC, 1995)
No regulation of violent programming	Rating code established for TV programming; TV sets must be equipped with V-chip (Congress, 1996)
Advertising	
Guidelines (not rules) on maximum amount of advertising time allowed	Guidelines dropped (FCC for radio, 1982; for TV, 1985)
Policy (no requirement) suggesting limitations on commercialization of children's programs (1974)	Specific limits on advertising in children's programming set (Congress, 1990)
Technical Rules	
Specific technical standards selected/mandated for each new service (e.g., color TV, FM stereo)	Starting with AM stereo decision, standards left to marketplace forces (FCC, 1982); Congress mandates FCC to select a specific standard (1992); DTV standard chosen (FCC, 1994, 1995, 1996)
Engineers ranked by class and specific classes required for certain jobs	Engineers no longer ranked by class (FCC, 1981); all engineer licensing requirements dropped (FCC, 1995)
Rules requiring signal quality and avoiding interference	Stations may use own methods (actions at various times by FCC)
Cable Television	
Cable systems must carry all local TV stations	Must-carry rule dropped (U.S. appeals court, 1985, 1987); rules reinstated (Congress, 1992)
Syndicated exclusivity rules allow a station to force local cable systems to black out syndicated programs on superstations that are also carried by the local station	Rules dropped (FCC, 1980), then reinstated (FCC, 1988, effective 1990)
Local communities may control rates charged by cable systems	Local regulatory role largely eliminated (Congress, 1984), then reinstated (Congress, 1992); upper-tier rate regulation eliminated (Congress, 1996)

EXHIBIT 12.h

Changing Cable Regulation

For more than five decades the FCC's regulation of cable television has undergone substantial changes in direction. From virtually no regulation in cable's earliest years to substantial and detailed FCC regulation of the medium in 1972, cable was steadily more controlled. For the next two decades, cable was progressively deregulated, especially with the 1984 cable act, which removed most remaining local franchise limitations. In 1992 Congress reversed course again and passed a strongly re-regulatory cable act, only to change course again four years later.

- The FCC's 1980s vision of all human activity reduced to simplistic economic terms (the marketplace) treated programs as the equivalent of manufactured goods with little concern for diversity, quality, or cultural content—what former chairman Mark Fowler called "a toaster with pictures."
- Sometimes deregulation simply does not deliver the results expected by Congress. Four years after passage of the Telecommunications Act of 1996, which was expected to open the way for new competition in the cable television industry, as well as the local and long-distance telephony markets, many of the expected changes had not happened. The telephone companies' ardor for entering the cable television business waned rapidly, though there were some exceptions (see Section 6.4, p. 174). And the regulatory requirements for permitting increased competition in telephony effectively delayed new entrants in those markets. As noted elsewhere, the act instead has led to increased consolidation of media ownership, which some would argue has actually decreased media competition.

12.11   OTHER REGULATION

Many other federal laws, as well as state and municipal statutes, affect the electronic media. The following sections briefly survey some of the most relevant.

Treeaties

The tendency of radio signals to ignore political boundaries necessitates international treaties to prevent interference. The United States and its neighbors have entered into separate regional treaties governing AM, FM, and television broadcasting. AM agreements cover the widest territory, because long-distance sky-wave propagation affects the scattered islands of the Caribbean as well as the two common-border nations, Canada and Mexico; agreements on simpler FM and television allocations have also long been in place with both countries.

On American international telecommunications policy, the FCC works closely with the U.S. State Department and the National Telecommunications and Information Administration (NTIA)—an agency of the Department of Commerce that acts as the president's chief adviser on telecommunications questions. The FCC provides technical expertise and helps coordinate private-sector participation.

Press Law

The electronic media share with the print media a body of laws, precedents, and privileges known as the law of the press. Typical areas include defamation, obscenity, trial coverage, freedom of access to information, right of privacy, labor laws, and copyright (many of which are discussed in Chapter 13).

One other area, *reporters' privilege*, concerns the asserted right of news personnel to withhold the identity of news sources and to refuse to surrender personal notes, including any audio and video "outtakes." This and many other aspects of press law fall under state jurisdiction, with results varying widely from state to state.

Regulation of Advertising

The FCC itself has no authority to punish licensees directly for unfair advertising. All it can do is cite fraudulent advertising as evidence that an applicant or licensee lacks the requisite character qualification to be a licensee and refer cases of deceptive advertising to a sister agency, the Federal Trade Commission (FTC).

The FTC settles most cases of alleged advertising deception by *stipulation*, an informal (and hence timesaving) way of getting advertisers to drop objectionable practices voluntarily. If a formal complaint becomes necessary, the FTC can seek a *consent order*, another nonpunitive measure under which the advertiser agrees to stop the offending practice without admitting guilt. Actual guilt has to be proved before the FTC can obtain a *cease and desist order* forcing compliance with the law. Such orders can be appealed in the courts. The appeal process can mean long delays in bringing the objectionable advertising to an end if the advertiser chooses to use it.

Even this clumsy machinery ground to a halt in the 1980s when the FTC chairperson declared that market forces rather than government regulations should protect consumers from exploitation. However, as the 1990s dawned, the FTC showed signs of renewed vigor. For example, in mid-1990 the FTC's Bureau of Consumer Protection obtained its first consent order canceling allegedly fraudulent program-length commercials. The offending company, which sold "cures" for baldness, impotency, and obesity, agreed to pay refunds to consumers and to identify future program-length commercials as paid advertisements.

The only legal product officially banned from broadcast and cable advertising is cigarettes and related tobacco products. A federal statute adopted in 1971 imposed the ban. Many broadcasters claim that they were unfairly singled out because cigarette ads continue to appear—even increase—in other media.

As noted elsewhere, broadcasters and distilled spirits producers traditionally have voluntarily refrained from airing ads for hard liquor products. This total abstinence ended in 1996 as several stations and cable networks began running ads for Seagram. This led to policy discussions in Congress, the FTC, and the FCC, which many broadcasters fear may result in a ban on all liquor advertising, including wine and beer spots.

The FCC does have jurisdiction over commercialization levels for children's television. A 1990 act of Congress limited the time that could be devoted to advertising in children's programs to 10.5 minutes per hour on weekends and 12 minutes per hour on weekdays. The FCC is in charge of enforcing these rules, and the Commission has assessed thousands of dollars in fines against stations violating these limits.

Lotteries

• • • • • • •

The advertising of illegal lotteries can subject a licensee to a fine and imprisonment for each day's offense, according to a provision of the U.S. Criminal Code (18 USC 1304). This statute concerns broadcasters because they frequently use contests in advertising and promotional campaigns. They must take care not to let an innocent promotional gimmick turn into an illegal lottery—defined as a *chance* to win a *prize* for a *price*.

A contest becomes a lottery if it requires participants to pay any kind of fee or go out of their way (*price* or *consideration*), chooses the winner by lot (*chance*), and awards the winner something of value (*prize*). Contests involving these elements can get a station into serious trouble. The advertising of legal city and state lotteries is exempted from the antilottery statute. Restrictions on broadcast advertising for casino gambling were declared unconstitutional by the Supreme Court in 1999. (US, 1999)

State and Local Laws

• • • • • • • • • • • • • •

Under the Constitution, federal laws prevail over state laws in subject areas designated as being of federal concern. This means that a state law cannot rise above the Communications Act. Nevertheless, state laws govern many electronic media activities that are not covered by federal statutes. Scores of state laws affect the print and electronic media, especially those concerned with these areas:

- individual rights (regarding defamation and privacy, as discussed in Section 13.4, p. 378)
- advertising of specific products and services
- noncommercial broadcasting (many states have commissions to coordinate statewide public radio and television activities)
- standard business operations (state taxes, for example)
- aspects of cable television—including franchising or cable right-of-way disputes, theft of service, and the attachment of cable lines to utility poles (Municipalities, too, have enacted many local regulations on cable franchising.)

Informal Controls

• • • • • • • • • • •

We have seen how all three branches of government—executive, legislative, and judicial—participate in the formal regulatory process. These forces divide the FCC's allegiance. The Commission finds itself under constant and often conflicting pressures not only from the White House, Congress, and the courts but also from the industries it regulates, lobbyists representing special interests, and the public.

Although Congress gave the FCC a mandate in the Communications Act of 1934 to act on its behalf, it continually brings the Commission up short if it wanders far afield. In addition to Senate approval of nominations to the FCC and congressional control of federal budgeting, Congress conducts frequent *oversight* hearings on the Commission's performance and plans.

Since the 1920s the White House has found ways, usually indirect, to influence electronic media policy. In the early 1990s, for example, the Bush administration slowed the pace of federal regulatory efforts—including new FCC rules—through the activities of the Council on Competitiveness, a pro-business group chaired by Vice President Dan Quayle and abolished by the Clinton administration in early 1993. The Clinton administration took a much more activist approach to telecommunications policy making. Several times President Clinton invited electronic media leaders to the White House for policy discussions on the topics of violence and children's television requirements. Vice President Al Gore was a leading administration spokesperson for telecommunication reform. The overall effect of White House pressure varies with each administration and its interest in communications matters.

Consumer Action

• • • • • • • • • • •

The highlight event that crystallized interest in broadcast reform came in the consumerism heyday of the 1960s. The WLBT case (see Exhibit 12.e) gave a station's audience the legal right to file complaints against the station and participate in license renewal proceedings. Hundreds did, though with relatively little lasting impact.

In recent years consumer action has increasingly taken the form of single-issue groups with specific complaints. The threat of a *boycott* has become the strongest action such groups can take to try to influence electronic media behavior. For example, in 1999 the NAACP threatened to boycott the major broadcast networks to protest the lack of minorities in network programs. Boycotts involve not watching or listening to targeted stations and refusing to buy advertised products. In a pluralistic society, boycotters have difficulty achieving sufficient consensus and discipline to do substantial economic damage.

Concern over abortion, gun control, prayer in the schools, sex and violence in the electronic media, and other "family-values" issues heightened after 1980. Programs and ads debated them all. Protests stimulated widespread news coverage.

Even noncommercial broadcasting, supposedly freer to present more points of view because it does not cater to advertisers, has succumbed to group pressure. Programs on gay lifestyles, avant-garde art, and Third World political movements have all been criticized—and sometimes temporarily removed from the air.

Self-Regulation
• • • • • • • • • •

Many professions and industries adopt voluntary codes of conduct to cultivate favorable public relations and forestall abuses that might otherwise bring on government regulation. But codes that restrict the freedom to compete run afoul of antitrust laws.

The National Association of Broadcasters (NAB) began developing a code for radio programming and advertising practices in the late 1920s and extended them to television in 1952. These codes used broad generalizations and many "shoulds" and "should nots" but left most decisions to the discretion of station management. Nonetheless, the codes reinforced generally observed standards, especially as to the amount of time broadcasters should devote to advertising.

Despite NAB precautions to avoid any hint of coercion, the Department of Justice brought suit against the Television Code in 1979. The suit alleged that code time standards "artificially curtailed" advertising, repressing price competition and "depriving advertisers of the benefits of free and open competition." In 1982 a federal district court approved a consent decree by which the NAB disbanded its code-making activities.

For a time, networks and many stations continued to follow their own internal standards—some even tougher than the former NAB codes. By the 1990s, the networks had cut back their program standards departments as part of more stringent budget controls. In an effort to restore a sense of industry standards, several members of Congress proposed legislation that would exempt the electronic media from antitrust laws so that industry codes of conduct could again be developed. To date, these proposals have not been enacted.

Much self-regulation results from media worries about externally imposed controls. As explained earlier, broadcasters generally refrained from carrying hard liquor ads because they feared a congressional reaction. Similarly, broadcast and cable's agreement to monitor violent program content and develop a television rating code resulted from outside pressure, not altruistic motives.

Press Criticism
• • • • • • • • • •

Coverage of the electronic media in both the trade and popular press often affects policy. Congress, the White House, the FCC, and other agencies closely follow reporting about the media, seeking reports of their actions.

Reviewers in major newspapers and magazines seem to have more impact on news and public-affairs programs than on entertainment shows. Further, they influence producers of such programs more than they do the general audience.

CHAPTER 13

Constitutional Issues

CHAPTER CONTENTS

The First Amendment to the U.S. Constitution, together with the Fourteenth Amendment (see Exhibit 13.a), prohibits government regulation of speech and press, yet the Communications Act imposes federal licensing and other limitations on those who own electronic media. This paradox isn't unique, as society often demands a balance between the ideal of *absolute* individual freedom and the practical need to limit speech that might harm others (for instance, one may be punished for yelling "fire" in a crowded theater if there is none). The essence of the constitutional controversies discussed here is how to compromise between these opposing goals.

13.1　FIRST AMENDMENT

The freedoms of speech and press guaranteed by the First Amendment were intended from the start—and the courts have construed them since—to encourage a wide-open *marketplace of ideas*. First Amendment theory holds that ideas and opinions from different sources should compete in such a marketplace. As the Supreme Court noted in a landmark electronic media decision, "It is the purpose of the First Amendment to preserve an uninhibited marketplace of ideas in which truth will ultimately prevail, rather than to countenance monopolization of the market" (US, 1969: 390).

Although freedom of expression is only part of one of the 10 amendments that make up the Bill of Rights, that freedom has played a pivotal role in the American political system (see Exhibit 13.a). In the words of Supreme Court Justice William O. Douglas, the First Amendment "has been the safeguard of every religious, political, philosophical, economic and racial group amongst us" (US, 1951: 584).

"No Such Thing as a False Idea"

"Under the First Amendment," said the Supreme Court, "there is no such thing as a false idea. However pernicious an opinion may seem, we depend for its correction not on the conscience of judges and juries, but on the competition of other

ideas" (US, 1974: 339)—again, the marketplace metaphor.

The First Amendment *encourages* disagreement. "A function of free speech under our system of government is to invite dispute," wrote Justice Douglas. "It may indeed best serve its highest purpose when it induces a condition of unrest, creates dissatisfaction with conditions as they are, or even stirs people to anger" (US, 1949: 4).

Anger certainly arose during the 1980s in Dodge City, Kansas, when station KTTL-FM broadcast daily hourlong sermons by two fundamentalist ministers attacking Jews, blacks, and other groups. They urged listeners to ignore police officers and attack such groups at will. The invective poured out in such abundance that several local groups protested renewal of KTTL's license when it expired in May 1983.

A huge media uproar resulted, with a congressional subcommittee hearing and extensive press coverage of the station owner's extremist conservative views favoring local armed vigilantes. In mid-1985 the FCC designated the license for a comparative hearing with another applicant for the same frequency. In a controversial decision to renew, the FCC said that the First Amendment protected the broadcasts, offensive though they might be to many listeners (FCCR, 1985a). The Commission found that because the material did not present a *clear and present danger* to the public, a test long established by the U.S. Supreme Court, it qualified as protected speech (US, 1919: 52). KTTL's exhortations to take illegal or dangerous actions did not, as the FCC had said in an earlier case, "rise far above public inconvenience, annoyance, or unrest" (FCCR, 1972c: 637).

Government Censorship

Many assume that the First Amendment also provides protection from private parties. But in fact the amendment aims at protecting people only from government (censorship), not from one another (broadcaster or cable editorial decisions). Station, system, and network officials who edit, cut, bleep, delete, revise, and otherwise mangle programs may be guilty of bad judgment, excessive timidity, and other faults, but editorial control is not censorship, and it does not violate the First Amendment.

EXHIBIT 13.a

• • • • • • • • • • •

"Congress Shall Make No Law . . ."

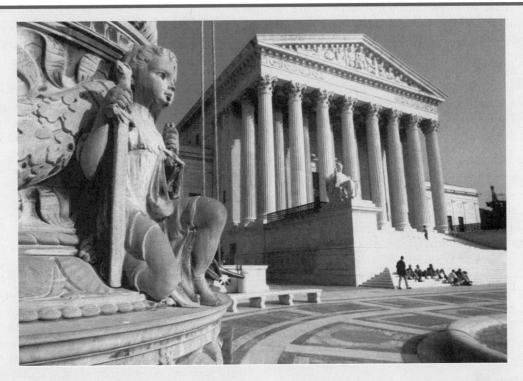

The First Amendment protects four fundamental rights of citizens that governments throughout history have had the most reason to fear and the greatest inclination to violate—freedom to *believe*, to *speak*, to *gather together*, and to ask rulers to *correct injustices*. The amendment conveys all this in only 45 words, of which just 14 guarantee freedom of expression:

> Congress shall make no law respecting an establishment of religion, or prohibiting the free exercise thereof, or abridging the freedom of speech, or of the press; or the right of the people peaceably to assemble, and to petition the Government for a redress of grievances.

These words limit not only Congress but also state and local governments, thanks to the Fourteenth Amendment, passed in 1868, which says, "No state shall make or enforce any law which shall abridge the privileges or immunities of citizens of the United States." Sec. 326 of the Communications Act of 1934 explicitly extends the First Amendment's protection to broadcasting:

> Nothing in this Act shall be understood or construed to give the Commission the power of censorship over the radio communications or signals transmitted by any radio station, and no regulation or condition shall be promulgated or fixed by the Commission which shall interfere with the right of free speech by means of radio communication.

Source: Bruce Young/Reuters-Archive Photos.

Such control becomes a violation only when it results from the kind of government intrusion known as *state action*. In promoting the "family viewing" concept in the 1970s, for example, the FCC attempted to reduce television violence not by rule but by "jawboning"—pressuring the television industry to regulate itself. When the National Association of Broadcasters (NAB) responded to this pressure by amending its Television Code, a court construed this private action as state action in violation of the First Amendment (F, 1979: 355).

Prior Restraint

The principal evil that First Amendment authors intended to address is *prior restraint*—that is, preventing someone in advance from saying, showing, or otherwise publishing something. Only in the most extreme and exceedingly rare case will a court issue an *injunction* barring, for example, a television program from broadcasting an allegedly libelous story (see Section 13.4). If libel does occur, the remedy is punishment after the fact, not censorship before.

Religious Freedom

Another First Amendment clause guarantees religious freedom. The active role of television evangelists in the national elections of 1980 and since (especially the candidacy of Pat Robertson for the GOP nomination in 1988 and some GOP platform planks in 1992) disturbed many people who were sensitive to First Amendment concerns about the establishment of any specific religion. Yet the same amendment protects the evangelists' right to have their say.

Doubly protected by the freedoms of speech and religion, stations owned by religious groups have often claimed near-immunity from FCC requirements because they regard their right of religious freedom (from any government action) as absolute. Though the FCC holds religious licensees to the same standards of regulation for all broadcasters, it has resisted close monitoring of any licensees. On the other hand, cable operator arguments that rules requiring them to carry religious television stations violate the systems' own First

Amendment rights to religious freedom have proved unsuccessful.

13.2 BROADCASTING'S LIMITED RIGHTS

Just what do the First Amendment words *speech* and *press* mean in today's context? Speech encompasses not only that amplified by public-address systems but also film, broadcasting, videos, cable, DBS, and the Internet. The press includes pornographic magazines. Can *all* ways of expressing oneself and *all* forms of the press claim equal First Amendment rights?

Broadcasting vs. Print

More specifically, should all regulated electronic media have First Amendment *parity* (equality) with unregulated media, especially the printed press? The answer has generally been no—the First Amendment protects some media more than others. Broadcasting, it is argued, has unique attributes that justify imposing certain regulations that would violate the First Amendment if imposed on newspapers. The three main arguments favoring this assumption follow.

Channel Scarcity

Not everyone who wants to own a station can do so, because intolerable interference would otherwise result. Because of this channel scarcity, the government (the FCC) has to choose among applicants. Where mutually exclusive applications seek facilities whose activation would cause interference, only one license can be granted—thereby abridging the freedom of other applicants. By contrast, anyone who can afford it can publish a newspaper or magazine without limit or license.

The development of new and improved services *has* decreased scarcity as a justification for regulating broadcasting. The huge increase in the number of stations since Congress wrote the 1927 Radio Act—from about 600 to more than 15,000, plus almost 11,000 cable systems—makes scarcity a relative factor at best. And although demand for channels in major markets continues, cable, MMDS, and DBS provide a multitude of additional channels—converting scarcity into abundance. Digital technology promises to increase the number of available channels dramatically.

By contrast, others point to the many applicants for desirable stations that become available—and the huge prices often paid for them—to suggest that scarcity continues. They argue that availability of cable and DBS channels does not fully relieve scarcity because they reach about three-fourths of the population, while broadcasting reaches virtually everyone. In any event, cable, MMDS, and DBS differ sharply from "free" broadcasting in requiring subscription fees over and above receiver expenses.

Others argue that even if spectrum scarcity does exist, that fact does not justify treating broadcasting differently from the printed press in terms of the First Amendment. Judge Robert Bork, in a case before the U.S. Court of Appeals, addressed the issue this way:

> The line drawn between the print media and the broadcast media, resting as it does on the physical scarcity of the latter, is a distinction without a difference.
>
> It is certainly true that broadcast frequencies are scarce but it is unclear why that fact justifies content regulation of broadcasting in a way that would be intolerable if applied to the editorial process of the print media. All economic goods are scarce, not the least the newsprint, ink, delivery trucks, computers, and other resources that go into the production and dissemination of print journalism. Not everyone who wishes to publish a newspaper, or even a pamphlet, may do so. Since scarcity is a universal fact, it can hardly explain regulation in one context and not another. (F, 1986: 508)

According to this argument, economic scarcity, not physical scarcity, is what drives media ownership. Anybody can speak via a broadcast station—as long as he or she has enough money to buy one.

Conflict in Licensing
• • • • • • • • • • • • • •

When writing the Communications Act, Congress explicitly confirmed "the right of free speech by means of radio communication" (see Exhibit 13.a).

Yet the Act also requires the FCC to grant or renew licenses only if they serve the public interest. It also requires the FCC to ensure that candidates for public office have equal opportunities to use broadcasting.

Any such requirements by definition place limits on licensees' freedom of speech. That fundamental contradiction came before the Supreme Court in the 1943 *NBC* case, where broadcasters argued that the FCC had the right to regulate only *technical* factors, and that any further regulation violated the First Amendment. But the Court emphatically rejected this argument, saying:

> We are asked to regard the Commission as a kind of traffic officer, policing the wave lengths to prevent stations from interfering with one another. But the Act does not restrict the Commission merely to supervision of the traffic. It puts upon the Commission the burden of determining the composition of that traffic. (US, 1943: 215)

Intrusiveness
• • • • • • • • •

Radio and television enter directly into virtually every U.S. household, readily available to people of all ages and types. Because of the intrusive nature of the medium, people never know exactly what they will see or hear when they first turn on their sets, and that initial exposure might be upsetting. The medium also is easily accessible to children; it requires no special training to turn on a television or radio and flip through the channels or stations. Children who are not yet old enough to read can use a broadcast receiver even when their parents are not around. Because of broadcasting's intrusiveness and easy accessibility to children, Congress, the courts, and the FCC have often imposed content regulation on broadcasters that would not be allowed in the print media. Some content acceptable in other media—indecency, for example—may be regarded as intolerable in broadcasting, at least when children form a substantial portion of the audience.

First Amendment purists respond that where such distinctions need to be made, the marketplace can make them better than government. If people object to a program, they can—and do—complain to station and system operators, networks, or advertisers, or they can simply hit the "off" button. Regarding protection of children, opponents of content regulation argue that parents, not the government, should be responsible for the viewing and listening behavior of children.

13.3   FIRST AMENDMENT STATUS OF OTHER ELECTRONIC MEDIA

Because of the factors noted in the previous section, broadcasting enjoys substantially less First Amendment protection than the printed press. Generally speaking, as new electronic media emerge, they are not regulated (and thus match the print model) until and unless government takes some specific action—congressional attempts to control indecency on the Internet, for example. Government must then discern how those media should be treated in terms of the free speech and free press guarantees. The Supreme Court has repeatedly stated that, for First Amendment purposes, each new medium must be judged based on its own unique characteristics. When judging new media, the government tends to analyze them in terms of what is already familiar. Often the analysis comes down to the question "Is the new service more like print or more like broadcasting?"

On a scale of First Amendment protection, cable television systems fall somewhere between the print and broadcast media. Courts consistently find that cable is much different than a broadcast station in terms of its physical characteristics. Cable systems do not use the electromagnetic spectrum in the same way broadcasters do, so spectrum scarcity typically is not an issue. Because cable is a subscription service, cable is not considered as intrusive as broadcasting; subscribers choose to bring the service into their homes every month as they pay their cable bills. For similar reasons, cable is not considered as accessible to children. Plus, cable systems have long provided special lock boxes that can lock out channels that parents may find unsuitable for children.

For these reasons, certain programming restrictions that have been upheld for broadcasters—indecency regulations, for example—have been judged invalid when applied to cable television systems. Critics consider this rather curious. Most consumers do not think much about how their television signals are delivered so long as they get a good picture with good service. But under the First Amendment rulings, television content is regulated differently based on how it is delivered to the home. On the other hand, courts have approved content-related regulations for cable systems—the must-carry rules offer a good example—that would never be approved if applied to newspapers or magazines.

Other electronic media whose First Amendment status has been considered by the courts have been treated similarly. Speech and press protections for telephone systems, DBS, and the Internet so far seem to fall somewhere between the protections afforded newspapers and broadcasters. In all cases, the courts must balance the interests of those seeking the broadest protection of speech and those arguing for protection from certain types of content.

13.4   THINGS YOU CAN'T SAY

Despite that seemingly strict First Amendment command "Congress shall make no law . . . ," government does in fact make laws that punish *unprotected* speech. This includes defamation (libel and slander), invasion of privacy, and obscenity.

Libel: Protecting Reputations

Libel is *defamation* by published untrue words that may expose their subject to public hatred, shame, or disgrace. Spoken defamation is called *slander,* but because electronic media spread spoken words far and wide, and because the words are often preserved on tape, broadcast or cable defamation is treated as libel.

Libel laws constitute another example of conflicting social interests. Although libel should be punishable because society has an interest in protecting its citizens, society also needs to expose official corruption and incompetence. Harassing libel suits can serve as a screen to protect dishonest politicians. In the United States all public figures,

EXHIBIT 13.b

. . . .

Times v. Sullivan **Libel Landmark**

The leading case establishing the relative protection of media from libel suits filed by public figures occurred during the civil rights protests of the 1960s. By chance it involved an instance of "editorial advertising," not investigative reporting. Supporters of the Montgomery, Alabama, bus boycott protesting segregation bought a large display advertisement in the *New York Times* that criticized Montgomery officials. Some of the statements in the advertisement were false, although they apparently were not *deliberate* lies. Sullivan, one of the officials, sued for libel in an Alabama court (all libel suits have to be brought at the state level). The court awarded Sullivan a half-million dollars in damages, and the Alabama Supreme Court affirmed the decision.

On appeal to the U.S. Supreme Court, the *Times* won a reversal. Criticism of public officials, said the Court, had broad First Amendment protection. Even though some of the allegations against the unnamed officials were untrue, they did not constitute libel. Argument over public officials, the Court continued,

should be "uninhibited, robust, and wide-open." It may include "vehement, caustic, and sometimes unpleasantly sharp attacks on government and public officials." Such freewheeling debate would be discouraged if, in the heat of controversy, the critic must pause to weigh every unfavorable word:

> The constitutional guarantees require, we think, a federal rule that prohibits a public official from recovering damages for a defamatory falsehood relating to his official conduct unless he proves that the statement was made with "actual malice"—that is, with knowledge that it was false or with reckless disregard of whether it was false or not. (US, 1964: 279)

Subsequent libel cases broadened the term *public officials* to include anyone who, because of notoriety, could be classed as a *public figure*. People so classified have little chance of bringing a successful libel suit against the media. Even when stories about public figures are false, plaintiffs find it exceedingly difficult to prove actual malice.

not just politicians, must be prepared to face harsh, sometimes even unfair and ill-founded, criticism from the media, because it is extremely difficult to prove the *actual malice* standard necessary for their libel suit to succeed (see Exhibit 13.b). By the 1990s the primary sources of libel trouble for electronic media were radio "shock jocks" such as Howard Stern and radio call-in shows (because licensees are liable for any defamation spoken over the air by those calling in).

A 1998 study by the Libel Defense Resource Center traced libel trials dating back to 1984 and found that the media won only a quarter of the jury trials in those 14 years—though less than 40 percent of the initial awards were finally upheld on appeal and judgments paid. The same study found that awards averaged more than $8 million in 1990 and 1991. Libel damage awards came down later in the decade. The average libel judgment in 1998 was less than $1.5 million.

In fact, about 90 percent of all such libel suits are resolved long before they reach trial: They are

dropped, settled, or dismissed by a judge during pretrial proceedings. Even when the media win, they spend a lot of money defending themselves. More important, such suits have a *chilling effect* on investigative reporting.

Preserving Privacy
. . . .

Privacy as an individual right, though not spelled out in the Constitution, is implied in the Fourth Amendment: "The right of the people to be secure in their persons, houses, papers, and effects . . . shall not be violated." The law of privacy is not as well developed as libel law, but several individual privacy rights have been identified:

- the right to physical solitude
- protection from intrusion on private property or publication of the details of one's personal life
- protection from being presented in a "false light" (for example, being photographed walking on campus and then being included in a visual

accompanying a television feature about drug use in college)

- protection from unauthorized use of one's name or image for commercial gain

Although the courts have held that public officials, performers, and people involved in news events have a lesser right to privacy because of legitimate public interest in those persons or events, privacy laws still limit the media by generally supporting individual rights.

Privacy advocates are concerned increasingly about loss of privacy on the Internet. The computer programs that allow web browsing and online purchasing also enable marketers to track an individual's web use. Opinions differ on whether such tracking is benign or pernicious, but many people do not like the fact that their web use can be monitored. Also, computer databases that contain private financial or medical records can easily be copied and transferred without an individual's knowledge or consent. Many groups concerned about privacy rights have lobbied Congress to enact legislation to prohibit the unauthorized use of private information on the Internet, but Congress seems unwilling to act, preferring that web companies engage in self-regulation.

Special privacy rules apply to children. For example, in 1998, Congress adopted the Children's Online Privacy Protection Act to regulate the collection of personal data from children by web site operators. (47 USC et seq.) Pursuant to the act, the Federal Trade Commission adopted rules requiring web sites that cater to children to post their privacy policies. The rules also restrict companies from collecting personal information from children without a parent's permission.

Restricting Court Coverage
. . . .

The due process clause of the Fifth Amendment ensures fair play to persons accused of crimes. The Sixth Amendment also spells out some of the elements of due process, among them the right to a fair and public trial in criminal cases. Ordinarily, news media freely cover trials, but that coverage can subject participants to such extensive publicity that a fair trial becomes impossible. In this *free press vs. fair trial* confrontation, the constitutional rights of the media sometimes have to give way to the constitutional rights of criminal defendants.

For decades there was virtually no live or recorded radio or television coverage of trials. After messy coverage of the Lindbergh kidnapping trial (see Exhibit 13.c), the American Bar Association (ABA) recommended in 1935 that judges discourage radio and photo coverage because it tended "to detract from the essential dignity of the proceedings, degrade the court, and create misconceptions with respect thereto in the mind of the public."

By the 1970s less intrusive equipment and seemingly more mature broadcast journalism judgment persuaded the ABA to recommend that judges be given wider latitude in allowing photo and video coverage of trials. In 1981 the Supreme Court—although it bans cameras and microphones in its own proceedings—noted the improved technology, holding that "the risk of juror prejudice . . . does not warrant an absolute constitutional ban on all broadcast coverage" (US, 1981: 560). By the early 1990s all but a handful of states either allowed cameras in their courtrooms or had conducted experiments to that end. That the American public has an apparently insatiable appetite for televised real-life trials is demonstrated by the arrival in 1991 of Court TV, a national 24-hour cable courtroom service, and by the unprecedented coverage given to the prosecution, beginning in 1994, of football star turned media personality O. J. Simpson for the murder of his former wife and her male friend. Wanting to avoid a second media circus, the judge in Simpson's subsequent civil trial refused to allow television coverage of the proceedings.

Federal courts have remained off-limits to such coverage, despite repeated media attempts to breach that barrier. Late in 1994, after a multiyear experiment with television coverage of civil cases in six federal district and two appeals courts, the Judicial Conference of the United States voted to restore its flat ban on cameras in federal courtrooms. For this reason, cameras and recording equipment are not allowed in the courtroom during even high-profile federal proceedings such as the 1997 Timothy McVeigh/Oklahoma City bombing trial.

Broadcast Hoaxes and Undercover Reporting
. . . .

While the broadcasting of "news" that isn't really news dates back at least to the 1938 Orson Welles

EXHIBIT 13.c
· · · · · · · · · · ·

Covering Courts—Changing Pictures

(A)

(B)

William Kennedy Smith Trial LIVE

(C)

Broadcast coverage of court trials began with the 1925 Scopes case and reached the height of overkill with the 1935 trial of Bruno Hauptmann for the kidnap and murder of Charles Lindbergh's son. Reaction to that intrusive coverage by radio and film led to the ABA's "Canon 35," which banned trial coverage for decades. (A) An experiment in television coverage in the 1960s, Billie Sol Estes's trial for fraud, led to a Supreme Court reversal because of television intrusiveness. (B) As restrictions on trial coverage declined in the 1980s, sometimes lurid trials came into people's homes—such as the 1992 rape trial of William Kennedy Smith. The jury found him not guilty, but the trial became known for its often graphic testimony and the media's use of a fuzzy dot over the victim's face to protect her identity. (C) Court coverage reached historic proportions in 1994 and 1995 when live cameras, hour after hour, watched every move by the prosecution and defense in the bizarre case of sports and media personality O. J. Simpson, charged with murdering his former wife and her friend.

Source: (A) UPI/Corbis Images; (B) AP-Wide World Photos; (C) Reed Saxon/AP-Wide World Photos.

broadcast of *War of the Worlds*, there have been relatively few cases of intentional news hoax broadcasts. However, a rash of cases in the early 1990s convinced the FCC that it needed stronger rules and a means of fining stations for hoaxes rather than just sending letters of concern. Several stations broadcast stories of false murders, one station said a nearby trash dump was about to explode because of methane gas buildup, another used the Emergency Alert System and said a nuclear attack was under way, and still another reported that a volcano had erupted in a suburban Connecticut town. All were presented as real events, and emergency forces responded accordingly.

In 1992 the Commission adopted a rule against "harmful" hoaxes—those that might divert police, fire, or other safety forces from real events. The FCC made clear that it had the First Amendment in mind as the rule was developed and hence limited its actions to cases where "broadcast of such information will cause substantial public harm, and broadcast of the information does in fact directly cause such harm." Such broadcasts can result in fines of up to $25,000 for each day on which a harmful hoax occurs—up to a maximum FCC fine of $250,000 for repeated violations. Several broadcasters agreed that the industry had not adequately policed itself and thus had brought about this action.

Improved technology has made another type of hoax possible. As cameras get smaller and operate at lower light levels, options for the use of hidden cameras increase. Of growing concern is the number of station and network or syndicated "undercover investigations" in which reporters using hidden cameras impersonate another person (*not* a reporter) to obtain information. This process of omission (not making clear that a reporter is asking the questions) raises substantial questions of ethics.

As noted in Section 9.11, p. 285, ABC News learned firsthand about the dangers of undercover reporting in 1996 when a North Carolina jury awarded $5.5 million to Food Lion supermarkets, which had sued ABC News over its news gathering techniques. ABC producers had misrepresented themselves to Food Lion in order to get jobs with the company. While on the job, they used hidden cameras to record the repackaging for sale of allegedly spoiled and outdated meat. Food Lion did

not sue ABC for libel or invasion of privacy; indeed, the accuracy of the report was not challenged in court. Instead, Food Lion sued on the basis of fraud, trespass, and breach of fiduciary duty by the reporters in getting their positions with the company under false pretenses. ABC appealed the decision, and the award to Food Lion was reduced to a mere $2 in 1999. Although ABC prevailed after three years of litigation, journalists publicly worried that the case might have a chilling effect on covering important stories.

13.5 OBSCENITY AND INDECENCY

One of the most controversial topics regarding electronic media concerns sexually explicit content. Although the terms "obscene" and "indecent" might be considered synonyms in polite conversation, in law they have different and specific meanings, though they both refer to sexually explicit material. If something is found to be "obscene," it has no First Amendment protection whatsoever. According to the Supreme Court,

> All ideas having even the slightest redeeming social importance—unorthodox ideas, controversial ideas, even ideas hateful to the prevailing climate of opinion—have the full protection of the [First Amendment] guarantees. . . . But implicit in the history of the First Amendment is the rejection of obscenity as utterly without redeeming social importance. . . . We hold that obscenity is not within the area of constitutionally protected speech or press. (US, 1957: 484)

"Indecent" speech, on the other hand, is protected expression under the First Amendment, but it can be regulated under certain conditions to protect children from being exposed to it. The major problem with this area of the law—for both students and the media—is defining and understanding the terms "obscene" and "indecent."

What Is Obscene?

The current definition of obscenity dates to the 1973 *Miller* case in which the Supreme Court upheld a California obscenity law. The decision reaffirmed the basic notion that obscene material does not receive protection under the First Amendment. The Court then emphasized that *community standards* vary from place to place: "It is neither realistic nor constitutionally sound to read the First Amendment as requiring that the people of Maine or Mississippi accept public depiction of conduct found tolerable in Las Vegas or New York City."

Nevertheless, the Court warned, state laws must carefully confine what they classify as *obscene* to "works which, taken as a whole, *appeal to the prurient interest in sex, which portray sexual conduct in a patently offensive way, and which, taken as a whole, do not have serious literary, artistic, political, or scientific value*" (US, 1973b: 24; emphasis added).

The *Miller* case, along with later decisions that added minor modifications, restricted obscenity censorship to "hard-core" materials, leaving to the states the difficult job of defining obscenity. Courts now interpret the First Amendment as preventing censors from taking such arbitrary actions as these:

- condemning an entire work because of a few isolated vulgar words
- using outdated standards no longer common to the local community
- applying as a standard the opinions of hypersensitive persons not typical of the general public
- ignoring serious artistic, scientific, literary, or political purpose in judging a work

Generally speaking, courts must consider the "average person" when applying contemporary community standards to potentially obscene material. Historically, those standards have been stricter for material available to children.

Limiting Indecency in Broadcasting

Sec. 1464 of the U.S. Criminal Code reads:

> Whosoever utters any *obscene, indecent,* or *profane* language by means of radio communication shall be fined not more than $10,000 or imprisoned for not more than two years, or both. (18 USC 1464; emphasis added)

Note the use of *indecent* and *profane* in the statute. As with *obscene*, definitions and degrees of control have varied for these terms over the years. In a 1992 policy statement, the Commission defined indecency as "language or material that, in context, depicts or describes, in terms patently offensive as measured by *contemporary community standards for the broadcast medium*, sexual or excretory activities or organs" (FCCR, 1992b: 6464, note 4, emphasis added).

The Commission long remained in doubt as to its power to enforce the Criminal Code provision. Because of ready availability of electronic media to children, material acceptable in other media might be regarded as unacceptable by many in the audience. Furthermore, because broadcast and cable network services have national reach, they confront a great variety of local standards.

Broadcasting's traditional conservatism delayed its response to the tolerant social climate of the 1960s, but the liberalization of standards in other media had its effect. In the 1970s a "topless radio" fad triggered thousands of complaints to the FCC. The format invited women to call in and talk on the air about intimate details of their sex lives. Although such talk shows are commonplace today—on television as well as radio—one Illinois radio broadcast in 1973 triggered a flood of complaints. The FCC imposed what was then a fairly steep $2,000 fine on the offending FM station (FCCR, 1973: 919). The Commission had actually hoped that the station would contest the fine, thus precipitating a test case, but the station dutifully mailed in a check instead (for an amount equal to only a few minutes' worth of advertising revenue in a major market).

Pacifica Decision

That same year, the FCC finally got its test case. A noncommercial station, WBAI-FM in New York, included in a discussion of social attitudes about language a recording of a nightclub act by comedian George Carlin. Called "Filthy Words," the monologue satirized society's hang-ups about seven sexually oriented words not likely to be heard on the air. This time, though, they were heard—no fewer than 106 times in 12 minutes. The single complaint came from a man who, as it later turned out, was associated with a group called Morality in Media. He heard the early-afternoon broadcast with his teenage son—a crucial element in the case.

On the basis of that lone complaint, the FCC advised station management that the broadcast appeared to violate the indecency statute. The licensee, Pacifica, challenged the ruling as a matter of First Amendment principles. The FCC received an initial setback in the appeals court but won Supreme Court approval of its reasoning. Focusing its argument on the Carlin monologue as *indecent* rather than obscene (as the Supreme Court had defined obscenity), the FCC stressed that the broadcast came when children would normally be in the audience.

The FCC argued that children need protection from indecency. Instead of meeting the First Amendment directly by flatly banning such material as the Carlin monologue, the FCC said it should be *channeled* to a part of the day when children are least likely to be in the audience.

The channeling concept has precedence in nuisance law, which recognizes that something acceptable in one setting could be an illegal nuisance in others. The Supreme Court agreed with this rationale when, recalling that a judge had once said that a nuisance "may be merely a right thing in a wrong place—like a pig in the parlor instead of the barnyard," the Court added that if the FCC "finds a pig has entered the parlor, the exercise of regulatory power does not depend on proof that the pig is obscene." The Court also accepted the FCC's narrowing of community standards in adding the words *for broadcasting*, saying, "We have long recognized that each medium of expression presents special First Amendment problems. . . . And of all forms of communication, it is broadcasting that has received the most limited First Amendment protection" (US, 1978: 748).

Political Football
. . . .

In 1987 a rising flood of complaints about perceived obscenity or indecency on the air caused the FCC to issue a statement announcing plans to extend enforcement of Sec. 1464 of the Criminal Code beyond Carlin's seven words at issue in the *Pacifica* case. The Commission reprimanded three radio stations for using overly explicit language, especially at times when children might be in the audience (FCCR, 1987a: 2726). The FCC argued that marginal program material (indecent, perhaps, but not obscene) might be scheduled during the

midnight-to-6:00 A.M. period. For some time the FCC had informally allowed such material to begin as early as 10:00 P.M. The narrower definition of allowable time led to a court appeal.

Several broadcasters joined in challenging the FCC decision and in 1988 won reversal of the Commission action on the grounds that limiting indecent language to the midnight-to-6:00 A.M. hours violated the First Amendment. The court remanded to the FCC the question of how best to "promote parental—as distinguished from government—control" of children's listening (F, 1988: 1332).

Thereupon Congress entered this hotly controversial arena. In approving the FCC budget for the 1989 fiscal year, legislators (mindful that they were in the midst of an election campaign) ordered the Commission to ban indecent material at *all hours* of the day.

While the controversy raged in the courts, the FCC took action against a score of radio stations, fining them for various indecency violations. In 1991 the Supreme Court declined to review an appeals court ruling that the 24-hour ban was too broad a limit on freedom of expression (F, 1991: 1504). In 1992 the Commission initiated a rule-making procedure to define and justify a "safe harbor" for such programming and stepped up its prosecution of indecency violations at a number of radio stations.

Meanwhile, the 1992 funding act for public broadcasting included a provision restricting indecent programming on all stations, commercial and public, to the midnight-to-6:00 A.M. period—right back where the FCC was in 1987. Adding to the pressure was an appropriations bill containing just such a definition of a safe harbor, signed by the president late in 1992. In early 1993 the FCC rule making concluded (sagely) that the midnight-to-6:00 A.M. safe harbor seemed the best compromise for all concerned. Yet another appeal, however, knocked the safe harbor back to 8:00 P.M. to 6:00 A.M. while arguments pro and con raged in court.

The battle continued until late 1993, when a three-judge panel of the U.S. Court of Appeals for the District of Columbia Circuit struck down the FCC's 6:00-A.M.-to-midnight prohibition of indecent programming as being unconstitutionally broad. In 1995 the full court of appeals apparently settled the issue by ordering the FCC to reinstate its original 10:00 P.M.-to-6:00 A.M. safe harbor

(F, 1995). The Commission appealed the decision, but the Supreme Court declined to hear the case.

Enforcement
• • • • • • • • •

The Commission has initiated 70 or so forfeiture actions against stations broadcasting indecent material outside the safe harbor. All but three of those actions were brought since the new enforcement standards were announced in 1987. All of the actions except one have been against radio stations, indicating that the Commission considers the fare of radio "shock jocks" more serious than the titillating discussions of daytime television talk shows or steamy afternoon soap operas.

Since 1987 fines have ranged from a low of $2,000 for a single offense to over $100,000 for repeated violations. The typical fine for a first offense is $10,000. Stations licensed to Infinity Broadcasting (now owned by CBS) have been fined the most, due to the repeated alleged violation of the indecency standards by its popular shock jock Howard Stern. In 1995 Infinity agreed to pay $1.715 million in fines rather than contest the FCC's actions in court. This did not spell the end of Stern's (or Infinity's) problems with the Commission; complaints about his program have continued.

Regulating Indecency on Cable
• •

The cable act of 1984 provided for fines or imprisonment for anyone who "transmits over any cable system any matter which is obscene or otherwise unprotected by the Constitution" (47 USC 639). Although the Supreme Court has held that cable television does have First Amendment protection, it has also said that such protection must be balanced against what it has called "competing social interests," without specifying what those interests may be or how that might be done (US, 1986: 488).

States had previously tried to go even further by applying the Supreme Court's *Pacifica* decision to questionable cable content. Four federal court decisions in the mid-1980s concluded that state laws banning cable indecency violated the First Amendment because they were too broad in scope. Because cable does not use open spectrum, the scarcity argument that had supported indecency limits in broadcasting could not be applied. The courts reasoned that cable is not as "uniquely intrusive" as broadcasting, nor as available to children, because subscribers pay for the service—and that *Pacifica* therefore does not apply to cable.

Congress also has attempted to regulate indecency on cable systems, with mixed results. The 1992 cable act for the first time gave system operators the authority to prohibit indecent programming on public and leased access channels and empowered system operators to enforce written policy on such programming. Additionally, the 1992 act required cable operators who did allow indecent programming on these channels to put all such programming on one channel and to provide that channel only to subscribers who wanted it (47 USC 612, 638). In 1996 the Supreme Court upheld the provision allowing cable operators to prohibit indecency on leased-access channels but declared the other requirements of the law unconstitutional infringements on cable programmers' and subscribers' First Amendment rights (US, 1996).

In 1996 Congress passed legislation requiring cable operators to block fully the audio and video of channels primarily dedicated to sexually oriented programming. The intent of this provision is to prevent nonsubscribers from receiving any part of the channel. The law further requires that until the full channel blocking is accomplished, the cable operator must protect the child audience by carrying the channel only at times of the day when children are not likely to be watching (47 USC 561). Several adult-oriented programmers, including the Playboy Channel, challenged the law in court. In May 2000 the U.S. Supreme Court declared the law unconstitutional, finding that there were less restrictive ways (the use of a channel lock-out device, for example) to protect children from indecent programming on cable (US, 2000).

In 1991 the FCC and the Justice Department agreed to divide up their overlapping concerns about obscenity and indecency. Justice would take primary responsibility for any actions against cable or pay-cable services, whereas the FCC would concentrate on broadcast cases (Ferris et al., 1983–1992: 8–38/9).

Indecency and Other Electronic Media
• •

Although no precedent exists with respect to indecency standards for other multichannel technologies such as DBS and wireless cable, they will most likely be treated similarly to cable television because they are primarily subscription services. The courts have spoken about the transmission of sexually explicit material on two other electronic media: the telephone system and the Internet.

Since the early 1980s the FCC had been trying to control the transmission of sexually explicit messages via the telephone, commonly known as "dial-a-porn," but several of its attempts at regulation were struck down by a court of appeals as being too restrictive. Finally, Congress acted in 1988 with legislation that sought to prohibit dial-a-porn by making it a crime to transmit obscene or indecent commercial telephone messages. The Supreme Court unanimously held the statute to be unconstitutional as applied to the transmission of indecent messages. The Court noted that indecent speech was constitutionally protected for adults and that as long as other techniques were available to block such messages from children, a total ban on telephone indecency violated the First Amendment (US, 1989). Consequently, so long as sexually explicit message providers take steps to make sure that children cannot access their services—through credit card sales, scrambling techniques, or verifiable passwords, for example—their transmissions of indecent material are protected.

The other great pornography battleground is the Internet. Spurred by press reports of easily accessible sexual material on the Internet, plus anecdotal evidence of children's harmful exposure to this type of content, Congress, as part of its revision of the Communications Act in 1996, passed the Communications Decency Act (CDA). The CDA made it a crime to make, create, solicit, initiate the transmission of, or display obscene or indecent material over the Internet to any person under the age of 18. Almost immediately after the CDA's passage, a coalition of free speech advocates, Internet users, and computer and telecommunications industry groups went to court, arguing that the CDA was unconstitutional.

In June of 1996 a special three-judge panel of the U.S. District Court in Pennsylvania agreed with the plaintiffs in the case. As a first step, the court noted that access to the Internet is much more complicated than access to broadcasting, implying that a different First Amendment standard should be applied. According to the U.S. District Court:

Internet communication, while unique, is more akin to telephone communication, at issue in *Sable*, than to broadcasting, at issue in *Pacifica*, because, as with the telephone, an Internet user must act affirmatively and deliberately to retrieve specific information online. Even if a broad search will, on occasion, retrieve unwanted materials, the user virtually always receives some warning of its content, significantly reducing the element of surprise or "assault" involved in broadcasting. (FS, 1996: 851–852)

The judges noted that while the CDA criminalized only transmissions to minors, there was no way to know the age of an Internet user. Thus content providers might stifle their messages not knowing who might receive them. Since indecent speech is fully protected when targeted to adults, the court concluded that this "chilling effect" on such content violated the First Amendment rights of the speakers. The decision was appealed directly to the Supreme Court. In June 1997 the Court, using similar arguments, upheld the lower court's decision, thus making the CDA unenforceable with respect to indecency.

In response to the Court's decision, Congress in 1998 again attempted to regulate sexually explicit programming on the Internet by passing the Child Online Protection Act. The act prohibits commercial Internet sites from making available to children material that is deemed "harmful to minors." (47 USC 201) Immediately after President Clinton signed the bill into law, several groups filed suit to stop its enforcement, arguing that the law, like the CDA before it, violates the First Amendment rights of adults. Subsequently, a federal judge enjoined the government from enforcing the act while the lawsuit is being considered. While the question of regulating indecency on the Internet is still being litigated, the government can punish more serious offenses. In late 1998, for example, enforcement agents in the United States and 13 other countries moved against members of an online club that allegedly distributed and traded child pornography.

Seeking a Balance
• • • • • • • • • • •

Obscenity/indecency law as applied to the electronic media will evolve as society's standards evolve. After three decades of increasing liberalization, during which many taboos fell and audiences grew more tolerant of explicit language and scenes, the pendulum for a while swung to a more conservative standard—restrictions aimed at preserving "family values." The country's trend toward political conservatism in the 1980s favored restrictions despite another conservative article of faith—reliance on the marketplace to set rules.

By contrast, no indication of a disappearing market for programs that pushed the limits was evident in the 1990s. Even network shows (such as ABC's *NYPD Blue*, with its partial nudity and strong language) suggested that the search for what was acceptable—how far one could go—was ongoing.

The 1990s trend toward a more permissive interpretation of laws dealing with sexuality for cable as compared to broadcast programming also has continued into the new century. This difference arises largely because cable lacks the universal reach of broadcasting and comes into the home only after a conscious decision to subscribe—a decision reviewed monthly when the bill comes in.

Although an all-out invasion of X-rated cable programs has not yet occurred, cable channels such as Playboy and Spice regularly offer only minimally sanitized pornographic videos. Indeed, the owner of one of the nation's largest producers and distributors of X-rated materials has predicted that if the much-touted 500-channel future ever arrives, "there's going to be an enormous new market for adult entertainment" (*Wall Street Journal*, 11 July 1994: A1).

Similarly, commercial Internet pornography sites experienced almost a 40 percent annual growth in the late 1990s. Several porn purveyors generate up to $200 million per year in revenue, accommodating up to 5 million viewers (and purchasers) annually.

13.6 OWNERSHIP

First Amendment theory stresses the value of many diverse and antagonistic sources of information and opinion. Unregulated competition can lead to monopoly control of media outlets, thus limiting the number of speakers in the market. Accordingly, to promote the goals of the First Amendment and prevent the monopolization of the electronic media, the government limits the number and kinds of media outlets that any one entity can own. This *diversification* of ownership and control has long been a major FCC goal.

Every broadcast station enjoys a limited monopoly—exclusive use of a given channel in a particular market. Similarly, wireless cable operators have monopoly control over specific frequencies in local markets. DBS operators have such control over their orbital slots. Cable, a monopoly service in most markets, has more far-reaching control. Once viewers subscribe, installers usually disconnect rooftop antennas so that homeowners receive even broadcast stations only (or at least primarily) by cable. A cable operator thus has control over all video coming into some homes, other than rented or purchased tapes. Government policies on media control and ownership are designed to lessen the impact of these monopoly situations.

Antitrust Law

The Sherman Act of 1890 and the Clayton Act of 1914 together aim to prevent excessive concentration of control by one or a few companies over a particular segment of the economy. They provide the basis for such government actions as the one that led to the 1984 breakup of the giant telephone monopoly AT&T.

Courts have long held that despite the First Amendment, business law—including antitrust statutes—may be applied to media. Sec. 313 of the Communications Act requires the FCC to consider revoking licenses held by companies found guilty of violating antitrust laws. However, the deregulatory atmosphere of the 1980s weakened the enforcement of antitrust principles. The Justice Department rarely questioned either the huge media mergers and takeovers during the 1980s or the trend toward ever-larger combinations in cable television.

This is changing, at least with respect to local radio concentration. Following the passage of the Telecommunications Act of 1996, which relaxed many of the media ownership restrictions, the Department of Justice stepped up its review of media acquisitions. Several radio mergers, which would have given one owner too much control over a local radio market, have been challenged by the Justice Department, and several stations had to be divested. According to the guidelines that seem to be emerging in such cases, if a proposed merger or acquisition would result in one owner controlling over about 40 percent of the radio advertising revenues in a market, the Department of Justice might object to the transaction based on antitrust principles.

Broadcast Ownership Defined

In general, the FCC defines ownership of a broadcast station as any active or participating equity interest in the station of 5 percent or more. Complex *attribution rules* pinpoint exactly what other kinds of interests may be deemed ownership under the FCC's rules (47 CFR Sec. 3555). In the 1990s several group owners (including ABC, CBS, Fox, and Tribune) attempted to avoid being considered owners of stations by acquiring passive (noncontrolling) interests, thus staying within the letter of the FCC's ownership limitations. In other situations broadcasters operate stations without owning them under *local marketing agreements* (LMAs). In 1999 the Commission ruled that operators of stations under LMAs generally will be considered owners of those stations.

Multiple Ownership

Prior to 1996 the FCC limited the total number of broadcast stations anyone could control. By 1994 those limits confined a single owner to no more than 20 AM, 20 FM, and 12 full-power television stations nationwide; for radio the cap for minority-owned stations was 25 AM and 25 FM; for television there was an additional limit of reaching markets totaling no more than 25 percent of the television homes in the country (UHF TV stations counted only one-half the homes they actually reached); and minority-owned TV station caps were 14 outlets or 30 percent coverage.

The Telecommunications Act of 1996 changed the national ownership rules substantially. Now there are no limits on the number of radio stations that one entity may own. As discussed in Section 6.2, this has led to an unprecedented number of radio station sales, with some owners now controlling hundreds of radio stations across the country. Congress also removed the limits on the number of television stations that can be controlled by one owner. Congress maintained the

national audience cap for television owners, though it raised that limit to 35 percent of television homes in the country. None of these regulations applies to ownership of noncommercial broadcast stations. Note also that while Congress removed these national ownership limits, local ownership restrictions on both radio and TV are still in effect (see Exhibit 13.d).

Neither DBS nor low-power TV has any ownership limitation, though the FCC has considered restricting the number of "full CONUS" orbital slots DBS operators can control. These slots are very valuable because they provide the ability to cover the entire continental United States (CONUS). There are only three full CONUS slots available, and the FCC has considered allowing DBS owners to control no more than one. Multichannel multipoint distribution service (MMDS) operators may control only one of the two sets of frequencies available per market and may not own more than 5 percent of co-located cable systems.

Duopoly

On the individual market level, a *duopoly* rule the FCC issued in 1940 and enforced for half a century held that no single owner could have more than one station of the same type (for example, more than one AM station) in the same market.

In 1992, while changing its radio ownership rules, the FCC struck down the duopoly limitation, allowing ownership of up to four stations in the largest radio markets. In 1996, as part of the Communications Act revisions, Congress went even further and codified new local ownership standards for radio depending on the size of the market (see Exhibit 13.d).

In 1999 the FCC amended the television duopoly rule to allow one entity to own two stations in a market under specific conditions. For one entity to own two television stations in one market, one of the stations cannot be among the top four stations in the market in terms of audience share. Also, the market must have at least eight other full-power television stations. The new rules also permit an existing owner to buy a second television station in the market if the station purchased is a "failed station"—one that is either off the air or is involved in involuntary bankruptcy proceedings. (47 C.F.R. Section 73.3555)

EXHIBIT 13.d
. . . .
Media Ownership Regulations

Multiple Ownership		
	Market Size	**Limitation**
Radio	45 or more commercial stations	8, no more than 5 in one service (AM or FM)
	30 to 44 commercial stations	7, no more than 4 in one service
	15–29 commercial stations	6, no more than 4 in one service
	14 or fewer commercial stations	5, no more than 3 in one service
		No more than 50 percent of stations in market
Television	May own 2 stations per market *if* one station is not among the top 4 stations in market *and* there are at least 8 other full-power stations in market	
	Nationally, group-owned stations may cover no more than 35 percent of U.S. television households	
Cable Television	MSOs may not serve more than 30 percent of U.S. video subscribers	
MMDS	Can control only one of two sets of frequencies available per market	

Crossownership	
Newspaper-Broadcast	Cannot own a newspaper and a broadcast station in the same market; existing combinations grandfathered
Cable-Broadcast	Cannot own a cable system and a broadcast station in the same market
TV-Radio	May own up to 2 television and 6 radio stations in markets with 20 additional independent media "voices" (full-power TV stations, radio stations, daily newspapers, cable systems)
	May own up to 2 television stations and 4 radio stations in markets with 10 additional voices
	May own a television and a radio station notwithstanding the number of independent voices in the market
Cable-MMDS-SMATV-LMDS	Cable operators cannot own MMDS, LMDS, or SMATV systems in the same market

Cable Systems
. . . .

Whereas the 1984 Cable Act *allowed* the FCC to establish cable ownership rules if it chose (it didn't), the 1992 Cable Act *mandated* the Commission to do so. In 1993 the FCC issued rules limiting ownership to systems that pass no more than 30 percent of all U.S. cable subscribers. The FCC revised this rule in 1999. Multiple system operators (MSOs) are now permitted to serve up to 30 percent of all multichannel video subscribers (which includes not only cable subscribers but also DBS and MMDS subscribers). (47 C.F.R. Section 76.503) A U.S. court of appeals upheld the constitutionality of this rule in May 2000 (F, 2000).

The Commission also has limited *vertical integration* by ruling that a cable system can devote no more than 40 percent of its channels to program services in which the system operator has an ownership interest (45 percent for minority-owned systems). The limit applies to only the first 75 channels.

Cross-Media Ties
• • • • • • • • • • •

Every instance of media *crossownership*, such as common control of newspapers and broadcast stations in the same market, reduces diversification. This reduction in alternative information sources (sometimes called media *voices*, as opposed to actual *outlets* such as stations or systems) is especially undesirable in small communities, in which the only newspaper might own the only broadcast station or cable system.

The FCC issued rules banning new daily newspaper–broadcasting crossownership in 1975 but allowed all but a very few existing combinations to continue (a process called *grandfathering*). Though both the FCC and Congress have considered limits on newspaper-cable crossownership on several occasions, no such limits have yet been adopted.

In 1970 the Commission prohibited broadcast television network ownership of cable systems as well as crossownership of telephone companies or television stations and cable systems in the same market (except in rural areas of fewer than 5,000 people). Congress eliminated both the network-cable and the cable-telco crossownership restrictions in the Telecommunications Act of 1996 (broadcast networks have always been allowed to own cable networks). The 1992 Cable Act banned crossownership of any MMDS or SMATV service by a cable system in its franchise area. The FCC, as usual, grandfathered combinations existing when the act was passed (Sec. 613[a][2]).

Common ownership of a television and a radio station in the same market had been prohibited since the mid-1970s (with existing combinations grandfathered). In 1999, as part of its ruling allowing television duopolies, the FCC amended its television-radio crossownership rules to permit common station ownership in the same market. In the largest markets, licensees may own up to two television and six radio stations. Even in the smallest markets, one owner can now control a television and a radio station in that market. (47 C.F.R. Section 73.3555)

Enter the Telcos
• • • • • • • • • • •

At the end of the 1980s, pushed in part by technical and structural changes in the telephone industry, a new crossownership controversy developed that found broadcasters and cable owners on the same side. Both groups feared that the FCC might allow local telephone companies to own cable systems—and possibly even broadcast stations—within their telephone service areas. A ban on cable-telco crossownership in the 1984 cable act had kept regulated common-carrier and media businesses apart in the same market, but telephone industry representatives increasingly desired a share in the information or content business.

In the early 1990s the FCC sought to increase telephone company participation in electronic media ownership for three reasons: to develop competition with largely monopoly cable systems; to hasten the installation of fiber-optic transmission lines, with their high-quality broadband capacity; and to encourage the development of two-way (interactive) video services.

The FCC's video dial tone (VDT) decision of 1992 allowed telephone companies to provide the means of audio and video program delivery, though they could control no more than a 5 percent interest in the programs themselves (FCCR, 1992a: 5781). The Commission prohibited telephone companies from buying an existing cable system in their service areas because that would simply replace one monopoly provider with another.

Meanwhile, a number of federal courts had ruled that the 1984 telephone-cable crossownership ban violated the telcos' First Amendment rights. Congress ended the debate once and for all by eliminating the cable-telco crossownership restriction in the 1996 Telecommunications Act. As detailed in Section 12.9, p. 362, telephone companies are now allowed to provide multichannel video services in competition with cable television systems and other video service providers. In order to foster competition at the local level, Congress expressly disallowed telephone companies from acquiring existing cable systems in their service areas and vice versa. Joint ventures between the local telephone company and the cable system also are not allowed.

Not to be outflanked, some large cable MSOs are investigating different kinds of telephone and other voice and data services they might provide.

Video compression and other technological developments aid the competing industries as they move onto one another's turf. AT&T, for example, plans to use Internet protocols to provide local telephone service through its cable systems. How regulators will control the shifting alliances and technologies promises to feed major electronic media policy debates well into the new century.

Minority Ownership
● ● ● ● ● ● ● ● ● ● ● ● ●

For years the FCC ignored minority status of owners as an aspect of diversifying media control. A series of court reversals in the 1960s and early 1970s forced it to reexamine its position, and it slowly began to give an advantage to minority applicants.

Since 1978 the Commission has taken several steps to enhance opportunities for members of ethnic and racial minority groups (African Americans, Hispanics, Asians, Native Americans) to become licensees.

- *Tax certificates* have encouraged the sale of stations and cable systems to minorities by allowing sellers to avoid or at least defer paying capital gains taxes on their profits.
- *Distress sales.* Normally, the Commission will not permit an owner whose license is in serious danger of nonrenewal to sell anything other than the station's physical assets. But to encourage sales to minority applicants, the FCC permits endangered licensees to recover some, though not all, of the market value of "intangible assets" (effectively, the station license and the operation's public image).
- *Controlling interests.* In 1982 the Commission sought to further encourage minority ownership by allowing members of a racial minority holding as little as 20 percent of the equity in a licensee to take advantage of both certificates and distress sale rules—provided that the minority owner had voting control. At the same time, the FCC made cable system sales eligible for tax certificate consideration.
- *Lotteries* used in choosing LPTV and MMDS licensees give preference to minority ownership.
- *Bidding credits* are awarded to companies owned by minorities or women participating in the FCC's frequency auctions.

Minority ownership of stations has increased from about 30 stations in 1976 to about 300 at the beginning of the new century (of which fewer than 10 percent are television stations). The National Black Media Coalition reported that to achieve ownership of stations in proportion to their actual numbers in society, some 1,250 broadcast stations would have to be owned by African Americans and about 450 by Hispanics. The industry has a long way to go before minorities achieve such ownership parity.

Despite this record, the deregulatory-minded FCC briefly suspended the distress sale minority preference option in the 1980s because of ideological opposition to "reverse discrimination," as opponents call special breaks for minorities. Pressure from Congress soon forced the Commission to reinstate it. Several legal attacks on this and other minority preference policies were consolidated into one case but rejected by the Supreme Court, which held that such preference policies "do not violate equal protection principles" (US, 1990: 547).

Minority preferences came under renewed attack in 1995. First, a Supreme Court decision striking down minority preferences for federal contracting cast doubt on the continuing constitutionality of the FCC's race-based policies (US, 1995). Next, a U. S. court of appeals barred the FCC from proceeding with a second planned auction of licenses for new wireless telephone and data services, under which small businesses owned by women or minorities would receive a 25 percent discount. At the same time, Congress—angered by reports that Viacom might avoid some $600 million in taxes by selling its cable systems to a black-owned company backed by TCI (now AT&T BIS), the nation's largest cable operator at the time—abolished the tax certificate program as part of its overall review of equal employment opportunity programs (see Section 12.5, p. 354).

In response to these actions, the FCC initiated two proceedings relating to minority and female ownership of electronic media. The first was a broad-based inquiry seeking insights into its existing minority policies and efforts to extend them (F, 1995). The second sought comments on identifying and eliminating market-entry barriers for small businesses, especially small businesses owned by minorities and females (F, 1996). A final outcome of these and related matters could take years.

Meanwhile, the National Association of Broadcasters spearheaded an effort to develop a minority and female ownership trust fund designed to provide seed money to minorities and females seeking to enter the broadcasting business. By the end of 1999 several companies, including CBS and Clear Channel, had committed $175 million to the effort.

Foreign Control
• • • • • • • • • •

Sec. 310(a) of the Communications Act forbids control of a U.S. broadcast or common-carrier radio station by a foreign entity. Specifically, no more than 25 percent of the station's stock or other means of control can be in the hands of foreign investors. Subscription-based electronic media are not restricted in this way: Canadian and other foreign interests, for example, own many U.S. cable systems.

The application of this rule was put in question in 1985 when Rupert Murdoch acquired the Metromedia group of stations that formed the basis of his Fox network. Murdoch, an Australian, became a U.S. citizen, and the FCC approved the ownership transfer. In 1994, however, both NBC and the NAACP raised the ownership question anew, arguing that the Fox owned-and-operated stations were actually owned by Murdoch's Australian-based News Corporation.

NBC later withdrew as a contestant (after negotiating carriage of its CNBC and NBC Super Channel on News Corporation's satellite-delivered Asian Star Television), but the underlying question remained. Fox argued that, although News Corporation—a foreign entity—holds 99 percent of the equity (ownership) in the stations, Murdoch—now an American citizen—owns 76 percent of the *voting* (and hence controlling) stock, and controls News Corporation through ownership in that entity as well. Fox also pointed out that the FCC may permit foreign ownership if it decides that doing so would be in the public interest and that creation of a viable fourth network, long encouraged by the Commission, met that test.

None of these arguments addressed the more serious issue of whether Fox had misrepresented to the FCC the ownership/control structure when it sought the Metromedia stations in 1985. A finding of *lack of candor* has historically been the most flagrant violation of Commission rules.

In May 1995 the FCC determined that Fox was in violation of Sec. 310 but that it had not lied about its foreign ownership. Fox filed a petition offering a partial restructuring and reasserting that a waiver of the rule would be in the public interest. In July, by a 5-to-0 vote, the Commission ruled that Fox could keep the company intact and that it had indeed served the public interest by competing with the Big Three networks, multiplying program choices, and creating thousands of jobs.

In 1997 the World Trade Organization passed a trade agreement that would allow foreign participation and often control of domestic common-carrier facilities. It is unclear at this time how this trade agreement will affect the enforcement of Sec. 310, but critics are worried that a similar trade agreement might permit foreign ownership of broadcast stations. Several concerned members of Congress circulated proposals that would require congressional approval of any trade agreement that changes U.S. law.

13.7   POLITICAL ACCESS

The Communications Act regulates programs most explicitly when they involve candidates for public office. Congress correctly foresaw in 1927 that broadcasting would one day exert a major influence on voters. If the party in power could monopolize the electronic media, opposing candidates would stand little chance of winning elections.

"Equal Time"
• • • • • • • • • •

The commonly but imprecisely used term *equal time* does not appear in the Communications Act. Indeed, the correct term, *equal opportunities*, requires more than equal time. A literal interpretation of equal time would, for example, permit a broadcaster to run a 30-second commercial for a favored candidate during prime time and run another for the opponent at 4:00 A.M.—a ploy not permitted under the equal opportunities rule.

Sec. 315 of the Act requires that

> if any licensee shall permit any person who is a legally qualified candidate for any public office to use a

broadcasting station, he shall afford equal opportunities to all other such candidates for that office in the use of such broadcasting station; *Provided,* That such licensee shall have no power of censorship over the material broadcast under the provisions of this section. No obligation is imposed under this subsection upon any licensee to allow the use of its station by any such candidate. (47 USC 315[a])

The Commission has interpreted "equal opportunities" to require not only equal amounts of time but also access to equivalent audiences and equal financial arrangements. Thus, if a broadcaster gives free airtime to his brother-in-law who is running for mayor, the licensee is obligated under Sec. 315 to provide a similar amount of free airtime, in a time slot calculated to reach a similar audience, to his relative's opponents as well. (47 USC 315)

A 1971 amendment to the Act mandates allowing time for *federal* candidates by adding a new basis for license revocation: the "willful or repeated failure to allow reasonable access to" federal candidates (47 USC 312[a][7]). The equal opportunities and reasonable access provisions of the Act apply only to broadcast stations and cable channels on which a cable operator originates programming. Cable networks such as Lifetime and MTV are not covered by these laws.

Buying Political Time

One of the trickiest problems in political broadcasting is defining what constitutes a station's "lowest unit charge." This phrase in Sec. 315(b) of the Communications Act defines the maximum rate that licensees may charge candidates who buy time for political purposes.

In effect, stations may charge a political candidate no more for a given commercial than would be charged to the lowest-paying advertiser for the same spot. For example, a commercial advertiser might have to buy several hundred spots to qualify for the maximum quantity discount, but a political candidate benefits from that discount even when buying only a single spot.

Even with the lowest unit rate, candidates spend hundreds of millions of dollars on broadcast advertising during political campaigns. This inevitably leads to calls for campaign reform. Following the 1996 elections, several members of Congress proposed legislation that would give federal candidates up to 30 minutes of free airtime during election campaigns. Under this proposal candidates could purchase additional time at a rate of 50 percent of the lowest unit cost. Many politicians suggest that this would be a small price to pay for the broadcasters' use of additional spectrum for digital television. President Clinton voiced his support for these proposals, as did FCC chairman William Kennard. Broadcasters for the most part oppose these plans, arguing that they should not be required to foot the bill for campaign finance reform. As of mid-2000, Congress had not passed any campaign finance reform proposals.

Candidates in the News

What about the presence of candidates in news stories? Do such "uses" also trigger Sec. 315's equal opportunities obligation? At first the FCC said no; later it said yes. Congress thereupon amended Sec. 315 to exempt news coverage of all candidates:

Appearance by a legally qualified candidate on any
(1) bona fide newscast,
(2) bona fide news interview,
(3) bona fide news documentary (if the appearance of the candidate is incidental to the presentation of the subject or subjects covered by the news documentary), or
(4) on-the-spot news coverage of bona fide news events (including but not limited to political conventions and activities incidental thereto),
shall not be deemed to be use of a broadcasting station within the meaning of this subsection. (47 USC 315[a]; emphasis added)

Although the amendment liberated news coverage from political equal opportunity claims, it also left the FCC with many knotty problems of interpretation. Questions regarding political candidates' rights are among those most frequently asked by licensees seeking interpretation of FCC rules. Some examples of the issues confronting stations are detailed in Exhibit 13.e. In addition to providing paid-for access, the broadcast media have a responsibility to inform the electorate about political issues in a nonpartisan way. This tricky combination of enforced cooperation, profit making, and the obligations of journalistic objectivity continues to create problems for station managers.

EXHIBIT 13.e
• • • • • • • • • • • •

What Does "Equal Opportunities" Mean?

- *Who gets equal opportunities?* Candidates for nomination in primary elections and nominees in general elections get equal opportunities. But equal opportunities can be claimed only by candidates for the same specific office; purchase of time by a candidate for Congress, for example, entitles all other candidates for that post in that district to equivalent opportunities, but not candidates for other districts or other offices.
- *Do presidential news conferences count as bona fide news programs?* Yes—they are exempt from equal opportunity claims.
- *Do presidential candidate debates count as news?* Though the rules have changed many times, the answer is yes. However, third-party candidates are less likely to be heard (networks or other sponsors exclude them as being of little interest—Ross Perot in 1992 being an exception, but even he was excluded in 1996)—a news exception that flies in the face of Sec. 315's intent to let voters hear all the candidates.
- *Are regularly scheduled interview and talk programs exempt from equal opportunity requests?* Yes—including "infotainment" programs such as *Oprah.*
- *How much time constitutes "reasonable access"?* The FCC refuses to provide any set amount, but stations are prohibited from limiting candidates for federal office to set numbers of spots or amounts of program time, even when their schedules do not easily allow extended political speeches.
- *May live news appearances be recorded for later playback and still be exempt?* Yes.
- *Is a candidate entitled to whatever time of day he or she wants?* No. Candidates cannot decree when they appear. Stations may keep political spots out of particular periods—such as the times scheduled for news programs.
- *May broadcasters require political ads to conform to standard advertising spot or program lengths?* No. Stations cannot ban campaign ads lasting for five minutes or other "nonstandard" lengths of time.
- *Must a station broadcast a candidate's use of obscene material?* No. When *Hustler* publisher Larry Flynt threatened to use arguably obscene material in a possible campaign for the presidency in 1984, licensees asked the FCC for policy guidance. The Commission held that the ban on obscenity overrode the "no censorship" provision of Sec. 315.
- *Are photos of aborted fetuses in political ads obscene and therefore not allowed?* No. During the 1992 and 1996 elections, several candidates ran on antiabortion platforms, and some used graphic pictures of aborted fetuses in their advertising. Yielding to both viewer and licensee concerns, the Commission held that such depictions, although well within the rights of candidates' uncensored access, might be indecent and thus could be restricted to the "safe harbor" hours of 10:00 P.M. to 6:00 A.M. A federal appeals court disagreed and ruled that broadcasters could not restrict such ads only to the safe harbor hours. The court argued that this would deprive many voters of hearing the candidates' message (F, 1996). Stations can run a disclaimer that federal law required the ad to be carried.
- *Must a station carry potentially libelous material by a candidate?* Yes. *Is the station legally liable?* No.
- *May electronic media endorse political candidates in editorials?* Yes, but if they do, they must notify opposing candidates for the same office and offer reply time, even if the editorial occurs within a newscast.
- *Are cable systems held to the same regulations?* Regulations apply only to channels on which the cable operator *originates* programming. Cable systems are not responsible for the political content of stations or services whose programs they carry but over which they have no content control.
- *May a licensee evade political broadcasting problems by banning all political advertising?* No, at least not for federal candidates (who must receive "reasonable access").
- *Is any appearance of a candidate a "use" of broadcast facilities?* Unless it qualifies as a news exemption, probably yes. Until 1992 anytime a candidate appeared on the air—even in an entertainment or other nonpolitical situation—that time constituted a use. In 1992 the FCC narrowed the definition to cover only time that is authorized or otherwise controlled by a candidate or his or her committee—in other words, overtly political use of time. In 1994 the Commission again reversed course and seemed to define "use" in the traditional way—any appearance by a candidate, whether purposely political or not. Stay tuned for further flip-flops.

13.8 PUBLIC ACCESS

Not everyone with an idea to express can own a station or cable system, nor can everyone expect access to stations or systems owned by others. The FCC long struggled with this problem by trying to provide access to broadcasting for *ideas* rather than for specific *people*. But even access for ideas has to be qualified. It would be impractical to force stations to give time for literally every idea that might be put forward. The FCC mandated access only for ideas about *controversial issues of public importance*.

Rise of the Fairness Doctrine

The FCC slowly elaborated its access ideas into a formalized set of procedures called the *fairness doctrine*. This doctrine obligated stations (a) to schedule time for programs on controversial issues of public importance and (b) to ensure expression of opposing views on those issues. Both stations and the FCC largely ignored (a), focusing their attention on (b). In practice, therefore, most fairness doctrine complaints came as reactions to ideas that had already been discussed on the air, rather than as complaints about the failure to initiate discussion of issues. Licensees had great latitude in deciding whether a subject qualified as both a controversial issue and one of public importance, how much time should be devoted to replies, when replies should be scheduled, and who should speak for opposing viewpoints.

In 1969 the Supreme Court upheld the FCC's fairness doctrine, as well as its related personal attack and political editorializing rules, in its landmark *Red Lion* decision (see Exhibit 13.f). By unanimously supporting the FCC in this case, the Court strongly affirmed the fairness doctrine concept in an opinion that emphasized four key principles relevant to broadcasting's First Amendment status (US, 1969: 367).

- *On the uniqueness of broadcasting:* "It is idle to posit an unabridgeable First Amendment right to broadcast comparable to the right of every individual to speak, write, or publish."
- *On the fiduciary principle:* "There is nothing in the First Amendment which prevents the Government from requiring a licensee to share his frequency with others and to conduct himself as a proxy or fiduciary."
- *On the public interest:* "It is the right of the viewers and listeners, not the right of the broadcasters, which is paramount."
- *On the scarcity factor:* "Nothing in this record, or in our own researches, convinces us that the [spectrum] resource is no longer one for which there are more immediate and potential uses than can be accommodated, and for which wise planning is essential."

Although the fairness doctrine is no longer in effect, these underlying constitutional principles remain intact.

End of the Doctrine

By 1985 the FCC, by then largely made up of Reagan administration appointees dedicated to deregulation, had joined a mounting chorus of opposition to the fairness doctrine. After a lengthy proceeding, the Commission concluded:

> [Based on] our experience in administering the doctrine and our general expertise in broadcast regulation, we no longer believe that the fairness doctrine, as a matter of policy, serves the public interest. . . . Furthermore, we find that the fairness doctrine, in operation, actually inhibits the presentation of controversial issues of public importance to the detriment of the public and in degradation of the editorial prerogatives of broadcast journalists. (FCCR, 1985b: 143)

Acting on a court remand requiring the FCC to reconsider an earlier fairness decision, the Commission instead abolished the doctrine entirely (FCCR, 1987b: 5043). The personal attack and political editorializing rules, however, remained in place, as discussed below.

That should have ended the matter, but the FCC's defiance in ending a practice that Congress had gone on record as favoring enraged key legislators. They vowed to reinstate the fairness mandate, in the meantime making their dissatisfaction known by holding up the approval of appointments to vacant FCC seats. Bills reinstating the doctrine were introduced and passed twice but were vetoed both times by President Reagan. When President

EXHIBIT 13.f

. . . .

A Place and a Case Called *Red Lion*

An unlikely small-town station formed the setting for one of the leading Supreme Court decisions on electronic media. During the 1960s right-wing preachers inundated radio with paid syndicated political commentary, backed by ultraconservative supporters such as Texas multimillionaire H. L. Hunt through tax-exempt foundations. Purchased time for these religious/political programs provided much-needed radio income in small markets.

The landmark case got its name from WGCB, a southeastern Pennsylvania AM/FM outlet licensed to John M. Norris, a conservative minister, under the name Red Lion Broadcasting. In 1964 one of the Reverend Billy James Hargis's syndicated broadcasts carried by the station attacked author Fred Cook, who had criticized defeated Republican presidential candidate Barry Goldwater and had written an article on what he termed the "hate clubs of the air," referring to the Hargis series *Christian Crusade*, among others. Hargis attacked Cook on the air, charging him with communist affiliations and with criticizing the FBI and the CIA—the standard litany of accusations Hargis routinely made against liberals.

Cook then accused the station of violating FCC rules by failing to inform him of a personal attack. When he wrote asking for time to reply, the station responded with a rate card, inviting him to buy time like anyone else. Cook appealed to the FCC, which

agreed that he had a right to free airtime for a reply. It ordered WGCB to comply.

It would have been easy for Norris to grant Cook a few minutes of time on the Red Lion station, but he refused on First Amendment grounds, appealing the Commission decision. The court of appeals upheld the FCC, but Norris took the case to the Supreme Court, which, in 1969, also upheld the FCC, issuing an opinion strongly defensive of the FCC's right to demand program fairness.

Several years later, Fred Friendly, the former head of CBS News but by then a Columbia University journalism professor, began looking into the background of this well-known case for a book about the fairness doctrine (Friendly, 1976). He discovered that Cook had been a subsidized writer for the Democratic National Committee and that his fairness complaint had been linked to a systematic campaign mounted by the Democrats to discredit right-wing extremists such as Hargis.

According to Friendly, the Democrats set out to exploit the fairness doctrine as a means of harassing stations that sold time for the airing of ultraconservative political programs. Cook and the Democratic National Committee claimed that Friendly had misinterpreted their activities, maintaining that Cook had acted as a private individual, not as an agent of the Democratic Party.

Clinton stated early in 1993 that he would sign a fairness doctrine reinstatement bill if it reached his desk, many predicted renewed congressional action. By the end of the decade, however, Congress had taken no action on the issue.

A *Right* of Reply?

. . . .

Two specific fairness-related requirements, the personal attack and political editorializing rules (both adopted in 1967 and retained even after the fairness doctrine's demise in 1987), continue to cause concern to opponents of FCC program interference.

The *personal attack rule* requires stations to inform individuals or groups of personal attacks on their "honesty, character, integrity or like personal qualities" that occur in the course of discussions of controversial public issues. Within a week after the offending broadcast, licensees must advise those attacked, explaining both the nature of the attack and how replies can be made. Specifically exempted from the right of reply are on-the-air attacks made against foreigners, those made by political candidates and their spokespersons during campaigns, and those occurring in news interviews, on-the-spot news coverage, and news commentaries (47 CFR 73.1920).

The *political editorializing rule* requires that all candidates or their representatives be given a chance to respond if a licensee endorses any of their opponents. If a station editorially opposes a candidate, that candidate also must be given an opportunity to respond. A station must inform a candidate of his or her right to respond within 24 hours of such an editorial. The rule does not apply to the use of a station's facilities by opposing candidates, a situation covered by the equal opportunities procedures.

The Radio-Television News Directors Association (RTNDA) has long urged the FCC to abolish the personal attack and political editorial rules, but the FCC consistently has refused to act. The RTNDA finally went to court and in 1996 obtained an order forcing the FCC to consider the issue. Early in 1997 the FCC invited public comment on the prospect of modifying or repealing the rules, but the commissioners could not reach a consensus on the issue. In 1999 the court asked the FCC for a formal justification of the rules, further delaying a final decision on the RTNDA's request.

Rights for Advertisers

Access by advertisers to broadcasting has been governed more by considerations of taste and public acceptance—and income potential—than by government regulation. However, cigarette advertising became the subject of a famous fairness doctrine complaint. In 1968 the FCC decided that the surgeon general's first report on the dangers of smoking, as well as Congress's 1965 act requiring a health warning on cigarette packages, justified treating cigarette advertising as a unique fairness doctrine issue. Therefore, stations would have to carry antismoking spots if they carried cigarette commercials. These counter-ads ended, of course, after Congress banned all broadcast advertising of cigarettes in 1971, though occasional public-service announcements concerning smoking still appear.

Editorial advertising, or "advertorials," posed a different kind of problem. Traditionally, the electronic media have declined to let advertisers use commercials as vehicles for comment on controversial issues, arguing that

- serious issues cannot be adequately discussed in short announcements

- selling larger blocks of time for editorializing by outsiders involves surrender of editorial responsibility
- not everyone can afford to buy time, so selling to those with funds is inherently unfair

The Supreme Court upheld the principle of licensee *journalistic discretion* in a case dealing with a fairness doctrine demand for access to editorial advertising:

> Since it is physically impossible to provide time for all viewpoints . . . the right to exercise editorial judgment was granted to the broadcaster. The broadcaster, therefore, is allowed significant journalistic discretion in deciding how best to fulfill the Fairness Doctrine obligations, although that discretion is bounded by rules designed to assure that the public interest in fairness is furthered. (US, 1973a: 111)

Despite deletion of the fairness doctrine, the question of editorial advertising continually reappears. Controversies arise periodically between advertisers and broadcasters. In several cases, ads that were turned down by broadcasters have appeared on cable or in print—with pointed commentary about the refusal of broadcasters to sell time for such advertising. The controversy over advertising of condoms is one example: Few broadcasters accepted condom advertising or editorial statements calling for their use to prevent AIDS, for fear that substantial parts of their audiences would be offended by such messages. Moreover, the refusal of many broadcasters to air the messages of antiabortion advocates led abortion foes to run for Congress. As detailed in Exhibit 13.e, broadcasters cannot censor the ads of political candidates; this gives antiabortion candidates an opportunity to put their views on the air.

Editorial Discretion

By their nature, news and public-affairs programs necessarily involve controversial issues, often leading partisans to bring charges of unfairness. It becomes difficult for broadcasting and cable journalists to deal with serious issues if their corporate bosses prefer to avoid controversy. Yet electronic media cannot win full public respect and First Amendment status without taking risks similar to those the printed press has always faced.

The FCC and courts generally assume that reporters and editors use editorial discretion, which calls for fair and considered news treatment of events, people, and controversies. No one believes that journalists always use the best judgment or that they totally lack bias or prejudice. First Amendment philosophy holds, however, that it is better to tolerate journalists' mistakes—and even their prejudice and incompetence—than to set up a government agency as an arbiter of truth. The Supreme Court reaffirmed this notion in confirming yet another FCC decision rejecting a fairness complaint:

> For better or worse, editing is what editors are for, and editing is selection and choice of material. That editors—newspapers or broadcast—can and do abuse this power is beyond doubt, but that is not reason to deny the discretion Congress provided. Calculated risks of abuse are taken in order to preserve higher values. (US, 1973a: 124)

News Bias

Critics often accuse the electronic media, especially television network news departments, of news bias. CNN discovered in an ironic way that it had "arrived" as a widely used news source when, in the wake of its 1991 Gulf War coverage, many complained of its alleged bias by keeping reporters in Baghdad.

Such charges usually come from political conservatives, many of whom regard all news media as too liberal in outlook. They argue that the cumulative effect of alleged liberal bias over time tends to build up one-sided perceptions of issues. Professional gadflies, such as the misnamed Accuracy in Media (which seems to find *in*accuracy and bias in every media report it investigates), regularly take up conservative causes against alleged television bias. Reagan and Bush administration elements lent support to such complaints, particularly after Bush's reelection defeat in 1992. More recently, conservative critics point to the disparity in media coverage of the sexual harassment allegations against Supreme Court nominee Clarence Thomas on the one hand and President Clinton on the other. Content-analysis research is used to support both sides of the bias questions, depending on who finances the studies. The debate continues.

Localism

The public needs access to diverse voices, but they also need voices of their own. The notion of a broad right of public access has two aspects—access to the ideas of others and to the means of expressing those ideas. The widespread availability of stations and cable systems can facilitate such two-way access, especially within local-market communities.

From its inception until well into the 1960s, the FCC encouraged *localism*—reflection of the local community's needs and interests—by distributing as many local outlets to as many localities as possible and by issuing guidelines that encouraged airing of programs reflecting community needs and interests. The Communications Act gives the Commission its own localism guidelines when it acts on applications for new stations:

> The Commission shall make such distribution of licenses, frequencies, hours of operation, and of power among the several States and *communities as to provide a fair, efficient, and equitable distribution of radio service to each of the same.* (47 USC 307[b]; emphasis added)

This policy of localizing stations has provided listeners in major markets 40 or more radio stations from which to choose, whereas people in rural areas have far fewer or, in some cases, none at all. As cable expanded, it equalized and increased television channel choices in most urban and suburban areas; but it, too, underserves rural areas.

Yet cable does take pressure off commercial broadcasters, inasmuch as coverage of meetings and local events has largely transferred to local cable government access channels. Most cable systems have public, educational, and government (PEG) channels as encouraged by federal law and often required by local franchise. Though budget problems facing most local governments and school systems have restricted their use, the potential of cable's growing number of channels to serve a variety of local concerns has increased greatly in the past decade.

Still, when threatened by changing technology—direct broadcast satellites or telephone industry takeover, for example—broadcasters immediately cite their unique public interest in providing locally relevant programs. They argued long and hard to the FCC and Congress, though to no avail, that cable and DBS are almost entirely national services, whereas broadcasters still hold to the ideal of station-based localism. However, as any discerning listener can attest, while broadcasting and cable do have local *outlets*, the *voices* they carry are increasingly the same nationwide. Evidence of this trend away from local programming abounds. Many radio stations eschew local news and public affairs programming altogether, preferring to program their stations with 24-hour-per-day syndicated music formats delivered via satellite. Some television stations are canceling local newscasts in favor of more profitable syndicated programs. Moreover, a March 2000 report spnsored by the Benton Foundation found that local public affairs programming is at an all-time low. According to the report, which profiled the public affairs programming in 24 television markets over a two-week period, there was an average of 6.52 hours of local public affairs programming per market. This translates to 2.3 hours of such programming per commercial station per month, or 0.3 percent of each station's total broadcast time (Napoli, 2000).

13.9   SERVING CHILDREN

Reacting to constant pressure from Action for Children's Television and others, Congress passed legislation in 1990 regulating the amount of advertising in children's programs as well as establishing a general obligation to serve child audiences. Its major provisions include the following:

- Advertising per hour of children's programs is limited to no more than 10½ minutes on weekends and 12 minutes on weekdays. This "limitation" applies to broadcast licensees and cable systems—for all programming produced and broadcast primarily for children.
- At license renewal, the FCC must consider whether stations have served the "educational and information needs of children" through their overall programming, which must include "some" (otherwise undefined) programming specifically designed to meet children's needs.
- Establishment of a National Endowment for Children's Educational Television was proposed, to be funded at $2 million for fiscal year 1991

and $4 million for fiscal 1992. The endowment was eventually funded at about half this rate and administered by the National Telecommunications and Information Administration (NTIA).

Licensee Duties

With the basic provisions of the Children's Television Act having set the stage, the FCC conducted rule making in 1990–1991 to put more specific regulations in place. After considerable input from the broadcast and cable industries and others, the Commission arrived at the following decisions:

- It took a very wide view of *programming* by defining "children" as those 16 and younger, thus allowing stations to include music and dance programs aimed at teens—not what Congress had in mind (the congressional limitation on *commercials*, for example, affects programs aimed at children 12 and under).
- It prohibited program-length commercials for children. A program built around a fantasy hero may no longer include advertising for products related to that hero—they must run in another, unrelated children's program. (Since 1974 the FCC had policies against "host selling" and the close intermixing of program and advertising elements.)
- It defined the "educational and cultural" programming called for in the Children's Television Act to mean "programming that furthers the positive development of the child in any respect."
- It declared that stations need not program for all children; if desired, they may select one or more age groups.
- It encouraged (but did not require) licensees to "assess" the needs of children in their communities.
- It required keeping of records, in a form chosen by the licensee, that must be made a part of the station's public file.
- It decided that although public broadcasters must meet the programming requirement, they are not held to the specific record-keeping regulations. Cable systems are not obligated to meet either the programming or record-keeping requirement, but they are bound by all advertising limits.

- Finally, it established a fine of $10,000 for each violation of any of the rules growing out of the Children's Television Act (47 CFR 73.660).

The Commission began to review compliance with this list of requirements in 1992 and found that 95 percent of monitored program hours met the commercial limits within weeks of their being put in place; a year later compliance stood at 98 percent. But several television stations and cable systems did run afoul of these limits and became widely publicized examples of the FCC's intent to enforce the new rules rigidly (*Broadcasting & Cable*, 25 April 1994: 10).

As for the programs themselves, one consumer group reported in late 1992 that some broadcasters were meeting the letter of the law's requirements (and those of the loose FCC reporting rules) simply by claiming positive aspects of existing cartoon or rerun comedy shows or by placing what few new programs were being produced at very early morning hours when few would watch (*New York Times*, 30 September 1992: 1, B8).

That study, and rising consumer group and congressional displeasure with broadcaster performance, led to 1993 hearings in Congress and inception of an FCC inquiry into whether its children's program requirements should be made more specific.

In 1996 President Clinton took the unusual step of becoming directly involved in this issue when he invited television industry and government leaders to the White House for a discussion of the children's television requirements. At this meeting a consensus arose among the participants that television broadcasters should, at a minimum, air three hours of educational programming for children per week. The FCC subsequently adopted this standard as a rule that all television broadcasters must satisfy.

The rule appears to be working, at least in part. A 1999 study by the University of Pennsylvania's Annenberg Public Policy Center found that the number of programs produced for children was increasing and that nearly three-quarters of all children's television shows had some "enriching content." The study also found that children's educational programming was less violent than other children's shows. On the downside, one in five shows listed by broadcasters as meeting the educational requirement contained little or no educational or informational elements. (Schmitt, 1999) Also, many West Coast affiliates regularly preempt their networks' weekend children's schedule to carry sporting events.

Television Violence

As detailed in Section 11.7 (p. 337), television violence and its effects on children have elicited more concern and study than any other entertainment program issue. The first congressional hearings on television violence were held in the 1950s, and there have been many more since then. Until relatively recently, however, the government has declined to regulate violent programming based primarily on First Amendment grounds. Throughout the controversy broadcasters have resisted regulation, arguing that parents and not the government should control children's television viewing. Not satisfied with broadcasters' efforts at self-regulation, Congress finally acted on this issue in 1996.

As part of the Telecommunications Act of 1996, Congress set in motion a process to help parents control the programming their children watch. The legislation requires the establishment of a television ratings system that rates television programs. The ratings are listed in television programming guides and displayed at the beginning of rated programs. The legislation also requires that new television sets be equipped with a device known as a V-chip (*V* for violence), which allows parents to program their televisions to block out all programs of a particular rating. Each program carries its rating in the signal's vertical blanking interval (see Section 4.8, p. 104) so that the V-chip can identify which programs could be shown and which should be blocked.

The act allows broadcasters to establish a ratings system to be submitted to the FCC for approval. After many months of study, an industry committee headed by Motion Picture Association of America (MPAA) President Jack Valenti unveiled its ratings system at the end of 1996. Similar to the MPAA movie ratings scheme, the initial television system rated programs according to age categories.

Almost immediately, critics of the proposal attacked the age-based system, arguing that the legislation intended a content-based ratings system instead. Parents' rights groups and government leaders alike charged that the age-based system did not provide enough information to help parents decide what was appropriate for their children. They demanded that the system be changed in such a way that the ratings reflect objectionable content: V for violence, S for sexual situation, and L for strong language. Broadcasters generally argued that a content-based rating system was too confusing and that advertisers would shy away from programs rated for objectionable content.

By mid-1997 Valenti and his committee succumbed to the pressure and added content descriptions to the ratings (see Exhibit 9.d, p. 267). In March 1998 the Commission approved the modified ratings system, but that has not ended the controversy surrounding objectionable programming. NBC does not include the content description in its ratings, believing that the content descriptors are too intrusive, and Black Entertainment Television refuses to carry any program ratings. Moreover, in May 1999 the Parents Television Council released a study showing that violence on television had increased dramatically since the adoption of the ratings code. Sexual content and adult language also had increased. The council suggested that the ratings system encourages some producers to push the envelope and include ever riskier elements in their programs. Other critics of the system suggest that the ratings do not provide enough meaningful differentiation, because the majority of shows are rated TV-14. They further argue that most children's cartoons containing violence fail to carry the FV rating for fantasy violence. (Parents Television Council, 1999)

13.10 COPYRIGHT

The Constitution recognizes the fundamental importance of encouraging national creativity. Article I, Section 8, calls on Congress to "promote the progress of science and the useful arts, by securing for limited times to authors and inventors the exclusive right to their respective writings and discoveries."

When broadcasting began, authors and composers had to rely on the copyright law of 1909, which dealt primarily with printed works and live performances. Despite amendments, the old act never caught up with the times. After long study and debate, Congress finally passed a new copyright law in 1976, effective for works published in 1978 and later.

Basics
•••••

Key provisions of the 1976 Copyright Act, administered by the Copyright Office (part of the Library of Congress), include the following:

- *Purpose.* Copyright holders license others to use their works in exchange for the payment of *royalties.* "Use" consists of making public by publishing, performing, displaying, or broadcasting.
- *Copyrightable works.* In addition to traditionally copyrightable works—books, musical compositions, motion pictures, and broadcast programs—works such as sculptures, choreographic notations, and computer programs can be copyrighted. Things *not* copyrightable include ideas, slogans, brand names, news events, and titles. (However, brand names, logos, and slogans can be protected under trademark regulations.)
- *Length of copyright.* In general, a copyright lasts for the life of the work's creator plus 70 years. Copyrights initially granted to corporations last 95 years from the first publication of the work or 120 years from the date the work was created, whichever comes first. After that, a work enters the *public domain* and can be used without securing permission or paying royalties.
- *Compulsory licensing.* Owners of copyrighted material who license television stations to use their work must grant *retransmission consent* to cable systems that lawfully pick up such programs off the air and deliver them to subscribers. Cable systems, in turn, must pay a preset small proportion of their revenue for these rights. Similar regulations apply to DBS operators who carry broadcast stations.
- *Fair use.* The new act retained the traditional copyright concept of fair use, which permits limited use of copyrighted works without payment or permission for certain educational and critical or creative purposes (such as reviews of books and musical works).

The advent of new media forms (electronic delivery of newspapers, digital recording, and the

Internet, for example) presents novel and difficult copyright questions. Royalty obligations to authors and others, whose works are prepared for single use by one medium but are then reused by others, will be the subject of negotiation, litigation, and legislation well into the future.

Music Licensing

Broadcast and cable programmers obtain rights to use recorded music by reaching agreements with the copyright licensing organizations ASCAP, BMI, and SESAC. Most stations hold *blanket licenses* from these agencies, which allow unlimited air play of any music in their catalogs in return for payment of an annual percentage of the station's gross income. Some radio stations, especially those with news and talk formats, pay on a per-use basis. These licenses, whether blanket or per use, permit the *broadcast* of the music by the stations. They do not cover other uses of the music, such as adding the music to a local commercial or public service announcement. Such uses require additional permission from the copyright owner.

Cable, Copyright, and the CRT

The Copyright Act of 1976 established the Copyright Royalty Tribunal (CRT). For cable television, the CRT was to establish rates that cable had to pay for the use of imported distant television broadcast signals, pool the resulting revenue, and divide the pooled money among copyright holders.

The process of apportioning payments became snarled in legal proceedings from the day the CRT began work. Throughout its short history, the CRT lacked sufficient staff and expertise to carry out its contentious assignments. In 1982 the agency's chairperson called the whole process unworkable and unfair. Finally, in 1993, Congress abolished the CRT, turning its duties over to a series of arbitration panels to be supervised by the Library of Congress and its copyright office.

Cable Piracy

Until the mid-1980s *signal piracy* chiefly involved illicit hookups to cable television feeder lines, enabling cable reception without payment of monthly subscriber fees. As pay cable developed, illegal "black boxes" (decoders) made possible the reception of even scrambled signals. The cable industry estimated that in the 1990s an equivalent of a quarter of cable revenue was lost to pirate activities.

Pirated cable reception violates Sec. 705 of the Communications Act, which defines penalties for the unauthorized "publication" of communications intended for reception only by subscribers paying to use special descrambling devices. Several cable systems prosecuted violators and publicized the resulting felony convictions, which entailed jail terms and fines of up to $25,000 for the first violation and $50,000 for subsequent violations. More than half the states have passed antipiracy laws prohibiting the manufacture, sale, or use of unauthorized decoders or antennas.

TVRO Piracy

Piracy increased when the prices of television receive-only (TVRO) Earth stations began to fall in the 1980s, making them affordable by ordinary households. Both pay- and basic-cable operators began to scramble satellite signals to protect their investments from pirates.

As a result, Congress modified Sec. 705 to allow individuals to pick up any *nonscrambled* satellite programming if they have obtained authorization, usually for a fee. By the early 1990s nearly all satellite channels were scrambled. Since the cable industry had long refrained from marketing its services to such scattered audiences, a number of third-party brokers had already begun to assemble packages of cable services for sale to TVRO users, although the use of illegal descramblers continued.

VCR Piracy

In 1976 a number of program producers brought suit against Sony, the pioneering manufacturer of home video recorders, for "indirect" copyright infringement ("indirect" because Sony provided the *means* of infringement—the machines that made recording possible—but did not itself do the copying). A district court found in favor of Sony, concluding that home recordings of broadcast programs not sold for a profit fall within the fair-use provision of the new copyright law. The Supreme Court affirmed the decision by a 5-to-4 vote (US, 1984a: 417). The Court cited audience research showing that people record broadcast signals primarily for time-shifting purposes, which it regarded as a fair use.

The decision left unresolved the legality of recording cable or pay programming; the *Sony* case covered only recording of over-the-air, or "free," broadcast material. The legality of showing a taped copy of a copyrighted film or program to a group outside the home remained in question, a problem intensified by increasing numbers of video rentals. Hollywood producers have pressured Congress— thus far unsuccessfully—to modify the Copyright Act once more, this time to extract from those who do home recording such indirect royalty payments as a surcharge on recorders or blank tapes at the time of purchase or on rental fees customers pay to video rental stores.

Digital Piracy
• • • • • • • •

Late in 1992 Congress enacted into law an industry compromise on royalties for recordings produced by the coming generation of digital audio home recording devices. The price of machines, digital audio tape (DAT), and digital compact cassettes (DCC), for example, includes a 2 percent royalty fee (from $1 to $8 per machine), whereas blank tapes and discs include a 3 percent fee on their wholesale price. The funds raised go to the Copyright Office to be disseminated to copyright holders.

The law also supports the industry's serial copy management system (SCMS) technical standard, which prevents a second-generation digital dub from being made on any of these machines. In other words, an owner can make a copy but not a copy of a copy (*Television Digest*, 28 September 1992: 10).

Similar issues involving video recording have arisen with the introduction of the personal video recorder (PVR; see Section 5.6, pp. 142). These devices can record and store perfect digital copies of television programs and can easily manipulate the programs to avoid commercials, for example. Concerned about the copyright implications of the devices, the Advanced Television Copyright Coalition is attempting to force the makers of PVRs to pay some sort of licensing fee. PVR manufacturers are resisting. Legislation may be needed to resolve this controversy.

Digital copying issues have slowed the production and distribution of cable-ready digital television receivers. The first generation of digital receivers was able to pick up over-the-air digital signals only; they were not equipped to receive digital signals from cable systems. Set manufacturers and the cable industry have been working on a standard to make digital sets compatible with cable systems, but a final agreement has not been reached (as of June 2000) because copyright holders insist that the standard include a mechanism for restricting the copying of digital signals. Noting that a final agreement on digital receiver standards is crucial to the rollout of digital broadcasting, the FCC stated in early 2000 that if the industries could not reach a voluntary agreement, it would set a standard by regulation. (FCCR, 2000)

Copyright and the Internet
• • • • • • • • • • • • • • • •

The ease and speed with which information travels on the Internet poses unique and thorny copyright problems. Anyone with an image scanner and a computer can easily put copyrighted material into the hands of thousands of Internet users in a matter of seconds. Technically, anytime a copyrighted work is transmitted from one computer to another and copied in the receiving computer's memory, a violation of copyright law has occurred. Acquiring music via easily downloaded MP3 files raises thorny copyright issues. Broadcast stations that simulcast copyrighted programs such as professional sports on their web sites face increased copyright liability. And commercial web sites that retransmit broadcast signals over the Internet may also be violating copyright law. Moreover, digitized information is easily manipulated, so copyright notices can be removed without much effort.

A further question involves the copyright liability of Internet service providers (ISPs) such as AOL. Because these services store information provided by their subscribers and make that information available to others, are the ISPs likewise liable for copyright infringement when the information is protected by copyright?

Internet user groups argue that the copyright laws should be interpreted as allowing the freest exchange of information possible. They contend that restrictive copyright enforcement will stifle the freewheeling and robust interchange of ideas that occurs on the Internet. Commercial interests and the Clinton administration, on the other hand, counter that copyright enforcement on the Internet should be increased in order to protect the economic interests of copyright owners.

These and other issues have been addressed both nationally and internationally. In the United States a task force looking into copyright and new technology produced a report in 1996 generally concluding that the use of copyrighted material on the Internet should come in line with the use of such material in more traditional media.

The very few courts that have addressed these issues thus far have reached mixed conclusions on how the copyright laws should apply to Internet users. A 1999 court case settled at least one music-recording issue: The use of hand-held MP3 players, such as the popular Rio, does not violate copyright law. The court ruled that the players are consistent with the intent of copyright law to ensure the right of consumers to make recordings of copyrighted music for their private noncommercial use. (F, 1999) In another case, a court enjoined the distribution of television broadcast signals by iCraveTV.com after several U.S. and Canadian television stations and other rights holders sued the company for copyright infringement. (FS, 2000) The case was settled in February 2000 when iCraveTV.com agreed to stop distributing the signals. Legal analysts suggest that because distribution of broadcast signals over the Internet is inevitable, Congress will have to address the issue soon.

Congress did settle some Internet copyright issues in 1998. The Online Copyright Infringement Liability Limitation Act shields telephone companies and ISPs from liability when subscribers illegally transmit or post infringing material using their facilities. The legislation protects telcos and ISPs only when they are unaware of the infringing activity. (17 USC 512) The Digital Millennium Copyright Act makes it illegal to break through the encryption codes that protect copyrighted material on the Internet. The act also makes the manufacture or sale of devices that circumvent encryption technology illegal. (17 USC 1201 et seq.)

On the international front, the World Intellectual Property Organization drafted treaties in 1996 dealing with copyright and the Internet that attempt to slow the flow of copyrighted information and make copyright violations easier to identify and prosecute. The Senate ratified the treaties in 1998, thus making them enforceable in the United States.

13.11   CHANGING PERSPECTIVES

Numerous media compete for attention and consumer dollars, but more can too often become less in a marketplace dominated by giant corporations. These well-financed communicators blanket the nation, reducing the diversity of competing voices.

Two totally different approaches to this problem are actively debated. Here stands the *deregulator*, confident that the marketplace will regulate itself, given maximum competition and unconstrained consumer choice. There stands the *regulator*, demanding that government intervene once more to oppose monopolistic media tendencies and protect the public interest from the effects of unrestrained competition.

Many practical examples of this fundamental clash of views exist:

- Should indecency in electronic media have the same First Amendment protections it has on newsstands and in movie theaters? If so, how best can children be protected from such content?

- Should electronic media have special responsibilities with regard to airing controversial issues of social importance? If so, why—and to what degree—should they be treated differently than print media that bear no such responsibilities?

- Does violence in the electronic media have such potentially damaging social effects as to justify tilting the regulatory playing field against it?

- As cable systems retransmit broadcast signals, as programming is delivered by satellite directly to the home, as cable systems move into telephony and telephone companies into cable, should a level regulatory field be ensured and how?

- How much concentration of control of the media is too much? As media conglomeration continues and more outlets increasingly are in the hands of fewer owners, can a freewheeling marketplace of ideas survive?
- What controls should—and realistically can—be imposed (and by whom) on the unlimited-channel universe of the superhighway and in cyberspace, with media that for some are mass and for others personal, with interactive, multiplayer computer video games and "private" bulletin boards?

As the twenty-first century begins, the debate on how these issues should be addressed continues.

Deregulation advocates argue that economic forces, competition, and technological advancements will resolve any problems. Proponents of government intervention counter that the public's interest in receiving diverse and relevant information are not likely to be advanced by commercial profit motives but are better guaranteed by well-crafted regulations. In fact, the electronic media will continue to be guided by a combination of both approaches, which, it is hoped, will further the First Amendment goals of guaranteeing free speech for all and promoting a free-flowing marketplace of diverse viewpoints.

CHAPTER **14**

Global View

ew Americans realize that most other countries operate electronic media quite differently than we do. Yet because radio waves ignore national boundaries (especially when sent from satellites) and telephone and computer networks reach almost all of the world, electronic media are truly international. Countless organizations here and abroad participate in global exchanges of equipment, programs, and training. Fully understanding electronic media in America requires at least a sense of how they operate elsewhere.

Although countries have the same theoretical potential for using electronic media, each adopts a system uniquely suited to serve its own conditions, economic abilities, and needs. Three attributes of these media promote a mirrorlike relationship between a nation's character and its electronic media:

- Broadcasting invites *government regulation* because it requires the use of electromagnetic spectrum, which all governments view as public property. Each government defines its responsibility to control spectrum in the context of its own history, geography, economics, culture, and, most important, political philosophy.
- Transmissions using spectrum can cause *interference*, creating a need for controlling the number and power of transmitters. Political realities affect how, and how much, a country chooses to regulate based on interference control.
- Electronic media have *political and cultural impact* because they can bypass government entities, going directly and instantly to the public. Whatever its politics, no nation can afford to leave such persuasive media totally unregulated.

14.1 CONTROLLING PHILOSOPHIES

Controls that a nation imposes on electronic media reflect the government's attitudes toward its own people. Despite dramatic political changes over the past two decades, worldwide electronic media structures still generally fall into three broad categories: permissive, paternalistic, and authoritarian.

Permissivism

The electronic media in America provide the classic example of a predominantly *permissive* or *laissez-faire* system. Our Constitution encourages free enterprise and makes freedom of communication a central article of faith. The industry argues that resulting commercialism creates more lively, popular, and expertly produced programs than are usually found elsewhere.

Generally speaking, countries now or once within American influence (such as the Philippines and many Latin American nations) have adopted similar permissive, profit-driven systems. Many other countries deplore American commercialism because it focuses almost exclusively on what people *want* rather than on what critics, experts, and government and religious leaders think they *need*.

Paternalism

Most countries perceive electronic media as playing a positive social role: the preservation and enhancement of predominant national cultures. They take a *paternalistic* approach, putting special emphasis on preserving national languages, religions, and social norms by ensuring a "balanced" program diet—meaning not too much light entertainment at the expense of information and culture. And many nations take special care with children's programs, ensuring positive and culturally relevant examples.

The British Broadcasting Corporation (BBC) is the classic example of paternalism. Growing out of a brief period of commercial operation, the BBC converted to a public-service role in 1926, seeking to avoid America's "mistake" of radio commercialization. BBC funds today come primarily from government-imposed license fees on television receivers, removing dependence on advertising. A well-articulated ideal emerged from the BBC experience. Adopted by many other countries, *public-service broadcasting* includes principles such as the following:

- balanced programming, representing all principal genres
- control by a public body, relatively independent of both political and commercial pressures
- relative financial autonomy, usually secured by partial or complete dependence on receiver

EXHIBIT 14.a

Broadcasting House, London

In pretelevision days, broadcasters from all over the world journeyed to this famous art deco building in the heart of London as a kind of broadcaster's mecca. The BBC moved here from its original quarters on the banks of the Thames in 1932. Though Broadcasting House tripled the corporation's previous space, the new building proved too small for BBC activities even before its completion. The giant BBC television center is located in a London suburb, and BBC radio and television own or lease many other buildings around the city. The British sometimes refer to the BBC as "Auntie" or the "Beeb."

Source: Photo courtesy BBC.

license fees, income from which is dedicated to support of broadcasting

- program services that can be received by and that hold interest for everyone, including rural dwellers and minorities
- strict impartiality in political broadcasts
- respect for the artistic integrity of program makers

This public-service broadcasting philosophy, adapted to varying national circumstances, soon spread worldwide. Thousands of foreign broadcasters have visited BBC's famous Broadcasting House in London (see Exhibit 14.a). Many take BBC training courses in all aspects of production, engineering, and management. Other national services modeled to some extent on BBC traditions include Australia's ABC, Canada's CBC, and Japan's NHK.

Authoritarianism

Traditionally, dictatorships of whatever political flavor take an authoritarian approach to electronic media. The state itself often finances and operates

media systems, along with other telecommunication services. Indeed, governments once owned and operated most broadcasting systems, although the number of privately owned services began to increase during the 1980s and at an accelerating pace in the 1990s.

Authoritarian approaches generally ignore popular taste, as programs have a propaganda or educational goal. Broadcasters pay little attention to marketing techniques that permissive and even paternalistic broadcasters employ to ensure attractive and cost-effective programs. Indeed, authoritarians look on broadcasting as a one-way medium—all give and no take. Either they do not understand or they actively suppress its potentially *democratic nature*—a medium that can offer access to many different ideas and that depends on freewill listening and viewing by audiences.

Third World Authoritarianism

The lack of purchasing power limits ownership of even simple radio sets in many Third World countries. The absence of rural electrical power as well as of relay facilities for networking further restricts growth—especially for line-of-sight television service. Neither receiver license fees nor advertising (alone or in combination) can bring in enough revenue to support broadcasting adequately. Therefore, most Third World governments own and operate their broadcasting systems. Dictators heading often-shaky Third World regimes dare not allow broadcasters to incite the illiterate masses. Prudence and ideology dictate tight control. (Note that in political upheavals, government-operated broadcast stations, usually in the capital city, are the initial targets of rebel takeovers.)

14.2 PLURALISTIC TREND

As we enter the new millennium, none of these three controlling philosophies—permissive, paternalistic, or authoritarian—exists anywhere in pure form. Compromise leads to combinations of features to create a fourth approach: *pluralism*.

Role of Competition

More than 80 years' worth of experience has proved that pluralistic electronic media systems work best in most national circumstances. Pluralism means more than simple competition among rival services. Given that similar motives drive all services, media would merely imitate one another in the absence of pluralistic competition.

Pluralism means putting a variety of motives to work, each with approximately equal status—usually a mix of commercial and public-service motives. Healthy competition between differently motivated broadcasting organizations can stimulate creativity, encourage innovation, and ensure variety. The result is usually a wider range of genuine program choices than any single-motive system could produce.

British Pluralism in Transition

The British eventually developed a widely admired pluralistic system based on control by two non-

commercial public authorities: the BBC and the Independent Broadcasting Authority (IBA), the latter set up in 1955. A nonprofit, government-chartered corporation, the IBA selected and supervised regional commercial television companies known collectively as Independent Television (ITV), as well as specialized national television networks and, eventually, local radio stations. Like the BBC, the IBA owned and operated its own transmitters, so commercial motives could not distort geographical coverage by outlets being placed only in major cities. Unlike the BBC, however, privately owned commercial companies provided programming and sold advertising time.

The BBC's national and regional radio services initially had little competition. Its networks feature pop music (Radio 1), middle-of-the-road programs (Radio 2), serious music and talk (Radio 3), and news/current affairs (Radio 4). In 1991 it added Radio 5 to bring instructional/educational programs to schools as well as to provide programming such as African music to underserved audiences. Today more than 40 BBC local radio stations provide regional programming of interest to audiences in England, Scotland, Wales, and Northern Ireland, while competing with an increasing number of local commercial radio stations.

This "comfortable duopoly" operated for 35 years, ending in 1990. A new broadcasting law, reflecting something of an American deregulatory approach, included the following features:

- The BBC continues as the "cornerstone of public-service broadcasting," deriving revenue from receiver license fees. The BBC's royal charter was extended in 1996, maintaining the BBC's special status in the U.K. until the end of 2006.
- The commercial network became Channel 3 at the beginning of 1993, with the BBC's two networks renamed Channels 1 and 2. Channel 4 (begun in 1982 as an "electronic publisher" to carry programs produced by others) continued and sold its own commercial time. The Channel 3 companies are responsible for Channel 4's financial health.
- A new commercial Channel 5, able to reach about 70 percent of the population, was authorized in 1994 with a 10-year franchise and program requirements similar to those for Channel 3. Regional franchises for Channel 3 and the new Channel 5 were auctioned to the highest

bidders, subject only to their ability to meet program-quality requirements. Channel 5 began broadcasting in 1997.

- The IBA and a separate Cable Authority merged into a new "light touch" regulatory authority called the Independent Television Commission (ITC). A separate Broadcasting Standards Council was created to develop a program code controlling sex and violence. And a new Radio Authority auctioned 200 to 300 new local commercial radio station franchises.

Britain's 1990 law reflected changes taking place to varying degrees worldwide: increases in competition, more local (in addition to national) services, greater private ownership, and deregulation.

14.3 DEREGULATION

Starting around 1980 U.S.-inspired deregulatory practices swept the telecommunications world. Although American influence alone did not create this trend, foreign experts studied American deregulation intensively, while U.S. government and media officials promoted both deregulation and private ownership abroad. Deregulation appealed overseas as it did here for another reason—no regulator could keep up with the fast-changing technology.

The Clinton administration negotiated aggressively in international discussions to further the deregulatory trend, seeking to stimulate global competition, avoid anticompetitive conduct, and open international markets to foreign participation. Talks began in 1995 and were concluded in 1997 with several international trade agreements that make it easier for companies from one country to participate in the telecommunications markets of other countries. These agreements are all subject to ratification by the individual nations involved, and not all countries agree with these goals. Similar agreements have been reached among the members of the European Union (EU); on January 1, 1998, all telecommunications services and infrastructure in the EU member nations were opened to full competition. Worldwide, we have come a long way from the monopolistic, nationally controlled media of the past.

Impact on Public-Service Broadcasting

With deregulation, traditional public-service operators faced new competition. Cable television, satellite-distributed programs for both cable and direct-to-home viewing, videocassettes, new commercial broadcast channels, and Internet sites now fight for audiences, advertisers, and programs. Some critics argue that excessive competition may force public-service broadcasters to lower program standards as they struggle to maintain viable audience shares. Others push for private ownership of government broadcasters.

Pressures on the Canadian Broadcasting Corporation (CBC) illustrate the problem. Created in the 1930s to overcome American radio's dominance among Canadian listeners, the CBC operates French and English services on both radio and television, the latter in part commercially supported. The French services are very popular, but many English-speaking Canadians tune in to American programs, either directly or as rebroadcast on commercial Canadian networks competing with the CBC. In the mid-1990s the CBC's huge federal subsidy (some $800 million, nearly three times what Congress gives to the Corporation for Public Broadcasting) came under concerted attack as unneeded in an era of multiple channels, commercial diversity, and government deficit. An early 1995 subsidy cut of 25 percent brought the resignation of the CBC's president and questions of how CBC services could be maintained (parallel again to similar events in American public broadcasting). The pressure to boost revenues and maintain its audience share continues to intensify as the service adjusts to competition from cable television and DBS.

Privatization

Converting state-owned facilities to private ownership (or creating new private outlets), known as *privatization*, rapidly took hold in France and Italy. Other European countries, as well as Australia, Canada, Japan, and New Zealand, plus some Latin American nations, have taken similar, though usually less drastic, deregulatory steps. Two European examples illustrate the process.

Government once operated all of France's broadcasting stations and networks. Private French stations and networks were first authorized in

1984. Two years later conservatives won control and took the unprecedented step of selling off France's leading public-service television broadcasting network, TF1, and of authorizing competition from new privately owned television networks. By 1994 TF1 controlled 40 percent of the audience and 55 percent of television advertising. A French educational over-the-air channel began daytime operation in 1994 on a fifth television channel, already used in the evening by Arte, a Franco-German cultural network.

Privatization came earlier in Italy and by a different route. In the 1970s the official Italian broadcasting monopoly, RAI (Radio Televisione Italia—the initials are traditional and date back to radio-only days) went to court to suppress unauthorized cable operations. The Italian constitutional court ruled in 1975 that RAI's legal monopoly covered only *national network* broadcasting, not local cable or local broadcast operations controlled by private owners. This decision opened the floodgates to thousands of private stations. At the end of the century these stations were still operating without benefit of formal regulation because the Italian parliament, though always seemingly about to pass a new law, could not agree on the form it should take.

Former Communist States

In the Soviet Union, and to an even greater degree in its Eastern European client states, democratization brought dramatic change. In 1985 new Soviet leader Mikhail Gorbachev inaugurated a policy of *glasnost*—roughly translated as "candor" or "openness." In addition, electronic media spillover signals from neighboring countries, insistent penetration by the Voice of America and other external services, satellite-borne programs from Western Europe, and videocassettes all helped push communist regimes toward greater openness.

Each of these media played an active part in the revolutions of 1989–1991. Eastern European regimes in 1989 toppled domino-like in the wake of Gorbachev's reforms. Nothing in history matches the startling suddenness with which four decades of rigid communist control, exemplified by the 30-year-old Berlin Wall, crumbled almost overnight in country after country—culminating with the collapse of the Soviet Union itself in late 1991.

Russian broadcasters soon scheduled more live shows, spontaneous interviews (in place of the obviously rehearsed recitals formerly used), and telephone call-in programs. They extended schedules into early-morning and late-night dayparts. For the first time, broadcasting began to please and inform Russians instead of lecturing them.

The first privately owned and commercially sponsored television station appeared in 1990, to be joined by two others in 1993. By the end of the decade all three were working toward developing national coverage. Their competition for—and success in gaining—audiences encouraged more advertising revenue and greater use of investigative reporting and often critical analyses (of the Russian invasion of Chechnya, for example) with vivid pictures contributing to sharp political debate. Resulting pressure to change the corrupt ways at one of the three state-run television stations led to the widely reported 1995 killing of a highly popular on-air personality who had just been named manager. Such setbacks underlined the growing pluralism in a system that for years had harbored no dissent. Further, the country's financial crisis in the late 1990s forced major cutbacks in television operations, thus wiping out many of the gains made earlier in the decade.

By the mid-1990s private broadcasting and cable was making headway in several formerly communist nations. TV NOVA took over a government station in Prague, providing 19 hours of daily programming to Czech listeners. Supported by advertising, the station expected to break even in about five years. In Poland, a quarter of the nation's viewers could receive Western-style satellite-delivered cable services, including a new one based in Warsaw. Home Box Office marketed itself actively in Hungary. Developing fiber-optic cable networks promised more broadband capacity. On the other hand, some old habits die hard. In several countries, including Hungary and Slovakia, new legislation placed strong controls on media content, including limits on antigovernment news or statements.

International Telecommunication Union

As with changes in individual countries, many international controls were revised during the 1980s. Most nations belong to the International Telecommunication Union (ITU), a United Nations affiliate headquartered in Geneva, Switzerland. As sovereign nations, members cannot be forced to obey ITU regulations—though they usually find it in their best interests to accept ITU recommendations and to adopt ITU standardized terms and procedures.

Acting together, ITU member states allocate frequency bands to specific services to avoid interference. Throughout the world, AM and FM radio, television, and broadcast satellite services have similar spectrum allocations. (The ITU also issues initial letters for station-identification call signs—hence the *K* or *W* used by American stations, whereas Canadian station calls begin with *C* and Mexican outlets with *X*.)

In the 1980s the ITU expanded its aid to help developing nations improve their telecommunication facilities. At the same time, with many newly independent Third World members, ITU conferences became more politicized. Third World countries have argued that ITU spectrum and orbital allocations should be planned for the long term. In other words, they want channels and slots *reserved* for their future use, even though decades might pass before most developing nations could activate them. The United States and most other heavy users of satellite communications, however, need immediate access to these resources to meet current needs and to foster technological development. Clashes on this issue occur at each ITU meeting.

In the early 1990s the ITU underwent its most dramatic reorganization in a half-century—evidence that its dozens of Third World members were having more impact on ITU decisions and processes alike. The ITU elevated assistance to Third World countries to equal status with its more traditional technical standard setting and frequency-allocation roles.

Competing Technical Standards
• • • • • • • • • • • • • • • • • • • •

Thanks to ITU uniform spectrum allocations and standards, taking a portable radio to listen abroad works in most places. But don't try traveling internationally with a television set. For although television has similar channel allocations throughout the world, its technical standards are not uniform. No fewer than 14 monochrome and 3 principal color technical standards require converters to interchange programs (as in Olympic or World Cup sports coverage) or to use one country's television equipment elsewhere.

Today's three basic analog color television systems—NTSC, PAL, and SECAM—reflect American, German, and French government promotion in the 1960s. Each country sought to persuade other governments to adopt its standard. Adoption meant not only national prestige but also huge revenues from international sales by manufacturers of equipment using the favored system. PAL is most widely used; the American NTSC system is nearly universal in North and South America and Japan; France, the former USSR, and their close allies or former colonies have adopted SECAM.

A second start on uniform world television standards could result from the ongoing ITU study of direct broadcast satellite (DBS) and high-definition television (HDTV) system technologies. Certainly, satellite-to-cable and DBS services designed to cover many countries simultaneously would benefit from such standardization.

By 1990 Europe, Japan, and the United States each had different and incompatible analog HDTV standards on the drawing board. The American breakthrough in digital HDTV (see Section 5.7, p. 146) led early in 1993 to the end of Europe's attempt to develop its own analog standard. Consequently, a consortium of European countries began developing a blanket video delivery standard that will apply to cable, satellite, and terrestrial distribution systems. In Japan, an analog version of satellite-delivered HDTV has been in service since 1991 and serves approximately 330,000 households. In 1997 the Ministry of Posts and Telecommunications announced that it will promote a new digital system that will closely resemble the U.S. standard. Additionally, the governments of Canada, South Korea, Taiwan, and Argentina have adopted the U.S. digital television standard.

PTT Deregulation
• • • • • • • • • • • • •

Deregulation has also had a widespread impact on common-carrier services, loosening control by highly centralized national post, telephone, and telegraph (PTT) monopolies, and spurring competition among new telecommunications service providers. In addition to their telephone operations, some PTTs have long held exclusive rights to install and operate broadcast transmitters and relays, as well as cable facilities. However, in many countries, transmission and programming have been regulated by separate government authorities.

For cable television, marketing and promotion play key roles: Consumers need to be convinced that this new service will benefit them. Lack of experience in facing competition generally means that monopolistic PTTs have little marketing expertise. Progress in cable installations has often languished under PTT leadership. Rising public and advertiser demand for cable and other services has helped to break down PTT dominance in many countries since the mid-1980s.

Privatization of the telephone and cable networks in Central and Eastern Europe led to some interesting business collaborations that would not have been thought possible a decade earlier. For example, Matav, the Hungarian telephone company, sold a 30 percent stake in its operations—worth $875 million—to a consortium of Ameritech International and Deutsche Telekom. The company hoped to increase the number of telephone lines in the country at least 15 percent per year for six years. The Slovak Republic and Romania also are in the process of privatizing their telephone networks. In the Czech Republic, a foreign consortium purchased a portion of the telephone system, SPT Telecom, and U.S. West acquired 28.6 percent of the largest Czech cable television company, Kabel Plus. U.S. West and Bell Atlantic hold 49 percent of the only Czech mobile telephone company, Eurotel. Similar arrangements exist throughout Latin America as telecommunications giants such as AT&T and Spain's Telefonica SA work with national service providers to build new broadband networks.

Privatization and foreign investment in telecommunications have been slower in other parts of the world. In China, for example, regulations stipulated that no foreign investment was allowed in the operation of telecommunications networks and services, although the country rarely enforced its rules. China changed its official policy in 1999, as a required component of its acceptance into the World Trade Organization. Under the new policy, China will allow foreign investors to hold up to 49 percent of certain telecommunications companies, including Internet firms.

The FCC, in its 2000 report on international telecommunications markets, concluded that liberalization and new competition have been boons to

consumers throughout most of the world. The report notes that rates for international services have declined, new entrants are making inroads in the telecommunications marketplace, and the satellite services industry is growing rapidly and continues to diversify its service offerings. (FCC, 2000)

14.4 ACCESS

That all electronic media can inform, persuade, and cultivate values makes access to them a jealously guarded prerogative. Traditionally, most countries have limited access to professional broadcasters, experts on topics of public interest, people currently in the news, entertainers, and politicians.

Politicians

Democracy requires fairness in the political use of the electronic media—without at the same time crippling the media's role as a means of informing voters. In America, weak parties and candidates have the same theoretical access rights as major parties—provided they can afford to buy advertising time. No other industrialized democracy permits such broad access for candidates—and such commercialization of elections. In most European countries, for example, despite strict fairness regulations on paper, ruling political parties often evade rules by controlling appointments to state broadcasting services and regulatory agencies.

Britain severely limits political broadcasts, while providing them free of charge. Campaign broadcasts focus on parties rather than on individual candidates (in keeping with the parliamentary system in which party membership is more important than in American elections). Given that British national campaigns last only 30 days, voters are spared the endless merchandising of candidates. Nor do candidates have to beg for donations and accept money from lobbyists to pay for expensive broadcast advertising. Britain is also one of the few countries that allow live televising of Parliament's debates—similar to C-SPAN's coverage of Congress here. Such a ready means of watching can bring the governing process closer to voters.

Citizens
• • • • •

During the 1960s people in many countries sought airway access. They argued that if spectrum is really a shared natural resource, then everybody should get a chance to use it. This access movement paralleled a widespread rise in ethnic and regional awareness.

The access movement at first had little impact on the centralized systems of Europe. Access seekers petitioned authorities to create regional and community stations. Many did. France, for example, legitimized more than a thousand small, privately owned FM radio stations following passage of a new broadcasting law in 1982. Many had started as pirate stations, which the French government had suppressed rigorously. The 1990 British broadcasting law placed only limited regulation on some 300 new local radio stations, as well as on new commercial networks first permitted under the new law. In Scandinavia, governments finance low-power FM *när radio* (neighborhood stations), inviting local groups to help program them, free of virtually all regulation.

Groups
• • • • •

Another way of dealing with access demands is to shift the emphasis from individuals to the groups to which those individuals belong. The now-defunct FCC fairness doctrine applied such a strategy (see Section 13.8, p. 395).

The uniquely structured access system of the Netherlands has gone farthest in assuring different social groups their own programs on national broadcast facilities. Most program time on government-run networks is turned over to citizen broadcasting associations. Some represent religious faiths; some have a nonsectarian outlook. Even very small constituencies, such as immigrant workers or people from Dutch colonies, can regularly obtain airtime. An umbrella organization, Netherlands Broadcasting Foundation (NOS), coordinates Dutch time-sharing and produces programs of broad interest, such as national and international news and major sporting events. Another organization handles all sales of advertising time, revenue from which goes to a central program fund.

Access can also work in a negative fashion. The once white-controlled *apartheid* government of South Africa for years limited black access to outside sources of news. Radio news directed at tribal groups in their own languages was channeled onto FM radio. As FM and shortwave tuning were not ordinarily available on the same receiver, the use of FM for controlled domestic news limited black listeners' access to external news sources.

The growth of the Internet worldwide provides a communications link for many people with no hope of access to traditional media. Although computer penetration in most countries is very low (see Exhibit 14.e on p. 428), people with a will often find a way. Users in Haiti often must use solar power, batteries, and generators to power their computers because there is no electricity most of the time. In the town of Goma in the Democratic Republic of Congo (formerly Zaire), people use high-frequency radio to send e-mail to an ISP based in Uganda because Congo has no ISPs of its own. In many countries, residents seeking access to the Internet must deal not only with a lack of infrastructure but also with a national government hostile to the kind of open communications the Internet provides. Some Middle Eastern countries, for example, suppress the use of the Internet because of its potential influence on local culture.

• • • • • • • • • • • • • • • •
14.5   ECONOMICS AND GEOGRAPHY

Economics is second only to politics in shaping a country's electronic media. National systems vary widely in audience size, facilities, revenue sources, and the ability to produce homegrown programs. Economic constraints account for these differences, though geography also plays an important role.

Audience Size
• • • • • • • • •

More than 200 countries and dependencies have their own radio broadcasting systems. Fewer than 40 countries (mostly small islands) lacked television stations by the end of the century. Television set penetration per household varies from nearly 100 percent in countries like the United States, the

United Kingdom, and Switzerland, to less than 10 percent in parts of Asia and Africa. High penetration levels depend only partly on economics; important as well are programs popular enough to motivate set purchase and a policy of licensing local stations. But set penetration does not tell the whole story of viewing and listening in other countries. Cultural differences can influence the use of electronic media even when set penetration levels are comparable. Television viewing time per household in Asia, for example, is much less than in the United States.

Former communist governments invested heavily in transmitters, relays, and production facilities, yet set penetration remains disproportionately low. The People's Republic of China did not begin television broadcasts until 1958, and set penetration lagged far behind that of most other Asian countries for many years. Recent reforms and the growth of a consumer culture have spurred the growth of television in China.

In tropical Third World countries, radio sets do not work because of humidity and a shortage of batteries. Government investment in transmitters and production facilities can therefore be a waste of money. It costs just as much in program and transmission expense to reach a few scattered individuals as to reach everyone within a transmitter's coverage range. Lack of communications infrastructure—electric power, telephones, and relay facilities for networks—further impedes Third World electronic media development. However, a few of the oil-rich Middle East states have achieved relatively high radio set penetration—for example, 69 per 100 people in Kuwait and 54 per 100 in Bahrain, compared with an average of fewer than 20 in most African nations. By contrast, the average person in the United States owns two or more radio sets.

Revenue Sources

Electronic media financial support comes from three main sources: government appropriations, receiver license fees, and advertising. Most countries' electronic media still depend at least partially on government funds.

In industrialized democracies, substantial support depends on user license fees—an annual payment per household or receiver (about $150 a year in most European countries by the late-1990s).

This somewhat insulates broadcasting organizations from inevitable biases caused by dependence on either direct government funding or advertising. But the protection is only partial, because governments create fee payment rules and set fee levels. As operational costs rise faster than audience fees increase, many formerly fee-supported systems turn to advertising, though theoretically advertising contributes too small a part to give advertisers any real influence over programs.

Program Economics

Television consumes expensive programs at such a rate that even developed nations with strong economies cannot afford to program several different television networks exclusively with home-grown productions. Britain is unusual in having five full-scale broadcast television networks, but even it imports some entertainment programs, though with a voluntary limit of 14 percent of total airtime for such imports.

In smaller countries, the lack of programs from domestic sources stimulated cable's growth even before satellite-to-cable program networks came into being. Community antennas in small countries can often pick up a half-dozen services from neighboring countries.

Most imported programs come from the United States (as discussed in Section 14.6), though many come from Britain. A few Third World countries (especially Brazil, Mexico, and India) are active in supplying international program markets.

International syndication and various forms of cost and talent sharing have helped increase the amount of programming available. For example, the European Broadcasting Union (EBU), an association of official broadcasting services in Europe and nearby countries, shares programs through Eurovision, which arranges regular exchanges (mainly sports and news) among its members. Similar associations exist in Asia, the Middle East, and the Caribbean, although they are not as active as the EBU in handling program exchanges.

Co-production is increasingly used to cover high program costs. Producers from two or more countries combine financial and other resources to co-produce television series or movies. They divide the capital expenses and profits from the distribution of programming in participating countries.

Coverage Problems

Cost-effective media coverage of a country depends on both its shape and size. Alaska and Hawaii, for example, had to await satellite transmission to enjoy same-time coverage with the U.S. mainland.

Indonesia's 6,000 or so widely scattered islands with diverse populations speaking many different languages present even more formidable coverage problems. And Russian territory extends so far east and west that national broadcast schedules have to be adapted to serve 10 different time zones (contrasted with only 4 in the continental United States). Program distribution needs prompted Russia and Indonesia to become early users of domestic satellites.

Signal Spillover

Geography (and a lack of shortwave radios) insulates most American listeners and viewers from programs spilling over from foreign countries. But spillover from the United States has strongly influenced not only broadcasting but also satellite and cable development in neighboring Canada. Most Canadians live within 100 miles of the American border and can easily receive American broadcast signals off-air and/or by cable and DBS (and vice versa, of course). As a result, Canada became the first heavily cabled country in the world.

To avoid being overwhelmed by American programs—and to ensure work for its own creative community—Canada imposes quotas limiting the amount of syndicated programming that the electronic media may import. To help fill the gap, Canada's government subsidizes Canadian production, leading to a vibrant Canadian popular music industry, for example. In addition, some American films and TV programs have been co-produced in Canada to lower their cost.

14.6 PROGRAMS

Interchangeability and sharing of programming simply reflects the fact that similar program formats exist throughout the world. News, commentary, public affairs, music, drama, variety, studio games, sports events—such program types appeal everywhere. National differences are evident only in the ways these genres are treated.

News and Public Affairs

Prime-time daily newscasts are universally popular, but their content and style differ among countries. Parochialism, chauvinism, and ideology affect the choice, treatment, and timing of news stories. Each country emphasizes its own national events, few of which hold interest elsewhere.

In authoritarian countries such as Iraq, Cuba, and North Korea, news focuses heavily on the national leader to the virtual exclusion of all else. Some broadcasting systems devote their news gathering facilities entirely to reporting on the head of state's every public move and developing a cult of personality.

Program Balance

Audiences everywhere prefer entertaining comedy and drama to more serious informational content. Accordingly, where audience appeal controls programs—as through audience ratings—entertainment dominates. Most industrialized democracies (including the United States, thanks to public broadcasting and news/culture-oriented cable channels) try to strike a balance among light entertainment, news, information, culture, and educational programs.

Those members of a society who buy most of the consumer products and services and enjoy access to political and social groups see and hear themselves through the electronic media far more than do the poor and powerless. Accordingly, the electronic media depict a largely urban, fairly well-educated, affluent people, whether fictional or real.

Third World nations can afford few local productions with popular appeal. Therefore, they import foreign entertainment that attracts audiences for a fraction of what it would cost to produce programs locally. But imports throw schedules out of balance (as discussed in the next section) by overemphasizing entertainment and playing up foreign cultures that are often quite different from the viewer's surroundings.

Schedules
• • • • • • •

Though common in America, broadcast days of 18 to 24 hours were rare even in other developed countries until quite recently. Traditionally, most foreign radio services would broadcast a short morning segment, take a break before a midday segment, then another break before evening programs, closing down for the night relatively early. Even in Britain, the BBC began 24-hour radio only in 1979, when Radio 2 filled in the previously silent hours of 2:00 to 5:00 A.M. Television in much of the world still begins only late in the afternoon, signing off by about 11:00 P.M. Extending programming into early-morning and late-night hours in the 1980s was one sign that new networks, stations, and cable systems were meaningful competition for long-time broadcasters.

International Syndication
• • • • • • • • • • • • • • • • •

Cultural differences often impact programming. One major difference between American and foreign television (especially in Europe) is the tolerance of nudity in programs and advertising. Full frontal nudes are not uncommon on some European television channels. What is acceptable in one country often is not in another, creating problems for program exchange or syndication. American programs, for example, are often criticized abroad for being excessively violent.

U.S. programs, long dominant in world syndication, came to new prominence with added demand from new satellite-distributed cable networks, direct broadcast satellite channels, and VCRs (see Sections 14.8 and 14.9). These services vacuum up programs from whatever sources they can find. Their enhanced demand has intensified old fears of American cultural domination. The low price for American programs (because their original cost has already been repaid through domestic sales) and their almost sure-fire mass appeal keep them in demand. Exhibit 14.b shows typical prices paid overseas for syndicated American programs.

In the 1990s American international syndicators' sales slowed down. In response, many American firms invested in overseas media, though such ventures may encounter legal and cultural opposition to American cost-efficient operating methods. Nor could investors count on quick or easy profits from newly introduced, advertiser-supported, privately owned media services. The rundown economies of former communist countries, for example, require cash for the most basic economic necessities, leaving audiences with little ability to purchase media-advertised goods. Despite that pressure, those countries also adopted local program requirements—the Czech Republic, for example, required 60 percent indigenous programming on a new commercial television station in Prague.

Cultural Imperialism
• • • • • • • • • • • • • • •

American dominance of international syndication has long fed Third World concern about cultural imperialism. Critics argue that images and values shown in imported television programs undermine local cultures. For example, American shows are criticized for encouraging excessive consumption, materialism, and disregard for other traditions. Further, every imported program denies an opportunity for locals to showcase their own talents. Cheaper imported programs can thus perpetuate dependence on foreigners.

Nor do such complaints come only from the Third World. Even developed nations with extensive production of their own limit the amount of entertainment their national systems may import. In 1992 the 12-nation European Community (representing a market larger than America) imposed program-import restrictions despite strong American objections. A year later France led Europe to protect its media market from American product often sold at bargain-basement prices (about a tenth of the cost to produce the same thing locally). There was interest—but no specific requirement—in increasing the amount of time given to French music on French stations.

American producers understandably strongly resist attempts by other countries to limit program imports. At the newspaper and broadcast industries' behest, U.S. diplomats successfully advocated a *free-flow* international communication policy when the United Nations was created in 1945. But more than 70 new nations, most extremely conscious of their histories as colonies of developed countries, have since joined the UN. They ask what value free flow has for them when it runs almost entirely in one direction—*from* the United States and other industrialized countries *to* the

EXHIBIT 14.b
. . . .

Program Bargains from U.S. Syndicators

Prices for U.S. syndicated material vary widely according to potential audience and ability to pay. Prices for feature films also fluctuate widely based on the quality of the movie. The figures here represent the average price, or range of prices, paid by mainstream broadcasters in various countries. Figures for dramas, documentaries, and music/arts programs are per hour; prices for sitcoms and children's programs are per half-hour.

Cost per Program (U.S. Dollars)

Purchasing Country	Feature Films	TV Movies	Dramas	Sitcoms	Documen- taries	Children's Programs	Music/Arts Programs
Argentina	$25,000	$6,000–12,000	$3,000	$1,000	$4,000	$1,000	$1,500
Australia	700,000	65,000	40,000	20,000	12,000	—	11,000
Austria	20,000	20,000	12,000	6,000	7,000	5,000	10,000
Belgium	18,000– 50,000	8,000	5,000	3,000	3,000	250	2,000
Brazil	7,000	4,000	2,000	1,000	1,500	1,500	1,000
Canada	50,000– 200,000	100,000	50,000	35,000	7,500	6,000	15,000
Czech Republic	5,000– 25,000	3,000	2,500	1,200	1,000	700	1,000
France	30,000– 2 million	90,000	55,000	30,000	13,000	12,000	—
Germany	200,000– 3.5 million	300,000	200,000	60,000	20,000	12,000	20,000
Greece	10,000– 60,000	5,000	5,000	2,000	—	2,000	2,000
Hong Kong	5,000– 100,000	2,500	1,000	450	2,500	400	1,500
Hungary	20,000	3,000	6,000	2,500	1,000	700	600
India	3,000– 5,000	4,000	2,000	1,200	—	2,500	500
Italy	250,000	75,000	10,000	5,000	6,000	6,000	—
Japan	80,000– 1.5 million	—	20,000	10,000	6,000	6,000	3,000
Korea	50,000– 200,000	15,000	4,000	1,500	1,500	1,000	2,000–2,500
Mexico	15,000– 40,000	20,000	9,000	4,000	2,000	1,500–2,000	—
New Zealand	15,000– 100,000	18,000	5,000	2,500	2,500	500	2,500
Philippines	20,000	3,000	6,000	2,500	1,000	700	600
Russia	7,000– 25,000	3,500	2,500	1,200	1,000	700	1,000
South Africa	15,000– 20,000	15,000	6,000	5,000	2,500	2,500	6,000
Spain	100,000– 2 million	55,000	40,000	15,000	7,000	5,000	6,000
United Kingdom	50,000– 2 million	200,000	150,000	80,000	25,000	30,000	10,000

Source: Based on data in "Global TV Price Guide," *Variety*, September 27, 1999: M18. Reprinted with permission of Variety, Inc. © 1997.

Third World. Instead, they call for *balanced* flow—defined as news and entertainment that treats Third World countries fairly and in proportion to their population significance. The debate between these polar positions continues in many international organizations today.

On the other hand, such American cultural creations as MTV reached more than 300 million households in 83 countries at the end of the 1990s. MTV International now operates several regional and country-specific services: MTV Europe, MTV Asia, MTV Latin America, MTV India, MTV Brazil, MTV Mandarin, MTV Australia, and MTV Japan. These channels, produced in Singapore, Taipei, and London, feature national talent and playlists designed to appeal to each channel's target market. In 1998 MTV announced that it would inaugurate a 24-hour music channel in Russia.

NBC Super Channel, also based in London and transmitting to much of Europe, offers a selection of the network's American domestic product to new audiences—apparently with considerable success. With similar aims, ABC has purchased part interest in several European production firms.

UNESCO's Role

Much of this debate came to a head in the 1970s and 1980s at the Paris-based UN Educational, Scientific, and Cultural Organization (UNESCO), which led in defining and calling for the establishment of a *new world information and communication order* (NWICO). NWICO called for, among other things, balanced media flow, training (and perhaps licensing) of journalists to operate in other cultures and languages, and support from rich nations to assist the development of poor nations' media infrastructure. American media leaders saw this development as an attack on journalistic independence and free-enterprise advertising. Claiming that UNESCO wasted money (a quarter of its budget came from the United States) and had become hopelessly politicized, the U.S. government withdrew as a member in 1984, leaving an agency that it had helped establish four decades earlier. By the mid-1990s America said it would rejoin UNESCO, as much of the agency's political and budget turmoil had been resolved. Ironically, that reentry has been delayed further—initially by tight budget conditions in Washington and subsequently by partisan politics.

14.7 TRANSBORDER BROADCASTING

The ability to use radio to penetrate political boundaries added a potent element to diplomatic relations. Never before had nations been able to talk directly to masses of foreigners, crossing even the most heavily defended national borders. By the 1990s more than 80 countries were operating official external services—programs aimed at foreign countries.

BBC World Service

Colonial commitments first prompted nations to broadcast radio internationally. The Dutch and Germans started in 1929, the French in 1931. After experimenting for several years, Britain's BBC began what was first called the Empire Service in 1932. It initially broadcast only in English, primarily seeking to maintain home-country ties with expatriates in colonies around the world. During World War II foreign listeners came to regard the BBC as having the most credibility among external broadcasters, and it has retained that reputation ever since. Today approximately 143 million listeners worldwide tune in to what is now called BBC World Service, especially when they doubt reports from other sources. It has more than a million American radio listeners; a few listen via short wave, but many tune in via rebroadcasts on Public Radio International or via the World Service web site. Europeans can also hear it as a direct-to-home satellite service, which is now broadcast in 44 languages.

BBC World Service Television (WSTV) began in 1991 as a cooperative effort of domestic television news and World Service radio. The TV service came under the control of BBC Worldwide, the commercial arm of BBC, in 1995. Currently, World Service provides news programming to the satellite-delivered 24-hour news and information channel BBC World. Video service also is available on the BBC Worldwide web site.

Voice of Russia (Radio Moscow)

Though the Soviet Union initially controlled no overseas colonial empire, it used radio to promote its revolutionary ideas. With Radio Moscow's inception in 1929, the Soviets recognized the importance of foreign-language radio as a means of gaining and influencing friends abroad.

Radio Moscow developed fewer overseas relay stations than major Western external broadcasters, though in the 1960s it built one such station in Cuba aimed at the Americas. Like radio services in all communist states, Radio Moscow was relentlessly propagandistic. It lightened its tone to gain wider appeal in the 1980s, even initiating a 24-hour English-language service. Renamed the Voice of Russia in 1995, the service was forced by budget problems to cut its staff by 30 percent and discontinue broadcasting in 16 foreign languages. Today the service broadcasts to listeners in 160 countries in 32 languages, and the English language service broadcasts to all continents 24 hours per day.

Voice of America

The United States joined the growing battle of words with Voice of America (VOA) early in 1942. Wary of creating a domestic propaganda agency, Congress forbids the broadcast of VOA programs in the United States. However, anyone with a shortwave radio can pick up VOA programs aimed at overseas listeners, and VOA scripts and recordings (as well as those from Radio/TV Martí and Radio Free Europe) can now be obtained on the Internet.

By the end of the century VOA was broadcasting more than 700 hours of programming per week in 52 languages. About 85 percent of the programming is devoted to American events and news. VOA programs originate in Washington, going overseas via leased satellite channels and 114 VOA shortwave transmitters located in Greenville, North Carolina, and several secondary U.S. sites. VOA also leases sites in a dozen foreign countries, where it maintains transmitters for rebroadcasting programs to listeners in nearby areas.

VOA news and public-affairs programs reflect official American policies. News commentaries are explicitly labeled as coming from the U.S. government. For the sake of credibility, VOA tries to observe the spirit of its original 1942 policy: "Daily at this time we shall speak to you about America. . . . The news may be good or bad. We shall tell you the truth." Truth telling continues to be VOA policy, despite occasional lapses when partisan officials bend facts to suit momentary political objectives.

Worldnet

The short range of broadcast television makes it useless for external services aimed at distant targets in the manner of shortwave radio. The United States Information Agency (USIA), VOA's parent organization, used to rely entirely on persuading foreign broadcasters to carry American television programs on their own domestic television services.

In 1983 USIA created Worldnet as a daily television service distributed abroad by satellite. To encourage foreign broadcasters to use its material, Worldnet offers interactive teleconferences with U.S. government officials responding to live questions from foreign news personnel. American interviewees speak from a Washington studio to questioners in other countries. Participation by another country's own broadcasters makes the American presence more acceptable on that country's television service. Additionally, Worldnet televises live call-in and public-affairs programs in Arabic, English, French, Mandarin, Russian, Polish, Serbian, Spanish, and Ukrainian.

Worldnet fills most of its schedule with "passive" programs—news and general information minus the interactive feature. American diplomatic posts throughout the world pick up Worldnet on some 230 TVROs, as do foreign cable systems and broadcast stations in about 200 cities in more than 125 countries. Worldnet is available via INTELSAT (see Section 14.8) to any foreign broadcasting or cable television operation that wishes to use it. Service costs, as well as research showing very small audiences, led to congressional budget cuts for Worldnet in the early 1990s, thus limiting its ability to provide original programming.

Radio Free Europe and Radio Liberty

In addition to conventional external broadcasting, the United States has long provided *surrogate* domestic services. Such broadcasts simulate domestic networks in target countries, bypassing censored domestic media.

Beginning in the early 1950s, Radio Liberty (RL) targeted the Soviet Union, while Radio Free Europe (RFE) aimed at Eastern European states then under Soviet control. CIA support of these services was revealed in the early 1970s, leading to their reorganization under the Board for International

Broadcasting. RFE and RL initially built studios and transmitters in Munich, Germany, and additional transmitters in Portugal and Spain at sites favorable for sky-wave transmission to their target countries. Broadcasting by radio, entirely in the languages of the target countries, RFE and RL provided domestic and foreign news from the listeners' perspective, often with strong messages encouraging resistance to communist thinking and governments. The United States spent more on operating these two surrogate services than on VOA.

The 1989–1991 collapse of communism in Russia and Eastern Europe led to major adjustments in RFE and RL. Both stepped up their activity once listening to them was no longer illegal, and they soon added transmitters and news bureaus right in their target countries—something undreamed of in Cold War days. Driven by budget concerns and Clinton administration proposals, Congress decided to merge RFE/RL into a new broadcasting bureau (including VOA) within USIA. As an indication of dramatic change among their former target countries, in 1994 RFE began relocating its broadcast functions from Munich to the less expensive city of Prague—at the invitation of the Czech government. From its new offices in Prague, RFE/RL broadcasts 800 hours of programming per week in 26 languages to a regular audience of more than 35 million listeners. More than 5 million people visit the RFE/RL web site every month to read news reports and hear RFE/RL programming.

Radio/TV Martí

In May 1985 the United States introduced Radio Martí, a surrogate service aimed at giving Cuba news and information free of Castro-regime bias. Supporters claim that Radio Martí has had a powerful effect, heightening dissatisfaction with the Castro regime by revealing facts it conceals from its own people. The expatriate Cuban community in south Florida, with solid political ties to Washington, strongly supports the operation. Critics suggest that some Cuban expatriots exert too much control over the station, thus damaging its credibility as a news source. When Cuban military jets shot down two civilian airplanes operated by a Cuban exile group in 1996, President Clinton responded in part by ordering Radio Martí to increase its power to provide an even stronger signal in Cuba.

Congress authorized funds in 1988 to start experiments with a television version of Radio Martí, using a transmitter hung from a balloon tethered 10,000 feet above Cudjoe Key in Florida. This height enabled television line-of-sight signals to reach Havana 110 miles away. Tests conducted in 1990 incited Cuba to *jam* TV Martí transmissions and to interfere with American AM radio stations. The Cuban government appealed to the ITU, claiming that TV Martí violated ITU allocations that designate television channels for domestic use only. Many American broadcasters also objected to the scheme, which was characterized in one *Broadcasting* editorial as a "huge, disastrous silliness" (9 April 1990: 98). By the end of the decade, due to technical problems with the transmitter and signal jamming by the Cuban government, it was unclear whether anyone in Cuba had ever viewed a complete program from TV Martí.

Radio Free Asia

The idea of bringing news reports to places otherwise unlikely to hear them also took hold on the other side of the world. In 1993 Congress authorized a Radio Free Asia (RFA) service aimed at potential audiences in the People's Republic of China and other Asian countries. Radio Free Asia began broadcasting in 1996 with an hourlong newscast in Mandarin. By 2000 RFA was broadcasting news and cultural programming in nine languages to China, Tibet, Burma (Myanmar), Laos, Cambodia, Vietnam, and Korea. Several of these countries, most notably China and Vietnam, have strongly criticized RFA operations.

Religious Broadcasters

A different ideological motivation led to transborder religious broadcasting. International radio gave evangelicals their first opportunity to deliver messages directly to potential converts in closed societies dominated by state religions. Official hostility—in some Moslem countries toward Evangelical Christians, for example—can prevent the establishment of on-site missions but cannot easily bar radio messages. Well-funded conservative religious broadcasters have so saturated shortwave bands that listeners can pick them up almost anywhere in the world, 24 hours a day.

Peripherals

Commercial motives account for transborder broadcasters known as *peripherals*. Several European ministates bordering large countries have long operated such stations, capitalizing on the unfulfilled demand for popular music and broadcast advertising. Both audiences and advertisers, frustrated by severely regulated domestic services, have welcomed these alternatives.

Peripherals beam commercial radio services in appropriate languages to neighboring countries. They specialize in popular music formats, sometimes supplemented by objective news programs, which are welcome where ruling political parties dominate broadcast news.

The Grand Duchy of Luxembourg, ideally located for peripheral transmitters at the intersection of Belgium, France, and Germany, receives much of its national income from international commercial television as well as radio broadcasting (see Exhibit 14.c on p. 424). Other notable transborder commercial radio stations operate in the German Saar (Europe No. 1), Monaco (Radio Monte Carlo), Cyprus (Radio Monte Carlo Middle East), Morocco (Radio Mediterranean International), and Gabon (Africa No. 1). Peripherals tend to be rather staid operations that are tolerated by their target countries, some of which even invest in them. They still leave some commercial and program demands unsatisfied, however, creating a vacuum filled by pirate radio outlets.

Pirates

The first offshore *pirate* stations began broadcasting from ships anchored between Denmark and Sweden in 1958. Lacking authorization from their target countries, they were often financed by American interests and copied American pop-music formats, advertising techniques, and promotional gimmicks. They quickly captured large and devoted youthful audiences—their very illegality adding spice to their attractiveness.

Some pirates made a lot of money but at considerable risk. They suffered from storms, raids by rival pirates, and stringent laws that penalized land-based firms for supplying or doing other business with them. The appetite for pop music whetted by pirate stations forced national systems to take no-

tice of formerly ignored musical tastes. The BBC, for example, reorganized its national radio network offerings, adding a pop-music network (Radio 1) imitative of the pirates. Some offshore DJs eventually worked for the BBC and other established broadcasters.

Controversy over pirate broadcasting caused the fall of one Dutch government years ago, and the country reorganized its broadcasting system as a result. It permitted two former pirate organizations to come ashore and develop into leading legitimate broadcasters. As national systems become more pluralistic in program appeal, and as more "virtual stations" stream on the Internet, the rationale for pirate stations has declined; few operate today.

14.8 INTERNATIONAL SATELLITES

Today commercial services increasingly cross borders by using satellite-to-cable networks and DBS facilities.

International Telecommunications Satellite Organization

The United States led the 1964 founding of the International Telecommunications Satellite Organization (INTELSAT). With relay satellites stationed above the Atlantic, Indian, and Pacific oceans, INTELSAT eventually made possible instant worldwide distribution not only of television programs but also of telephone conversations, news agency stories, and business data.

More than 140 countries have part ownership of INTELSAT, but the United States owns the largest block of its shares (about 20 percent), houses its headquarters in Washington, DC, and has provided most of its director generals. Communications Satellite Corporation (Comsat) initially operated INTELSAT's network under contract on behalf of the rest of the consortium until INTELSAT itself took over in 1979. In the early 1990s the former Soviet Union and its onetime client states of Eastern Europe abandoned their Intersputnik rival network and became members of

EXHIBIT 14.c
• • • • • • • • • • • •

Luxembourg: Home of Peripherals

The tiny Grand Duchy of Luxembourg granted a broadcasting monopoly to a hybrid government/private corporation, now known as RTL (Radio-Télé-Luxembourg), back in 1930. A year later it began operating as what the French called a *radio peripherique* (peripheral radio). In those days, when official European radio services tended to be rather highbrow and stuffy, listeners far and wide avidly tuned in to its pop-music programs. Legend has it that Radio Luxembourg, received in Liverpool, England, gave the Beatles their first taste of pop music.

Today RTL has high-power long-wave (2,000 kW), medium-wave (1,200 kW), and shortwave (500 kW) radio transmitters radiating across the borders, carrying programs in Dutch, French, and German. After 60 years of service, RTL ended its English-language radio programming in 1992.

Television's short range made it impossible for the Grand Duchy of Luxembourg to repeat its radio coverage with the newer medium, but it overcame this problem in part by setting up jointly owned broadcasting services in neighboring countries, notably RTL Plus in Germany (both a broadcast and a satellite-to-cable service), TV (French-speaking Belgium's first commercial broadcast channel), and RTL-Veronique (Holland's first privately owned commercial broadcasting channel).

In 1997 RTL's parent company, Compagnie Luxembourgeoise de Télédiffusion (CLT), merged with the Hamburg-based Universum Film AG (UFA), a subsidiary of publishing giant Bertelsmann AG. In addition to interests in German TV and radio, UFA is involved in the production of films and television programming and holds the broadcast rights to many sporting events. CLT-UFA has expanded RTL's television activities into Sweden, Poland, Hungary, the United Kingdom, and Austria. In 1999 CLT-UFA was the largest TV and radio group owner in Europe, owning 22 TV stations and 18 radio stations.

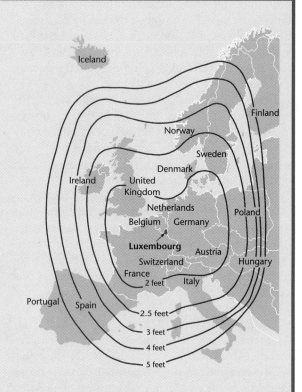

The ITU's 1977 allocation of DBS orbital slots to European countries offered Luxembourg a chance to extend its television coverage to the whole of Europe. SES (Societé Européenne des Satellites), founded in 1985, launched ASTRA—the first privately owned European communication satellite—in 1988. The numbers on the map above indicate the diameters of home TVRO antennas needed to receive ASTRA in the zones defined by the contour lines.

ASTRA contracted with a variety of direct-to-home and satellite-to-cable services to occupy its 16 channels, including Disney and MTV-Europe. It also downlinked the first British DBS service, Sky Television, later known as BSkyB.

EXHIBIT 14.d
. . . .
INTELSAT and "Separate Systems"

The globe shows an example of INTELSAT coverage. INTELSAT 801, located above the Atlantic Ocean, beams both to the east and to the west. Note that such satellites can project beams to the entire region or to a specific zone or spot within a region.

Beginning in the 1970s American satellite firms, strongly backed by the American government, argued that INTELSAT's monopoly was inefficient, that its rates did not reflect actual costs, and that its sheer size made it inflexible. Competition from smaller, nimbler satellite firms, they contended, would lower prices for all, enhance services, and encourage innovation. The first private American "separate system" to launch a satellite, Pan American Satellite Corporation (PANAMSAT), offers both domestic and international satellite services. It links the United States and countries of the Hispanic world—Central and South America and the Caribbean—with Europe and serves Hispanic domestic needs as well.

Peru became the first PANAMSAT customer, followed by the Dominican Republic and Costa Rica on this side of the Atlantic and by Britain, Ireland, Luxembourg, Sweden, and the former West Germany on the other, with more yet to come. CNN and other U.S. satellite-to-cable networks contracted with PANAMSAT to relay their programs to countries to

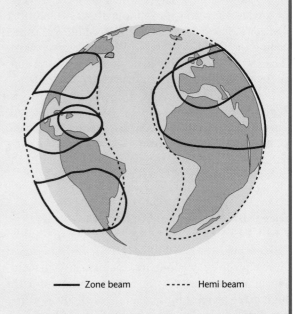

——— Zone beam   - - - - - Hemi beam

the south. In 1997 PANAMSAT announced that it was merging with another international satellite operator, Hughes Galaxy.

Source: Satellite footprints used by permission of INTELSAT.

INTELSAT. Exhibit 14.d offers more details on INTELSAT and its competition.

Though primarily an international carrier, INTELSAT also leases satellite capacity at reasonable rates to Third World countries for their *domestic* use. Thus INTELSAT enables many such nations to vault directly into the satellite era, avoiding costly construction of microwave and coaxial cable circuits throughout their territories.

Satellite Launching
. . . .

The U.S. National Aeronautics and Space Administration (NASA) long monopolized Western capacity to launch communication satellites. In 1984 a consortium of European countries began to challenge NASA's monopoly with its own launch facility, Arianespace (named for *Ariane*, the rocket

used). Ariane rockets operate from a site in French Guiana on the northern coast of South America, near the equator. That location gives Arianespace better conditions for attaining equatorial orbit than does NASA's more northerly location at Cape Canaveral, Florida.

After 1986 NASA launches were confined to government projects. This decision opened the launch market to private U.S. and foreign rocket makers. China and Russia made their government launch facilities available to foreign commercial users—although American national security concerns limited U.S. participation (we did not want to give away expensive technology information to potentially hostile countries). By the end of the 1990s, however, U.S. companies were using the former Soviet facilities in Kazakhstan to launch their communications satellites.

Domsats

The Soviet Union initiated domestic satellite (domsat) development. In 1965 it began launching the *Molniya* satellite series, which enabled relaying of Soviet television throughout its vast territory. Canada's *Anik* domsat series followed in 1972, preceding the first U.S. domsat, Westar, by two years. The 1970s also saw the launching of domestic and regional satellite services elsewhere: Indonesia's *Palapa*, the first Third World satellite; Europe's *Eutelsat*, the first of a regional satellite series; and Japan's first satellite. France, first among the European countries with its own domsat, launched *Telecom* in 1985, in part to enable relays to French overseas territories.

Direct Broadcast Satellite Service

Direct broadcast satellite (DBS) services have shown promise in countries, such as Australia and Japan, where they do not have to compete with high cable penetration. Japan forestalled cable expansion by concentrating on DBS experiments. It led the world with the first full-time DBS service in 1987. By the end of the 1990s three commercial DBS operators were competing for Japanese customers—PerfeTV, Direct TV Japan, and JSkyB, a joint venture among Rupert Murdoch's News Corporation, Sony, and Fuji TV Network. Australia's AUSSAT, though a general-purpose satellite rather than a specialized DBS vehicle, transformed Australian broadcasting in the late 1980s. It brought television for the first time to remote areas too thinly populated to support either terrestrial relay or cable systems.

ASTRA and BSkyB

In 1988 a Luxembourg corporation launched AS-TRA, Europe's first privately owned satellite suitable for DBS service, described in Exhibit 14.c on p. 424. It is used by BSkyB, a DBS service resulting from a 1990 merger of two competing British firms. The first direct-to-home satellite venture in Britain, Rupert Murdoch's Sky Television, began in 1989; a second service launched a satellite in 1990. The two could not sustain the costly competition for home subscribers—a competition costing Murdoch

$3.2 million a week. The merged BSkyB service provides sports, general programming, films, and an all-news channel. By the turn of the century more than 28 million Europeans received some form of satellite-delivered television service.

Asia's StarTV

In 1991 StarTV, more formally known as Satellite Television Asian Region Ltd., began satellite delivery of television programming to all of Asia. In July 1993 Rupert Murdoch's News Corporation purchased nearly two-thirds of the operation for $525 million. Based in Hong Kong, the service offers more than 20 channels featuring news, movies, music, sports, and general entertainment programming. Two of the channels are entirely in the Mandarin language; the others feature programs in English and several indigenous languages. The service is viewed in 53 countries by some 300 million people, who welcome a wider choice of viewing options. Large advertisers support the programming, which is free to viewers. StarTV has a potential audience in the billions as receiving dishes become more widely available.

Radio by Satellite

Americans tend to think of DBS services exclusively in television terms, but Europeans can receive some 60 radio services by satellite. In addition to such major public-service broadcasters as the BBC, a number of colorfully named private radio stations can be received by satellite—Kiss FM, Radio Radio, Fun FM, and Skyrock, for example. Europeans receive satellite radio services directly by means of home TVROs or indirectly via cable systems and FM radio stations.

14.9   CABLE AND NEWER MEDIA

Many developed countries had primitive community antenna television (CATV) for years before modern cable emerged. These early systems merely extended domestic broadcast station coverage, sometimes adding a few channels for the purpose of carrying neighboring foreign broadcast networks. Most operated noncommercially, often owned by municipalities. They offered few channels, no local origination, and no pay television.

Cable
• • • •

Modern cable has developed most extensively in Canada, the United States, and industrialized smaller countries (see Exhibit 14.e). Although cable penetration continues to increase, public demand in many larger countries remains sluggish. If basic-cable systems do not offer a wide variety of popular programs, subscribership remains low. High-appeal cable services delivered by satellite did not begin in Europe until 1982. After that, satellite networks grew rapidly. More than 100 were serving European cable systems by the end of the century. Some targeted only a single country; others were aimed at several nations that shared common languages.

VCRs
• • • •

Both as an alternative and as a supplement to cable and DBS, videocassette recorders (VCRs) have proved a boon to viewers in countries where broadcast services fail to satisfy demand. They offer a relatively inexpensive shortcut to programs banned from, or otherwise not available on, national broadcast or cable channels.

In formerly communist countries, rigid political censorship created appetites that only VCRs could help satisfy. Clandestine VCR tapes hastened the demise of those repressive regimes. VCRs are common in Britain, used primarily to time-shift desired programs to more convenient viewing hours and to view feature films. By the end of the century there were an estimated 102 million VCRs in all of Europe. But penetration varies from a high in the U.K. of about 75 percent of television households to a low in Russia of only 5 percent of TV households (see Exhibit 14.e).

VCRs help compensate for often inadequate domestic television schedules and poor program quality. Few individuals in Third World countries can afford to buy a VCR outright, but rentals, club purchases, and group viewing in bars, coffeehouses, and even on buses resolve cost problems. In some cases heavy censorship encourages VCR growth—as in Saudi Arabia, where conservative Moslem standards severely limit broadcast television. A worldwide underground market in VCRs and tapes defeats most government attempts to limit sales and rentals. It also undermines the profitability of video sales.

Teletext and Videotex
• • • • • • • • • • • • • •

The BBC and the IBA invented teletext, which has proved far more successful overseas than in the United States (see Section 3.3, p. 66). Though not a runaway success, it has nevertheless found greater acceptance in Britain and Europe than in this country, where too many alternative media choices exist.

France leads the world in videotex development with its *Minitel* system. The government telecommunications monopoly promotes Minitel, renting terminals to consumers for a few dollars a month. Telephone subscribers forgo printed telephone directories, instead paying about six cents a minute for Minitel's many services, which include, in addition to telephone directory assistance, more than 25,000 independent electronic services such as transportation schedules, banking, and personal "chat lines." By the year 2000 more than 20 million users accessed Minitel services using 6.5 million dedicated Minitel terminals. More than 600,000 French households now access Minitel services via their PCs.

The Internet
• • • • • • • • •

With millions of users around the world and thousands more logging on every month, the Internet is the fastest-growing electronic medium in the world. The popularity of the Internet and the ability of net users to communicate effortlessly across national boundaries make the medium at once a very liberating and a dangerous technology, depending on one's point of view. Totalitarian governments in particular worry about the uncontrolled access of their citizens to ideas and opinions contrary to the party line. Some countries have even outlawed the private ownership of personal computers in an effort to quell the use of the Internet inside their borders. As

EXHIBIT 14.e
• • • • • • • • • • • •

Telecommunications Around the World

In an age of nearly ubiquitous telecommunications availability in the United States, it is useful to remember that half of the world's adult population has never placed or received a telephone call. The world's richest nations have about 20 percent of the world's population but 91 percent of its Internet users and 74 percent of its telephone users. There are more computers in the United States than in the rest of the world combined. And whereas a personal computer costs less than one month's wages for a typical U.S. citizen, that same computer costs about eight *years* wages for a citizen of Bangladesh.

As this exhibit shows, the penetration of telecommunication services varies greatly around the world. Note that the first four columns measure availability per 1,000 people. The last two columns show the percentage penetration of the technology in television households (TVHH).

Country	Main Telephone Lines (per 1,000 people)	Televisions (per 1,000 people)	Personal Computers (per 1,000 people)	Internet Hosts (per 1,000 people)	Cable Penetration (% of TVHH)	VCR Penetration (% of TVHH)
Algeria	44	68	3.40	—	—	—
Australia	519	666	311.30	42.70	—	79%
Belgium	465	464	167.30	16.00	95%	59%
Brazil	96	289	18.40	1.04	—	11%
Canada	602	709	243.60	53.50	77%	69%
Chile	156	277	45.10	2.07	2%	12%
Guinea	2	8	0.30	—	—	—
India	15	64	1.50	0.01	10%	17%
Israel	441	300	116.30	14.20	—	—
Japan	489	700	167.30	16.00	20%	84%
Kenya	8	19	1.60	0.04	—	—
Mexico	95	193	29.00	0.92	7%	38%
Norway	555	569	284.50	71.80	44%	60%
Russia	175	386	23.70	1.05	95%	5%
Saudi Arabia	106	263	37.20	0.01	—	—
Singapore	513	361	216.80	15.10	2%	84%
Slovenia	333	375	47.80	9.85	—	—
South Africa	100	123	37.70	3.82	—	—
Syria	82	91	1.40	—	—	—
United States	640	806	362.40	88.90	67%	80%
Uruguay	209	305	22.00	5.02	—	17%

Sources: United Nations, *Human Development Report 1999*, Appendix A1.3: "Information Flows"; National Center for Accessible Media, *International Captioning Report 1998*. Photo © Steve Starr/Index Stock Imagery.

with any new medium, the international repercussions of the Internet will not be known entirely for many years. It is definitely a story worth watching.

Conclusion
● ● ● ● ● ● ● ●

And so, as we begin a new millennium, the electronic media in America—now a huge industry of many competing technologies and owners, all sharing the common goal of providing entertainment and news to American homes—continue to thrive and, in many cases, to set benchmarks to which other nations aspire. Even the newest electronic media services owe much to traditional broadcasting because their programming draws directly from genres developed and still provided by radio and television networks and stations. Audiences rarely care about how program content reaches their home screens or radios, and as the information in this book demonstrates, fewer and fewer citizens access many of these services via traditional forms of broadcasting. Although our roots will remain forever in broadcasting, perhaps a more accurate title for future editions of this book might be *Telecommunications in America*.

BIBLIOGRAPHY

AAAA (American Association of Advertising Agencies) and ANA (Association of National Advertising). 2000. *1999 Television Commercial Monitoring Report.* New York, NY.

Abramson, Albert. 1995. *Zworykin: Pioneer of Television.* U. of Illinois Press, Urbana.

Adams, R. C. 1989. *Social Survey Methods for Mass Media Research.* Erlbaum, Hillsdale, NJ.

Advertising Age. Weekly. Crain, Chicago.

Alten, Stanley R. 1994. *Audio in Media,* 4th ed. Wadsworth, Belmont, CA.

American Academy of Pediatrics. 1999. "Policy Statement, Media Education." *Pediatrics,* 104 (August): 341.

Anderson, James. 1987. *Communication Research: Issues and Methods.* McGraw-Hill, New York.

———, and Meyer, Timothy. 1988. *Mediated Communication: A Social Action Perspective.* Sage, Beverly Hills, CA.

APA. 1992. "APA Task Force Explores Television's Positive and Negative Influences on Society." News Release. American Psychological Association, Washington, DC.

Arbitron Internet Information Services. 2000. Infostream Headlines. http://internet.arbitron.com.

Arlen, Michael J. 1969. *Living-Room War.* Viking, New York.

ATSC Digital Television Standard (A/53). 1995. Advanced Television Systems Committee.

Auletta, Ken. 1991. *Three Blind Mice: How the TV Networks Lost Their Way.* Random House, New York.

Avery, Robert K., and Pepper, Robert. 1979. *The Politics of Interconnection: A History of Public Television at the National Level.* National Association of Educational Broadcasters, Washington, DC.

Baker, W. J. 1970. *A History of the Marconi Company.* Methuen, London.

Baldwin, Thomas F., and Lewis, Colby. 1972. "Violence in Television: The Industry Looks at Itself," in Comstock and Rubenstein, 290–365.

———, McVoy, D. Stevens, and Steinfield, Charles. 1996. *Convergence Integrating Media, Information & Communication.* Sage, Thousand Oaks, CA.

Banning, William P. 1946. *Commercial Broadcasting Pioneer: The WEAF Experiment, 1922–1926.* Harvard U. Press, Cambridge, MA.

Barnouw, Erik. 1978. *The Sponsor: Notes on a Modern Potentate.* Oxford U. Press, New York.

Baudino, Joseph E., and Kittross, John M. 1977. "Broadcasting's Oldest Station: An Examination of Four Claimants," *Journal of Broadcasting* 21 (Winter): 61.

Berger, Arthur Asa. 1991. *Media Research Techniques.* Sage, Newbury Park, CA.

BIB (Board for International Broadcasting). Annual. *Annual Report.* GPO, Washington, DC.

Bilby, Kenneth. 1986. *The General: David Sarnoff and the Rise of the Communications Industry.* Harper and Row, New York.

Birch Scarborough. 1991. "How America Found Out About the Gulf War: A Birch Scarborough Study of Media Behavior," Birch Scarborough Research, Coral Springs, FL.

Bishop, John. 1988. *Making It in Video: An Insider's Guide to Careers in the Fastest-Growing Industry of the Decade.* McGraw-Hill, New York.

Black, Uyless. 1997. *Emerging Communications Technologies.* 2nd ed. Prentice Hall PTE, Upper Saddle River, NJ.

Bliss, Edward, Jr. 1991. *Now the News: The Story of Broadcast Journalism.* Columbia U. Press, New York.

Block, Alex Ben. 1990. *Outfoxed: Marvin Davis, Barry Diller, Rupert Murdoch, Joan Rivers, and the Inside Story of America's Fourth Television Network.* St. Martin's, New York.

Block, Eleanor S., and Bracken, James K. 1991. *Communication and the Mass Media: A Guide to the Reference Literature.* Libraries Unlimited, Englewood, CO.

Blum, Eleanor, and Wilhoit, Frances Goins. 1990. *Mass Media Bibliography: An Annotated Guide to Books and Journals for Research and Reference.* U. of Illinois Press, Urbana.

Boorstin, Daniel J. 1964. *The Image: A Guide to Pseudo-Events in America.* Harper and Row, New York.

———. 1978. "The Significance of Broadcasting in Human History," in Hoso-Bunka Foundation, *Symposium on the Cultural Role of Broadcasting,* The Foundation, Tokyo.

Bower, Robert T. 1973. *Television and the Public.* Holt, Rinehart and Winston, New York.

———. 1985. *The Changing Television Audience in America.* Columbia U. Press, New York.

Broadcast Engineering. Monthly. Intertec, Overland Park, KS.

Broadcasting (after 1992, *Broadcasting & Cable*). Weekly. Cahners, Washington, DC.

Broadcasting & Cable, September 27, 1999. p.38.

————, February 7, 2000. p. 34.

Broadcasting & Cable Yearbook. (Title varies.) Annual. Broadcasting, Washington, DC (to 1991); Bowker, New Providence, NJ (1992–present).

Brooks, John. 1976. *Telephone: The First Hundred Years*. Harper and Row, New York.

Brooks, Tim, and Marsh, Earle. 1992. *The Complete Directory to Prime Time Network TV Shows: 1946–Present*, 5th ed. Ballantine, New York.

Brotman, Stuart N. 1990. *Telephone Company and Cable Television Competition: Key Technical, Economic, Legal and Policy Issues*. Artech, Norwood, MA.

Brown, Les. 1971. *Televi$ion: The Business Behind the Box*. Harcourt Brace Jovanovich, New York.

————. 1992. *Les Brown's Encyclopedia of Television*, 3d ed. Facts on File, New York.

Browne, Steven E. 1992. *Film Video Terms and Concepts*. Focal, Stoneham, MA.

Buckley, Tom. 1978. "Popularity of '60 Minutes' Based on Wide-Ranging Reports," *New York Times* (December 17): 99.

Buxton, Frank, and Owen, Bill. 1972. *The Big Broadcast: 1920–1950*. Viking, New York.

CAB (Cable Advertising Bureau). 1999. http://www.cabletvbureau.com/ news/040299news.htm

Cablevision. Weekly. International Thomsom Communications, Denver.

Cal. (California Reporter). 1981. *Olivia N. v. National Broadcasting Co.*, 178 Cal. Rptr. 888 (California Court of Appeal, First District).

Cantril, Hadley. 1940. *The Invasion from Mars: A Study of the Psychology of Panic*. Princeton U. Press, Princeton, NJ.

Carothers, Diane Foxhill. 1991. *Radio Broadcasting from 1920 to 1990: An Annotated Bibliography*. Garland, New York.

Carroll, Raymond L., and Davis, Donald M. 1993. *Electronic Media Programming: Strategies and Decision Making*. McGraw-Hill, New York.

Carter, T. Barton, et al. 1991. *The First Amendment and the Fourth Estate: The Law of Mass Media*, 5th ed. Foundation, Westbury, NY.

————. 1996. *The First Amendment and the Fifth Estate: Regulation of Electronic Mass Media*, 4th ed. Foundation, Westbury, NY.

CCET (Carnegie Commission on Educational Television). 1967. *Public Television: A Program for Action*. Harper and Row, New York.

CCFPB (Carnegie Commission on the Future of Public Broadcasting). 1979. *A Public Trust*. Bantam, New York.

CEMA. 1997. http://www.cemacity.org

————. 1999. http://www.cemacity.org/gazette/today/1999/1997nb.htm

Center for Research on Electronic Commerce, University of Texas at Austin, Graduate School of Business. 1999. *Internet Economy Indicators*, Austin, TX.

CFR (Code of Federal Regulations). Annual. *Title 47: Telecommunications*, 5 vols. GPO, Washington, DC.

Christians, Clifford, et al. 1991. *Media Ethics: Cases and Moral Reasoning*, 3d ed. Longman, White Plains, NY.

Clifford, Martin. 1992. *Modern Audio Technology: A Handbook for Technicians and Engineers*. Prentice Hall, Englewood Cliffs, NJ.

Comstock, George, et al. 1978. *Television and Human Behavior*. Columbia U. Press, New York.

————, and Rubenstein, F., eds. 1972. *Television and Social Behavior: Media Content and Control I*. GPO, Washington, DC.

Consumer Electronics Manufacturers Association. 1999. http://www.cemacity.org/ gazette/files3/dtvsales.htm

Coolidge, Calvin. 1926. "Message to Congress," 68 *Congressional Record* 32.

Cooper, Eunice, and Jahoda, Marie. 1947. "The Evasion of Propaganda: How Prejudicial People Respond to Anti-Prejudice Propaganda." *Journal of Psychology* (January 23): 15.

CPB (Corporation for Public Broadcasting, Washington, DC).

Annual. *Annual Report*.

Annual. *Public Broadcasting Revenue* (title varies).

CTAM. 1999. *Multichannel News*, May 31, 1999, No. 23, Vol. 20, p.16.

DirecTV. 1999. http://www.directv.com/press/presscurrent/1,11331,1,00.html

Donald, Ralph, and Spann, Thomas. 2000. *Fundamentals of Television Production*. Iowa State University Press, Ames, IA.

Dorsey, Tom. 1998. "Cable Was the Worst," *Louisville Courier-Journal* (August 19): 2C.

Doyle, Marc. 1992. *The Future of Television: A Global Overview of Programming, Advertising, Technology and Growth*. NTC Business Books, Lincolnwood, IL.

Drake-Chenault Enterprises Inc. 1978. "History of Rock and Roll," Drake-Chenault, Canoga Park, CA.

Duncan, James. Five times a year. *American Radio*. Duncan Media, Indianapolis, IN.

Eastman, Susan Tyler, ed. 1993. *Broadcast/Cable Programming: Strategies and Practices*, 4th ed. Wadsworth, Belmont, CA.

————, and Newton, Gregory D. 1995. "Delineating Grazing: Observations of Remote Control Use," *Journal of Communication* 45 (Winter): 77.

EBUR. 1987. "Morning Has Broken: An Idea Whose Time Has Come," *European Broadcasting Union Review* (Sept.): 12.

EIA (Electronic Industries Association). Annual. *The U.S. Consumer Electronics Industry in Review* (title varies). EIA, Washington, DC (after 1994: Arlington, VA).

Ellerbee, Linda. 1986. *"And So It Goes": Adventures in Television*. Putnam, New York.

Ennes, Harold E. 1953. *Principles and Practices of Telecasting Operations*. Sams, Indianapolis.

F (*Federal Reporter, 2d and 3d Series*). West, St. Paul, MN.
  1926. *U.S. v. Zenith Radio Corp.*, 12 F2d 614.
  1969. *Office of Communication v. FCC*, 425 F2d 543.
  1972. *Brandywine-Main Line Radio v. FCC*, 473 F2d 16.
  1975. *Natl. Assn. of Independent TV Distributors v. FCC*, 516 F2d 526.
  1977. *Home Box Office v. FCC*, 567 F2d 9.
  1979. *Writers Guild v. ABC*, 609 F2d 355.
  1986. *Telecommunications Research and Action Center v. FCC*, 801 F2d 501, *rehearing denied*, 806 F2d 1115 (D.C. Cir. 1986).
  1987. *Century Communications Corp. v. FCC*, 835 F2d 292.
  1988. *Action for Children's Television et al. v. FCC*, 852 F2d 1332.
  1991. *Action for Children's Television v. FCC*, 932 F2d 1504.
  1995. *Action for Children's Television v. FCC*, 58 F3d 654.
  1996. *Becker v. FCC*, 95 F3d 75, 4 CR 882.
  1998. *Lutheran Church—Missouri Synod v. FCC*, 141 F3d 344.
  1999. *Recording Industry Association of America v. Diamond Multimedia Systems, Inc.*, 180 F3d 1072 (9th Cir).
  2000. *Time Warner Entertainment Co. v. United States*. ___F3d___, 2000 U.S. App. LEXIS 11032 (D.C. Cir.).

FCC (Federal Communications Commission). GPO, Washington, DC.
  Annual Report. Annual.
  Broadcast and Cable Employment Report. Annual.
  1980a. (Network Inquiry Special Staff). *New Television Networks: Entry, Jurisdiction, Ownership and Regulation*. Final Report, Vol. 1. *Background Reports*, Vol. 2.
  1980b. *Staff Report and Recommendations in the Low-Power Television Inquiry*. FCC, Washington, DC.
  1998. *Bechtel v. FCC*, 10 F.3d 875.
  47 C.F.R. Section 73.3555
  47. C.F.R. Section 76.503
  47. C.F.R. Section 73.3555
  2000. *Report on International Telecommunications Markets for 1999*.

FCCR (*FCC Reports*, 1st and 2d Series; and *FCC Record*). GPO, Washington, DC.
  1952. *Amendment of Sec. 3.606* [adopting new television rules] . . . Sixth Report and Order. 41 FCC 148.
  1965. *Comparative Broadcast Hearings*. Policy Statement. 1 FCC2d 393.
  1972a. *DOMSAT*. Report and Order. 35 FCC2d 844.
  1972b. *Cable Television*. Report and Order. 36 FCC2d 143.
  1972c. *Complaint by Atlanta NAACP*. Letter. 36 FCC2d 635.
  1973. *Apparent Liability of Stations WGLD-FM* [Sonderling Broadcasting]. News Release. 41 FCC2d 919.
  1985a. *Cattle Country Broadcasting [KTTL-FM]*. Hearing Designation Order and Notice of Apparent Liability. 58 RR 1109.
  1985b. *Inquiry into . . . the General Fairness Doctrine Obligations of Broadcast Licensees*. Report. 102 FCC2d 143.
  1985c. *Application of Simon Geller*. Memorandum Opinion and Order. 102 FCC2d 1443.
  1987a. *New Indecency Enforcement Standards to Be Applied to All Broadcast and Amateur Radio Licenses*. Public Notice. 2 FCC Rcd 2726.
  1987b. *Complaint of Syracuse Peace Council Against Television Station WTVH*. Memorandum Opinion and Order. 2 FCC Rcd 5043.
  1992a. *Telephone Company-Cable Television Cross-Ownership Rules*. Second Report and Order, Recommendation to Congress and Second Further Notice of Proposed Rulemaking. 7 FCC Rcd 5781. ("Video Dial Tone")
  1992b. *Enforcement of Prohibitions Against Broadcast Indecency in 18 USC 1464*. 7 FCC Rcd 6464.
  1995. *Minority and Female Ownership*, 10 FCC Rcd 2788.
  1996. *Section 257 Proceeding to Identify and Eliminate Market Entry Barriers for Small Businesses*, 11 FCC Rcd 6280.
  2000. Review of Commission's Rules and Policies Affecting the Conversion to Digital Television. Notice of Proposed Rule Making. ___FCC Rcd ___, 65 FR 15600, http://www.fcc.gov/Bureaus/Mass_Media/Notices/2000/fcc00083.txt

Federal Trade Commission. 1999. "FTC to Conduct Study on Marketing Practices of Entertainment Industry." Press Release, June 1.

Ferris, Charles, et al. 1983–present. *Cable Television Law: A Video Communications Guide*, 3 vols. Matthew Bender, New York.

The Freedom Forum, May 1994: 4

Friendly, Fred. 1976. *The Good Guys, the Bad Guys, and the First Amendment: Free Speech vs. Fairness in Broadcasting*. Random House, New York.

FS (*Federal Supplement*). West, St. Paul, MN.
  1996. *American Civil Liberties Union v. Reno*, 929 F. Supp. 824.
  2000. *Twentieth Century Fox Film Corp. v. ICRAVETV*. ___F. Supp. ___, 2000 U.S. Dist. LEXIS 1013 (W.D. Pa.).

Goldenson, Leonard H., with Wolf, Marvin J. 1991. *Beating the Odds: The Untold Story Behind the Rise of ABC—The Stars, Struggles and Egos That Transformed Network Television*. Scribner's, New York.

The Graphic, Visualization and Usability Center. 1998. Georgia Tech Research Corporation, Atlanta,

GA. http://www.gvu.gatech.edu/gvu/user_surveys/survey-1998-10/

Great Britain, Command (Cmnd.) Papers. 1986. *Report of the Committee on Financing the BBC* ["Peacock Report"], Cmnd. 9824. Her Majesty's Stationery Office, London.

Greenfield, Jeff. 1977. *Television: The First Fifty Years.* Abrams, New York.

Greenfield, Thomas A. 1989. *Radio: A Reference Guide.* Greenwood, Westport, CT.

Halberstam, David. 1979. *The Powers That Be.* Knopf, New York.

Harris Interactive Surveys. http://www.harrisinteractive.com/news/index.asp?NewsID=55&HI_election=other

Head, Sydney W. 1985. *World Broadcasting Systems: A Comparative Analysis.* Wadsworth, Belmont, CA.

Henry J. Kaiser Family Foundation. 1999. *Kids & Media @ the New Millenium.* Menlo Park, CA.

Hurly, Paul, et al. 1985. *The Videotex and Teletext Handbook.* Harper & Row, New York, NY.

Intellectual Property and the National Information Infrastructure: The Report of the Working Group on Intellectual Property Rights. 1996. GPO, Washington, DC.

Isailovic, Jordan. 1985. *Videodisc and Optical Memory Systems.* Prentice-Hall, Englewood Cliffs, NJ.

Johnson, Leland L. 1992. *Telephone Company Entry into Cable Television: Competition, Regulation, and Public Policy.* Rand, Santa Monica, CA.

Jowett, Garth. 1976. *Film: The Democratic Art.* Little, Brown, Boston.

Kaltenborn, H. V. 1938. *I Broadcast the Crisis.* Random House, New York.

Katz, Elihu, and Lazarsfield, Paul F. 1955. *Personal Influence: The Part Played by People in the Flow of Mass Communications.* Free Press, Glencoe, IL.

Katz Television Group. 1997. *Katz Syndicated Programming Guide, 1997.* Katz Television Group, New York.

Kendrick, Alexander. 1969. *Prime Time: The Life of Edward R. Murrow.* Little, Brown, Boston.

Klapper, Joseph T. 1960. *The Effects of Mass Communication.* Free Press, Glencoe, IL.

Klein, Paul. 1971. "The Men Who Run TV Aren't That Stupid . . . They Know Us Better Than You Think," *New York* (January 25): 20.

Koch, Tom. 1991. *Journalism for the 21st Century: Online Information, Electronic Databases and the News.* Praeger, New York.

Lichter, S. Robert, Lichter, Linda S., and Amundson, Dan. 1999. *Merchandising Mayhem: Violence and Popular Culture.* Center for Media and Public Affairs, Washington, DC.

McClellan, Steve. 1999. "This News Hound Still Hunts." *Broadcasting & Cable* (August 8): 18.

Mallinson, John C. 1993. *The Foundations of Magnetic Recording.* Academic, San Diego, CA.

Markey, Edward J. 1988. "Statement Accompanying Congressional Research Service Letter to House Subcommittee on Telecommunications and Finance. Cases Involving the Federal Communications Commission That Were Reversed . . ." (March 21). Library of Congress, Washington, DC.

Mayo, John S. 1992. "The Promise of Networked Multimedia Communications." Speech delivered at Bear Sterns Sixth Annual Media & Communications Conference, Coronado, CA (October 28).

Media International. February 1999.

MediaWeek, April 14, 1997. No. 15, Vol 1, p.78.

Metz, Robert. 1975. *CBS: Reflections in a Bloodshot Eye.* Playboy Press, Chicago.

———. 1977. *The Today Show: An Inside Look . . .* Playboy Press, Chicago.

Meyersohn, Rolf B. 1957. "Social Research in Television," in Rosenberg and White, 345–357.

Miami Herald, 2 April 1994: A2.

———. 9 May 1994: A2.

Minow, Newton N. 1964. *Equal Time: The Private Broadcaster and the Public Interest.* Atheneum, New York.

———. 1991. "How Vast the Wasteland Now?" Gannett Foundation Media Center, Columbia University, New York (May 9).

Montgomery, Kathryn C. 1989. *Target: Prime Time: Advocacy Groups and the Struggle Over Entertainment Television.* Oxford U. Press, New York.

Mothersole, Peter L., and White, Norman W. 1990. *Broadcast Data Systems: Teletext and RDS.* Butterworth & Co. Ltd., London.

NAB (National Association of Broadcasters). 1999. "More Women Fill Top Spots Despite Lack of EEO Rules." *Billboard*, December 18 (111): 90.

———. Washington, DC. Annual. *Radio Financial Report* (to 1992). Annual. *Television Financial Report.* Semiannual. *Broadcast Regulation* (title varies). 1978. *The Television Code.* http://www.nab.org/PressRel/Dtvstations.asp

Napoli, Philip M. 2000. *Market Conditions and Public Affairs Programming: Implications for Digital Television Policy.* Benton Foundation: Washington, DC.

National Association of Broadcasters. 1999. http://www.nab.org/Newsroom/Issues/digitaltv/DTVstations.asp

National Cable Television Association. Quarterly. *Cable Television Developments.* Washington, DC.

National Retail Federation/Forrester Online Retail Index. January 2000 report. http://www.forrester.com/ER/Press/Release/0,1769253,FF.htm

National Telecommunications and Information Administration (a division of the U.S. Department of Commerce). 8 July 1999. *Falling Through the Net: Defining the Digital Divide.* http://ntia.doc.gov/ntiahome/fttn99/execsummary.html

NCTA. 1999. *Cable Television Developments.* Summer 1999. National Cable Television Association, Washington, DC. http://www.ncta.cyberserv.com/qs/user_pages/Dev(servicetype).cfm

New Century Net. 1997. http://www.newcentury.net

NewMedia. Monthly. HyperMedia, San Mateo, CA.

Newseum. 1997. http://www.newseum.org

Newspage. 1997. http://www.newspage.com

The New Yorker. 6 January 1997: 76.

New York Times. 30 September 1992: 1, B8.

———. 20 January 1995: C3.

Nielsen. 1996. *Television Audience*.

Nielsen Media Research. 1993. *1992–1993 Report on Television*. Nielsen Media Research, New York.

———. 1998. *1998 Report on Television*. Nielsen Media Research, New York, NY. p.45

Nielsen//NetRatings. 2000. Hot off the Net. http://nielsen.netratings.com

Ohio State University. School of Journalism. Annual. *Journalism and Mass Communications Graduate Survey: Summary Report*. Ohio State University, Columbus, OH.

Owen, Bruce M. 1999. *The Internet Challenge to Television*. Harvard University Press, Cambridge, MA.

Paley, William S. 1979. *As It Happened: A Memoir*. Doubleday, New York.

Pareles, Jon. 1989. "After Music Videos, All the World Has Become a Screen," *New York Times* (10 December): E6.

Parents Television Council. 1999. *Unintended Consequences: With Ratings in Place, TV More Offensive Than Ever*. Alexandria, VA.

Paterson, Richard. Annual. *TV and Video International Guide*. Tantivy, London.

Paul Kagan Associates. Annual. *The Cable TV Financial Databook*. Carmel, CA. http://www.kagan.com/screen/kmarket/press11.html

———. 1999. http://www.ncta.com/yearend98_3.html#std

Persico, Joseph E. 1988. *Edward R. Murrow, An American Original*. McGraw-Hill, New York, NY.

Pichaske, David. 1979. *A Generation in Motion: Popular Music and Culture of the Sixties*. Schirmer Books, New York.

Pike and Fischer. *Radio Regulation* (after 1995 *Communications Regulation*). Bethesda, MD.

Pohlmann, Ken C. 1989. *The Compact Disc: A Handbook of Theory and Use*. A-R Editions, Madison, WI.

Radio World. Bi-weekly. IMAS, Falls Church, VA.

Rather, Dan. 1987. "From Murrow to Mediocrity?" *New York Times* (March 10): A27.

Roper Organization. 1993. *America's Watching: Public Attitudes Toward Television*. National Association of Broadcasters, Washington, DC, and Network Television Association, New York, NY.

Roper Starch Worldwide. 1998. *America's Watching*. New York, NY.

Rosenberg, Bernard, and White, David Manning, eds. 1957. *Mass Culture: The Popular Arts in America*. Free Press, Glencoe, IL.

Rubin, Bernard. 1967. *Political Television*. Wadsworth, Belmont, CA.

Rumsey, Francis. 1990. *Tapeless Sound Recording*. London.

Schmitt, Kelly L. 1999. *The Three-Hour Rule: Is It Living Up to Expectations?* Annenberg Public Policy Center, University of Pennsylvania: Philadelphia, PA.

Senate CC (U.S. Congress, Senate Committee on Commerce). 1972. *Surgeon General's Report by Scientific Advisory Committee on Television and Social Behavior*. Hearings. 92nd Cong., 2d Sess. GPO, Washington, DC.

Settel, Irving. 1967. *A Pictorial History of Radio*, 2d ed. Grosset and Dunlap, New York.

———. 1983. *A Pictorial History of Television*, 2d ed. Ungar, New York.

Setzer, Florence, and Levy, Jonathan. 1991. *Broadcast Television in a Multichannel Marketplace*. FCC Office of Plans and Policy (Staff Study No. 26, June).

SGAC (Senate Government Affairs Committee). 1991. *Pentagon Rules on Media Access to the Persian Gulf War*. Hearing. 102nd Cong., 1st Sess. GPO, Washington, DC.

Shannon and Weaver. 1949. *The Mathematical Theory of Communication*.

Shawcross, William. 1992. *Murdoch*. Simon and Schuster, New York.

Sieber, Robert. 1988. "Industry Views on the People Meter: Cable Networks," *Gannett Center Journal* 2 (Summer): 70.

Silbergleid, Michael, and Pescatore, Mark J., eds. 1999. *The Guide to Digital Television*. 2nd edition. Miller Freeman PSN Inc., New York, NY.

Sinclair Broadcast Group. 1999. http://www.sbgi.net/press_releases.html

Singleton, Loy A., and Copeland, Gary A. 1997. "FCC Broadcast Indecency Enforcement Actions: Changing Patterns Since *Pacifica*," Law and Policy Division, BEA Annual Conference.

Smith, Bruce Lannes, et al. 1946. *Propaganda, Communication and Public Opinion*. Princeton U. Press, Princeton, NJ.

Smith, Richard A. 1985. "TV: The Light That Failed," *Fortune* (December): 78.

Smith, Sally Bedell. 1990. *In All His Glory: The Life of William S. Paley—The Legendary Tycoon and His Brilliant Circle*. Simon and Schuster, New York.

Sperber, A.M. 1986. *Murrow: His Life and Times*. Freundlich Books, New York, NY.

Steiner, Gary A. 1963. *The People Look at Television: A Study of Audience Attitudes*. Knopf, New York.

Stephenson, William. 1967. *The Play Theory of Mass Communication*. U. of Chicago Press, Chicago.

Sterling, Christopher H. 1984. *Electronic Media: A Guide to Trends in Broadcasting and Newer Technologies, 1920–1983*. Praeger, New York.

———, and Kittross, John M. 1990. *Stay Tuned: A Concise History of American Broadcasting*, 2d ed. Wadsworth, Belmont, CA.

Strong, William S. 1993. *The Copyright Book: A Practical Guide*, 4th ed. MIT Press, Cambridge, MA.

TCAF (Temporary Commission on Alternative Financing for Public Telecommunications). 1982–1983. *Alternative Financing Options for Public Broadcasting* (Vol. 1); *Final Report* (Vol. 2). FCC, Washington, DC.

Television Broadcast. Monthly. Miller Freeman, New York, NY.

Television Digest. 28 September 1992: 10.

Toffler, Alvin. 1970. *Future Shock.* Random House, New York.

———. 1980. *The Third Wave.* Morrow, New York.

TV Technology. Semi-monthly. IMAS, Falls Church, VA.

US (*United States Reports*). GPO, Washington, DC.

1919. *Schenk v. U.S.,* 249 US 47.

1942. *Marconi Wireless Telegraph Co. of America v. U.S.* 320 US 1.

1943. *NBC v. U.S.,* 319 US 190.

1949. *Terminiello v. Chicago,* 337 US 1.

1951. *Dennis v. U.S.,* 341 US 494.

1957. *Roth v. United States,* 354 US 476.

1964. *New York Times v. Sullivan,* 376 US 254.

1968. *U.S. v. Southwestern Cable,* 392 US 157.

1969. *Red Lion v. FCC,* 395 US 367.

1973. *Miller v. California,* 413 US 15.

1974. *Miami Herald v. Tornillo,* 418 US 241.

1978. *FCC v. Pacifica Foundation,* 438 US 726.

1981. *Chandler v. Florida,* 449 US 560.

1984. *Universal Studios v. Sony,* 464 US 417.

1986. *City of Los Angeles and Department of Water and Power v. Preferred Communications, Inc.,* 476 US 488.

1989. *Sable Communications of California, Inc. v. FCC,* 492 US 115.

1990. *Metro Broadcasting v. FCC,* 497 US 547.

1995. *Adarand Constructors, Inc. v. Pena,* 515 US 200, 115 SCt 2097.

1996. *Denver Area Educational Telecommunications Consortium v. FCC,* 116 SCt 2374.

1999. *Greater New Orleans Broadcasting Association, Inc. v. United States,* 527 US 173.

2000. *United States v. Playboy Entertainment Group, Inc.* __U.S.__, 120 S. Ct. 1878, 2000 U.S. LEXIS 3427.

USC (*United States Code*). Regularly revised. GPO, Washington, DC.

47 USC 315.

1998. *Children's Online Privacy Protection Act.* 47 USC 6501 et seq.

1998. *Children's Online Protection Act.* 47 USC 201.

1998. *Digital Millennium Copyright Act.* 17 USC 1201 et seq.

1998. *Online Copyright Infringement Liability Limitation Act.* 17 USC 512.

U.S. Commission on Civil Rights. 1977, 1979. *Window Dressing on the Set: Women and Minorities in Television,* 2 vols. GPO, Washington, DC.

USDOS (U.S. Department of State). 1990. *Eastern Europe: Please Stand By—Report of the Task Force on Telecommunications and Broadcasting in Eastern Europe.* GPO, Washington, DC.

U.S. Education Department's National Center for Education Statistics (NCES). 1998. http://cyberatlas/internet.com/big_picture/demographics/education/article/0,1323,5951_308911,00.html

Variety. Weekly. Variety, New York.

The Veronis, Suhler & Associates Communications Industry Forecast: Historical and Projected Expenditures for 9 Industry Segments. Annual. Veronis, Suhler, New York.

Wall Street Journal, 11 July 1994: A1.

Walsh, David. 1999. *Video and Computer Game Report Card.* National Institute on Media and the Family, Minneapolis, MN:. http://www.mediaandthefamily.org/research/ugrc/index.shtml

Washington Post, 12 October 1992: D5.

Webster, James, and Lichty, Lawrence W. 1991. *Ratings Analysis: Theory and Practice.* Erlbaum, Hillsdale, NJ.

Wiener, Norbert. 1950. *The Human Use of Human Beings.* Houghton Mifflin Company, Boston, MA.

Williams, Christian. 1981. *Lead, Follow, or Get Out of the Way: The Story of Ted Turner.* Times Books, New York.

Williams, Margorie. 1989. "MTV as Pathfinder for Entertainment," *The Washington Post* (December 13): A1.

Wood, James. 1992a. *History of International Broadcasting.* Peter Peregrinus/Science Museum, London.

———. 1992b. *Satellite Communications and DBS Systems.* Focal, Stoneham, MA.

Woolfe, Roger. 1980. *Videotex.* Heyden & Son Ltd., London.

World Radio-TV Handbook. Annual. Billboard, New York.

Wurtzel, Alan, and Acker, Stephen R. 1989. *Television Production,* 3d ed. McGraw-Hill, New York.

Zettl, Herbert. 1996. *Television Production Handbook,* 6th ed. Wadsworth, Belmont, CA.

INDEX

ABC (American Broadcasting
Company)
and affiliates, 173
changing ownership of, 80
children's programs, 264
competition with NBC, 77
Disney/ABC, 80–81, 256, 261
Food Lion suit, 286, 382
game show success of, 258
Internet programming and,
283
news programming, 268
origins of, 36
public affairs programming,
271
radio networks, 275–276
recording ban by, 38
sports programming on,
260–263
Accommodation theory, 319
Action for Children's Television,
341
Active-matrix LCDs, 115
Addressability, and PPV, 65
Addressable converters, 121, 192
Administrative law judges (ALJs),
358
Adult Learning Service (ALS),
222
Advanced Television Systems
Committee (ATSC), 147,
148, 152
Advertising
access rights of, 397
audience targeting and, 228
before and after deregulation,
367
on cable, 179, 184, 185, 187
children and, 341
cigarette, 182, 190, 397
on early radio, 31–32
editorial, 189, 397
election campaign, 330
impact of, 340–341
infomercials, 189, 235
integration of, 182–184
Internet, 159, 166, 188, 189,
313

liquor, 182, 190
local, 180, 181, 187
national-spot, 180
network, 180, 183, 184, 187,
191
promotion, 167, 183, 245–247
proof of performance, 189
pros and cons of, 180, 182
rates for, 184–187
regulation of, 369
sale of, 187–189
sports sponsorship and, 240
standards, 189–191
top budgets in, 183 (exhibit)
unethical practices, 190–191
volume of, in media, 181
Advertising agencies, 188
Advertorials, 189, 397
Affiliates, television
networks and, 171–173, 190,
201
news origination by, 277–278
schedule strategies of, 243
African Americans
in electronic media, 198
in FCC, 350
ownership by, 391
portrayal of, on radio, 36–37
portrayal of, on TV, 253, 255
talk radio, 282
See also Stereotypes
Agency for Instructional Televi-
sion (AIT), 222
Agenda setting, 323
Allen, Paul, 11
Allied Business Intelligence, 161
All in the Family, 250, 319
All News Channel, 240
All Things Considered, 222
Ally McBeal, 229
Alphastar, 69
American Experience, The, 220
American Marconi, 23–24, 27
American Public Radio. *See* Public
Radio International
American Society of Composers,
Authors, and Publishers
(ASCAP), 37–38

American Telephone and Telegraph Company. *See* AT&T
American Top 40, 276
America Online (AOL), 8, 77,
159, 160
AMFM, Inc., 81, 171, 276
Amos 'n' Andy, 36–37
Ampex Corporation, 38, 48
AM radio
antennas, 98, 99 (exhibit), 101
channels for, 90, 101
classes of, 101, 102 (exhibit)
decline of, 53, 81
digital radio and, 155, 156
vs. FM, 93–94
FRC frequencies for, 34
public, 208
shortwave, 103
stereo, 102–103, 365
transmission, 101–102
wave propagation in, 97
Analog format, *vs.* digital, 92–93,
94
Anderson, Gillian, 252 (exhibit)
Annenberg, Walter, 215, 246, 247
Annenberg/CPB project, 222
Announcements (spots), 182–184
Antennas
and contour grades, 114
DARS, 154
DBS service and, 64 (exhibit)
directional propagation and,
98, 100
direct waves and, 95–96
FM/television, 103
propagation by, 112
satellite, 131, 132, 133, 136
transmitting, 98, 99 (exhibit)
TVRO, 59–60, 69, 120 (ex-
hibit), 133 (exhibit), 136, 192,
402
Antiques Roadshow, 220
Antitrust law, 387–388
Arafat, Yassir, 329 (exhibit)
Arbitron, 292, 294 (exhibit), 296,
297, 299 (exhibit), 302, 314
Area of dominant influence (ADI),
302